Endothelial Signaling in Development and Disease

Mirko HH Schmidt • Stefan Liebner
Editors

Endothelial Signaling in Development and Disease

Springer

Editors
Mirko HH Schmidt
Microscopic Anatomy and Neurobiology
Johannes Gutenberg University
School of Medicine
Mainz, Germany

Stefan Liebner
Institute of Neurology
Goethe University Frankfurt
School of Medicine
Frankfurt, Germany

ISBN 978-1-4939-2906-1 ISBN 978-1-4939-2907-8 (eBook)
DOI 10.1007/978-1-4939-2907-8

Library of Congress Control Number: 2015944714

Springer New York Heidelberg Dordrecht London

Printed on acid-free paper

Springer Science+Business Media LLC New York is part of Springer Science+Business Media (www.springer.com)

Foreword

How blood and lymphatic vessels form functional networks that are adapted to effectively meet the metabolic and homeostasis requirement of organs and the organism as a whole is a most fascinating question that stimulates the rapidly growing field of vascular biology research. As we appreciate the complexity of cellular and molecular interactions, the intricacies of localized functional requirements, and the heterogeneity of the cellular constituents that form blood and lymphatic vessels, the growing number of open questions may seem overwhelming. Advanced genetic model systems to selectively study gene functions in the endothelial cells lining blood and lymphatic vessels, as well as the supporting smooth muscle cells, have helped to chart key pathways controlling the formation and differentiation of vascular networks. Rapidly improving imaging methodology, functional labeling, and organotypic dynamic in vitro assays help to unravel morphogenic principles of sprouting, lumen formation, stabilization, and regression. Signaling networks that have traditionally been mapped in cell culture now take shape in vivo as we see genetic models dissect individual residues in signaling receptors, and quantitative assessment of protein–protein complexes allow glimpses into when and where, in the context of vessel formation, receptor signaling takes place.

As in many disease areas, cardiovascular research also uncovers the power of noncoding RNAs and genetic, as well as epigenetic, modulation of vascular patterning and risk factors. At the same time, concepts of vascular functions in tumor growth and metastasis are evolving as we begin to understand the reasons for the disappointing clinical success of antiangiogenic therapy. Exciting developments also stem from studies into the cellular metabolism of endothelial cells, and the interplay of hypoxia with multiple endothelial cell responses and functions.

The editor(s) and authors of the present book have done a terrific job in collating a comprehensive overview of current concepts and insight into the development and differentiation of vascular networks (Part I), mechanisms controlling vascular homeostasis (Part II), and timely topics in pathophysiology of the vascular system (Part III). Although the individual chapters are stand-alone pieces by individuals or teams of leaders in the field, the collection of chapters will provide interesting and timely reading and extended educational value for specialists and broader-interest readers alike.

The collection of chapters covers the most prominent signaling pathways that impact on a vast array of vascular and endothelial functions, such as vascular endothelial growth factor-A, Notch, hypoxia, fibroblast growth factor, nitric oxide, and the angiopoietins/Tie2 system. In addition, the most prominent and powerful model systems, such as zebrafish for in vivo analysis and retina for its accessibility and clinical relevance, are covered by experts. The reader will further gain an up-to-date understanding of the heterogeneity of vascular networks and specialization in vascular beds, including the lung and ocular vasculature. Most of the current knowledge on vascular formation and function focuses on the inner lining of vessels—the endothelium. The endothelium functions as a collective of cells that are tightly connected by special junctions, functionally connected by signaling mechanisms, mechanically embedded in extracellular matrix lending stability and providing additional signaling input, and yet highly specialized and diversified, both within organs and between organs, to serve the functional complexity of the organism. Coordination of endothelial signaling and cellular behavior is therefore of particular relevance, and several chapters shed light on emerging concepts and principles.

Finally, our current understanding of the dynamic regulation of vascular permeability and fluid homeostasis, as well as the control of transendothelial migration for immunosurveillance and tissue homeostasis, is discussed in several chapters, including an expert account of the latest developments in lymph vessel formation and regulation.

I hope this work will stimulate more young scientists to embark on the exciting discovery science that can be followed in vascular biology, and potentially kindle new ideas that may help us to overcome limitations that currently hamper our progress in clinical translation. I wish you enjoyable and stimulating reading.

Berlin, Germany Holger Gerhardt

Contents

About the Editors

Mirko HH Schmidt, Univ.-Prof. Dr. Dr., is a professor of anatomy and biochemistry at the Institute of Microscopic Anatomy and Neurobiology, Mainz, Germany. He received his PhD in biochemistry from the University of Lübeck, Germany, and studied molecular signaling processes as a postdoc at the Hermelin Brain Tumor Center, Detroit, MI, USA, and at the Institute of Biochemistry II, Frankfurt/Main, Germany. His research group 'Molecular Signal Transduction' focuses on the degeneration of signaling pathways in the diseased brain, and was located at the Institute of Neurology (Edinger Institute), Frankfurt, before moving to Mainz. Prof. Schmidt has received numerous fellowships and awards for his research on translational molecular medicine.

Stefan Liebner, Ph.D., is currently a research group leader at the Institute of Neurology (Edinger Institute), Johann Wolfgang Goethe University Clinic, Frankfurt/Main, Germany. After receiving his PhD in cell biology at the Eberhard Karls University Tübingen, he investigated the role of endothelial β-catenin for vascular development at the Italian Foundation for Cancer Research (FIRC) Institute of Molecular Oncology (IFOM), Milan, Italy, as a postdoctoral fellow. His research focus is on molecular mechanisms of blood–brain barrier development, maintenance, and dysfunction. Dr. Liebner has received numerous fellowships and awards, is an active member of the International Brain Barriers Society (IBBS) and European Vascular Biology Society (EVBM), and acts as an academic editor of PLoS ONE and ISRN Vascular Medicine.

Contributors

Jeremiah Bernier-Latmani Department of Oncology, University of Lausanne, Lausanne, Switzerland

Frank Bicker Molecular Signal Transduction Laboratories, Institute for Microscopic Anatomy and Neurobiology, Johannes Gutenberg University School of Medicine, Mainz, Germany

Luca Bravi FIRC Institute for Molecular Oncology Fondazione, Milan, Italy

Gustaf Christoffersson Department of Medical Cell Biology, Uppsala University, Uppsala, Sweden

Lena Claesson-Welsh Department of Immunology, Genetics and Pathology, Uppsala University, Uppsala, Sweden

Richard Daneman Departments of Neuroscience and Pharmacology, University of California, San Diego, La Jolla, CA, USA

Henricus J. Duckers Erasmus MC, Rotterdam, The Netherlands

Nevenka Dudvarski Stanković Molecular Signal Transduction Laboratories, Institute for Microscopic Anatomy and Neurobiology, Johannes Gutenberg University School of Medicine, Mainz, Germany

German Cancer Consortium (DKTK), Heidelberg, Germany

German Cancer Research Center (DKFZ), Heidelberg, Germany

Ingrid Fleming Institute for Vascular Signalling, Goethe University, Frankfurt/am Main, Germany

Ulrich Förstermann Department of Pharmacology, Johannes Gutenberg University Medical Center, Mainz, Germany

Hans-Peter Hammes 5th Medical Department, UMM – University of Heidelberg, Mannheim, Germany

Anne-Theres Henze Laboratory of Molecular Oncology and Angiogenesis, Vesalius Research Center, VIB, Leuven, Belgium

Laboratory of Molecular Oncology and Angiogenesis, Vesalius Research Center, Department of Oncology, KU Leuven, Leuven, Belgium

Dorien M. A. Hermkens Hubrecht Institute Utrecht, Utrecht, The Netherlands

Erasmus MC, Rotterdam, The Netherlands

Meike Hoffmeister Institute for Biochemistry II, Goethe University Frankfurt, Medical School, Frankfurt/Main, Germany

Annika Keller Division of Neurosurgery, University Hospital Zürich, Zürich, Switzerland

Igor Kovacevic Department of Molecular Cell Biology, Sanquin Research and Landsteiner Laboratory, Academic Medical Center, University of Amsterdam, Amsterdam, CX, The Netherland

Institute for Biochemistry II, Goethe University Frankfurt, Medical School, Frankfurt/Main, Germany

Maria Grazia Lampugnani FIRC Institute for Molecular Oncology Fondazione, Milan, Italy

Mario Negri Institute for Pharmacological Research, Milan, Italy

Huige Li Department of Pharmacology, Johannes Gutenberg University Medical Center, Mainz, Germany

Robert Luck Biochemistry Center, Heidelberg University, Heidelberg, Germany

Davit Manukyan Center for Thrombosis and Hemostasis, University Medical Center Mainz, Mainz, Germany

Institute of Clinical Chemistry and Laboratory Medicine, University Medical Center Mainz, Mainz, Germany

Massimiliano Mazzone Laboratory of Molecular Oncology and Angiogenesis, Vesalius Research Center, VIB, Leuven, Belgium

Laboratory of Molecular Oncology and Angiogenesis, Vesalius Research Center, Department of Oncology, KU Leuven, Leuven, Belgium

Iva Nikolic Cancer Research Division, The Garvan Institute of Medical Research, Sydney, NSW, Australia

Stefanie Oess Institute for Biochemistry II, Goethe University Frankfurt, Medical School, Frankfurt/Main, Germany

Tatiana V. Petrova Department of Oncology, University of Lausanne, Lausanne, Switzerland

Swiss Institute for Cancer Research, Ecole Polytechnique Federale de Lausanne, Lausanne, Switzerland

Mia Phillipson Department of Medical Cell Biology, Uppsala University, Uppsala, Sweden

Karl H. Plate Institute of Neurology (Edinger Institute), Goethe University Medical School, Frankfurt, Germany

German Cancer Consortium (DKTK), Partner Site Frankfurt/Mainz, Frankfurt, Germany

German Centre of Cardiovascular Research (DZHK), Partner Site Rhein/Main, Frankfurt, Germany

Christoph Reinhardt Center for Thrombosis and Hemostasis, University Medical Center Mainz, Mainz, Germany

Yvonne Reiss Institute of Neurology (Edinger Institute), Goethe University Medical School, Frankfurt, Germany

German Cancer Consortium (DKTK), Partner Site Frankfurt/Mainz, Frankfurt, Germany

German Centre of Cardiovascular Research (DZHK), Partner Site Rhein/Main, Frankfurt, Germany

Wolfram Ruf Center for Thrombosis and Hemostasis, University Medical Center Mainz, Mainz, Germany

Department of Immunology and Microbial Science, The Scripps Research Institute, La Jolla, CA, USA

Carmen Ruiz de Almodovar Biochemistry Center, Heidelberg University, Heidelberg, Germany

Amélie Sabine Department of Oncology, University of Lausanne, Lausanne, Switzerland

Mirko HH Schmidt Molecular Signal Transduction Laboratories, Institute for Microscopic Anatomy and Neurobiology, Johannes Gutenberg University School of Medicine, Mainz, Germany

Focus Program Translational Neuroscience, Mainz, Germany

German Cancer Consortium (DKTK), Heidelberg, Germany

German Cancer Research Center (DKFZ), Heidelberg, Germany

Alexander Scholz Institute of Neurology, (Edinger Institute), Goethe University Medical School, Frankfurt, Germany

Laboratory of Immunology and Vascular Biology, Department of Pathology, Stanford University School of Medicine, Stanford, CA, USA

Stefan Schulte-Merker Cells-in-Motion Cluster of Excellence, University of Münster, Münster, Germany

Hubrecht Institute, Utrecht, Utrecht, The Netherlands

Institute for Cardiovascular Organogenesis and Regeneration, Westfälische Wilhelms-University, Münster, Germany

Michael Seimetz Excellence Cluster Cardio-Pulmonary System, Giessen, Germany

Universities of Giessen and Marburg Lung Center, Giessen, Germany

German Center for Lung Research, Giessen, Germany

Justus-Liebig-University, Giessen, Germany

Marcin Teodorczyk Molecular Signal Transduction Laboratories, Institute for Microscopic Anatomy and Neurobiology, Johannes Gutenberg University School of Medicine, Mainz, Germany

Norbert Weissmann Excellence Cluster Cardio-Pulmonary System, Giessen, Germany

Universities of Giessen and Marburg Lung Center, Giessen, Germany

German Center for Lung Research, Giessen, Germany

Justus-Liebig-University, Giessen, Germany

Michael Welsh Department of Medical Cell Biology, Uppsala University, Uppsala, Sweden

Ning Xia Department of Pharmacology, Johannes Gutenberg University Medical Center, Mainz, Germany

Part I
Development and Differentiation of the Vascular System

Chapter 1
VEGF and Notch Signaling in Angiogenesis

Marcin Teodorczyk, Nevenka Dudvarski Stanković, Frank Bicker and Mirko HH Schmidt

1.1 Overview of Vascular Morphogenesis

The vascular system, consisting of a three-dimensional (3D) and hierarchical tubular network, functions as the main supply system of the body, as it carries nutrients, circulating cells, gasses, fluids and hormones to almost every tissue and organ in higher metazoans. The development of this elaborate system is spatially and temporally tightly regulated and relies on the orchestration of different cells, growth factors and components of the extracellular matrix (ECM). A functional blood vessel network is important for maintaining homeostasis of the healthy vertebrate body, and dysregulation of vascular morphogenesis is a hallmark of many diseases. This is the case in deadly tumors, where cancer cells promote neovascularization leading to tumor progression and metastasis, while insufficient blood flow is the underlying cause of ischemic diseases [1].

Marcin Teodorczyk and Nevenka Dudvarski Stanković contributed equally to this work

---

M. HH. Schmidt (✉) · M. Teodorczyk · N. D. Stanković · F. Bicker
Molecular Signal Transduction Laboratories, Institute for Microscopic Anatomy and Neurobiology, Johannes Gutenberg University School of Medicine, Langenbeckstr. 1, 55131 Mainz, Germany
e-mail: mirko.schmidt@unimedizin-mainz.de

Focus Program Translational Neuroscience (FTN), Mainz, Germany

M. Teodorczyk
e-mail: marcin.teodorczyk@unimedizin-mainz.de

N. D. Stanković
e-mail: nevenka.dudvarski-stankovic@unimedizin-mainz.de

F. Bicker
e-mail: frank.bicker@unimedizin-mainz.de

M. HH. Schmidt, S. Liebner (eds.), *Endothelial Signaling in Development and Disease,* DOI 10.1007/978-1-4939-2907-8_1

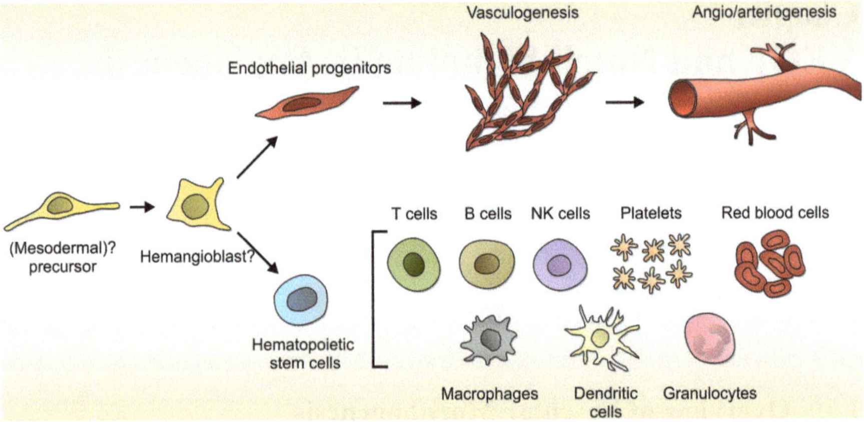

Fig. 1.1 Schematic of the development of the vascular and hematopoietic systems. *In vitro* embryonic stem-cell differentiation assays led to the identification of a common mesoderm-derived endothelial/hematopoietic progenitor, the so-called hemangioblast which commits to more differentiated cell lineages. Endothelial cell (*EC*) precursors, angioblasts, undergo sequential maturation and eventually express markers of mature ECs. Early in organogenesis a primitive vascular network forms from ECs in the process of vasculogenesis. All other blood vessels formed during embryogenesis and in adults arise by sprouting from this pre-existing vasculature in a process defined as angiogenesis. The growth of pre-existing interlocking arterioles towards functionally mature collateral arteries is denoted as arteriogenesis. Hematopoietic stem cells give rise to precursors of all blood cells lineages, e.g., erythrocytes, leukocytes and thrombocytes

In vitro embryonic stem cell (ESC) differentiation assays led to the identification of a common endothelial/hematopoietic progenitor, the so-called hemangioblast (Fig. 1.1) [2–4]. The existence of the hemangioblast *in vivo*, however, is still under debate. Within the yolk sac of the mouse embryo, equivalent cells have been identified that also give rise to smooth muscle cells (SMCs) [5]. Thus, the question remains whether they are tri-potent progenitors or mesodermal cells with a broader differentiation potential [6]. The discovery of the hemogenic endothelium in recent years offers an alternative and/or complementary model for the generation of hematopoietic stem cells (HSCs) [7–12]. The cells that cover the ventral part of the dorsal aorta display an endothelial phenotype and morphology. However, both generate hematopoietic as well as endothelial progeny [13–15]. Two transcription factors play a major role in this process: Runx1 promotes hematopoiesis while HoxA3 drives endothelial differentiation [11, 16]. According to Rybtsov et al., HSC differentiation from the hemogenic endothelium proceeds *via* at least two non-EC intermediates and segregation of the endothelial and HSC lineages occurs prior to E10.5 [17, 18].

At approximately the same time during mouse embryonic development (E9.5), lymphatic EC (LEC) progenitors begin to bud off embryonic veins into the surrounding mesenchyme. Interestingly, the progenitors remain connected by adherens junctions (AJs) expressing high levels of vascular endothelial (VE)-cadherin [19]. At around E11.5, they start forming hollow structures called lymph sacs along the embryo's anterior-posterior axis [19, 20]. Lymph sacs become the main source of

LECs required for the formation of the entire lymphatic vasculature [21]. As the lymph sac forms, the LECs continue to proliferate and migrate into the mesenchymal tissue. The lymphatic plexus further differentiates to form two types of vessels: larger collecting lymphatic vessels and smaller lymphatic capillaries [22]. Lymphatic capillaries are thin-walled, blind-ended vessels that collect interstitial fluid and transport it to the larger collecting vessels, which are surrounded by SMCs and contain intraluminal valves in order to assure the movement of lymph against hydrostatic pressure and prevent its backflow [23–25]. For more information on lymphangiogenesis and lymphatic vessel maturation, please refer to Chap. 5.

The development of the blood vasculature occurs by two mechanisms: (1) vasculogenesis, which represents the *de novo* formation of endothelium, and (2) angiogenesis, which is defined as the growth of blood vessels from the pre-existing vasculature. For a long period of time it was believed that only differentiation of the primitive vascular network, occurring early in organogenesis, proceeds by vasculogenesis [26], while all other blood vessels are generated by angiogenesis, but in the last 20 years, various studies have reported the occurrence of vasculogenesis during postnatal blood vessel growth [27, 28]. However, whether or not vasculogenesis occurs in the adult organism remains under debate.

Initial formation of the primitive vasculature starts when the early mesoderm has been formed by gastrulation. Migrating mesodermal cells form aggregates in the yolk sac, referred to as blood islands [29]. Soon, two types of progeny cells can be distinguished in each blood island, namely the internally localized hematopoietic precursor cells (HPCs) and the peripherally positioned angioblasts, termed endothelial progenitor cells (EPCs) [30]. These blood islands undergo fusion and the outcome is the differentiation of the primitive capillary plexus. Further, it has been described that migrating mesodermal cells can give rise to angioblasts without differentiation of HPCs. Aside of differentiating into the primitive blood plexus, these intraembryonic angioblasts can also directly form vessels [31]. Interestingly, even at the earliest stages of blood vessel formation, vascular endothelial growth factor (VEGF) is a key signaling molecule. It has been confirmed that VEGF is strongly expressed by the extraembryonic endoderm and mesoderm. Further, in mice with the deletion of one of VEGF receptors, VEGFR2, yolk-sac blood islands could not be formed, indicating that VEGF is indeed a key player in vasculogenesis [32].

In general, two different, but non-exclusive mechanisms of angiogenic the vessel growth are known, namely intussusception and endothelial sprouting [30, 33]. In adulthood, the vascular network is mostly quiescent, except in the cycling ovary and uterus during pregnancy. However, the mature and resting vasculature can be reactivated in numerous pathological conditions by stimuli released from tumor cells or in inflammatory processes [34].

1.1.1 Sprouting

For a long time, sprouting angiogenesis was thought to be the sole process of the vessel formation in adulthood. Several distinguishing phases during this process can be recognized. Initially, sprouting starts with an increased vessel permeability, local degradation of the basement membrane, detachment of mural cells such as SMCs and pericytes (PCs) from blood vessels, and liberation of ECs. The next stage is characterized by EC migration and proliferation. Guided sprout outgrowth is followed by elongation, fusion and lumen formation. Finally, blood starts to perfuse the newborn vessel, which coincides with its maturation [35, 36].

When the quiescent endothelium senses a pro-angiogenic stimulus, a cascade of activating events occurs. One of the most important mediators of the vessel sprouting is VEGF. The remodeling endothelium releases angiopoietin 2 (Ang-2), which inhibits signaling *via* the receptor tyrosine kinase (RTK) Tie2 and as a result triggers the delocalization of SMCs and PCs [37] (please also refer to Chap. 13). VEGF induces nitric oxide (NO)-dependent local vasodilatation and escalation of the vessel's permeability. This allows extravasation of matrix metalloproteinases that degrade the basement membrane and further serve as a specific scaffold for the prospective sprout [36]. Eventually, ECs achieve a motile and invasive phenotype.

One of the ECs from an activated vessel is chosen to be a so-called tip cell with the function of sensing VEGF gradients and directing sprout growth. This is a highly dynamic process and tip cells are continuously replaced; however, there is always only a single tip cell [38]. The tip cell is a non-proliferative, highly sensitive and motile cell, characterized by numerous and dynamic filopodia covered with VEGFR2, the major RTK responsible for VEGF-induced EC sprouting. The tip cell is followed by cells possessing another special phenotype termed stalk cells, which are highly proliferative and thus facilitate the outgrowth of the nascent sprout [39]. Cells that are just behind the growing sprout help the tip cell to "choose" its direction away from the parental endothelium, as those cells express and secrete high levels of soluble VEGFR1 which binds VEGF, and in that way models the gradient of VEGF [40]. The molecular mechanisms that underlie the process of tip cell selection and tip-stalk cell communication have been studied thoroughly and the Notch signaling pathway is considered to be the major regulator of these processes [41]. Furthermore, parallels have been reported between the growth of the axonal cone and the nascent sprout. The same ligand-receptor interactions are involved in both processes and the result is either attraction or repulsion of the growing sprout. Molecules with a known role in angiogenesis are semaphorins and their plexin and neuropilin receptors (Nrp1 and Nrp2) [42]. Further, SLIT ligands and their Robo receptors [43], netrins and their DCC and UNC5B receptors [44], as well as ephrins and their Eph receptors [45]. In addition, Nrp1 and ephrin-B2 have an additional function as modulators of VEGF-induced sprouting, as Nrp1 is a co-receptor of VEGFR2 [46], while ephrin-B2 is involved in the internalization of VEGFR2 and VEGFR3 [47]. For more information on the role of axon guidance factors in angiogenesis, please refer to Chap. 11.

Migration of the tip cell is based on repetitive, stepwise cycles of the actin cytoskeleton reorganization. In short, the tip cell protrudes in the direction of migration via actin polymerization. CDC42, a member of the family of small Rho GTPases, controls whether or not the protrusions become filopodia, which harbor a core of long, bundled actin filaments that explore the cellular surroundings and use VEGFR2 enrichment to sense VEGF gradients. At the growing end, Rac1, another member of the Rho GTPase family, regulates the formation of lamellipodia, which harbor a core of a cross-linked actin mesh [48]. Lamellipodia are specialized structures that attach to the substratum *via* integrins linked to actin filaments at the intracellular side of the cell membrane. Integrins activate RhoA which regulates the contraction of the actin cytoskeleton [49]. The opposing side of the tip cell retracts in order to reduce the tension that has arisen after attachment and contraction.

Stalk cells, aside from their divisions that elongate the sprout, can form a lumen. *In vitro* studies suggested a "cord-hollowing" mechanism of lumen formation based on integrin and Rho GTPase-dependent intracellular vacuolization and their further intercellular fusion [50]. This hypothesis was confirmed *in vivo* when outgrowth of intersegmental vessels in zebrafish occurred *via* fusion of small and numerous vacuoles in larger ones and further into multicellular lumen (for more information on vascular development in the zebrafish, please refer to Chap. 2) [51]. More recently, the report of Blum et al. suggested an alternative mechanism of the lumen formation using the same approach [52, 53]. In this scenario, the rearrangement of tight junctions led to a continuous increase in the extracellular space between two stalk cells until the lumen was formed [53]. It is possible that the two described mechanisms of the lumen formation might not be exclusive, but rather context specific [54]. When tip cells from two growing sprouts come into contact, a new vascular connection is formed. At that moment, the motile phenotype of the tip cell is suppressed and VE-cadherin, the major protein of AJs, becomes highly expressed [55].

In the last step of the angiogenic sprouting, vessels mature and become functional. Several distinguishing events can be recognized in the transition of the activated endothelium to quiescence. While migration and proliferation of ECs is reduced, more and more AJs are formed. This is possibly due to increased expression of VE-cadherin [56], while N-cadherin is involved in the recruitment of mural cells and thereby in the essential step of maturation of the newly formed sprout [57]. For more information on junctional signaling in ECs, please refer to Chap. 6, and VEGF-induced permeability and leukocyte extravasation are described in Chap. 8.

As mentioned above, SMC and PCs have been described as mural cells. While PCs cover immature vessels and capillaries where they have intimate contact with ECs, SMCs are spatially separated and regulate the deposition of their own basement membrane in all large-diameter vessels [58]. The origin of mural cells is still under debate. There are data that demonstrate the mesenchymal origin [59] as well as possible differentiation from bone marrow-derived HPCs [60]. On the other hand, PCs in the central nervous system and cardiac tract can arise from neural crest cells [59, 61, 62] (please refer to Chap. 3). Even the transdifferentiation of ECs into SMCs has been reported [63]. PCs themselves are multipotent cells able to differentiate into fibroblasts, osteoblasts, adipose cells and SMCs [64]. Nicosia et al. showed that SMCs can give rise to PCs *in vitro* [65]. No matter which cell is the actual progenitor, it is known that transforming growth factor β (TGF-β) is a key regulator of the number of mural cells [66].

PCs and SMCs have several supportive roles. They stabilize vessels regulating the deposition of ECM components and the secretion of various factors. With their contractile capability they facilitate blood flow [67]. The platelet-derived growth factor B (PDGF-B) pathway has an essential role in the recruitment of mural cells. The ligand PDGF-BB is mainly expressed by vascular ECs and promotes proliferation and migration of mural cells towards the nascent sprout after binding to its receptor PDGFR-β expressed on the latter cell type. Furthermore, genetic ablation of components of this key pathway led to hemorrhages and a dysfunctional vasculature, explaining the embryonic lethal phenotype [68]. Another pathway with a pivotal role in angiogenesis is TGF-β. Although the function of TGF-β is undoubtedly important, it is also antagonistic as it can either promote or inhibit angiogenesis. After binding of TGF-β to its receptor TGF-β R-II, two other molecules can be recruited, namely the activin-like kinase receptors ALK-1 and ALK-5. While ALK-1 is mostly expressed on ECs, where it mediates proliferation and migration, ALK-5 is expressed by PCs. There, it has an opposing function, as it promotes PC differentiation and migration towards the growing vessel. Here, TGF-β regulates the deposition of components of the ECM *via* ALK-5 and thus stabilizes the vasculature [69]. When mural cells are recruited to ECs, they secrete Ang-1, which increases their attachment to the vasculature and at the same time reduces vascular permeability [70]. For more information on PCs in vascular development, please refer to Chap. 3. The role of ECM in vascular signaling has been outlined in Chap. 7. Lastly, the onset of the blood flow is considered to be highly important for the finalization of angiogenic sprouting as effective blood flow increases the local oxygen level which leads to a decrease in VEGF expression [35].

1.1.2 Intussusceptive Angiogenesis

Intussusception or growth within itself was described in the late 1980s [71], but it remains largely underrepresented in the current literature for a technical reason: there are only few experimental approaches for studying intussusceptive angiogenesis both *in vivo* and *in vitro*.

Non-sprouting angiogenesis occurs in a stepwise fashion [72]. It starts with invaginations of the opposing walls of a single EC until they come into contact. In the second step, AJs are reorganized and the central pore develops. In the third phase, PCs and myofibroblasts invade the pore and further deposit components of the ECM. Thereby, the transluminal pillar with a diameter of up to 2.5 μm is formed [73]. Finally, the newly formed pillar enlarges its diameter. The whole process is completed within a few hours.

Depending on the result of this type of angiogenesis, there are three possible "phenotypes": intussusceptive microvascular growth (IMG), intussusceptive arborization (IAR) and intussusceptive branching remodeling (IBR) [72]. The IMG begins with the protrusion of opposing capillary walls into the lumen. The contact zone between them is perforated, followed by the invasion of the supporting cells and the formation of transluminal pillars. In this way, the network enlarges the diffusing surface several times [73–76]. IAR is characterized by a series of pillars that eventually fuse and shape the hierarchical architecture of the vasculature. The outcome of the third "face" of intussusception can oppose IAR as pruning of the unneeded branch is classified as IBR. Here, fusion of pillars cut off blood flow in the extra branch leading to branch regression. Nevertheless, a more frequent effect of IBR is a change in vasculature geometry in order to optimize hemodynamic forces. The relocalization of a branching point and a change in the branching angle are usual events characteristic of IBR [76].

Intussusception is a delicate and metabolically saving phenomenon occurring in the vasculature developed through either vasculogenesis or sprouting angiogenesis. It is a mechanism designed to optimize the supply of blood vessels. Virtually, intussusception happens in the absence of proliferation or migration of EC and without disruption of the basement membrane. Thus, it occurs without interference of the continuous blood flow and is thereby a faster and lower-energy process as compared to sprouting angiogenesis. The total number of ECs remains unchanged in the vasculature undergoing intussusceptive growth but ECs increase their size and surface by flattening [75].

The regulation of the non-sprouting growth of blood vessels differs from that in sprouting. Still, it remains poorly understood what drives intussusception. Since it has been noticed in chick chorioallantoic membrane assays that blood flow modifies the existing vasculature [77], hemodynamic forces are considered to play a key role in the vascular remodeling. Furthermore, *in silico* models have suggested that shear stress is essential for intussusception [78]. However, it should be noted that there are two additional forces exerted toward blood vessels, namely hydrostatic pressure of blood onto the vasculature walls and cycling stretch, which occurs when the blood pressure changes. It is possible that a combination of all three factors regulates intussusceptive angiogenesis [79].

1.1.3 Shear Stress

Shear stress is a tangential force exerted by flowing blood on the vessel surface [80] that depends on three factors: vessel diameter, flow velocity and blood viscosity. Increased blood flow potentiates the shear stress and thus the vessel's diameter enlarges in order to compensate for this increase and return it to normal levels [81, 82]. Arteries that experience increased flow increase their caliber, whereas those experiencing decreased flow rates decrease their caliber, and in extreme cases, completely regress [83]. This mechanism plays an important role during molding of the embryonic vasculature as the blood flow is established before the transport of oxygen and nutrients is required [84]. Further, shear stress has been postulated to promote angiogenesis. Whereas the evidence for stress-induced sprouting is still lacking, intussusceptive angiogenesis has been shown to be significantly modulated by the shear stress [79, 85].

Mechanistically, shear stress causes the glycocalyx, a glycoprotein-polysaccharide network coating the endothelial surface, to bend in the direction of flow to transduce this force *via* the cortical actin skeleton to AJs [86]. Continuous high laminar shear stress promotes EC quiescence and protection from atherosclerosis, while disturbed flow and consequently low or acute shear stress elicits an inflammatory response in ECs [87]. Krüppel-like factor (KLF2) has been shown to play a major role in the former case [88]. KLF2 upregulates expression of genes involved in vasodilation (widening of blood vessels), such as NO synthase (NOS) [89, 90]. Furthermore, it protects the vessels by inhibiting the expression of pro-inflammatory genes regulated by nuclear factor kappa B (NF-κB) [91].

1.1.4 EPCs

Traditionally, the *de novo* formation of blood vessels by vasculogenesis has been considered to be restricted to the embryonic development [92]. In the case of vessel injury, proangiogenic signals activate ECs to proliferate and migrate. At the end of the last century, the so-called circulating EPCs (cEPC) were identified. Those $CD34^+$ cells had the ability to differentiate into endothelial-like cells *ex vivo* and to form cord-like structures when seeded on fibronectin-coated dishes. Further, they had the capacity to repair the injured endothelium [27]. Several subsequent studies reported that cEPC have the ability to be incorporated into ischemic limbs [28], and are recruited into tumors where they play a considerable role in tumor neovascularization [93]. Despite the latter function having been challenged [94], putative EPCs continue to receive much attention owing to their promising therapeutic use in endothelial recovery and replacement of damaged ECs [95]. The work of Asahara et al. importantly solidified the notion that EPCs exist in adults, although the population of cells that the authors isolated was later shown not to consist of true EPCs, rather cells having a similar phenotype and a common origin. From then on, the criteria for the identification of EPCs have been under debate. Different cell populations have been suspected to be genuine vascular precursor cells. Among them, cells derived from bone marrow or circulating in peripheral blood were the most frequently observed. These cells share many of the hallmarks of ECs. They express the same surface markers—CD31, CD34 as well as von Willebrand factor (vWF)—and show VEGF-induced migration and proliferation [26]. The work of Auerbach et al. suggested that the above-mentioned circulating endothelial-like cells can even integrate into existing endothelial networks *in vitro* [96]. However, only cells that have a clonogenic potential, high self-renewal and proliferative capacity, the capability to differentiate into mature EC and that interact with components of the ECM can be considered as EPCs [97]. Despite the reparative potential of bone marrow-derived cells and circulating cells in peripheral blood not being negligible [98], they do not resemble EPCs in maintaining the vasculature under physiological conditions and during pathological remodeling of vessels [99].

1.1.5 Embryoid Bodies

Isolation of EC lines and establishment of *in vitro* culture conditions was a significant milestone in the field of vascular biology [100]. This system, however, is limited as it lacks 3D interactions between ECs, neighboring cells and ECM.

Therefore, the development of embryoid bodies (EBs) greatly facilitated vascular research. EBs are in principle aggregates of pluripotent stem cells cultured in suspension and capable of the differentiation into all three germ lineages [101]. Although postnatal stem and progenitor cells are capable to form spherical aggregates *in vitro*, EBs are mostly derived from ESCs. The most common EB culture methods, hanging drop and static suspension culture, were adopted from *in vitro* differentiation methods originally used for embryonic carcinoma cells or ESCs [101, 102]. The hanging drop method is performed by dispersing a fixed number of ESCs in droplets of media hanging from the lid of a Petri dish. Hanging drops provide a good environment for EB formation as the round bottom of the drop facilitates aggregation of cells and allows control of the number of clustered ESCs [103–105]. A static suspension culture, on the other hand, is simply produced by adding a suspension of ESCs to a bacteriological grade Petri dish or a similar vessel that does not support cell adhesion (e.g., agar- or other hydrophilic-coated substrates) [106–108].

The EB differentiation begins with the formation of an ESC aggregate, whose size depends on the number of cells clustering *via* cell-cell adhesion receptors [109–111]. The first sign of the EB differentiation is the formation of a primitive endoderm on the exterior surface [112]. As the EB development progresses, differentiated cell phenotypes of all three germ layers begin to arise [113]. The first signs of vasculogenesis emerge on day 3 of the differentiation as the common precursors for endothelial and hematopoietic cells, the so-called hemangioblasts, appear [2, 3]. Subsequently, hemangioblasts commit to a more differentiated lineage. The EC precursors, the angioblasts, undergo sequential maturation and eventually express markers of mature ECs such as VEGFR2, CD31 or VE-cadherin [114]. The vessel formation, however, strongly depends on the culture conditions and is induced by the addition of VEGF. The EB cultured in 2D form a primary vascular plexus, which is remodeled from day 6 onwards, by sprouting angiogenesis. Invasive angiogenesis in 3D collagen gels, on the other hand, is manifested by the formation of EC sprouts projecting outward from the central core of the EB [115] and involves the formation of tip and stalk cells. This process strongly resembles sprouting angiogenesis observed in zebrafish and mouse retinae [39, 116]. Eventually, the sprout's branch and the tip cells occasionally fuse with adjacent vessels to form a vascular network surrounded by perivascular cells. Lumen formation occurs between post-differentiation day 10 and 12. Interestingly, these lumenized vessels exhibit features of arterial or venous specification as they specifically express either ephrin-B2 (arterial marker) or EphB4 (venous marker) [117].

The EB model offers a valuable research tool since it mimics developmental and morphological features of ECs *in vivo* in several genetic models. For example, deletion of *Vegfa* results in the arrest of the vascular development and remodeling, both *in vitro* and *in vivo* [118]. Moreover, the exclusive contribution of different VEGF-A isoforms to specific stages of the vessel development can be tested in the EB model. Another advantage of EBs is the possibility of investigating mutations that are lethal in early development stages, especially if they are exhibited before the onset of vasculogenesis [115]. The EB paradigm has, however, limitations as well. Although the development of vascular ECs is faithfully reproduced as in various *in vivo* models, the subsequent differentiation of lymphatic ECs from blood vessels is difficult to control [119]. Moreover, EB vessels lack blood flow—an aspect that might be crucial as flow-induced shear stress influences the remodeling of the vascular system [120].

1.2 Signaling Pathways Governing Blood Vessel Formation

1.2.1 Molecular Structure of VEGFs and VEGFRs

The VEGF pathway is the essential regulator of vascular morphogenesis and implicated in every aspect of the vascular development. Already in vasculogenesis, the earliest phase of the embryonic development, components of the VEGF pathway coordinate the interaction of EPCs/angioblasts [30]. Further, VEGF controls angiogenesis as well as lymphangiogenesis, and even when the vasculature is quiescent, VEGF signals are crucial for EC survival and maintenance of the intact endothelium [121]. For more information on VEGF signaling and vascular permeability, please refer to Chap. 8.

VEGF was isolated for the first time and identified from the medium of a tumor cell line [122]. Soon, its importance in the formation and maintenance of the vasculature was described. In mammals, VEGF ligands comprise a family of five members of homodimeric disulfide-bound glycoproteins, namely, VEGF-A, also referred to as VEGF, VEGF-B, VEGF-C, VEGF-D and the placenta growth factor (PlGF). Among them, VEGF is the most studied. There are three VEGFRs, VEGFR1 (Flt-1), VEGFR2 (Flk-1, KDR) and VEGFR3, all of which are classified as type-V RTKs as they harbor seven immunoglobulin homology domains in their extracellular regions that are involved in the ligand recognition and receptor dimerization. The intracellular domains of the receptors display tyrosine kinase activity, which causes receptor autophosphorylation and subsequently, signal transduction [123] *via* signaling modules such as Erk1/2 in the case of VEGFR2 or Erk1/2 in addition to Akt when VEGFR3 is activated [124]. While VEGFR3 and its primary ligand VEGF-C play a role in lymphangiogenesis, the other two VEGFRs are implicated in the formation and survival of blood vessels. Furthermore, VEGFR1 and VEGFR2 exist as secreted forms which are a result of alternative splicing [125]. VEGF-B and PlGF bind exclusively to VEGFR1, whereas VEGF associates with VEGFR1 and VEGFR2. VEGF-C and VEGF-D show a high affinity for VEGFR2 when proteolytically processed and exhibit an important role in angiogenesis [126]. All VEGF ligands bind to the non-tyrosine kinase receptors Nrp1 and Nrp2. Genetic ablation of *Nrp1* showed defects in EC migration with a fatal outcome, which indicated that Nrp1 acts as a coreceptor that supports the VEGF-VEGFR2 interaction [127].

1.2.2 Signaling for Angiogenesis: VEGF and VEGFR2

VEGF is considered to be the strongest proangiogenic stimulus. Even loss of a single allele resulted in embryonic death with the undeveloped vasculature [128].

This suggested that the concentration of VEGF is tightly controlled in both spatial and temporal sense. VEGF exists in three isoforms, which is a consequence of the alternative splicing causing a loss of one or two exons. Aside from the differences in size, the human VEGF isoforms VEGF121, VEGF165 and VEGF189 (*nota bene*: one amino acid less in the respective isoforms in mice) have a distinctive binding affinity to the ECM [129], making VEGF121 the most soluble form and VEGF189 having the strongest ECM affinity [130]. The existence of three isoforms suggests that all form independent VEGF gradients. Indeed, experiments where either VEGF121 or the heparin-binding isoform VEGF189 were genetically deleted led to lethality. Moreover, upon loss of VEGF121, dilated vessels with low levels of branching were observed, while the VEGF189 knockout resulted in a thin and over-branched vasculature [131]. Previous data suggest that proper availability of VEGF is crucial for vascular morphogenesis as VEGF has a high affinity for VEGFR2. This receptor is highly expressed in tip cells, where the activation of VEGFR2 upregulates the expression of Delta-like ligand 4 (Dll4). In turn, Dll4 activates its receptor Notch1 on the neighboring cell and as a consequence downregulates the expression of VEGFR2 and upregulates the expression of VEGFR1 in the adjacent EC [132]. This leads to the suppression of the tip and promotion of the stalk cell phenotype (Fig. 1.2) [133].

1.2.3 Angiogenic Modifiers: VEGFR1, VEGF-B and PlGF

The stalk cell phenotype is promoted even more by upregulation of VEGFR1. Although VEGFR1 shows an approx. 10 times higher binding affinity for VEGF, it has a much weaker kinase activity. These VEGFR1 features suggest it acts as a decoy receptor that limits VEGFR2 signaling. In addition, VEGFR1 exists in a soluble form. Stalk cells primarily synthesize sVEGFR1 and in that way shape the VEGF gradient. Previous indications have been confirmed in experiments where *Vegfr1* was deleted, leading to uncontrolled angiogenesis [134]. Similar observations were collected in mice with genetic loss of both VEGFR1-specific ligands, VEGF-B and PlGF [135, 136]. The main function of VEGFR1, VEGF-B and PlGF is to control the VEGF-induced activation of VEGFR2 and modulate angiogenic sprouting [40] and proliferation of ECs [137].

1.2.4 Fibroblast Growth Factor (FGF)

In contrary to the relatively restricted function and expression pattern of VEGF and its receptors, FGF receptors and ligands regulate a plethora of cell types [138]. There are at least 18 mammalian FGFs (FGF1-FGF10 and FGF16-FGF23), which are divided into 6 subfamilies. Most FGFs are paracrine factors, which play an important role in tissue patterning and organogenesis during embryogenesis [139]. FGF receptors (FGFRs) are RTKs encoded by four genes in mammals (*FGFR1-FGFR4*) [140]. Additional isoforms are generated by alternative splicing (e.g., the

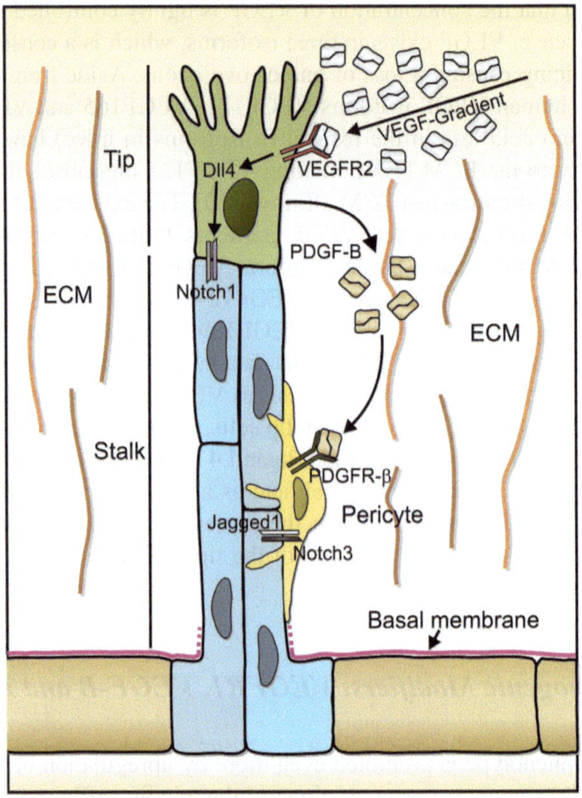

Fig. 1.2 Schematic illustration of a growing sprout. A fraction of endothelial cells (*shown in green*) extends long filopodia and acquires motile and invasive behavior in response to tissue-derived vascular endothelial growth factor (*VEGF*). These tip cells guide new sprouts, whereas other endothelial cells form the sprout stalk (*shown in blue*) and stay behind to maintain tissue perfusion (*shown in brown*). The VEGF gradient induces the expression of Delta-like ligand 4 (*Dll4*) in the tip cell to activate Notch signaling in neighboring stalk cells. Stalk cells attract supporting PCs (*orange*) by releasing PDGF-B

epithelial variants FGFR1b-3b or the mesenchymal variants FGFR1c-3c) [141]. All of them are single-pass transmembrane proteins that require binding of both FGF and heparan sulfate glycosaminoglycan (HSGAG) for full stimulation [142]. This results in the activation of the classical RTK signaling pathways mediated among others by phospholipase Cγ1 (PLCγ1), mitogen-activated protein kinase (MAPK) or phosphatidylinositol 3-kinase (PI3K)/Akt [143].

FGF2 has a pro-angiogenic effect on ECs and stromal cells, which is mediated by the induction of VEGFR expression [144–146]. However, once VEGFR is expressed, its ability to induce angiogenesis becomes FGFR-independent [147]. FGF9 plays an important role in the formation of the coronary vascular plexus as it induces the expression of VEGF-A, VEGF-B, VEGF-C and Ang-2 *via* Hedgehog in cardiomyoblasts [148]. Moreover, inhibition of FGFR results in the vessel disintegration, suggesting that low levels of FGF are required for the maintenance of vascular integrity [138]. It should be noted, however, that due to high redundancy in the FGF superfamily, the knock-out of *Fgf* genes does not result in vascular defects [139]. Thus, it is challenging to dissect the specific role of FGF/FGFR pathway components in developmental angiogenesis. For more information on FGF in the vascular development, please refer to Chap. 4.

1.2.5 Ang/Tie Signaling

Angs are secreted glycoproteins that regulate the blood vessel development and stability [149]. Three human paralogues exist: Ang-1, Ang-2 and Ang-4 (an orthologue of murine Ang-3) [150]. Ang-1 and Ang-2 form dimers, trimers and tetramers while Ang-1 is able to form multimers of higher order [151, 152]. Only tetrameric and higher multimeric forms of Ang-1 are capable of activating its receptor Tie-2 (tyrosine kinase with immunoglobulin and epidermal growth factor (EGF) homology domains 2). Oligomeric Ang-2 is either a weak agonist or an antagonist of Tie-2 depending on the cellular context. Another Tie paralog, Tie-1, is considered to be an orphan receptor. Yet, Tie-1 has been shown to form heterodimers with Tie-2 and regulate its activity [150, 153].

Angs are expressed in the vascular system in a complementary manner, as Ang-1 is mostly detected in PCs, SMCs and fibroblasts, whereas Ang-2 is restricted to ECs. The transcription of both isoforms is dependent on hypoxia and VEGF-A [150, 154]. In angiogenic sprouting, Ang-1-mediates activation of Tie-2, expressed mostly in stalk cells [155], promotes EC survival and quiescence [156]. Ang-2, on the other hand, is stored in the so-called Weibel-Palade bodies (WPBs) of quiescent cells and released in response to angiogenic stimulators [157]. Recently, vWF has been identified as an important component of WPBs since its knock-down leads to the release of Ang-2 [158] (for more information on the role of coagulation factors in angiogenesis, please refer to Chap. 10). Once released by sprouting ECs, Ang-2 acts as an Ang-1 antagonist and causes destabilization of the endothelium in an autocrine manner by promoting mural cell detachment and vascular permeability [153]. Ang-2 primes the vasculature for a robust response to growth factors like VEGF by enabling the penetration of proteases, cytokines and angiogenic myeloid cells. Cell migration and sprouting angiogenesis have both been shown to be promoted in a Tie-2-independent manner, as Ang-2 is able to bind to several integrin family members and activate focal adhesion kinase [159]. For more information on the Ang/Tie system in angiogenesis, please refer to Chap. 13.

1.2.6 Hypoxia-inducible Factor (HIF)

HIF-1 is a heterodimeric protein composed of two subunits: HIF-1α and HIF-1β. Both subunits are encoded by several paralogous genes and members of the basic helix-loop-helix family of transcription factors [160]. HIF-2α is closely related to HIF-1α and both proteins bind to hypoxia-responsive elements (HREs) located within the regulatory sites of the HIF target genes. In contrast to the ubiquitously expressed HIF-1α, HIF-2α is restricted to the lungs, carotid body and endothelium. Moreover, its specific targets are not directly involved in the pro-angiogenic hypoxic response [161]. The primary function of HIF-3α remains elusive [162].

The stability of HIF-1α is dependent on the intracellular concentration of oxygen. In the presence of O_2, one of the three prolyl hydroxylases (PHD1, PHD2, PHD3) hydroxylizes HIF-1α at Pro402 and Pro564, amino acid residues situated in the oxygen-dependent degradation domain [163]. This post-translational modification targets HIF-1α for ubiquitination and subsequent proteasomal degradation [164]. Another hydroxylase, called factor inhibiting HIF-1 (FIH-1), utilizes Asn803 at the C-terminal transactivation domain as a substrate. The hydroxylation of Asn803 does not result in the protein degradation but interferes with the binding of HIF-1 to the transcriptional co-activator CBP/p300 [165]. Therefore, hypoxia leads to a stabilization of HIF-1α and its translocation from the cytoplasm to the nucleus, where it undergoes dimerization with HIF-1β. The resulting protein complex binds to HRE and activates hypoxia-induced target genes.

Both VEGFR1 and VEGFR2 are upregulated by hypoxia: while the promoter of *VEGFR1* contains an HRE, VEGFR2 is regulated at the post-transcriptional level [166, 167]. Other pro-angiogenic transcriptional targets of HIF-1 include VEGF, Ang-1, Ang-2, PlGF, PDGF-B and their different receptors as well as ECM-modifying proteins such as matrix metalloproteinases, plasminogen activator receptors and inhibitors, and procollagen prolyl hydroxylase. The processes regulated by these genes are of utmost importance in ECs and contribute to angiogenesis, ECM invasion and lumen formation [168]. For more information on oxygen signaling in angiogenesis, please refer to Chap. 14.

1.2.7 NO Signaling

NO is a gas that regulates blood-vessel diameter and consequently blood pressure [169]. In addition to its role in vasodilation, it has been shown to regulate inflammation, angiogenesis and tumor progression [170]. NO is generated by NOS from L-arginine and oxygen [171]. There are three different NOS isoforms: neuronal NOS (nNOS or NOS1), inducible NOS (iNOS or NOS2) and endothelial NOS (eNOS or NOS3) [172]. nNOS and eNOS are constitutively expressed and mostly restricted to neurons and ECs, respectively. NO is predominately generated by eNOS in the endothelium and regulates vessel function and maturation [172]. There are several factors that regulate its expression *in vivo* including hypoxia, shear stress, cytokines and inflammation [173]. The promoter of *NOS3* contains an HRE and is strictly dependent on HIF-2 *in vitro* [174]. Moreover, hypoxia-induced VEGF enhances NO production *via* eNOS and VEGF-induced vascular permeability is significantly reduced in *Nos3*-deficient mice [175, 176]. One of the mechanisms decreasing the tightness of AJs is NO-dependent S-nitrosylation of VE-cadherin and β-catenin.

The binding between these two proteins is reduced upon S-nitrosylation and the weakened cell-cell contact allows angiogenesis to occur [177]. Two crucial proteins involved in the oxygen-sensing machinery, namely HIF-1α and PHD2, are also modulated by S-nitrosylation. This chemical modification results in the increased stability of HIF-1α and inhibits PHD2 activity, resulting in the accumulation of HIF-1α in normoxia [178–181]. The interplay between HIFs and NOSs has been investigated in macrophages, and it strongly depends on the cell subtype. In M1 macrophages, cytokine-induced HIF-1α drives expression of iNOS and results in NO-mediated VEGF secretion and thus angiogenesis. In M2 macrophages, on the other hand, HIF-2α is upregulated and promotes expression of arginase, an enzyme that reduces NO synthesis by converting L-arginine to ornithine. Eventually M2 macrophages promote arteriogenesis and vessel maturation rather than angiogenesis [182]. For more information on NO synthesis in vascular physiology and pathophysiology, please refer to Chap. 16 and for information on vasodilators and vasoconscrictors dependent on cytochrome P450 enzymes, please refer to Chap. 9.

1.2.8 The Notch Signaling Pathway

Notch signaling is a cell communication pathway that is evolutionarily conserved in virtually all metazoans. In contrast to *Drosophila*, whose genome encodes only a single Notch gene, there are four paralogs (Notch 1–4) in mammals [183]. Although every Notch receptor is translated as a single polypeptide, the mature protein consists of two non-covalently associated polypeptide chains forming single-pass type I transmembrane protein. The precursor is split in the trans-Golgi network by furin-like proteases. This so-called S1 cleavage results in the formation of two polypeptides: N-terminal extracellular domain (NECD) and a C-terminal portion consisting of the transmembrane (NTM) and intracellular (NICD) domains [7].

The NECD contains predominately EGF-like repeats, which are necessary for ligand binding [184]. Many of them bind calcium ions, which influence the structure of the receptor and ligand binding affinity [185]. In addition, EGF-like repeats are fucosylated on specific serine and threonine residues by O-fucosyltransferases, a process required for the efficient ligand binding [184, 186, 187]. These O-fucose moieties can be further extended by the addition of N-acetylglucosamine mediated by the Fringe family of 1,3 N-acetylglucosamintransferases [188]. Such modifications fine-tune the affinity of Notch binding towards specific ligands [188, 189]. The EGF-like repeats are followed by a negative regulatory region (NRR), responsible for the auto-inhibition of the Notch receptor [190, 191] and binding to a short extracellular region of NTM [192]. The NRR consists of three cysteine-rich Lin12/Notch repeats (LNR) [193] and is adjacent to a juxtamembrane heterodimerization domain.

The intracellular domain NICD is a mediator of Notch signaling and includes the recombination signal-binding protein Jκ (RBP-Jκ) associated module (RAM) [194] just after the NTM. This domain forms a high-affinity binding module centered on a conserved Trp-X-Pro motif. Seven ankyrin (ANK) repeats [195] are linked to RAM *via* an unstructured motif containing a nuclear-localization signal (NLS) and followed by a bipartite NLS [196] in addition to a loosely defined transactivation domain [197]. Finally, the C-terminus of Notch contains a PEST sequence (rich in proline, glutamic acid, serine and threonine) [198] that harbors a degradation signal responsible for NICD stability.

The canonical Notch ligands are type I transmembrane proteins and orthologues of *Drosophila* Delta and Serrate. The so-called DSL (Delta-Serrate-Lag2) family includes five mammalian ligands: Dll1 [199], Dll3 [200], Dll4 [201], Jagged1 [202] and Jagged2 [203]. The N-terminal region, the DSL domain and the first two EGF-like repeats, called the DOS (Delta and OSM-11-like proteins) domain, are essential for the interaction with EGF-like repeats of Notch receptors [204, 205]. These ligands contain a variable total number of EGF repeats and can be classified depending on the presence (Jagged) or absence (Delta) of a cysteine-rich domain. It should be noted that Dll3 is the most structurally divergent DSL ligand as it lacks the DOS domain [200], is incapable of inactivating Notch receptors *in trans* [206] and is rarely, if ever, present at the cell surface [207, 208]. Moreover, soluble forms of Notch ligands have been shown to inhibit receptor activation [209, 210]. It has been even suggested that exosomal release of free Notch ligands is a physiological process involved in blood-vessel outgrowth [211]. In addition to the above-mentioned proteins, several transmembrane and soluble proteins containing EGF-like repeats and lacking a DSL domain have been described as non-canonical ligands, e.g., Delta-like 1 (Dlk1), Dlk2, Delta and Notch-like EGF-related receptor (DNER) and the EGF-like protein 7 (EGFL7) [209, 212, 213]. Dlk1, Dlk2 and DNER are transmembrane proteins (although Dlk1 and Dlk2 also exist in soluble forms), while EGFL7 is a *bona fide* secreted factor. According to available data, Dlk1/2 and EGFL7 are Notch inhibitors while DNER activates Notch signaling [209].

Canonical Notch signaling requires binding of two membrane-bound proteins, which can be expressed by the same or two adjacent cells. The interaction between neighboring cells, referred to as *in trans* interaction, switches the Notch signaling on (Fig. 1.3) while *in cis* interaction, i.e. between a receptor and a ligand expressed on the same cell, inhibits the Notch pathway [214–216]. *In trans* activation leads to the ubiquitination and internalization of the interacting ligand. It has been proposed that this internalization acts as a "pulling" mechanism to disrupt the hydrophobic interactions between NECD and NTM in the Notch receptor. Subsequently, NTM becomes exposed to extracellular S2 cleavage by "a disintegrin and metalloprotease" 10 (ADAM10) or ADAM17 [217] creating a membrane-tethered intermediate called Notch extracellular truncation (NEXT). NEXT is consequently processed by the γ-secretase complex, a multi-subunit protease complex containing presenilin, nicastrin, presenilin enhancer 2 (Pen2) and anterior pharynx-defective 1 (Aph1) [192, 218, 219]. In turn, the intracellular Notch domain NICD, which is released upon this so-called S3 cleavage, translocates into the nucleus [220]. NICD cannot directly bind to DNA but heterodimerizes with a member of the CSL protein family (RBP-Jκ/CBF-1/KBF2 in mammals). In the absence of an activator such as NICD, RBP-Jκ represses transcription of Notch target genes by recruiting co-repressor

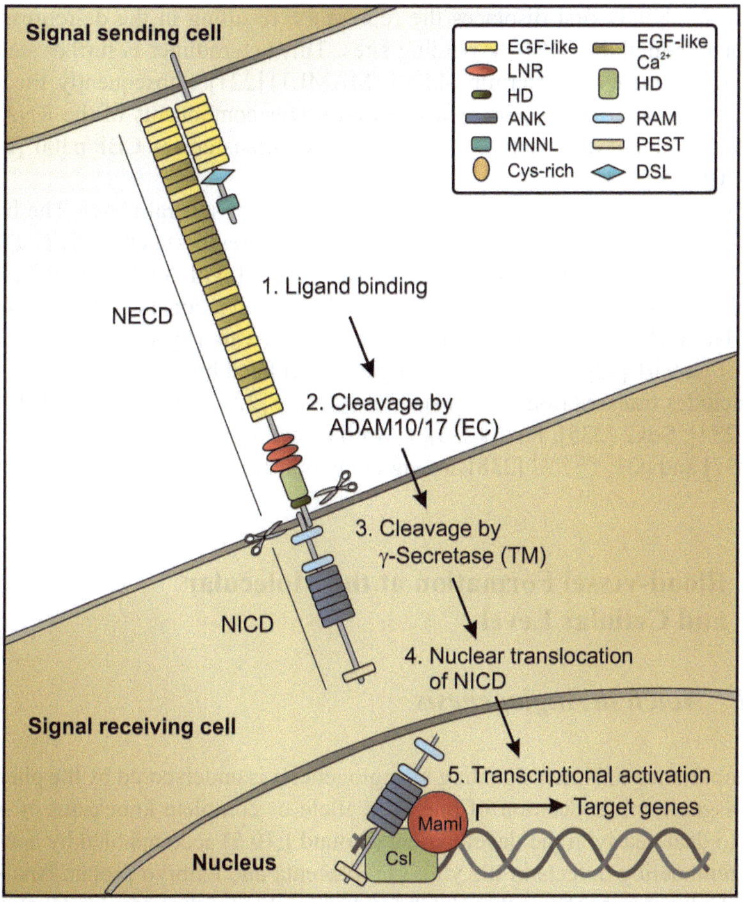

Fig. 1.3 Canonical Notch signaling. All four mammalian Notch receptors consist of two non-covalently bound polypeptide chains. The Notch extracellular domain (*NECD*) contains a variable number of epidermal growth factor (*EGF*)-like repeats (many EGF-like repeats bind calcium ions, which influence the structure of the receptor and ligand binding affinity) followed by a negative regulatory region (*NRR*) composed of three cysteine-rich (*Cys-rich*) Lin12/Notch repeats (*LNR*) and a juxtamembrane heterodimerization domain. The N-terminus of the second subunit begins with a single transmembrane domain (*NTM*) extending into the intracellular domain (*NICD*), which is involved in cellular signaling and includes the recombination signal-binding protein Jκ (*RBP-Jκ*) associated module (*RAM*), seven ankyrin (*ANK*) repeats, two nuclear localization sequences (*NLS*), a transactivation domain and a C-terminal PEST sequence (rich in proline, glutamic acid, serine and threonine). The so-called canonical Notch ligands are transmembrane molecules composed of an N-terminal domain of Notch ligands (*MNNL*; except *Dll3*), a Delta-Serrate-Lag2 (*DSL*) domain, several EGF-like domains and a cysteine-rich domain, which is specific only for Jagged ligands. The interaction between Delta/Jagged-type ligands and Notch receptors leads to S2 cleavage on the extracellular site by "a disintegrin and metalloprotease" 10 (*ADAM10*) or ADAM17, which is followed by S3 cleavage by the γ-secretase-presenilin complex. The S3 cleavage gives rise to an intracellular Notch fragment (*NICD*) that translocates into the nucleus, where NICD binds to a protein complex containing recombination signal-binding protein Jκ (*RBP-Jκ*). This mediates the conversion of RBP-Jκ from a repressor to a transcriptional activator and is followed by the recruitment of the co-activator mastermind-like 1 (*MAML1*). These events lead to the de-repression of transcription of Hairy/Enhancer of Split (*Hes*) and Hey. NECD—Notch extracellular domain, NTM—Notch transmembrane domain, HD—heterodimerization domain

complexes. NICD first displaces the repressors, resulting in the de-repression of promoters containing RBP-Jκ binding sites. This heterodimer is further stabilized by the recruitment of mastermind-like 1 (MAML1) [221]. Subsequently, the NICD-RBP-Jκ-MAML1 ternary complex recruits further components of the RNA polymerase II holoenzyme such as the histone acetyltransferases CBP/p300 [222] or PCAF/GCN5 [223].

Ultimately, these events lead to the transcription of several genes. The best described group of Notch targets are transcriptional repressors such as Hairy/Enhancer of Split (Hes) and Hey (subfamily of Hes, related with YRPW motif) proteins [224–226]. Most of the family members are direct Notch targets as the promoters of Hes1, Hes5, Hes7, Hey1, Hey2 and HeyL can be activated by a constitutively active form of Notch1 [226–228]. The list of genes regulated by Notch is still expanding and includes transcription factors such as NF−κB [229, 230], PPAR [231], c-Myc [232–234], Sox2 [235], Pax6 [236], as well as cell-cycle regulators such as cyclin D1 [237] and $p21^{CIP1/WAF1}$ [238] among many others.

1.3 Blood-vessel Formation at the Molecular and Cellular Level

1.3.1 Notch in Angiogenesis

The importance of Notch signaling in angiogenesis is underscored by the phenotype of transgenic mice: deletion of one *Dll4* allele or complete knock-out of *Notch1* leads to death early in the development (around E10.5) accompanied by acute vascular remodeling defects in the yolk sac, placenta and embryo proper. Analysis of *Notch4* mutant mice shows that Notch4 is dispensable for the vascular development. The double *Notch1/Notch4* knock-out, however, results in a more severe phenotype than *Notch1* knock-out including the abnormal development of intersomitic vessels and cardinal veins as well as of the dorsal aorta [239, 240]. The knock-out of *Jag1* or *Dll1* leads to a less severe outcome, yet such mice also perish early during gestation due to vascular defects and intense degree of hemorrhage [241, 242]. Notch pathway mutations are known to cause cardiovascular diseases in humans: cerebral autosomal dominant arteriopathy and subcortical infarcts and leukoencephalopathy (CADASIL) as well as Alagille syndrome have been associated with *NOTCH3* and *JAG1* mutations, respectively. Finally, the relevance of Notch receptors and ligands is reflected by their ample expression either in the endothelium or in the surrounding mural cells [243–245]. Moreover, expression of Notch4 and Dll4 is mostly restricted to the vascular system [201, 246–248].

One of the most important functions of Notch in the vascular development is to regulate the formation of an appropriate numbers of tip cells. The specification of ECs into either tip or stalk cells represents a binary decision process, which is regulated by the so-called lateral inhibition: a process in which a cell that stochastically acquires enhanced ligand expression stimulates a neighboring cells. The *in cis* inhibition of Notch on the ligand-expressing cells renders this interaction unilateral. The classical example of lateral inhibition is the neural-epidermal choice in *Drosophila*: the signal-sending cell will differentiate into a neuronal precursor while the signal receiving cell will adopt the epidermal fate instead [183, 249]. In mammals, Dll4 is prominently expressed in endothelial tip cells [132, 250] while the strongest Notch activity is most frequently observed in stalk cells [132, 251]. Stalk cells also express Jagged1, which has a lower binding affinity to Notch than Dll4, but nevertheless is able to compete for a receptor binding. Fringe modification of Notch renders it more prone for activation by Dll4 over Jagged1 and thus Jagged1 binding is antagonistic in this situation [252]. Genetic inactivation or pharmacological inhibition of either Dll4 or Notch1 results in augmented sprouting, branching and hyperperfusion of the capillary network. Excessive tip cell formation is responsible for this phenotype as evidenced by widespread filopodia formation and enhanced expression of tip cell-specific genes such as PDGF-B, UNC5B, VEGFR1 and VEGFR2 [132, 133, 253]. The Notch requirement for the stalk cell differentiation by suppressing tip cell fate was first determined *in vitro* in 3D EC culture, where blocking Notch resulted in increased branching [254]. This observation has been confirmed by the mosaic analysis of ECs deficient in Notch signaling as the majority adopted the tip cell phenotype. Ectopic activation of Notch signaling or injection of the Jagged1 peptide, on the other hand, reduced the number of tip cells and filopodia protrusions in the mouse retina [132]. These data indicate that the tip–stalk cell specification is a highly dynamic process. ECs stimulated by VEGF-A compete for the tip cell position *via* Dll4/Notch signaling. A cell that expresses more Dll4 will become a tip cell and suppress the neighbors *via* lateral inhibition. It appears that in the absence of Notch signaling the default phenotype is that of a tip cell, while the differentiation into stalk cells is activated upon Notch stimulation [132, 255]. The exact mechanism of this phenotypic switch remains elusive but the regulation of genes encoding VEGFR 1 and VEGFR2 and their co-receptors such as Nrp1 and VEGFR3 plays an important role in this process [256].

1.3.2 Negative Regulators of Notch Signaling in ECs

Notch signaling is regulated at multiple levels in ECs starting with the competition between membrane-bound and soluble ligands such as EGFL7. This secreted Notch interacting protein contains an N-terminal Emilin-like (EMI) domain and 2 EGF-like repeats harboring a putative DSL and a Ca^{2+}-binding domain [202, 257–259]. EGFL7 is highly expressed in the proliferative vasculature but absent from mature blood vessels [260]. EGFL7 knock-down in human umbilical vein ECs (HUVECs) leads to suppression of proliferation, migration and capillary sprouting [258]. *Egfl7* loss-of-function mouse models exhibit partial embryonic lethality and vascular abnormalities such as edema and reduced vascular coverage of the head and retina [261]. However, this phenotype was not fully assigned to the lack of EGFL7 protein as the *Eglf7* locus also encodes the EC-relevant miR126, located in intron 7 [262–264] and an *Egfl7* knock-out with unaltered miR126 expression did not result in an overt vascular malfunction [265]. Nevertheless, the function of EGFL7 in blood vessel formation has been presented by other means: overexpression of EGFL7 in the postnatal mouse resulted in a phenotype resembling *Dll4*$^{+/-}$ retinas, with elevated number of sprouting filopodia and a denser vascular network [258]. Further, EGFL7 has been shown to promote angiogenic vessel growth: it decreased adhesion of HUVECs, and increased random cell migration on fibronectin in an integrin $\alpha v\beta_3$-dependent manner. Deregulation of EGFL7 in zebrafish embryos led to severe integrin-dependent malformations of the caudal venous plexus. Moreover, the pro-angiogenic activity of EGFL7 was confirmed *in ovo* and *in vivo* [266].

Another level of fine-tuning Notch signaling is the regulation of NICD stability by ubiquitination and proteasomal degradation. The F-box protein, Fbxw7, is the NICD-recognizing component in the SCF-E3 type ubiquitin ligase complex [267]. Fbxw7 has been shown to promote angiogenesis by inducing the NICD degradation in an EC-specific manner. Fbxw7 inactivation in the postnatal mouse resulted in reduced sprouting, proliferation and tip cell formation as well as increased Dll4 staining and decreased VEGFR3 expression consistent with augmented Notch activity [268]. Sirtuin-1 (SIRT1) is another protein regulating NICD proteolysis by mediating NAD^+-dependent deacetylation as acetyl groups prevent proteasomal degradation. Deletion of *Sirt1* in ECs resulted in a decrease in angiogenic sprouting and increased expression of Notch target genes, leading to a less dense vascular network [269]. Furthermore, RBP-Jκ protein stability is also regulated by ubiquitination and it has been shown in ECs that the zinc finger protein BAZF, which is upregulated upon VEGF stimulation, promoted proteasomal degradation of RBP-Jκ by recruiting CUL3 E3 ubiquitin ligase. Consequently, *Bazf* knock-out mice exhibited reduced sprouting angiogenesis in the retina and increased Hey1 expression [270].

Negative regulators of Notch signaling can themselves be downstream targets of Notch, like Notch-regulated ankyrin repeat protein (Nrarp). This protein has been shown to form a ternary complex with NICD and RBP-Jκ *in vitro* and to inhibit NICD-mediated transcription in *Xenopus* and zebrafish [271]. Nrarp is expressed in stalk cells at the branch points, where it restricts Notch signaling in addition to enhancing Wnt signaling to promote EC proliferation and vessel stability. This has been confirmed *in vivo* as the loss of *Nrarp* reduced vessel density, and led to the formation of poorly lumenized vessels, remodeling of endothelial junctions and vessel regression [272].

1.3.3 Notch and Cellular Signaling in ECs

Notch signaling regulates several processes in ECs and has been shown to inhibit cell proliferation in several systems: 3D sprouting assays [254], developing and adult mouse retina [273, 274], and during tumor angiogenesis [275, 276]. Cell culture-based studies indicate that the regulation of proliferation is mediated at the transcriptional level downstream of NICD/RBP-Jκ/MAML targets and involves inhibition of the MAPK and PI3K/Akt pathways [277]. Furthermore, Notch signaling has been linked to cell cycle regulators: Notch1 and Notch4 cause downregulation of $p21^{CIP1/WAF1}$ expression, which in turns leads to enhanced translocation of cyclin

D-CDK4 to the nucleus. Cyclin D-CKD4-mediated Rb phosphorylation results in cell-cycle progression [278]. These data have been confirmed *in vivo* in reverse as the endothelial deletion of RBP-Jκ in adult mice resulted in augmented EC proliferation [274]. Data supporting the anti-proliferative function of Notch mostly reflected the situation *in vitro*, in adult tissue or in pathological situations and did not take into account the heterogeneity of ECs in angiogenic sprouting. Stalk cell proliferation is required for shaft elongation *in vivo*. Thus, Nrarp is expressed in these cells upon Notch activation, which limits Notch signaling and promotes Wnt signaling. In this way, Nrarp increases the levels of the Wnt/β-catenin target gene cyclin D1 and thereby promotes stalk cell proliferation [272].

Formation of filopodia, which are cell protrusions involved in cell adhesion, chemotaxis and cell migration, is one of the defining morphological features of a tip cell. Thereby, consistent with suppressing tip cell features, Notch has been shown to inhibit EC migration. In mice overexpressing Dll4, there is decreased EC migration and sprouting from the dorsal aorta to form intersomitic vessels. A similar response has been observed *in vitro* as HUVECs expressing full-length Dll4 exhibited decreased motility in the presence of exogenous VEGF [279]. Several pathways are responsible for the regulation of migration but the available data point towards the VEGF co-receptor Nrp1 as a crucial mediator of motility. Nrp1 is strongly downregulated by Notch signaling [280, 281] and the *Nrp1* knock-out resulted in reduced EC migration and EC guidance in mice, while proliferation remained unaffected [282, 283].

Migration of a cell is dependent on its adhesion to the surrounding environment, underscoring the importance of ECM molecules, whose expression is regulated by the Notch signaling pathway. The mRNA levels of fibronectin, laminin and collagen I and IV were increased in mouse embryos overexpressing Dll4. As a consequence, increased deposition of ECM (including fibronectin and laminin) around the dorsal aorta was observed in these transgenic animals [279]. Haploinsufficiency of *Dll4* led to the opposite effect, further supporting this model, as *Dll4*$^{+/-}$ mutant mice exhibit decreased expression and irregular deposition of collagen IV and laminin [284]. The adhesion to ECM is often regulated by integrins and it has been shown that the overexpression of intracellular domain of Notch4 (NICD4) in ECs augmented their adhesion to collagen in an integrin β_1-dependent manner [285]. Along these lines, NICD1 was able to activate β_1-integrins in a non-transcriptional manner and promote EC adhesion to fibronectin via $\alpha_5\beta_1$integrin. R-Ras, activated by NICD1 in these cells, antagonized H-Ras mediated integrin suppression and in turn increased integrin affinity [286]. Recent evidence supported the notion that expression of Notch pathway components can be induced by integrins in ECs. Combined $\alpha_2\beta_1$ and $\alpha_6\beta_1$ integrin signaling has been shown to stimulate laminin, enhance Dll4 expression and involve Foxc2-dependent transcription. Interestingly, none of the other endothelial Notch ligands was regulated by integrins [287].

1.3.4 Arterial Specification: Notch Meets Wnt

According to current knowledge, arterial-venous differentiation is specified before the initiation of blood flow and governed by genetic factors involving Notch signaling. Dll4 expression is restricted to large arteries and, consequently, Dll4 heterozygous knock-out mice exhibited defects in the arterial development. However, veins were also malformed suggesting that Notch signaling indirectly affected vein formation [239, 288, 289]. The genetic studies also provided evidence that arteriogenesis is dependent on Foxc transcription factors. Mouse embryos double mutant for *Foxc1* and *Foxc2* displayed arteriovenous malformations and a lack of arterial markers. This defect in arterial specification is probably due to disrupted regulation of Dll4 transcription since upregulation of either Fox transcription factors results in increased expression of Dll4, Notch1, Notch4 and ephrin-B2 [290]. During zebrafish and mouse development, Notch signaling affected the specification of immature vessels into the arterial or venous lineage by regulating the expression of specific markers [291–293]. For example, homozygous deletion of *Rbpj* led to the loss of arterial markers such as ephrin-B2 and CD44 [289]. Along these lines, upregulation of Dll4 led to an increased expression of arterial markers such as Hey2 and ephrin-B2 and a decrease in the expression of venous markers such as COUP-TFII [279]. Inhibition of Notch had the opposite effect [294, 295], suggesting that the venous phenotype is the default one and it has to be actively repressed by Notch signaling in order to induce arterial differentiation. Murine ES cell differentiation studies demonstrated that high VEGF-A levels induced an arterial EC phenotype, whereas low intermediate concentrations were associated with a venous identity [296]. The genetic predisposition for vein formation is demonstrated by the repression of Notch signaling by COUP-TFII in the venous compartment [297].

ECs express various types of Wnt ligands and their frizzled (Fzd) receptors. The role of Wnt pathway is most apparent in the specification of arterial ECs; gain-of-function of β-catenin, a transcriptional co-activator of the Wnt signaling pathway, resulted in the impairment of arteriovenous specification, namely the loss of venous markers and acquisition of arterial markers. Such a phenotype closely resembles consequences of Dll4 overexpression and indeed the Dll4 promoter has been shown to be directly bound by β-catenin [298]. Moreover, β-catenin has been shown to form a complex with NICD and RBP-Jκ in arterial ECs and bind regulatory regions of arterial-specific genes, such as *EphrinB2*, *Nrp1*, *Hes1*, *Dll4* and *Cxcr4* [299]. Recently, Sox17 has been proposed to act downstream of Wnt and promote Notch signaling in ECs by upregulating Hey1, Dll4, Dll1 and Notch4. Moreover, endothelial-specific *Sox17* deletion impaired arterial specification [300]. Another group, however, reported that Notch acts upstream of Sox17 by suppressing its expression in ECs [301]. Thus, additional data is required to fully understand this process. Interestingly, Notch regulation of arterial-venous specification of blood vessels is not restricted to the embryonic development since activated Notch4 in adult mice led to hepatic vascular shunting, arterialization and induction of other Notch pathway genes, culminating in lethality within weeks [302]. In addition, the induction of adult arteriogenesis following ischemia has been shown to be dependent on Dll1 and Notch1 [303, 304].

1.4 VEGF and Notch: the Interplay

1.4.1 The Physiological Interplay: EC Sprouting

It is impossible to describe the involvement of Notch signaling pathway in endothelial sprouting without mentioning components of the VEGF pathway since they are interconnected at several levels. VEGF regulates Notch signaling by inducing the expression of Dll4 *via* VEGFR2 as observed in mouse retina [253, 273, 291] and endothelial cultures [291]. These observations have been confirmed by elegant mosaic studies with EBs derived from wild-type, *Vegfr1*$^{+/-}$ and *Vegfr2*$^{+/-}$ mouse cells: ECs expressing less *Vegfr1* or more *Vegfr2* were more likely to occupy the tip position in angiogenic sprouts. Moreover, Dll4 was expressed at a higher level in *Vegfr1*$^{+/-}$ cells and at a lower level in *Vegfr2*$^{+/-}$ cells relative to wild-type cells [38]. When it comes to the signaling mechanism, it has been shown in the arterial endothelium that the induction of Dll4 expression is mediated by PI3K and ERK together with Foxc transcription factors [290, 305]. Another link between VEGF and Dll4, which involves disassembly of the transcriptionally repressive Tel/Cebp complexes at the Dll4 promoter, has been recently proposed [306]. Other reports offer an amplification loop for VEGF stimulation and describe the involvement of integrins. VEGF was found to stimulate the expression of laminin-γ1, which in turn activated the integrins $\alpha_2\beta_1$ and $\alpha_6\beta_1$ and augmented Dll4 expression. Additionally, the knock-down of α_2 or α_6 integrin subunits resembled Dll4 deletion by increasing the number of branching points in a HUVEC 3D culture [287].

The interaction between VEGF and Notch pathways is bilateral as Notch signaling regulates expression of several VEGFRs and their co-receptors. Hey1 has been shown to bind to the *Vegfr2* promoter and suppress the expression of the receptor [307, 308]. Nrp1 and VEGFR3 are also downregulated by Dll4/Notch [281, 292, 309, 310], the latter one possibly post-transcriptionally since the *Rbpj* knock-out led to a slight increase in VEGFR3 mRNA levels but a strong upregulation of the protein [311]. The role of VEGFR3 in endothelial sprouting is still not fully understood as pharmacological inhibition of VEGFR3 or VEGF-C reduced tip cell formation [133], while the *Vegfr3* knock-out caused the opposite phenotype [312]. The most plausible model implies that ligand-dependent VEGFR3 signals are pro-angiogenic whereas ligand-independent signaling activates Notch [312]. Another level of Notch–VEGF interaction is the increased expression of VEGFR1 by activated Notch. VEGFR1 is a decoy receptor for VEGF-A, which can quantitatively and spatially regulate VEGFR2 signaling [313]. Further, the soluble splice variant of VEGFR1 is upregulated in ECs upon Notch activation [280], suggesting that sequestration of VEGF-A is one of the means of decreasing VEGF signaling by Notch.

All in all, the negative feedback loop between VEGF and Notch pathways can be simplified to three steps: (i) VEGF induces endothelial Dll4 expression followed by (ii) Dll4 activating Notch in neighboring cells ultimately leading to (iii) reduced expression of activating receptors such as VEGFR2 as well as enhanced expression of decoy receptors (VEGFR1 and sVEGFR1) and consequently lowered VEGF signaling (Fig. 1.4). The cell expressing VEGFR2 and Dll4 becomes a non-proliferative tip cell, whereas the adjacent Notch-expressing cell becomes a stalk cell that proliferates and contributes to the elongation of the nascent sprout [41]. Sprout elongation is facilitated by EGFL7, which is released by stalk cells into the ECM to convey spatial information about the position of their neighbors [314].

Once the tip cell of the growing sprout makes contact with another sprout or vessel, their lumens begin connecting by anastomosis. On the molecular level the process starts with the downregulation of Dll4 in the tip cell that in turn relieves Notch signaling in the stalk cell. VEGF signaling in the anastomosing cell switches from driving migration to upregulating Notch expression and VEGFR3 seems to play an important role in this process. According to available data, the VEGFR3-positive tip cells are induced by macrophage-secreted VEGF-C resulting in the upregulation of Notch target genes. Ultimately, these events lead to decreased VEGF sensitivity and downregulation of VEGFR3 in tip cells [133, 252, 310, 312]. In addition, macrophages can interact with tip cells *via* Notch1 or Nrp1 [315, 316]. The process of sprout fusion, however, is not yet fully understood.

1.4.2 Pathological Interplay: Tumor Angiogenesis

Tumors larger than 2 mm^3 are unlikely to survive without the proper vasculature and thus upregulate proangiogenic signaling. It should be noted, however, that tumors can become vascularized by mechanisms not observed under physiological conditions: vessel co-option—tumor cells hijacking the existing vasculature; vascular mimicry—tumor cells lining blood vessels; or the differentiation of stem-like cancer cells into ECs [154]. Nevertheless, tumor angiogenesis remains vital for many cancer types, and VEGF is indeed overexpressed by the vast majority of human tumors. Although tumor cells represent the major source of VEGF, it is not a useful predictive marker because tumor-associated stromal cells such as ECs, muscle cells, macrophages and platelets also produce it [317]. VEGF expression is upregulated by numerous factors but HIF-1α, induced in hypoxic conditions, plays a crucial role in solid tumors. Elevated VEGF induces endothelial proliferation, migration, survival and vessel formation in tumors. VEGFRs (VEGFR1, VEGFR2 and VEGFR3) are also upregulated in tumors, and the expression of VEGFR1 and VEGFR2 has been shown to be augmented by tumor hypoxia [318].

The role of Notch signaling in tumor angiogenesis is also well-supported and it was first described *in vivo* in head and neck squamous cell carcinoma (HNSCC); Jagged1 expressed by HNSCC cells activated Notch signaling in human dermal microvascular ECs and consequently, promoted tumor angiogenesis and tumor growth in a SCID mouse model [319]. According to accumulating evidence, however, Dll4 plays a much more vital role in tumor angiogenesis than any other Notch ligand. Dll4 is strongly upregulated in the tumor vasculature in mouse models and human breast, kidney and bladder cancers [318]. This protein plays an important role in

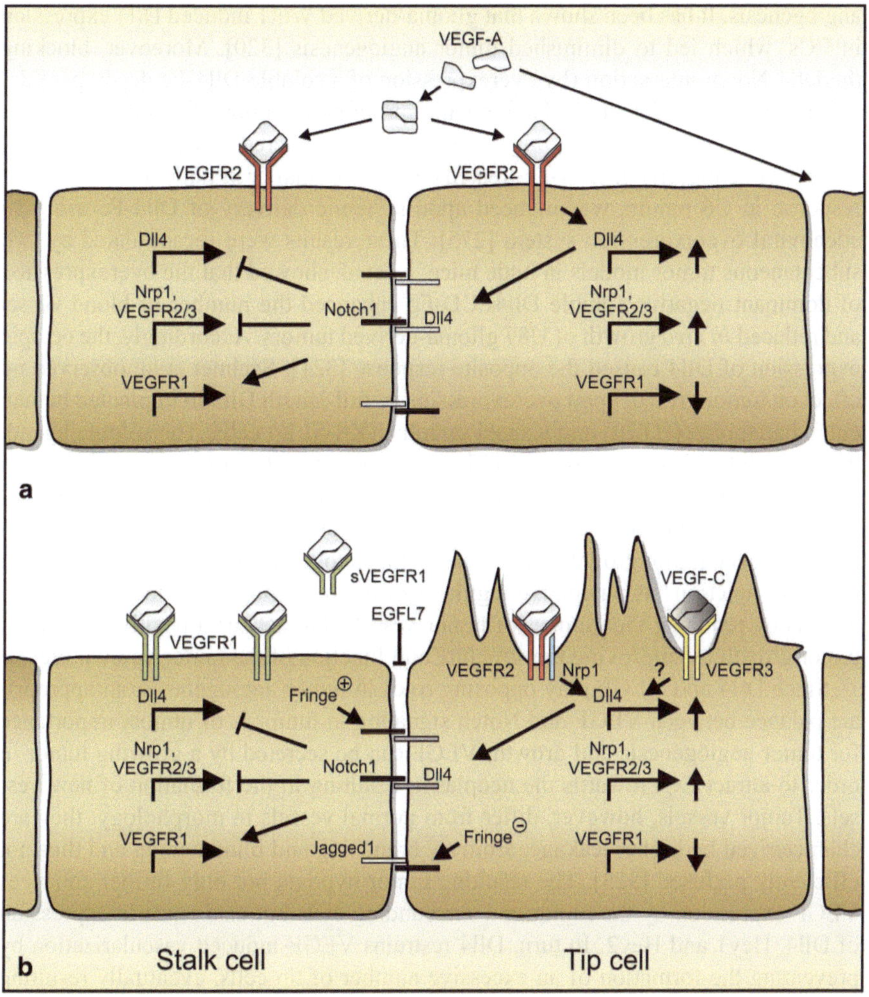

Fig. 1.4 **a** VEGF/Notch crosstalk during endothelial cell sprouting **a** Vascular endothelial growth factor A (*VEGF-A*) is released by hypoxic tissue and binds to VEGF receptor 2 (*VEGFR2*) expressed by endothelial cells, which compete for the tip cell position *via* Delta-like ligand 4 (*Dll4*)-induced Notch signaling. The cell that expresses more Dll4 will become a tip cell and suppress the neighbor cell *via* lateral inhibition. Future tip cells continue to express VEGF-A target genes, such as Dll4, VEGFR2, VEGFR3 and Nrp1. In contrast, future stalk cells cell become subject to Notch activation, which inhibits the expression of Dll4, VEGFR2, and VEGFR3 but promotes transcription of the decoy receptor VEGFR1. **b** The differences in signaling between two cells become amplified by additional mechanisms. Fringe modification of Notch renders it more prone to activation by Dll4 over Jagged1. Thus, Jagged1 binding is antagonistic in this situation and further inhibits Notch in the Dll4-expressing cell. Moreover, epidermal growth factor-like protein 7 (EGFL7), a Notch inhibitor released by stalk cells into the ECM, serves a similar purpose. The cell expressing VEGFR2 and Dll4 becomes a non-proliferative tip cell, whereas the adjacent Notch-expressing cell becomes a sprout cell that proliferates and contributes to the elongation of the nascent sprout

angiogenesis. It has been shown that glioma-derived Wnt1 induced Dll4 expression in ECs, which led to diminished tumor angiogenesis [320]. Moreover, blocking the Dll4-Notch interaction (by overexpression of a soluble Dll4-Fc decoy peptide) enhanced vascular density and angiogenic sprouting in tumors derived from the rat glioma line C6. Surprisingly, this vasculature was rendered non-functional as subcutaneous xenografts of Dll4-Fc-treated cells grew smaller in nude mice. A similar response in C6 tumors was induced upon systemic delivery of Dll4-Fc using an adenoviral overexpression system [275]. These results were recapitulated by two subcutaneous tumor models in nude mice. Li et al. showed that the overexpression of dominant negative soluble Dll4ECD-Fc enhanced the number of blood vessel and reduced *in vivo* growth of U87 glioma-derived tumors. Accordingly, the ectopic expression of Dll4 caused the opposite response [321]. Scehnet et al. observed no effect on tumor growth upon overexpression of full-length Dll4 in implanted human colon carcinoma (HT29) and Kaposi sarcoma (KS-SLK) cells. The soluble ligand, though, inhibited tumor growth and increased the density of thin poorly perfused vessels in both tumors [322]. Treatment of the tumor by pre-mixing soluble Dll4 with either HT29 or KS-SLK cells just before implantation also exhibited significantly reduced tumor growth over 2 weeks. Taken together, all these studies indicate that Dll4 functions as a negative regulator of tumor angiogenesis in mouse tumor models by reducing the number of tumor vessels, but acts as a positive driver for tumor growth by improving the structure and function of the tumor vasculature.

Since Dll4 and VEGF play opposing roles in tumor angiogenesis, an appropriate balance between VEGF and Notch signaling in tumor is of utmost importance for tumor angiogenesis and growth. VEGF can be secreted by a growing tumor in order to attract ECs towards the neoplasm, resulting in the formation of new vessels. Tumor vessels, however, differ from normal vessels in morphology: they are characterized by higher leakage, arbitrary branching and blind ending and thus not efficiently perfused [323]. The resulting tumor hypoxia not only further enhances VEGF expression by the tumor and surrounding cells but also leads to expression of Dll4, Hey1 and Hey2. In turn, Dll4 restrains VEGF-induced vascularization by preventing the formation of an excessive number of tip cells, eventually resulting in the establishment of a functional vasculature [276, 280, 302, 318, 321, 324]. The VEGF/Dll4 nexus, however, is just one of multiple signaling interplays in tumor angiogenesis as both VEGF and Notch interact with several pro-angiogenic pathways such as TGF-β, Wnt and hypoxic signaling [183, 318, 325].

Cancer is not the only disorder connected to pathological angiogenesis, as deregulated production of vessels contributes to the progression of diseases such as diabetic retinopathy and rheumatoid arthritis [326]. For more information on the vasculature in the diseased eye, please refer to Chap. 12, and the pulmonary vasculature in chronic obstructive lung disease is covered in Chap. 15.

1.4.3 Targeting Tumor Angiogenesis

In 1971, Judah Folkman proposed that blocking angiogenesis could be an effective anticancer therapy [327]. The first proof of principle for an antiangiogenic therapy in tumors came with the approval of bevacizumab, a monoclonal antibody directed against VEGF. The most promising results obtained with bevacizumab were in the treatment of metastatic colorectal cancer (CRC) and non-small cell lung cancer (NSCLC) patients, and it has also been used to treat renal cell carcinoma (RCC) and glioblastoma. All approved treatments involve a combination of chemotherapeutics. Other drugs used in clinics are aflibercept, a chimeric VEGF/PlGF neutralizing receptor, and several small molecule RTK inhibitors: sorafenib, sunitinib, pazopanib, vandetanib, vatalanib, cediranib and axitinib [328]. Despite the initial success of bevacizumab, responses are often transient and tumors become unresponsive [329]. Many cancers that clearly require vascularization are highly resistant to VEGF-A blockade from the start or acquire resistance in the course of treatment [329, 330]. Therefore, clinical data indicate that anti-VEGF therapy is ineffective in most cases [318].

VEGF, however, is not the only tumor-relevant proangiogenic factor. Therefore, approaches targeting several signaling pathways, such as Dll4/Notch may prove more efficacious. It has been shown in various preclinical models that the disruption of Dll4/Notch signaling remarkably inhibited tumor growth *in vivo* [275, 276, 318, 321, 322]. Although enhanced angiogenesis has been long connected with tumor growth, these reports suggest an opposite correlation when modulating Dll4 signaling. Blocking Dll4 inhibited tumor growth by increasing vessel density. Thus, it can be concluded that functionality of the vessels is more important for the tumor growth than vessel density [321]. The human glioma cell line U87 subcutaneously implanted in the nude mice tumor model was applied to identify Dll4 as a mediator of the tumor resistance to anti-VEGF therapy. According to data presented, the Dll4-mediated formation of larger vessels, insensitive to anti-VEGF treatment, was responsible for the observed resistance [331]. Notably, blocking Dll4/Notch signaling was effective in growth inhibition of VEGF-resistant tumors. The inhibitory role of Dll4 was confirmed in a pharmacological treatment model, i.e., using defined amounts of a systematically delivered recombinant agent. Administration of recombinant Dll4-Fc or anti-Dll4 polyclonal antibody in an HT1080-RM (generated from a bevacizumab-resistant human fibrosarcoma) tumor model caused an increase in vessel density and smaller tumors volumes [275]. In another model, generated using mouse leukemia WEHI3 cells, treatment with anti-Dll4 antibodies (YW152F) significantly suppressed the growth of tumors, which were highly resistant to the anti-VEGF monoclonal antibody therapy [276]. Along these lines, soluble Dll4ECD-Fc, secreted by co-transplanted cells, decreased *in vivo* growth of bevacizumab-unresponsive PC3 tumors [321]. Taken together, targeting Dll4/Notch signaling may become an alternative therapy for anti-VEGF-resistant cancers. Such compounds are already in clinical trials but studies are still at relatively early stages and thus no drugs have been approved for the cancer treatment as of yet [332, 333].

1.5 Concluding Remarks

The tip–stalk cell model presents an elegant paradigm for EC sprouting. This model, however, was established in EBs, zebrafish and mouse retinas. Thus, its relative simplicity does not always translate to the intricacy of angiogenesis in more complex tissues or organs. ECs exhibit great plasticity and are strongly influenced by external factors such as hypoxia, ECM, and macrophages. On the molecular level, blood vessel formation and function relies heavily on the interaction between VEGF and Notch pathways. This crosstalk is of equal importance in health and disease, e.g., neoplasms. Since bevacizumab was approved for anti-cancer treatment in 2004, several challenges facing an anti-VEGF treatment in humans, as well as possible improvements, have been identified. There is a need for reliable predictive biomarkers, optimization of drug regimens, and identification of factors that render the tumor vasculature non-responsive to the VEGF blockage. Animal experiments indicate that Dll4 belongs to the latter category and thus is a potential therapeutic target. Moreover, based on the results of Dll4 modulation, it has become evident that there is no direct correlation between tumor growth and vasculature density *in vivo*; the rapid outgrowth of a non-functional blood vessel network led to tumor shrinkage. Therefore, uncontrolled angiogenesis might actually enhance tumor hypoxia. However, it should be noted that the opposite approach, i.e., improving the functional vasculature, has also been proposed as an anti-tumor treatment. The rationale behind vessel normalization is that a dysfunctional vasculature hinders drug delivery and promotes metastases [334]. As every tumor represents a single entity, each approach might suit a different subgroup of malignancies or be applicable at different stages of a medical intervention. The future of the cancer treatment lies in personalized medicine and targeted cancer therapies; Notch/Dll4-targeting represents a promising tool belonging to the latter category and offers a means of modulating VEGF-induced pathological angiogenesis.

Acknowledgements We thank Darragh O'Neill and Stefan Liebner for proofreading the manuscript. This work was supported by the German Research Foundation (DFG) *via* the collaborative research center 1080, project A3 and grant SCHM 2159/2-1.

References

1. Herbert SP, Stainier DY. Molecular control of endothelial cell behaviour during blood vessel morphogenesis. Nat Rev Mol Cell Biol. 2011;12:551–64.
2. Choi K, Kennedy M, Kazarov A, Papadimitriou JC, Keller G. A common precursor for hematopoietic and endothelial cells. Development. 1998;125:725–32.
3. Kennedy M, D'Souza SL, Lynch-Kattman M, Schwantz S, Keller G. Development of the hemangioblast defines the onset of hematopoiesis in human ES cell differentiation cultures. Blood. 2007;109:2679–87.
4. Keller G. Embryonic stem cell differentiation: emergence of a new era in biology and medicine. Genes Dev. 2005;19:1129–55.
5. Huber TL, Kouskoff V, Fehling HJ, Palis J, Keller G. Haemangioblast commitment is initiated in the primitive streak of the mouse embryo. Nature. 2004;432:625–30.
6. Medvinsky A, Rybtsov S, Taoudi S. Embryonic origin of the adult hematopoietic system: advances and questions. Development. 2011;138:1017–31.
7. Blaumueller CM, Qi H, Zagouras P, Artavanis-Tsakonas S. Intracellular cleavage of Notch leads to a heterodimeric receptor on the plasma membrane. Cell. 1997;90:281–91.
8. Speck NA, Iruela-Arispe ML. Conditional Cre/LoxP strategies for the study of hematopoietic stem cell formation. Blood Cells Mol Dis. 2009;43:6–11.

9. Zovein AC, Hofmann JJ, Lynch M, French WJ, Turlo KA, Yang Y, Becker MS, Zanetta L, Dejana E, Gasson JC, et al. Fate tracing reveals the endothelial origin of hematopoietic stem cells. Cell Stem Cell. 2008;3:625–36.
10. Boisset JC, van Cappellen W, Andrieu-Soler C, Galjart N, Dzierzak E, Robin C. In vivo imaging of haematopoietic cells emerging from the mouse aortic endothelium. Nature. 2010;464:116–20.
11. Lancrin C, Sroczynska P, Stephenson C, Allen T, Kouskoff V, Lacaud G. The haemangioblast generates haematopoietic cells through a haemogenic endothelium stage. Nature. 2009;457:892–95.
12. Oberlin E, Hafny B E, Petit-Cocault L, Souyri M. Definitive human and mouse hematopoiesis originates from the embryonic endothelium: a new class of HSCs based on VE-cadherin expression. Int J Dev Biol. 2010;54:1165–73.
13. Nishikawa SI, Nishikawa S, Hirashima M, Matsuyoshi N, Kodama H. Progressive lineage analysis by cell sorting and culture identifies FLK1 + VE-cadherin + cells at a diverging point of endothelial and hemopoietic lineages. Development. 1998;125:1747–57.
14. Nishikawa SI, Nishikawa S, Kawamoto H, Yoshida H, Kizumoto M, Kataoka H, Katsura Y. In vitro generation of lymphohematopoietic cells from endothelial cells purified from murine embryos. Immunity. 1998;8:761–9.
15. Taoudi S, Medvinsky A. Functional identification of the hematopoietic stem cell niche in the ventral domain of the embryonic dorsal aorta. Proc Natl Acad Sci U S A. 2007;104:9399–403.
16. Iacovino M, Chong D, Szatmari I, Hartweck L, Rux D, Caprioli A, Cleaver O, Kyba M. HoxA3 is an apical regulator of haemogenic endothelium. Nat Cell Biol. 2011;13:72–8.
17. Rybtsov S, Sobiesiak M, Taoudi S, Souilhol C, Senserrich J, Liakhovitskaia A, Ivanovs A, Frampton J, Zhao S, Medvinsky A. Hierarchical organization and early hematopoietic specification of the developing HSC lineage in the AGM region. J Exp Med. 2011;208:1305–15.
18. Swiers G, Rode C, Azzoni E, de Bruijn MF. A short history of hemogenic endothelium. Blood Cells Mol Dis. 2013;51:206–12.
19. Yang Y, Garcia-Verdugo JM, Soriano-Navarro M, Srinivasan RS, Scallan JP, Singh MK, Epstein JA, Oliver G. Lymphatic endothelial progenitors bud from the cardinal vein and intersomitic vessels in mammalian embryos. Blood. 2012;120:2340–8.
20. Hagerling R, Pollmann C, Andreas M, Schmidt C, Nurmi H, Adams RH, Alitalo K, Andresen V, Schulte-Merker S, Kiefer F. A novel multistep mechanism for initial lymphangiogenesis in mouse embryos based on ultramicroscopy. EMBO J. 2013;32:629–44.
21. Yang Y, Oliver G. Development of the mammalian lymphatic vasculature. J Clin Invest. 2014;124:888–97.
22. Jurisic G, Detmar M. Lymphatic endothelium in health and disease. Cell Tissue Res. 2009;335:97–108.
23. Sacchi G, Weber E, Agliano M, Raffaelli N, Comparini L. The structure of superficial lymphatics in the human thigh: precollectors. Anat Rec. 1997;247:53–62.
24. Scallan JP, Davis MJ. Genetic removal of basal nitric oxide enhances contractile activity in isolated murine collecting lymphatic vessels. J Physiol. 2013;591:2139–56.
25. Casley-Smith JR. The fine structure and functioning of tissue channels and lymphatics. Lymphology. 1980;13:177–83.
26. Kaushal S, Amiel GE, Guleserian KJ, Shapira OM, Perry T, Sutherland FW, Rabkin E, Moran AM, Schoen FJ, Atala A, et al. Functional small-diameter neovessels created using endothelial progenitor cells expanded ex vivo. Nat Med. 2001;7:1035–40.
27. Asahara T, Murohara T, Sullivan A, Silver M, van der Zee R, Li T, Witzenbichler B, Schatteman G, Isner JM. Isolation of Putative Progenitor Endothelial Cells for Angiogenesis. Science. 1997;275:964–6.
28. Takahashi T, Kalka C, Masuda H, Chen D, Silver M, Kearney M, Magner M, Isner JM, Asahara T. Ischemia- and cytokine-induced mobilization of bone marrow-derived endothelial progenitor cells for neovascularization. Nat Med. 1999;5:434–8.
29. Flamme I, Risau W. Induction of vasculogenesis and hematopoesis *in vitro*. Development. 1992;116:435–9.
30. Risau W. Mechanisms of angiogenesis. Nature. 1997;386:671–4.

31. Coultas L, Chawengsaksophak K, Rossant J. Endothelial cells and VEGF in vascular development. Nature. 2005;438:937–45.
32. Shalaby F, Rossant J, Yamaguchi T, Gertsenstein M, Wu X, Breitman M, Schuh A. Failure of blood-island formation and vasculogenesis in Flk-1-deficient mice. Nature. 1995;376:62–6.
33. Patan S. Vasculogenesis and angiogenesis as mechanisms of vascular network formation, growth and remodeling. J Neurooncol. 2000;50:1–15.
34. Carmeliet P. Angiogenesis in life, disease and medicine. Nature. 2005;438:932–6.
35. Adams RH, Alitalo K. Molecular regulation of angiogenesis and lymphangiogenesis. Nat Rev Mol Cell Biol. 2007;8:464–78.
36. Potente M, Gerhardt H, Carmeliet P. Basic and therapeutic aspects of angiogenesis. Cell. 2011;146:873–87.
37. Maisonpierre PC, Suri C, Jones PF, Bartunkova S, Wiegand SJ, Radziejewski C, Compton D, McClain J, Aldrich TH, Papadopoulos N, et al. Angiopoietin-2, a natural antagonist for Tie2 that disrupts in vivo angiogenesis. Science. 1997;277:55–60.
38. Jakobsson L, Franco CA, Bentley K, Collins RT, Ponsioen B, Aspalter IM, Rosewell I, Busse M, Thurston G, Medvinsky A, et al. Endothelial cells dynamically compete for the tip cell position during angiogenic sprouting. Nat Cell Biol. 2010;12:943–53.
39. Gerhardt H, Golding M, Fruttiger M, Ruhrberg C, Lundkvist A, Abramsson A, Jeltsch M, Mitchell C, Alitalo K, Shima D, et al. VEGF guides angiogenic sprouting utilizing endothelial tip cell filopodia. J Cell Biol. 2003;161:1163–77.
40. Chappell JC, Taylor SM, Ferrara N, Bautch VL. Local guidance of emerging vessel sprouts requires soluble Flt-1. Dev Cell. 2009;17:377–86.
41. Phng LK, Gerhardt H. Angiogenesis: a team effort coordinated by notch. Dev Cell. 2009;16:196–208.
42. Toyofuku T, Yabuki M, Kamei J, Kamei M, Makino N, Kumanogoh A, Hori M. Semaphorin-4 A, an activator for T-cell-mediated immunity, suppresses angiogenesis via Plexin-D1. EMBO J. 2007;26:1373–84.
43. Bedell VM, Yeo SY, Park KW, Chung J, Seth P, Shivalingappa V, Zhao J, Obara T, Sukhatme VP, Drummond IA, et al. roundabout4 is essential for angiogenesis in vivo. Proc Natl Acad Sci USA. 2005;102:6373–8.
44. Larrivee B, Freitas C, Trombe M, Lv X, Delafarge B, Yuan L, Bouvree K, Breant C, Del Toro R, Brechot N, et al. Activation of the UNC5B receptor by Netrin-1 inhibits sprouting angiogenesis. Genes Dev. 2007;21:2433–47.
45. Pasquale EB. Eph-ephrin bidirectional signaling in physiology and disease. Cell. 2008;133:38–52.
46. Soker S, Takashima S, Miao HQ, Neufeld G, Klagsbrun M. Neuropilin-1 is expressed by endothelial and tumor cells as an isoform-specific receptor for vascular endothelial growth factor. Cell. 1998;92:735–45.
47. Sawamiphak S, Seidel S, Essmann CL, Wilkinson GA, Pitulescu ME, Acker T, Acker-Palmer A. Ephrin-B2 regulates VEGFR2 function in developmental and tumour angiogenesis. Nature. 2010;465:487–91.
48. De Smet F, Segura I, De Bock K, Hohensinner PJ, Carmeliet P. Mechanisms of vessel branching: filopodia on endothelial tip cells lead the way. Arterioscler Thromb Vasc Biol. 2009;29:639–49.
49. Lamalice L, Boeuf F L, Huot J. Endothelial cell migration during angiogenesis. Circ Res. 2007;100:782–94.
50. Davis GE, Bayless KJ. An integrin and Rho GTPase-dependent pinocytic vacuole mechanism controls capillary lumen formation in collagen and fibrin matrices. Microcirc. 2003;10:27–44.
51. Kamei M, Saunders WB, Bayless KJ, Dye L, Davis GE, Weinstein BM. Endothelial tubes assemble from intracellular vacuoles in vivo. Nature. 2006;442:453–6.
52. Blum Y, Belting HG, Ellertsdottir E, Herwig L, Luders F, Affolter M. Complex cell rearrangements during intersegmental vessel sprouting and vessel fusion in the zebrafish embryo. Dev Biol. 2008;316:312–22.

53. Strilic B, Kucera T, Eglinger J, Hughes MR, McNagny KM, Tsukita S, Dejana E, Ferrara N, Lammert E. The molecular basis of vascular lumen formation in the developing mouse aorta. Dev Cell. 2009;17:505–15.
54. Iruela-Arispe ML, Davis GE. Cellular and molecular mechanisms of vascular lumen formation. Dev Cell. 2009;16:222–31.
55. Montero-Balaguer M, Swirsding K, Orsenigo F, Cotelli F, Mione M, Dejana E. Stable vascular connections and remodeling require full expression of VE-cadherin in zebrafish embryos. PLoS One. 2009;4:e5772.
56. Lampugnani MG, Orsenigo F, Gagliani MC, Tacchetti C, Dejana E. Vascular endothelial cadherin controls VEGFR-2 internalization and signaling from intracellular compartments. J Cell Biol. 2006;174:593–604.
57. Gerhardt H, Wolburg H, Redies C. N-cadherin mediates pericytic-endothelial interaction during brain angiogenesis in the chicken. Develop Dyn. 2000;218:472–9.
58. Jain RK. Molecular regulation of vessel maturation. Nat Med. 2003;9:685–93.
59. Creazzo TL, Godt RE, Leatherbury L, Conway SJ, Kirby ML. Role of cardiac neural crest cells in cardiovascular development. Annu Rev Physiol. 1998;60:267–86.
60. Rajantie I, Ilmonen M, Alminaite A, Ozerdem U, Alitalo K, Salven P. Adult bone marrow-derived cells recruited during angiogenesis comprise precursors for periendothelial vascular mural cells. Blood. 2004;104:2084–6.
61. Bergwerff M, Verberne ME, DeRuiter MC, Poelmann RE, Gittenberger-de Groot AC. Neural crest cell contribution to the developing circulatory system: implications for vascular morphology? Circ Res. 1998;82:221–31.
62. Diaz-Flores L, Gutierrez R, Madrid JF, Varela H, Valladares F, Acosta E, Martin-Vasallo P, Diaz-Flores L, Jr. Pericytes. Morphofunction, interactions and pathology in a quiescent and activated mesenchymal cell niche. Histol Histopathol. 2009;24:909–69.
63. DeRuiter MC, Poelmann RE, VanMunsteren JC, Mironov V, Markwald RR, Gittenberger-de Groot AC. Embryonic endothelial cells transdifferentiate into mesenchymal cells expressing smooth muscle actins in vivo and in vitro. Circ Res. 1997;80:444–51.
64. Farrington-Rock C, Crofts NJ, Doherty MJ, Ashton BA, Griffin-Jones C, Canfield AE. Chondrogenic and Adipogenic Potential of Microvascular Pericytes. Circulation. 2004;110:2226–32.
65. Nicosia RF, Villaschi S. Rat aortic smooth muscle cells become pericytes during angiogenesis in vitro. Lab Invest. 1995;73:658–66.
66. Armulik A, Abramsson A, Betsholtz C. Endothelial/Pericyte Interactions. Circ Res. 2005;97:512–23.
67. Hall AP. Review of the Pericyte during Angiogenesis and its Role in Cancer and Diabetic Retinopathy. Toxic Pathol. 2006;34:763–75.
68. Hellstrom M, Kalén M, Lindahl P, Abramsson A, Betsholtz C. Role of PDGF-B and PDGFR-beta in recruitment of vascular smooth muscle cells and pericytes during embryonic blood vessel formation in the mouse. Development. 1999;126:3047–55.
69. Chantrain CF, Henriet P, Jodele S, Emonard H, Feron O, Courtoy PJ, DeClerck YA, Marbaix E. Mechanisms of pericyte recruitment in tumour angiogenesis: a new role for metalloproteinases. Eur J Cancer. 2006;42:310–8.
70. Asahara T, Chen D, Takahashi T, Fujikawa K, Kearney M, Magner M, Yancopoulos GD, Isner JM. Tie2 Receptor Ligands. Angiopoietin-1 and Angiopoietin-2, Modulate VEGF-Induced Postnatal Neovascularization. Circ Res. 1998;83:233–40.
71. Caduff JH, Fischer LC, Burri PH. Scanning electron-microscope study of the developing microvasculature in the postnatal rat lung. Anat Rec. 1986;216:154–64.
72. Burri PH, Djonov V. Intussusceptive angiogenesis—the alternative to capillary sprouting. Mol Aspects Med. 2002;23:1–27.
73. Djonov VG, Galli AB, Burri PH. Intussusceptive arborization contributes to vascular tree formation in the chick chorio-allantoic membrane. Anat Embryol. 2000;202:347–57.
74. Makanya AN, Stauffer D, Ribatti D, Burri PH, Djonov V. Microvascular growth, development, and remodeling in the embryonic avian kidney: the interplay between sprouting and intussusceptive angiogenic mechanisms. Microsc Res Tech. 2005;66:275–88.

75. Djonov V, Baum O, Burri P. Vascular remodeling by intussusceptive angiogenesis. Cell and Tissue Res. 2003;314:107–17.
76. Djonov VG, Kurz H, Burri PH. Optimality in the developing vascular system: Branching remodeling by means of intussusception as an efficient adaptation mechanism. Dev Dyn. 2002;224:391–402.
77. le Noble F, Fleury V, Pries A, Corvol P, Eichmann A, Reneman RS. Control of arterial branching morphogenesis in embryogenesis: go with the flow. Cardiovasc Res. 2005;65:619–28.
78. Lee GS, Filipovic N, Miele LF, Lin M, Simpson DC, Giney B, Konerding MA, Tsuda A, Mentzer SJ. Blood flow shapes intravascular pillar geometry in the chick chorioallantoic membrane. J Angiogenes Res. 2010;2:11.
79. Styp-Rekowska B, Hlushchuk R, Pries AR, Djonov V. Intussusceptive angiogenesis: pillars against the blood flow. Acta Physiol (Oxf). 2011;202:213–23.
80. Hoefer IE, den Adel B, Daemen MJ. Biomechanical factors as triggers of vascular growth. Cardiovasc Res. 2013;99:276–83.
81. Kamiya A, Togawa T. Adaptive regulation of wall shear stress to flow change in the canine carotid artery. Am J Physiol. 1980;239:H14–H21.
82. Langille BL, Bendeck MP, Keeley FW. Adaptations of carotid arteries of young and mature rabbits to reduced carotid blood flow. Am J Physiol. 1989;256:H931–9.
83. Roman BL, Pekkan K. Mechanotransduction in embryonic vascular development. Biomechanics modeling in mechanobiology. 2012;11:1149–68.
84. Burggren WW. What is the purpose of the embryonic heart beat? Or how facts can ultimately prevail over physiological dogma. Physiol Biochem Zool. 2004;77:333–45.
85. Egginton S. In vivo shear stress response. Biochem Soc Trans. 2011;39:1633–8.
86. Thi MM, Tarbell JM, Weinbaum S, Spray DC. The role of the glycocalyx in reorganization of the actin cytoskeleton under fluid shear stress: a "bumper-car" model. Proc Natl Acad Sci U S A. 2004;101:16483–8.
87. Garcia-Cardena G, Comander J, Anderson KR, Blackman BR, Gimbrone MA Jr. Biomechanical activation of vascular endothelium as a determinant of its functional phenotype. Proc Natl Acad Sci USA. 2001;98:4478–85.
88. Dekker RJ, van Soest S, Fontijn RD, Salamanca S, de Groot PG, VanBavel E, Pannekoek H, Horrevoets AJ. Prolonged fluid shear stress induces a distinct set of endothelial cell genes, most specifically lung Kruppel-like factor (KLF2). Blood. 2002;100:1689–98.
89. Dekker RJ, Boon RA, Rondaij MG, Kragt A, Volger OL, Elderkamp YW, Meijers JC, Voorberg J, Pannekoek H, Horrevoets AJ. KLF2 provokes a gene expression pattern that establishes functional quiescent differentiation of the endothelium. Blood. 2006;107:4354–63.
90. Parmar KM, Larman HB, Dai G, Zhang Y, Wang ET, Moorthy SN, Kratz JR, Lin Z, Jain MK, Gimbrone MA Jr., et al. Integration of flow-dependent endothelial phenotypes by Kruppel-like factor 2. J Clin Invest. 2006;116:49–58.
91. SenBanerjee S, Lin Z, Atkins GB, Greif DM, Rao RM, Kumar A, Feinberg MW, Chen Z, Simon DI, Luscinskas FW, et al. KLF2 Is a novel transcriptional regulator of endothelial proinflammatory activation. J Exp Med. 2004;199:1305–15.
92. Schmidt A, Brixius K, Bloch W. Endothelial precursor cell migration during vasculogenesis. Circ Res. 2007;101:125–36.
93. Lyden D, Hattori K, Dias S, Costa C, Blaikie P, Butros L, Chadburn A, Heissig B, Marks W, Witte L, et al. Impaired recruitment of bone-marrow-derived endothelial and hematopoietic precursor cells blocks tumor angiogenesis and growth. Nat Med. 2001;7:1194–201.
94. Machein MR, Renninger S, De Lima-Hahn E, Plate KH. Minor Contribution of Bone Marrow-Derived Endothelial Progenitors to the Vascularization of Murine Gliomas. Brain Pathol. 2003;13:582–97.
95. Roura S, Gálvez-Montón C, Bayes-Genis A. The challenges for cardiac vascular precursor cell therapy: lessons from a very elusive precursor. J Vasc Res. 2013;50:304–23.
96. Auerbach R, LewisR, Shinners B, Kubai L, Akhtar N. Angiogenesis assays: a critical overview. Clin Chem. 2003;49:32–40.
97. Timmermans F, Plum J, Yöder MC, Ingram DA, Vandekerckhove B, Case J. Endothelial progenitor cells: identity defined? J Cell Mol Med. 2009;13:87–102.

98. Kawamoto A, Gwon H-C, Iwaguro H, Yamaguchi J-I, Uchida S, Masuda H, Silver M, Ma H, Kearney M, Isner JM, et al. Therapeutic potential of ex vivo expanded endothelial progenitor cells for myocardial ischemia. Circulation. 2001;103:634–7.
99. Psaltis PJ, Simari RD. Cell therapy for refractory angina: time for more ACTion. Stem Cell Res Ther. 2011;2:43–3.
100. Gimbrone MA Jr, Cotran RS, Folkman J. Endothelial regeneration: studies with human endothelial cells in culture. Ser Haematol. 1973;6:453–5.
101. Bratt-Leal AM, Carpenedo RL, McDevitt TC. Engineering the embryoid body microenvironment to direct embryonic stem cell differentiation. Biotechnol Prog. 2009;25:43–51.
102. Martin GR, Evans MJ. Differentiation of clonal lines of teratocarcinoma cells: formation of embryoid bodies in vitro. Proc Natl Acad Sci USA. 1975;72:1441–5.
103. Hopfl G, Gassmann M, Desbaillets I. Differentiating embryonic stem cells into embryoid bodies. Methods Mol Biol. 2004;254:79–98.
104. Maltsev VA, Wobus AM, Rohwedel J, Bader M, Hescheler J. Cardiomyocytes differentiated in vitro from embryonic stem cells developmentally express cardiac-specific genes and ionic currents. Circ Res. 1994;75:233–44.
105. Kurosawa H. Methods for inducing embryoid body formation: in vitro differentiation system of embryonic stem cells. J Biosci Bioeng. 2007;103:389–98.
106. Doetschman TC, Eistetter H, Katz M, Schmidt W, Kemler R. The in vitro development of blastocyst-derived embryonic stem cell lines: formation of visceral yolk sac, blood islands and myocardium. J Embryol Exp Morphol. 1985;87:27–45.
107. Klug MG, Soonpaa MH, Koh GY, Field LJ. Genetically selected cardiomyocytes from differentiating embronic stem cells form stable intracardiac grafts. J Clin Invest. 1996;98:216–24.
108. Zhang SC, Wernig M, Duncan ID, Brustle O, Thomson JA. In vitro differentiation of transplantable neural precursors from human embryonic stem cells. Nat Biotechnol. 2001;19:1129–33.
109. Dang SM, Gerecht-Nir S, Chen J, Itskovitz-Eldor J, Zandstra PW. Controlled, scalable embryonic stem cell differentiation culture. Stem Cells. 2004;22:275–82.
110. Dasgupta A, Hughey R, Lancin P, Larue L, Moghe PV. E-cadherin synergistically induces hepatospecific phenotype and maturation of embryonic stem cells in conjunction with hepatotrophic factors. Biotechnol Bioeng. 2005;92:257–66.
111. Larue L, Antos C, Butz S, Huber O, Delmas V, Dominis M, Kemler R. A role for cadherins in tissue formation. Development. 1996;122:3185–94.
112. Maurer JNelson B, Cecena G, Bajpai R, Mercola M, Terskikh A, Oshima RG. Contrasting expression of keratins in mouse and human embryonic stem cells. PLoS One. 2008;3:e3451.
113. Keller GM. In vitro differentiation of embryonic stem cells. Curr Opin Cell Biol. 1995;7:862–9.
114. Vittet D, Prandini MH, Berthier R, Schweitzer A, Martin-Sisteron H, Uzan G, Dejana E. Embryonic stem cells differentiate in vitro to endothelial cells through successive maturation steps. Blood. 1996;88:3424–31.
115. Jakobsson L, Kreuger J, Claesson-Welsh L. Building blood vessels–stem cell models in vascular biology. J Cell Biol. 2007;177:751–5.
116. Lawson ND, Weinstein BM. In vivo imaging of embryonic vascular development using transgenic zebrafish. Dev Biol. 2002;248:307–18.
117. Muller-Ehmsen J, Schmidt A, Krausgrill B, Schwinger RH, Bloch W. Role of erythropoietin for angiogenesis and vasculogenesis: from embryonic development through adulthood. Am J Physiol Heart Circ Physiol. 2006;290:H331–40.
118. Ng YS, Ramsauer M, Loureiro RM, D'Amore PA. Identification of genes involved in VEGF-mediated vascular morphogenesis using embryonic stem cell-derived cystic embryoid bodies. Lab Invest. 2004;84:1209–18.
119. Kreuger J, Nilsson I, Kerjaschki D, Petrova T, Alitalo K, Claesson-Welsh L. Early lymph vessel development from embryonic stem cells. Arterioscler Thromb Vasc Biol. 2006;26:1073–8.

120. Nguyen TH, EichmannA, Noble F L, Fleury V. Dynamics of vascular branching morphogenesis: the effect of blood and tissue flow. Phys Rev E Stat Nonlin Soft Matter Phys. 2006;73:061907.
121. Jeltsch M, Leppänen V-M, Saharinen P, Alitalo K. Receptor Tyrosine Kinase-Mediated Angiogenesis. Cold Spring Harb Perspect Biol. 2013;5:a009183.
122. Senger DR, Galli SJ, Dvorak AM, Perruzzi CA, Harvey RS, Dvorak HF. Tumor cells secreta a vascular permeability factor that promotes accumulation of ascites fluid. Sci. 1983;219:983–5.
123. Koch S, Tugues S, Li X, Gualandi L, Claesson-Welsh L. Signal transduction by vascular endothelial growth factor receptors. Biochem J. 2011;437:169–83.
124. Tvorogov D, Anisimov A, Zheng W, Leppanen VM, Tammela T, Laurinavicius S, Holnthoner W, Helotera H, Holopainen T, Jeltsch M, et al. Effective suppression of vascular network formation by combination of antibodies blocking VEGFR ligand binding and receptor dimerization. Cancer Cell. 2010;18:630–40.
125. Kendall RL, Thomas KA. Inhibition of vascular endothelial cell growth factor activity by an endogenously encoded soluble receptor. Proc Natl Acad Sci USA. 1993;90:10705–9.
126. Joukov V, Kaipainen A, Jeltsch M, Pajusola K, Olofsson B, Kumar V, Eriksson U, Alitalo K. Vascular endothelial growth factors VEGF-B and VEGF-C. J Cell Physiol. 1997;173:211–5.
127. Kawasaki T, Kitsukawa T, Bekku Y, MatsudaY, Sanbo M, Yagi T, Fujisawa H. A requirement for neuropilin-1 in embryonic vessel formation. Development. 1999;126:4895–902.
128. Carmeliet P, Mackman N, Moons L, Luther T, Gressens P, Van Vlaenderen L, Demunck H, Kasper M, Breier G, Evrard P, et al. Role of tissue factor in embryonic blood vessel development. Nature. 1996;383:73–5.
129. Park JE, Keller GA, Ferrara N. The vascular endothelial growth factor (VEGF) isoforms: differential deposition into the subepithelial extracellular matrix and bioactivity of extracellular matrix-bound VEGF. Mol Biol Cell. 1993;4:1317–26.
130. Ferrara N. Binding to the extracellular matrix and proteolytic processing: two key mechanisms regulating vascular endothelial growth factor action. Mol Biol Cell. 2010;21:687–90.
131. Ruhrberg C, Gerhardt H, Golding M, Watson R, Ioannidou S, Fujisawa H, Betsholtz C, Shima DT. Spatially restricted patterning cues provided by heparin-binding VEGF-A control blood vessel branching morphogenesis. Genes Dev. 2002;16:2684–98.
132. Hellstrom M, Phng LK, Hofmann JJ, Wallgard E, Coultas L, Lindblom P, Alva J, Nilsson AK, Karlsson L, Gaiano N, et al. Dll4 signalling through Notch1 regulates formation of tip cells during angiogenesis. Nature. 2007;445:776–80.
133. Tammela T, Zarkada G, Wallgard E, Murtomaki A, Suchting S, Wirzenius M, Waltari M, Hellstrom M, Schomber T, Peltonen R, et al. Blocking VEGFR-3 suppresses angiogenic sprouting and vascular network formation. Nature. 2008;454:656–60.
134. Ho VC, Duan L-J, Cronin C, Liang BT, Fong G-H. Elevated vascular endothelial growth factor receptor-2 abundance contributes to increased angiogenesis in vascular endothelial growth factor receptor-1–deficient mice. Circ. 2012;126:741–52.
135. Carmeliet P, Moons L, Luttun A, Vincenti V, Compernolle V, De Mol M, Wu Y, Bono F, Devy L, Beck H, et al. Synergism between vascular endothelial growth factor and placental growth factor contributes to angiogenesis and plasma extravasation in pathological conditions. Nat Med. 2001;7:575–83.
136. Bellomo D, Headrick JP, Silins GU, Paterson CA, Thomas PS, Gartside M, Mould A, Cahill MM, Tonks ID, Grimmond SM, et al. Mice lacking the vascular endothelial growth factor-B gene (Vegfb) have smaller hearts, dysfunctional coronary vasculature, and impaired recovery from cardiac ischemia. Circ Res. 2000;86:e29–e35.
137. Kearney JB, Ambler CA, Monaco KA, Johnson N, Rapoport RG, Bautch VL. Vascular endothelial growth factor receptor Flt-1 negatively regulates developmental blood vessel formation by modulating endothelial cell division. Blood. 2002;99:2397–407.
138. Murakami M, Simons M. Fibroblast growth factor regulation of neovascularization. Curr Opin Hematol. 2008;15:215–20.

139. Beenken A, Mohammadi M. The FGF family: biology, pathophysiology and therapy. Nat Rev Drug Discov. 2009;8:235–53.
140. Mohammadi M, Olsen SK, Ibrahimi OA. Structural basis for fibroblast growth factor receptor activation. Cytokine Growth Factor Rev. 2005;16:107–37.
141. Johnson DE, Lu J, Chen H, Werner S, Williams LT. The human fibroblast growth factor receptor genes: a common structural arrangement underlies the mechanisms for generating receptor forms that differ in their third immunoglobulin domain. Mol Cell Biol. 1991;11:4627–34.
142. Schlessinger J, Plotnikov AN, Ibrahimi OA, Eliseenkova AV, Yeh BK, Yayon A, Linhardt RJ, Mohammadi M. Crystal structure of a ternary FGF-FGFR-heparin complex reveals a dual role for heparin in FGFR binding and dimerization. Mol Cell. 2000;6:743–50.
143. Dailey L, Ambrosetti D, Mansukhani A, Basilico C. Mechanisms underlying differential responses to FGF signaling. Cytokine Growth Factor Rev. 2005;16:233–47.
144. Claffey KP, Abrams K, Shih SC, Brown LF, Mullen A, Keough M. Fibroblast growth factor 2 activation of stromal cell vascular endothelial growth factor expression and angiogenesis. Lab Invest. 2001;81:61–75.
145. Seghezzi G, Patel S, Ren CJ, Gualandris A, Pintucci G, Robbins ES, Shapiro RL, Galloway AC, Rifkin DB, Mignatti P. Fibroblast growth factor-2 (FGF-2) induces vascular endothelial growth factor (VEGF) expression in the endothelial cells of forming capillaries: an autocrine mechanism contributing to angiogenesis. J Cell Biol. 1998;141:1659–73.
146. Tsunoda S, Nakamura T, Sakurai H, Saiki I. Fibroblast growth factor-2-induced host stroma reaction during initial tumor growth promotes progression of mouse melanoma via vascular endothelial growth factor A-dependent neovascularization. Cancer Sci. 2007;98:541–8.
147. Magnusson P, Rolny C, Jakobsson L, Wikner C, Wu Y, Hicklin DJ, Claesson-Welsh L. Deregulation of Flk-1/vascular endothelial growth factor receptor-2 in fibroblast growth factor receptor-1-deficient vascular stem cell development. J Cell Sci. 2004;117:1513–23.
148. Lavine KJ, White AC, Park C, Smith CS, Choi K, Long F, Hui CC, Ornitz DM. Fibroblast growth factor signals regulate a wave of Hedgehog activation that is essential for coronary vascular development. Genes Dev. 2006;20:1651–66.
149. Eklund L, Olsen BR. Tie receptors and their angiopoietin ligands are context-dependent regulators of vascular remodeling. Exp Cell Res. 2006;312:630–41.
150. Fagiani E, Christofori G. Angiopoietins in angiogenesis. Cancer Lett. 2013;328:18–26.
151. Davis S, Papadopoulos N, Aldrich TH, Maisonpierre PC, Huang T, Kovac L, Xu A, Leidich R, Radziejewska E, Rafique A, et al. Angiopoietins have distinct modular domains essential for receptor binding, dimerization and superclustering. Nat Struct Biol. 2003;10:38–44.
152. Kim KT, Choi HH, Steinmetz MO, Maco B, Kammerer RA, Ahn SY, Kim HZ, Lee GM, Koh GY. Oligomerization and multimerization are critical for angiopoietin-1 to bind and phosphorylate Tie2. J Biol Chem. 2005;280:20126–31.
153. Augustin HG, Koh GY, Thurston G, Alitalo K. Control of vascular morphogenesis and homeostasis through the angiopoietin-Tie system. Nat Rev Mol Cell Biol. 2009;10:165–77.
154. Carmeliet P, Jain RK. Molecular mechanisms and clinical applications of angiogenesis. Nature. 2011;473:298–307.
155. Yana I, Sagara H, Takaki S, Takatsu K, Nakamura K, Nakao K, Katsuki M, Taniguchi S, Aoki T, Sato H, et al. Crosstalk between neovessels and mural cells directs the site-specific expression of MT1-MMP to endothelial tip cells. J Cell Sci. 2007;120:1607–14.
156. Saharinen P, Eklund L, Miettinen J, Wirkkala R, Anisimov A, Winderlich M, Nottebaum A, Vestweber D, Deutsch U, Koh GY, et al. Angiopoietins assemble distinct Tie2 signalling complexes in endothelial cell-cell and cell-matrix contacts. Nat Cell Biol. 2008;10:527–37.
157. Fiedler U, Scharpfenecker M, Koidl S, Hegen A, Grunow V, Schmidt JM, Kriz W, Thurston G, Augustin HG. The Tie-2 ligand angiopoietin-2 is stored in and rapidly released upon stimulation from endothelial cell Weibel-Palade bodies. Blood. 2004;103:4150–6.
158. Starke RD, Ferraro F, Paschalaki KE, Dryden NH, McKinnon TA, Sutton RE, Payne EM, Haskard DO, Hughes AD, Cutler DF, et al. Endothelial von Willebrand factor regulates angiogenesis. Blood. 2011;117:1071–80.

159. Felcht M, Luck R, Schering A, Seidel P, Srivastava K, Hu J, Bartol A, Kienast Y, Vettel C, Loos EK, et al. Angiopoietin-2 differentially regulates angiogenesis through TIE2 and integrin signaling. J Clin Invest. 2012;122:1991–2005.
160. Michel G, Minet E, Ernest I, Roland I, Durant F, Remacle J, Michiels C. A model for the complex between the hypoxia-inducible factor-1 (HIF-1) and its consensus DNA sequence. J Biomol Struct Dyn. 2000;18:169–79.
161. Semenza GL. Targeting HIF-1 for cancer therapy. Nat Rev Cancer. 2003;3:721–32.
162. Maynard MA, Qi H, Chung J, Lee EH, Kondo Y, Hara S, Conaway RC, Conaway JW, Ohh M. Multiple splice variants of the human HIF-3 alpha locus are targets of the von Hippel-Lindau E3 ubiquitin ligase complex. J Biol Chem. 2003;278:11032–40.
163. Hopfl G, Ogunshola O, Gassmann M. HIFs and tumors–causes and consequences. Am J Physiol Regul Integr Comp Physiol. 2004;286:R608–23.
164. Kaelin WG. Proline hydroxylation and gene expression. Annu Rev Biochem. 2005;74:115–28.
165. Hewitson KS, McNeill LA, Riordan MV, Tian YM, Bullock AN, Welford RW, Elkins JM, Oldham NJ, Bhattacharya S, Gleadle JM, et al. Hypoxia-inducible factor (HIF) asparagine hydroxylase is identical to factor inhibiting HIF (FIH) and is related to the cupin structural family. J Biol Chem. 2002;277:26351–5.
166. Tang N, Wang L, Esko J, Giordano FJ, Huang Y, Gerber HP, Ferrara N, Johnson RS. Loss of HIF-1alpha in endothelial cells disrupts a hypoxia-driven VEGF autocrine loop necessary for tumorigenesis. Cancer Cell. 2004;6:485–95.
167. Koop EA, Lopes SM, Feiken E, Bluyssen HA, van der Valk M, Voest EE, Mummery CL, Moolenaar WH, Gebbink MF. Receptor protein tyrosine phosphatase mu expression as a marker for endothelial cell heterogeneity; analysis of RPTPmu gene expression using LacZ knock-in mice. Int J Dev Biol. 2003;47:345–54.
168. Yang Y, Sun M, Wang L, Jiao B. HIFs, angiogenesis, and cancer. J Cell Biochem. 2013;114:967–74.
169. Knowles RG, Moncada S. Nitric oxide as a signal in blood vessels. Trends Biochem Sci. 1992;17:399–402.
170. Fukumura D, Kashiwagi S, Jain RK. The role of nitric oxide in tumour progression. Nat Rev Cancer. 2006;6:521–34.
171. Moncada S, Higgs A. The L-arginine-nitric oxide pathway. N Engl J Med. 1993;329:2002–12.
172. Forstermann U, Sessa WC. Nitric oxide synthases: regulation and function. Eur Heart J. 2012;33:829–37, 837a-d.
173. Fraisl P. Crosstalk between oxygen- and nitric oxide-dependent signaling pathways in angiogenesis. Exp Cell Res. 2013;319:1331–9.
174. Coulet F, Nadaud S, Agrapart M, Soubrier F. Identification of hypoxia-response element in the human endothelial nitric-oxide synthase gene promoter. J Biol Chem. 2003;278:46230–40.
175. Fukumura D, Gohongi T, Kadambi A, Izumi Y, Ang J, Yun CO, Buerk DG, Huang PL, Jain RK. Predominant role of endothelial nitric oxide synthase in vascular endothelial growth factor-induced angiogenesis and vascular permeability. Proc Natl Acad Sci USA. 2001;98:2604–9.
176. Ziche M, Morbidelli L, Choudhuri R, Zhang HT, Donnini S, Granger HJ, Bicknell R. Nitric oxide synthase lies downstream from vascular endothelial growth factor-induced but not basic fibroblast growth factor-induced angiogenesis. J Clin Invest. 1997;99:2625–34.
177. Thibeault S, Rautureau Y, Oubaha M, Faubert D, Wilkes BC, Delisle C, Gratton JP. S-nitrosylation of beta-catenin by eNOS-derived NO promotes VEGF-induced endothelial cell permeability. Mol Cell. 2010;39:468–76.
178. Berchner-Pfannschmidt U, Yamac H, Trinidad B, Fandrey J. Nitric oxide modulates oxygen sensing by hypoxia-inducible factor 1-dependent induction of prolyl hydroxylase 2. J Biol Chem. 2007;282:1788–96.
179. Chowdhury R, Flashman E, Mecinovic J, Kramer HB, Kessler BM, Frapart YM, Boucher JL, Clifton IJ, McDonough MA, Schofield CJ. Studies on the reaction of nitric oxide with the hypoxia-inducible factor prolyl hydroxylase domain 2 (EGLN1). J Mol Biol. 2011;410:268–79.

180. Li F, Sonveaux P, Rabbani ZN, Liu S, Yan B, Huang Q, Vujaskovic Z, Dewhirst MW, Li CY. Regulation of HIF-1alpha stability through S-nitrosylation. Mol Cell. 2007;26:63–74.
181. Metzen E, Zhou J, Jelkmann W, Fandrey J, Brune B. Nitric oxide impairs normoxic degradation of HIF-1alpha by inhibition of prolyl hydroxylases. Mol Biol Cell. 2003;14:3470–81.
182. Takeda N, O'Dea EL, Doedens A, Kim JW, Weidemann A, Stockmann C, Asagiri M, Simon MC, Hoffmann A, Johnson RS. Differential activation and antagonistic function of HIF-{alpha} isoforms in macrophages are essential for NO homeostasis. Genes Dev. 2010;24:491–501.
183. Bray SJ. Notch signalling: a simple pathway becomes complex. Nat Rev Mol Cell Biol. 2006;7:678–89.
184. Rebay I, Fleming RJ, Fehon RG, Cherbas L, Cherbas P, Artavanis-Tsakonas S. Specific EGF repeats of Notch mediate interactions with Delta and Serrate: implications for Notch as a multifunctional receptor. Cell. 1991;67:687–99.
185. Cordle J, Redfieldz C, Stacey M, van der Merwe PA, Willis AC, Champion BR, Hambleton S, Handford PA. Localization of the delta-like-1-binding site in human Notch-1 and its modulation by calcium affinity. J Biol Chem. 2008;283:11785–93.
186. Okajima T, Irvine KD. Regulation of notch signaling by o-linked fucose. Cell. 2002;111:893–904.
187. Okajima T, Xu A, Irvine KD. Modulation of notch-ligand binding by protein O-fucosyltransferase 1 and fringe. J Biol Chem. 2003;278:42340–5.
188. Moloney DJ, Panin VM, Johnston SH, Chen J, Shao L, Wilson R, Wang Y, Stanley P, Irvine KD, Haltiwanger RS, et al. Fringe is a glycosyltransferase that modifies Notch. Nature. 2000;406:369–75.
189. Haines N, Irvine KD. Glycosylation regulates Notch signalling. Nat Rev Mol Cell Biol. 2003;4:786–97.
190. Sanchez-Irizarry C, Carpenter AC, Weng AP, Pear WS, Aster JC, Blacklow SC. Notch subunit heterodimerization and prevention of ligand-independent proteolytic activation depend, respectively, on a novel domain and the LNR repeats. Mol Cell Biol. 2004;24:9265–73.
191. Gordon WR, Vardar-Ulu D, Histen G, Sanchez-Irizarry C, Aster JC, Blacklow SC. Structural basis for autoinhibition of Notch. Nat Struct Mol Biol. 2007;14:295–300.
192. Mumm JS, Schroeter EH, Saxena MT, Griesemer A, Tian X, Pan DJ, Ray WJ, Kopan R. A ligand-induced extracellular cleavage regulates gamma-secretase-like proteolytic activation of Notch1. Mol Cell. 2000;5:197–206.
193. Aster JC, Simms WB, Zavala-Ruiz Z, Patriub V, North CL, Blacklow SC. The folding and structural integrity of the first LIN-12 module of human Notch1 are calcium-dependent. Biochem. 1999;38:4736–42.
194. Tamura K, Taniguchi Y, Minoguchi S, Sakai T, Tun T, Furukawa T, Honjo T. Physical interaction between a novel domain of the receptor Notch and the transcription factor RBP-J kappa/Su(H). Curr Biol. 1995;5:1416–23.
195. Lubman OY, Korolev SV, Kopan R. Anchoring notch genetics and biochemistry; structural analysis of the ankyrin domain sheds light on existing data. Mol Cell. 2004;13:619–26.
196. Lieber T, Kidd S, Alcamo E, Corbin V, Young MW. Antineurogenic phenotypes induced by truncated Notch proteins indicate a role in signal transduction and may point to a novel function for Notch in nuclei. Genes Dev. 1993;7:1949–65.
197. Kurooka H, Kuroda K, Honjo T. Roles of the ankyrin repeats and C-terminal region of the mouse notch1 intracellular region. Nucleic Acids Res. 1998;26:5448–55.
198. Rechsteiner M. Regulation of enzyme levels by proteolysis: the role of pest regions. Adv Enzyme Regul. 1988;27:135–51.
199. Bettenhausen B, Hrabe de Angelis M, Simon D, Guenet JL, Gossler A. Transient and restricted expression during mouse embryogenesis of Dll1, a murine gene closely related to Drosophila Delta. Development. 1995;121:2407–18.
200. Dunwoodie SL, Henrique D, Harrison SM, Beddington RS. Mouse Dll3: a novel divergent Delta gene which may complement the function of other Delta homologues during early pattern formation in the mouse embryo. Development. 1997;124:3065–76.

201. Shutter JR, Scully S, Fan W, Richards WG, Kitajewski J, Deblandre GA, Kintner CR, Stark KL. Dll4, a novel Notch ligand expressed in arterial endothelium. Genes Dev. 2000;14:1313–8.
202. Lindsell CE, Shawber CJ, Boulter J, Weinmaster G. Jagged: a mammalian ligand that activates Notch1. Cell. 1995;80:909–17.
203. Shawber C, Boulter J, Lindsell CE, Weinmaster G. Jagged2: a serrate-like gene expressed during rat embryogenesis. Dev Biol. 1996;180:370–6.
204. Parks AL, Stout JR, Shepard SB, Klueg KM, Dos Santos AA, Parody TR, Vaskova M, Muskavitch MA. Structure-function analysis of delta trafficking, receptor binding and signaling in Drosophila. Genetics. 2006;174:1947–61.
205. Shimizu K, Chiba S, Kumano K, Hosoya N, Takahashi T, Kanda Y, Hamada Y, Yazaki Y, Hirai H. Mouse jagged1 physically interacts with notch2 and other notch receptors. Assessment by quantitative methods. J Biol Chem. 1999;274:32961–9.
206. Ladi E, Nichols JT, Ge W, Miyamoto A, Yao C, Yang LT, Boulter J, Sun YE, Kintner C, Weinmaster G. The divergent DSL ligand Dll3 does not activate Notch signaling but cell autonomously attenuates signaling induced by other DSL ligands. J Cell Biol. 2005;170:983–92.
207. Chapman G, Sparrow DB, Kremmer E, Dunwoodie SL. Notch inhibition by the ligand DELTA-LIKE 3 defines the mechanism of abnormal vertebral segmentation in spondylocostal dysostosis. Hum Mol Genet. 2011;20:905–16.
208. Geffers I, Serth K, Chapman G, Jaekel R, Schuster-Gossler K, Cordes R, Sparrow DB, Kremmer E, Dunwoodie SL, Klein T, et al. Divergent functions and distinct localization of the Notch ligands DLL1 and DLL3 in vivo. J Cell Biol. 2007;178:465–76.
209. Dikic I, Schmidt MH. Notch: Implications of endogenous inhibitors for therapy. Bioessays. 2010;32:481–7.
210. Small D, Kovalenko D, Kacer D, Liaw L, Landriscina M, Serio C D, Prudovsky I, Maciag T. Soluble Jagged 1 represses the function of its transmembrane form to induce the formation of the Src-dependent chord-like phenotype. J Biol Chem. 2001;276:32022–30.
211. Sheldon H, Heikamp E, Turley H, Dragovic R, Thomas P, Oon CE, Leek R, Edelmann M, Kessler B, Sainson RC, et al. New mechanism for Notch signaling to endothelium at a distance by Delta-like 4 incorporation into exosomes. Blood. 2010;116:2385–94.
212. D'Souza B, Meloty-Kapella L, Weinmaster G. Canonical and non-canonical Notch ligands. Curr Top Dev Biol. 2010;92:73–129.
213. Meister J, Schmidt MH. miR-126 and miR-126*: new players in cancer. Sci World J. 2010;10:2090–100.
214. de Celis JF, Bray S. Feed-back mechanisms affecting Notch activation at the dorsoventral boundary in the Drosophila wing. Development. 1997;124:3241–51.
215. Li Y, Baker NE. The roles of cis-inactivation by Notch ligands and of neuralized during eye and bristle patterning in Drosophila. BMC Dev Biol. 2004;4:5.
216. Miller AC, Lyons EL, Herman TG. cis-Inhibition of Notch by endogenous Delta biases the outcome of lateral inhibition. Curr Biol. 2009;19:1378–83.
217. Zolkiewska A. ADAM proteases: ligand processing and modulation of the Notch pathway. Cell Mol Life Sci. 2008;65:2056–68.
218. Jorissen E, De Strooper B. Gamma-secretase and the intramembrane proteolysis of Notch. Curr Top Dev Biol. 2010;92:201–30.
219. Andersson ER, Sandberg R, Lendahl U. Notch signaling: simplicity in design, versatility in function. Development. 2011;138:3593–612.
220. De Strooper B, Annaert W, Cupers P, Saftig P, Craessaerts K, Mumm JS, Schroeter EH, Schrijvers V, Wolfe MS, Ray WJ, et al. A presenilin-1-dependent gamma-secretase-like protease mediates release of Notch intracellular domain. Nature. 1999;398:518–22.
221. Wu L, Aster JC, Blacklow SC, Lake R, Artavanis-Tsakonas S, Griffin JD. MAML1, a human homologue of Drosophila mastermind, is a transcriptional co-activator for NOTCH receptors. Nat Genet. 2000;26:484–9.
222. Wallberg AE, Neely KE, Hassan AH, Gustafsson JA, Workman JL, Wright AP. Recruitment of the SWI-SNF chromatin remodeling complex as a mechanism of gene activation by the glucocorticoid receptor tau1 activation domain. Mol Cell Biol. 2000;20:2004–13.

223. Kurooka H, Honjo T. Functional interaction between the mouse notch1 intracellular region and histone acetyltransferases PCAF and GCN5. J Biol Chem. 2000;275:17211–20.
224. Iso T, Sartorelli V, Chung G, Shichinohe T, Kedes L, Hamamori Y. HERP, a new primary target of Notch regulated by ligand binding. Mol Cell Biol. 2001;21:6071–9.
225. Iso T, Sartorelli V, Poizat C, Iezzi S, Wu HY, Chung G, Kedes L, Hamamori Y. HERP, a novel heterodimer partner of HES/E(spl) in Notch signaling. Mol Cell Biol. 2001;21:6080–9.
226. Jarriault S, Brou C, Logeat F, Schroeter EH, Kopan R, Israel A. Signalling downstream of activated mammalian Notch. Nature. 1995;377:355–8.
227. Maier MM, Gessler M. Comparative analysis of the human and mouse Hey1 promoter: Hey genes are new Notch target genes. Biochem Biophys Res Commun. 2000;275:652–60.
228. Nishimura M, Isaka F, Ishibashi M, Tomita K, Tsuda H, Nakanishi S, Kageyama R. Structure, chromosomal locus, and promoter of mouse Hes2 gene, a homologue of Drosophila hairy and Enhancer of split. Genomics. 1998;49:69–75.
229. Cheng P, Zlobin A, Volgina V, Gottipati S, Osborne B, Simel EJ, Miele L, Gabrilovich DI. Notch-1 regulates NF-kappaB activity in hemopoietic progenitor cells. J Immunol. 2001;167:4458–67.
230. Oswald F, Liptay S, Adler G, Schmid RM. NF-kappaB2 is a putative target gene of activated Notch-1 via RBP-Jkappa. Mol Cell Biol. 1998;18:2077–88.
231. Nickoloff BJ, Qin JZ, Chaturvedi V, Denning MF, Bonish B, Miele L. Jagged-1 mediated activation of notch signaling induces complete maturation of human keratinocytes through NF-kappaB and PPARgamma. Cell Death Differ. 2002;9:842–55.
232. Palomero T, Lim WK, Odom DT, Sulis ML, Real PJ, Margolin A, Barnes KC, O'Neil J, Neuberg D, Weng AP, et al. NOTCH1 directly regulates c-MYC and activates a feed-forward-loop transcriptional network promoting leukemic cell growth. Proc Natl Acad Sci USA. 2006;103:18261–6.
233. Sharma VM, Calvo JA, Draheim KM, Cunningham LA, Hermance N, Beverly L, Krishnamoorthy V, Bhasin M, Capobianco AJ, Kelliher MA. Notch1 contributes to mouse T-cell leukemia by directly inducing the expression of c-myc. Mol Cell Biol. 2006;26:8022–31.
234. Weng AP, Millholland JM, Yashiro-Ohtani Y, Arcangeli ML, Lau A, Wai C, Del Bianco C, Rodriguez CG, Sai H, Tobias J, et al. c-Myc is an important direct target of Notch1 in T-cell acute lymphoblastic leukemia/lymphoma. Genes Dev. 2006;20:2096–109.
235. Ehm O, Goritz C, Covic M, Schaffner I, Schwarz TJ, Karaca E, Kempkes B, Kremmer E, Pfrieger FW, Espinosa L, et al. RBPJkappa-dependent signaling is essential for long-term maintenance of neural stem cells in the adult hippocampus. J Neurosci. 2010;30:13794–807.
236. Kumar JP, Moses K. EGF receptor and Notch signaling act upstream of Eyeless/Pax6 to control eye specification. Cell. 2001;104:687–97.
237. Ronchini C, Capobianco AJ. Induction of cyclin D1 transcription and CDK2 activity by Notch(ic): implication for cell cycle disruption in transformation by Notch(ic). Mol Cell Biol. 2001;21:5925–34.
238. Rangarajan A, Talora C, Okuyama R, Nicolas M, Mammucari C, Oh H, Aster JC, Krishna S, Metzger D, Chambon P, et al. Notch signaling is a direct determinant of keratinocyte growth arrest and entry into differentiation. EMBO J. 2001;20:3427–36.
239. Duarte A, Hirashima M, Benedito R, Trindade A, Diniz P, Bekman E, Costa L, Henrique D, Rossant J. Dosage-sensitive requirement for mouse Dll4 in artery development. Genes Dev. 2004;18:2474–8.
240. Krebs LT, Xue Y, Norton CR, Shutter JR, Maguire M, Sundberg JP, Gallahan D, Closson V, Kitajewski J, Callahan R, et al. Notch signaling is essential for vascular morphogenesis in mice. Genes Dev. 2000;14:1343–52.
241. Hrabe de Angelis M, McIntyre J 2nd, Gossler A. Maintenance of somite borders in mice requires the Delta homologue DII1. Nature. 1997;386:717–21.
242. Xue Y, Gao X, Lindsell CE, Norton CR, Chang B, Hicks C, Gendron-Maguire M, Rand EB, Weinmaster G, Gridley T. Embryonic lethality and vascular defects in mice lacking the Notch ligand Jagged1. Hum Mol Genet. 1999;8:723–30.
243. Hofmann JJ, Luisa Iruela-Arispe M. Notch expression patterns in the retina: An eye on receptor-ligand distribution during angiogenesis. Gene Expr Patterns. 2007;7:461–70.

244. Karsan A. The role of notch in modeling and maintaining the vasculature. Can J Physiol Pharmacol. 2005;83:14–23.
245. Lovschall H, Mitsiadis TA, Poulsen K, Jensen KH, Kjeldsen AL. Coexpression of Notch3 and Rgs5 in the pericyte-vascular smooth muscle cell axis in response to pulp injury. Int J Dev Biol. 2007;51:715–21.
246. Mailhos C, Modlich U, Lewis J, Harris A, Bicknell R, Ish-Horowicz D. Delta4, an endothelial specific notch ligand expressed at sites of physiological and tumor angiogenesis. Differentiation. 2001;69:135–44.
247. Uyttendaele H, Marazzi G, Wu G, Yan Q, Sassoon D, Kitajewski J. Notch4/int-3, a mammary proto-oncogene, is an endothelial cell-specific mammalian Notch gene. Development. 1996;122:2251–9.
248. Yoneya T, Tahara T, Nagao K, Yamada Y, Yamamoto T, Osawa M, Miyatani S, Nishikawa M. Molecular cloning of delta-4, a new mouse and human Notch ligand. J Biochem. 2001;129:27–34.
249. Artavanis-Tsakonas S, Rand MD, Lake RJ. Notch signaling: cell fate control and signal integration in development. Science. 1999;284:770–6.
250. Claxton S, Fruttiger M. Periodic Delta-like 4 expression in developing retinal arteries. Gene Expr Patterns. 2004;5:123–7.
251. Hofmann JJ, Iruela-Arispe ML. Notch signaling in blood vessels: who is talking to whom about what? Circ Res. 2007;100:1556–68.
252. Benedito R, Roca C, Sorensen I, Adams S, Gossler A, Fruttiger M, Adams RH. The notch ligands Dll4 and Jagged1 have opposing effects on angiogenesis. Cell. 2009;137:1124–35.
253. Suchting S, Freitas C, le Noble F, Benedito R, Breant C, Duarte A, Eichmann A. The Notch ligand Delta-like 4 negatively regulates endothelial tip cell formation and vessel branching. Proc Natl Acad Sci USA. 2007;104:3225–30.
254. Sainson RC, Aoto J, Nakatsu MN, Holderfield M, Conn E, Koller E, Hughes CC. Cell-autonomous notch signaling regulates endothelial cell branching and proliferation during vascular tubulogenesis. FASEB J. 2005;19:1027–9.
255. Roca C, Adams RH. Regulation of vascular morphogenesis by Notch signaling. Genes Dev. 2007;21:2511–24.
256. Siekmann AF, Covassin L, Lawson ND. Modulation of VEGF signalling output by the Notch pathway. Bioessays. 2008;30:303–13.
257. Fitch MJ, Campagnolo L, Kuhnert F, Stuhlmann H. Egfl7, a novel epidermal growth factor-domain gene expressed in endothelial cells. Dev Dyn. 2004;230:316–24.
258. Nichol D, Shawber C, Fitch MJ, Bambino K, Sharma A, Kitajewski J, Stuhlmann H. Impaired angiogenesis and altered Notch signaling in mice overexpressing endothelial Egfl7. Blood. 2010;116:6133–43.
259. Schmidt MH, Bicker F, Nikolic I, Meister J, Babuke T, Picuric S, Muller-Esterl W, Plate KH, Dikic I. Epidermal growth factor-like domain 7 (EGFL7) modulates Notch signalling and affects neural stem cell renewal. Nat Cell Biol. 2009;11:873–80.
260. Parker LH, Schmidt M, Jin SW, Gray AM, Beis D, Pham T, Frantz G, Palmieri S, Hillan K, Stainier DY, et al. The endothelial-cell-derived secreted factor Egfl7 regulates vascular tube formation. Nature. 2004;428:754–8.
261. Schmidt M, Paes K, De Maziere A, Smyczek T, Yang S, Gray A, French D, Kasman I, Klumperman J, Rice DS, et al. EGFL7 regulates the collective migration of endothelial cells by restricting their spatial distribution. Development. 2007;134:2913–23.
262. Fish JE, Santoro MM, Morton SU, Yu S, Yeh RF, Wythe JD, Ivey KN, Bruneau BG, Stainier DY, Srivastava D. miR-126 regulates angiogenic signaling and vascular integrity. Dev Cell. 2008;15:272–84.
263. Wang S, Aurora AB, Johnson BA, Qi X, McAnally J, Hill JA, Richardson JA, Bassel-Duby R, Olson EN. The endothelial-specific microRNA miR-126 governs ascular integrity and angiogenesis. Dev Cell. 2008;15:261–71.
264. Nikolic I, Plate KH, Schmidt MH. EGFL7 meets miRNA-126: an angiogenesis alliance. J Angiogenes Res. 2010;2:9.

265. Kuhnert F, Mancuso MR, Hampton J, Stankunas K, Asano T, Chen CZ, Kuo CJ. Attribution of vascular phenotypes of the murine Egfl7 locus to the microRNA miR-126. Development. 2008;135:3989–93.
266. Nikolic I, Stankovic ND, Bicker F, Meister J, Braun H, Awwad K, Baumgart J, Simon K, Thal SC, Patra C, et al. EGFL7 ligates alphavbeta3 integrin to enhance vessel formation. Blood. 2013;121:3041–50.
267. Welcker M, Clurman BE. FBW7 ubiquitin ligase: a tumour suppressor at the crossroads of cell division, growth and differentiation. Nat Rev Cancer. 2008;8:83–93.
268. Izumi N, Helker C, Ehling M, Behrens A, Herzog W, Adams RH. Fbxw7 controls angiogenesis by regulating endothelial Notch activity. PLoS One. 2012;7:e41116.
269. Guarani V, Deflorian G, Franco CA, Kruger M, Phng LK, Bentley K, Toussaint L, Dequiedt F, Mostoslavsky R, Schmidt MH, et al. Acetylation-dependent regulation of endothelial Notch signalling by the SIRT1 deacetylase. Nature. 2011;473:234–8.
270. Ohnuki H, Inoue H, Takemori N, Nakayama H, Sakaue T, Fukuda S, Miwa D, Nishiwaki E, Hatano M, Tokuhisa T, et al. BAZF, a novel component of cullin3-based E3 ligase complex, mediates VEGFR and Notch cross-signaling in angiogenesis. Blood. 2012;119:2688–98.
271. Lamar E, Deblandre G, Wettstein D, Gawantka V, Pollet N, Niehrs C, Kintner C. Nrarp is a novel intracellular component of the Notch signaling pathway. Genes Dev. 2001;15:1885–99.
272. Phng LK, Potente M, Leslie JD, Babbage J, Nyqvist D, Lobov I, Ondr JK, Rao S, Lang RA, Thurston G, et al. Nrarp coordinates endothelial Notch and Wnt signaling to control vessel density in angiogenesis. Dev Cell. 2009;16:70–82.
273. Lobov IB, Renard RA, Papadopoulos N, Gale NW, Thurston G, Yancopoulos GD, Wiegand SJ. Delta-like ligand 4 (Dll4) is induced by VEGF as a negative regulator of angiogenic sprouting. Proc Natl Acad Sci USA. 2007;104:3219–24.
274. Dou GR, Wang YC, Hu XB, Hou LH, Wang CM, Xu JF, Wang YS, Liang YM, Yao LB, Yang AG, et al. RBP-J, the transcription factor downstream of Notch receptors, is essential for the maintenance of vascular homeostasis in adult mice. FASEB J. 2008;22:1606–17.
275. Noguera-Troise I, Daly C, Papadopoulos NJ, Coetzee S, Boland P, Gale NW, Lin HC, Yancopoulos GD, Thurston G. Blockade of Dll4 inhibits tumour growth by promoting non-productive angiogenesis. Nature. 2006;444:1032–7.
276. Ridgway J, Zhang G, Wu Y, Stawicki S, Liang WC, Chanthery Y, Kowalski J, Watts RJ, Callahan C, Kasman I, et al. Inhibition of Dll4 signalling inhibits tumour growth by deregulating angiogenesis. Nature. 2006;444:1083–7.
277. Liu ZJ, Xiao M, Balint K, Soma A, Pinnix CC, Capobianco AJ, Velazquez OC, Herlyn M. Inhibition of endothelial cell proliferation by Notch1 signaling is mediated by repressing MAPK and PI3K/Akt pathways and requires MAML1. FASEB J. 2006;20:1009–11.
278. Noseda M, Chang L, McLean G, Grim JE, Clurman BE, Smith LL, Karsan A. Notch activation induces endothelial cell cycle arrest and participates in contact inhibition: role of p21Cip1 repression. Mol Cell Biol. 2004;24:8813–22.
279. Trindade A, Kumar SR, Scehnet JS, Lopes-da-Costa L, Becker J, Jiang W, Liu R, Gill PS, Duarte A. Overexpression of delta-like 4 induces arterialization and attenuates vessel formation in developing mouse embryos. Blood. 2008;112:1720–9.
280. Harrington LS, Sainson RC, Williams CK, Taylor JM, Shi W, Li JL, Harris AL. Regulation of multiple angiogenic pathways by Dll4 and Notch in human umbilical vein endothelial cells. Microvasc Res. 2008;75:144–54.
281. Williams CK, Li JL, Murga M, Harris AL, Tosato G. Up-regulation of the Notch ligand Delta-like 4 inhibits VEGF-induced endothelial cell function. Blood. 2006;107:931–9.
282. Gerhardt H, Ruhrberg C, Abramsson A, Fujisawa H, Shima D, Betsholtz C. Neuropilin-1 is required for endothelial tip cell guidance in the developing central nervous system. Dev Dyn. 2004;231:503–9.
283. Jones EA, Yuan L, Breant C, Watts RJ, Eichmann A. Separating genetic and hemodynamic defects in neuropilin 1 knockout embryos. Development. 2008;135:2479–88.
284. Benedito R, Trindade A, Hirashima M, Henrique D, da Costa LL, Rossant J, Gill PS, Duarte A. Loss of Notch signalling induced by Dll4 causes arterial calibre reduction by increasing endothelial cell response to angiogenic stimuli. BMC Dev Biol. 2008;8:117.

285. Leong KG, Hu X, Li L, Noseda M, Larrivee B, Hull C, Hood L, Wong F, Karsan A. Activated Notch4 inhibits angiogenesis: role of beta 1-integrin activation. Mol Cell Biol. 2002;22:2830–41.
286. Hodkinson PS, Elliott PA, Lad Y, McHugh BJ, MacKinnon AC, Haslett C, Sethi T. Mammalian NOTCH-1 activates beta1 integrins via the small GTPase R-Ras. J Biol Chem. 2007;282:28991–9001.
287. Estrach S, Cailleteau L, Franco CA, Gerhardt H, Stefani C, Lemichez E, Gagnoux-Palacios L, Meneguzzi G, Mettouchi A. Laminin-binding integrins induce Dll4 expression and Notch signaling in endothelial cells. Circ Res. 2011;109:172–82.
288. Gale NW, Dominguez MG, Noguera I, Pan L, Hughes V, Valenzuela DM, Murphy AJ, Adams NC, Lin HC, Holash J, et al. Haploinsufficiency of delta-like 4 ligand results in embryonic lethality due to major defects in arterial and vascular development. Proc Natl Acad Sci USA. 2004;101:15949–54.
289. Krebs LT, Shutter JR, Tanigaki K, Honjo T, Stark KL, Gridley T. Haploinsufficient lethality and formation of arteriovenous malformations in Notch pathway mutants. Genes Dev. 2004;18:2469–73.
290. Seo S, Fujita H, Nakano A, Kang M, Duarte A, Kume T. The forkhead transcription factors, Foxc1 and Foxc2, are required for arterial specification and lymphatic sprouting during vascular development. Dev Biol. 2006;294:458–70.
291. Liu ZJ, Shirakawa T, Li Y, Soma A, Oka M, Dotto GP, Fairman RM, Velazquez OC, Herlyn M. Regulation of Notch1 and Dll4 by vascular endothelial growth factor in arterial endothelial cells: implications for modulating arteriogenesis and angiogenesis. Mol Cell Biol. 2003;23:14–25.
292. Lawson ND, Scheer N, Pham VN, Kim CH, Chitnis AB, Campos-Ortega JA, Weinstein BM. Notch signaling is required for arterial-venous differentiation during embryonic vascular development. Development. 2001;128:3675–83.
293. Zhong TP, Childs S, Leu JP, Fishman MC. Gridlock signalling pathway fashions the first embryonic artery. Nature. 2001;414:216–20.
294. Diez H, Fischer A, Winkler A, Hu CJ, Hatzopoulos AK, Breier G, Gessler M. Hypoxia-mediated activation of Dll4-Notch-Hey2 signaling in endothelial progenitor cells and adoption of arterial cell fate. Exp Cell Res. 2007;313:1–9.
295. Iso T, Maeno T, Oike Y, Yamazaki M, Doi H, Arai M, Kurabayashi M. Dll4-selective Notch signaling induces ephrinB2 gene expression in endothelial cells. Biochem Biophys Res Commun. 2006;341:708–14.
296. Lanner F, Sohl M, Farnebo F. Functional arterial and venous fate is determined by graded VEGF signaling and notch status during embryonic stem cell differentiation. Arterioscler Thromb Vasc Biol. 2007;27:487–93.
297. You LR, Lin FJ, Lee CT, DeMayo FJ, Tsai MJ, Tsai SY. Suppression of Notch signalling by the COUP-TFII transcription factor regulates vein identity. Nature. 2005;435:98–104.
298. Corada M, Nyqvist D, Orsenigo F, Caprini A, Giampietro C, Taketo MM, Iruela-Arispe ML, Adams RH, Dejana E. The Wnt/beta-catenin pathway modulates vascular remodeling and specification by upregulating Dll4/Notch signaling. Dev Cell. 2010;18:938–49.
299. Yamamizu K, Matsunaga T, Uosaki H, Fukushima H, Katayama S, Hiraoka-Kanie M, Mitani K, Yamashita JK. Convergence of Notch and beta-catenin signaling induces arterial fate in vascular progenitors. J Cell Biol. 2010;189:325–38.
300. Corada M, Orsenigo F, Morini MF, Pitulescu ME, Bhat G, Nyqvist D, Breviario F, Conti V, Briot A, Iruela-Arispe ML, et al. Sox17 is indispensable for acquisition and maintenance of arterial identity. Nat Commun. 2013;4:2609.
301. Lee SH, Lee S, Yang H, Song S, Kim K, Saunders TL, Yoon JK, Koh GY, Kim I. Notch pathway targets proangiogenic regulator Sox17 to restrict angiogenesis. Circ Res. 2014;115:215–26.
302. Carlson TR, Yan Y, Wu X, Lam MT, Tang GL, Beverly LJ, Messina LM, Capobianco AJ, Werb Z, Wang R. Endothelial expression of constitutively active Notch4 elicits reversible arteriovenous malformations in adult mice. Proc Natl Acad Sci USA. 2005;102:9884–9.

303. Limbourg A, Ploom M, Elligsen D, Sorensen I, Ziegelhoeffer T, Gossler A, Drexler H, Limbourg FP. Notch ligand Delta-like 1 is essential for postnatal arteriogenesis. Circ Res. 2007;100:363–71.
304. Takeshita K, Satoh M, Ii M, Silver M, Limbourg FP, Mukai Y, Rikitake Y, Radtke F, Gridley T, Losordo DW, et al. Critical role of endothelial Notch1 signaling in postnatal angiogenesis. Circ Res. 2007;100:70–8.
305. Hayashi H, Kume T. Foxc transcription factors directly regulate Dll4 and Hey2 expression by interacting with the VEGF-Notch signaling pathways in endothelial cells. PLoS One. 2008;3:e2401.
306. Roukens MG, Alloul-Ramdhani M, Baan B, Kobayashi K, Peterson-Maduro J, van Dam H, Schulte-Merker S, Baker DA. Control of endothelial sprouting by a Tel-CtBP complex. Nat Cell Biol. 2010;12:933–42.
307. Henderson AM, Wang SJ, Taylor AC, Aitkenhead M, Hughes CC. The basic helix-loop-helix transcription factor HESR1 regulates endothelial cell tube formation. J Biol Chem. 2001;276:6169–76.
308. Holderfield MT, Henderson Anderson AM, Kokubo H, Chin MT, Johnson RL, Hughes CC. HESR1/CHF2 suppresses VEGFR2 transcription independent of binding to E-boxes. Biochem Biophys Res Commun. 2006;346:637–48.
309. Fischer A, Schumacher N, Maier M, Sendtner M, Gessler M. The Notch target genes Hey1 and Hey2 are required for embryonic vascular development. Genes Dev. 2004;18:901–11.
310. Siekmann AF, Lawson ND. Notch signalling limits angiogenic cell behaviour in developing zebrafish arteries. Nature. 2007;445:781–4.
311. Benedito R, Rocha SF, Woeste M, Zamykal M, Radtke F, Casanovas O, Duarte A, Pytowski B, Adams RH. Notch-dependent VEGFR3 upregulation allows angiogenesis without VEGF-VEGFR2 signalling. Nature. 2012;484:110–4.
312. Tammela T, Zarkada G, Nurmi H, Jakobsson L, Heinolainen K, Tvorogov D, Zheng W, Franco CA, Murtomaki A, Aranda E, et al. VEGFR-3 controls tip to stalk conversion at vessel fusion sites by reinforcing Notch signalling. Nat Cell Biol. 2011;13:1202–13.
313. Kappas NC, Zeng G, Chappell JC, Kearney JB, Hazarika S, Kallianos KG, Patterson C, Annex BH, Bautch VL. The VEGF receptor Flt-1 spatially modulates Flk-1 signaling and blood vessel branching. J Cell Biol. 2008;181:847–58.
314. Benedito R, Hellstrom M. Notch as a hub for signaling in angiogenesis. Exp Cell Res. 2013;319:1281–8.
315. Fantin A, Vieira JM, Plein A, Denti L, Fruttiger M, Pollard JW, Ruhrberg C. NRP1 acts cell autonomously in endothelium to promote tip cell function during sprouting angiogenesis. Blood. 2013;121:2352–62.
316. Outtz HH, Tattersall IW, Kofler NM, Steinbach N, Kitajewski J. Notch1 controls macrophage recruitment and Notch signaling is activated at sites of endothelial cell anastomosis during retinal angiogenesis in mice. Blood. 2011;118:3436–9.
317. Kut C, Mac Gabhann F, Popel AS. Where is VEGF in the body? A meta-analysis of VEGF distribution in cancer. Br J Cancer. 2007;97:978–85.
318. Li JL, Harris AL. Crosstalk of VEGF and Notch pathways in tumour angiogenesis: therapeutic implications. Front Biosci (Landmark. Ed). 2009;14:3094–110.
319. Zeng Q, Li S, Chepeha DB, Giordano TJ, Li J, Zhang H, Polverini PJ, Nor J, Kitajewski J, Wang CY. Crosstalk between tumor and endothelial cells promotes tumor angiogenesis by MAPK activation of Notch signaling. Cancer Cell. 2005;8:13–23.
320. Reis M, Czupalla CJ, Ziegler N, Devraj K, Zinke J, Seidel S, Heck R, Thom S, Macas J, Bockamp E, et al. Endothelial Wnt/beta-catenin signaling inhibits glioma angiogenesis and normalizes tumor blood vessels by inducing PDGF-B expression. J Exp Med. 2012;209:1611–27.
321. Li JL, Sainson RC, Shi W, Leek R, Harrington LS, Preusser M, Biswas S, Turley H, Heikamp E, Hainfellner JA, et al. Delta-like 4 Notch ligand regulates tumor angiogenesis, improves tumor vascular function, and promotes tumor growth in vivo. Cancer Res. 2007;67:11244–53.

322. Scehnet JS, Jiang W, Kumar SR, Krasnoperov V, Trindade A, Benedito R, Djokovic D, Borges C, Ley EJ, Duarte A, et al. Inhibition of Dll4-mediated signaling induces proliferation of immature vessels and results in poor tissue perfusion. Blood. 2007;109:4753–60.
323. Dimova I, Popivanov G, Djonov V. Angiogenesis in cancer—general pathways and their therapeutic implications. J BUON. 2014;19:15–21.
324. Shawber CJ, Funahashi Y, Francisco E, Vorontchikhina M, Kitamura Y, Stowell SA, Borisenko V, Feirt N, Podgrabinska S, Shiraishi K, et al. Notch alters VEGF responsiveness in human and murine endothelial cells by direct regulation of VEGFR-3 expression. J Clin Invest. 2007;117:3369–82.
325. Hurlbut GD, Kankel MW, Lake RJ, Artavanis-Tsakonas S. Crossing paths with Notch in the hyper-network. Curr Opin Cell Biol. 2007;19:166–75.
326. Jakobsson L, Bentley K, Gerhardt H. VEGFRs and Notch: a dynamic collaboration in vascular patterning. Biochem Soc Trans. 2009;37:1233–6.
327. Folkman J. Tumor angiogenesis: therapeutic implications. N Engl J Med. 1971;285:1182–6.
328. Welti J, Loges S, Dimmeler S, Carmeliet P. Recent molecular discoveries in angiogenesis and antiangiogenic therapies in cancer. J Clin Invest. 2013;123:3190–200.
329. Jain RK, Duda DG, Clark JW, Loeffler JS. Lessons from phase III clinical trials on anti-VEGF therapy for cancer. Nat Clin Pract Oncol. 2006;3:24–40.
330. Shojaei F, Wu X, Malik AK, Zhong C, Baldwin ME, Schanz S, Fuh G, Gerber HP, Ferrara N. Tumor refractoriness to anti-VEGF treatment is mediated by CD11b + Gr1 + myeloid cells. Nat Biotechnol. 2007;25:911–20.
331. Bogdahn U, Hau P, Stockhammer G, Venkataramana NK, Mahapatra AK, Suri A, Balasubramaniam A, Nair S, Oliushine V, Parfenov V, et al. Targeted therapy for high-grade glioma with the TGF-beta2 inhibitor trabedersen: results of a randomized and controlled phase IIb study. Neuro Oncol. 2011;13:132–42.
332. Andersson ER, Lendahl U. Therapeutic modulation of Notch signalling–are we there yet? Nat Rev Drug Discov. 2014;13:357–78.
333. Teodorczyk M, Schmidt MHH. Notching on cancer's door: Notch signaling in brain tumors. Front Oncol. 2015;4:431.
334. Carmeliet P, Jain RK. Principles and mechanisms of vessel normalization for cancer and other angiogenic diseases. Nat Rev Drug Discov. 2011;10:417–27.

Chapter 2
Vascular Development in the Zebrafish

Dorien M. A. Hermkens, Henricus J. Duckers and Stefan Schulte-Merker

Abbreviations

DA	Dorsal aorta
DLAV	Dorsal longitudinal anastomotic vessel
HLT	Hypotrichosis-lymphedema-telangiectasia
HMG	High-mobility group
HPF	Hours post-fertilization
ISV	Intersegmental vessels
LDA	Lateral dorsal aorta
LPM	Lateral plate mesoderm
MO	Morpholino
NICD	Notch intracellular domain
PCV	Posterior cardinal vein
PHBC	Primordial hindbrain channel
TAD	Transactivation domain

S. Schulte-Merker (✉)
Cells-in-Motion Cluster of Excellence (EXC 1003—CiM), University of Münster, Münster, Germany
e-mail: schultes@ukmuenster.de

D. M. A. Hermkens · S. Schulte-Merker
Hubrecht Institute—KNAW & UMC Utrecht, 3584 CT Utrecht, The Netherlands

D. M. A. Hermkens · H. J. Duckers
Erasmus MC, Rotterdam, The Netherlands

S. Schulte-Merker
Institute for Cardiovascular Organogenesis and Regeneration, WWU, Mendelstr. 7, 48149 Münster, Germany

M. HH. Schmidt, S. Liebner (eds.), *Endothelial Signaling in Development and Disease,* DOI 10.1007/978-1-4939-2907-8_2

2.1 Zebrafish Vasculogenesis and Arteriovenous Specification

The first organ system that becomes functional in most vertebrates is the cardiovascular system. Given the transparency of the early zebrafish embryo, the development of this particular organ system has been traditionally a subject of interest. Large offspring numbers and external development of zebrafish embryos makes them very well suited for developmental research, aided by the fact that circulation within the zebrafish embryo commences very quickly; the first signs of heart beat can be appreciated at approximately 22 h post-fertilization (hpf), while the first erythrocytes move through the embryonic body only a few hours later [89]. The appeal of the zebrafish embryo for *in vivo* observation [43], the availability of transgenic lines that mark precursor cells of the vascular endothelium and that allow distinguishing between arterial and venous cells within the same embryo, the possibility of interfering with this process experimentally, and the generation of mutants that affect the process [73] have resulted in significant insight into various aspects of vascular development [84]. In this chapter, we will mainly focus on vasculogenesis and the early events of arteriovenous specification. The ensuing processes of arterial and venous differentiation, and the behavior of endothelial cells that contribute to later aspects of angiogenesis and lymphangiogenesis have recently been reviewed [65, 77] (see also Chap. 5) and will only be touched on here where necessary.

It was previously thought that the differentiation between arteries and veins is mainly established by the difference in hemodynamic forces such as blood pressure [22]; however, in recent years more and more evidence has emerged that molecular differences between arterial and venous precursor cells regulate the formation of the arteriovenous system irrespective of hemodynamic forces [81]. Specifically, recent studies have shown that molecular differences between arterial and venous cells have already been established in the early stages when endothelial precursors begin to arise [88].

Formation of the vascular system starts with vasculogenesis, which is defined as the *de novo* formation of vessels from individual mesenchymal cells. In the zebrafish trunk, this process results in the formation of the dorsal aorta (DA) and the posterior cardinal vein (PCV). How do these vessels arise during early development? It has been apparent for some time that there is a set of bilaterally aligned cells in the posterior lateral plate mesoderm (LPM) which constitute a precursor pool for these axial vessels. At this point in time, these cells stain positive for a variety of endothelial markers, and are commonly referred to as angioblasts, a term used by Sabin as early as 1917 when studying equivalent processes in the chicken embryo. Angioblasts migrate over the endodermal layer to the embryonic midline, where they form a vascular cord [60, 61, 63]. The migration of these angioblasts and their importance for DA and PCV formation is undisputed, but where these cells are initially localized in the LPM and whether they are actually already specified at the onset of migration is far less clear. In addition, the exact events upon reaching the embryonic midline still need to be resolved. Both issues will be discussed herein.

Independent reports showed two distinct waves of angioblast migration during vascular development in the zebrafish trunk. The first wave of migration starts at approximately 14 hpf, and the second wave starts at approximately 16 hpf. This has been confirmed by a number of groups, most recently by Kohli et al. [17, 34, 39]. Before the actual onset of migration, and as early as the 4-somite stage (12 hpf), a subset of medial angioblasts can be found to express the marker *etv2/etsrp*, a transcription factor that has key roles during vasculogenesis. Three hours later, a second and more laterally positioned line of *etv2/etsrp*-expressing cells becomes apparent [58]. The heterochrony of *etv2/etsrp* expression among these two angioblast populations is further substantiated when looking at other marker genes; *kdrl* (formerly known as *flk1* [6]) highlights medial angioblasts at the 10-somite stage but gets expressed in lateral angioblasts only a few hours later. Similarly, the widely used pan-endothelial marker *fli1a* [76] is first expressed in medial angioblasts, before then becoming expressed in the lateral angioblast population [39]. The arterial marker *gridlock/hey2* can only be found to be expressed in the medial population, not in the lateral population [87], suggesting that the medial angioblast population constitutes a pool of arterial precursor cells, while the lateral population contributes to the PCV.

This notion was further confirmed by *in vivo* observations that tracked individual cells and addressed the question as to whether angioblasts, irrespective of their position within the LPM, can contribute to both DA or PCV, or whether early positioning within either the medial or lateral angioblast population is largely, or even entirely, predictive for cells to become part of the DA or PCV. The zebrafish embryo is very suitable for this type of analysis as one can mark individual cells *in vivo* and follow the fate of labeled cells over time. Of course, *in vivo* imaging tracks cell movements, while in situ hybridizations (such as the ones discussed above) provide static representations of gene expression, which makes the direct comparison between these different modes of acquiring data difficult. That notwithstanding, lineage tracing has provided a number of important insights. When labeling a single cell at the margin of a gastrula stage embryo (6 hpf), Vogeli et al. [80] reported that a few of these cells contribute exclusively to the endothelial and the hematopoietic lineage, providing direct demonstration of the existence of hemangioblasts. However, only part of the hematopoietic and the endothelial lineage arise from these bi-potential cells, indicating that there are other cellular sources for the vasculature [80]. Another study labeled cells of the LPM at the 7- to 12-somite stage, and only observed contribution to either the arterial or venous lineage (consistent with *gridlock/hey2* expression) [87]. Hence, based on both marker expression studies and *in vivo* tracing experiments, it appears that there are lineage-restricted angioblasts in the zebrafish LPM during the early stages of somitogenesis that have been specified to become either arteries or veins (Fig. 2.1a).

How about the movement of these angioblasts from the posterior LPM to the midline? As mentioned above, it has been reported a number of times that first the medial, then the lateral, angioblast population migrates to the midline. Both migration waves occur in an anterior to posterior manner. The use of a transgenic line that expresses the photo-convertible fluorophore *kaede* from the *etv2* promoter has enabled elegant studies which demonstrated that the medial angioblasts give rise to

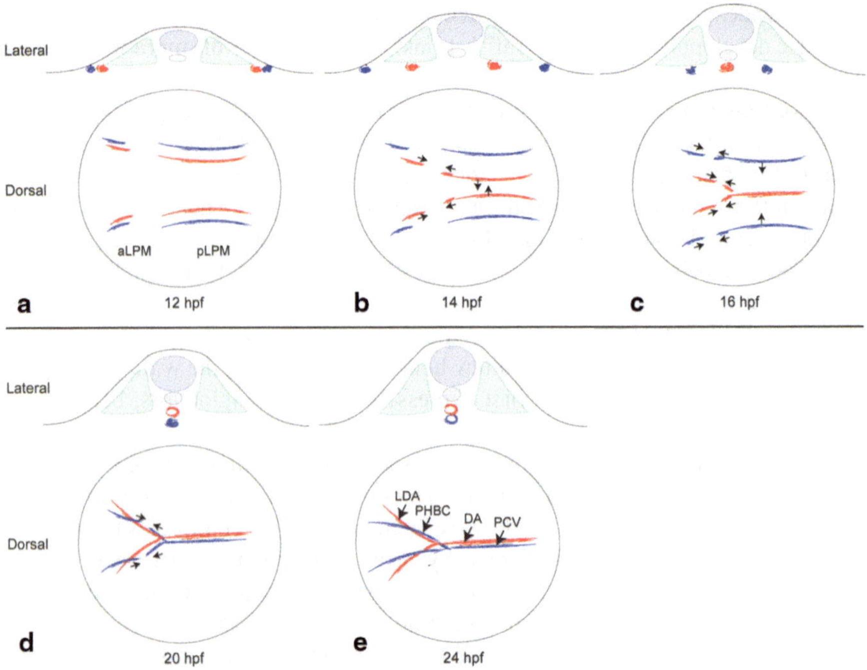

Fig. 2.1 Arterial and venous precursor cell migration. **a** At 12 hpf, angioblasts are located within the anterior and posterior lateral plate mesoderm (*LPM*) in two bilateral stripes. Presumably, even at this early stage, the medial angioblast population consists of arterial precursor cells (*red*), while the lateral population consists of venous precursor cells (*blue*). **b** At approximately 14 hpf, the precursor cells located at more medial positions within the posterior LPM start to migrate to the midline. At the same time, some arterial cells of the posterior LPM start to migrate anteriorly, whereas arterial cells from the anterior LPM start to migrate posteriorly. **c** At 16 hpf, arterial precursor cells will form the first axial vessel, the dorsal aorta. Cells within the more lateral located stripes of the posterior LPM start to migrate to the midline. In addition, some of these cells will migrate anteriorly, whereas in the anterior LPM some venous progenitor cells start to migrate posteriorly. **d** At 20 hpf, the venous precursor cells have migrated to the midline and form the posterior cardinal vein (PCV). The migration of arterial cells in the anterior region results in formation of the lateral dorsal aortae. **e** At 24 hpf, venous precursor cells in the anterior region have migrated to form the Primordial hindbrain channel (*PHBC*). *Green* indicates somites, *purple* indicates notochord and hypochord, *yellow* indicates embryonic tissue

the DA, and that lateral angioblasts constitute the PCV [39] (Fig. 2.1). Furthermore, these studies support a model where angioblasts move directly to a more dorsal location and form the DA, while lateral angioblasts migrate directly to a more ventral position and form the PCV. This is in contrast to a study by Herbert et al. [26], who suggested a ventral sprouting mechanism, during which angioblasts from the DA contribute directly to the PCV [26]. Whether this discrepancy is possible due to imaging in different regions of the embryo needs to be resolved.

As a consequence of early specification events and a highly orchestrated (both in time and space) array of cell movements, two cords of angioblasts/endothelial cells align along the embryonic axis. These cells establish cell–cell junctional complexes among each other, express markers of apical and basal polarity, and eventually form a luminized DA and PCV, even before the onset of circulation (reviewed by Schuermann et al. [65]).

2.2 Molecular Cues During Vasculogenesis

There is a plethora of genes and factors that have been connected to the genetic control of vasculogenesis and angiogenesis, but only a few for which zebrafish mutants are available and for which we fully understand the mechanistic implications. One of the first cardiovascular mutants described presents with a near-complete failure to specify blood and endothelial lineages, and these *cloche* mutants have been very instructive to understand many aspects of early vascular development [72]. It has been suggested that mutations in the *lycat* gene, encoding an acyl transferase, are causative for the phenotype [86]. The *cloche* mutant phenotype can be rescued via forced expression of the ETS1-related protein, placing this key transcription factor downstream of cloche [74]. As mentioned above, the ETS-domain transcription factor Etv2/Etsrp/ER71 is one of the earliest markers specifically expressed in angioblasts, and Etsrp is required for the expression of *vegfr2/kdrl* in early development. In *etsrp* zebrafish morphants, angioblasts are unable to differentiate, migrate, or form functional axial vessels. Overexpression of *etsrp* causes the induction of vascular endothelial markers in several cell types. Etsrp is thus a key regulator in the induction of vascular endothelial fate in early development [58, 74].

Overexpression studies showed that vascular endothelial growth factor A (*vegf-a*) and Sonic Hedgehog (Shh) are involved in the localization of the medial and lateral angioblasts to the midline, with high levels of *vegf-a* or *shh* resulting in a random distribution of medial and lateral angioblasts at the midline [39]. Similarly, *vegf-a* morphants showed a single vessel consisting of lateral and medial angioblasts [39], suggesting that Vegf and Shh are critical factors for arteriovenous specification at the time point of angioblast positioning at the midline. In zebrafish, *shh* is expressed in the notochord and floorplate [16], whereas *vegf-a* is expressed in the somites [49]. It is hypothesized that the medially located angioblasts receive a higher concentration of Shh and Vegf signaling than the angioblasts localized at the lateral positions [39]. This difference in concentration could induce the distinct pathways for arterial or venous specification. This model is intriguingly simple but important questions remain. Are these morphogens (Shh, Vegf) sufficient to both serve as chemoattractants for cell migration and for specifying arterial and venous cell fates? Or is there a stochastic initiation of some angioblasts in the LPM to migrate medially in a first wave, and these 'front runners' inhibit arterial fates in the trailing cells of the second wave? A thorough fate mapping of migration events might clarify some of these issues; if all angioblasts are equally naïve before the onset of migration, then fate mapping should reveal that the cells located most medially in the LPM (and which should therefore perceive the highest levels of Vegf and Shh) should invariably end up in the aorta. However, should more medially located cells be overtaken by more distally located cells, this would argue that the cells might not be naïve.

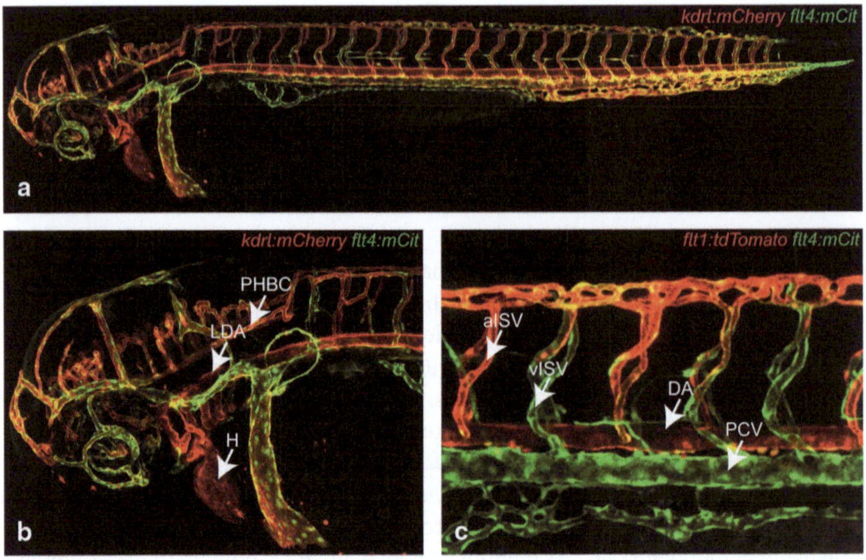

Fig. 2.2 Transgenic zebrafish vasculature. **a** Overview of *kdrl:mCherry;flt4:mCit* zebrafish embryo at 2 days post-fertilization (dpf), with arteries depicted in *red* and veins depicted in *yellow*. **b** Enlarged head region of (a), including the heart (*H*), lateral dorsal aorta (*LDA*), and the primordial hindbrain channel (PHBC). **c** Trunk region of *flt1:tdTomato;flt4:mCit* embryo at 3 dpf, with arteries depicted in *red* and veins depicted in *green*, including the arterial intersegmental vessel (*aISV*), venous intersegmental vessel (*vISV*), dorsal aorta (*DA*), and posterior cardinal vein (*PCV*)

While the axial vasculature is defective in *vegf*-deficient embryos and *shh* mutants, the anterior vasculature is unaffected, suggesting a different mechanism of vessel formation in the anterior LPM of the zebrafish. Time-lapse analysis showed that the anterior part of the DA, the lateral DA (LDA) develops in a different manner. At 14 hpf, a subset of angioblasts localized at the anterior LPM migrate posteriorly, while a subset of angioblasts from the posterior LPM start to migrate anteriorly. These cell populations migrate towards each other, eventually connecting and forming the LDA. This process of angioblast migration is suggested to be mediated by the chemokine Cxcr4 [70]. The expression of *cxcr4* is restricted to the angioblasts that will form the anterior-most LDA. Knockdown of *cxcr4* resulted in inappropriate fusion of anterior and posterior angioblasts, and consequently formed a disrupted LDA [70]. At approximately 17 hpf, the same process of migration happens for venous angioblasts that form the largest anterior vein in the head, the primordial hindbrain channel (PHBC) [7, 70] (Fig. 2.1 and 2.2a, b).

After the initiation of vasculogenesis, angiogenesis will further remodel the vascular network. Angiogenesis is the formation of blood vessels from pre-existing blood vessels by sprouting and remodeling of endothelial cells. In the zebrafish trunk, endothelial cells start to sprout dorsally from the DA at the somite boundaries and form the intersegmental vessels (ISVs). This primary sprouting results in ISVs that connect at the dorsal site to form the dorsal longitudinal anastomotic vessel (DLAV). After establishing this arterial network, secondary sprouting will occur in which endothelial cells sprout dorsally from the PCV [32]. These sprouts can connect to the ISVs and remodel the arterial ISV into a venous ISV. Statistically, only every second sprout becomes a venous ISV. The other endothelial sprouts will migrate further to the horizontal myoseptum and constitute a population of parachordal lymphangioblasts, which then migrate either dorsally or ventrally and start to form the lymphatic vasculature [29] (reviewed by van Impel and Schulte-Merker [77, 78]) (Fig. 2.2c).

One of the prevalent questions concerning the above angiogenic processes is how venous sprouts make the decision on whether to connect to an intersegmental artery and to remodel it into a vein in the process, or whether not to do this and contribute to the pool of lymphatic precursor cells. One might presume that this process might involve Notch and Delta but, until now, no members of the Notch/Delta signaling pathway have been detected to be expressed in the PCV or in venous sprouts.

2.3 Molecular Regulation of Arterial-Venous Specification

2.3.1 EphrinB2/Eph Receptor B4

Two of the most widely referred to markers for arteriovenous specification are Ephrin-B2 (Efnb2) and Eph receptor B4 (EphB4), based on the finding that Efnb2 and the EphB4 receptor are differentially expressed in arteries versus veins, respectively [81]. The EphB4 receptor belongs to the receptor tyrosine kinase family and is the only Ephrin receptor that specifically binds to Efnb2, a membrane-bound ligand of the Ephrin ligand family [20]. The Ephrin ligands and Eph receptors are both transmembrane proteins, and signaling requires cell-to-cell contact, which can be bidirectional [5]. Eph receptors and their ligands are often, but not always, localized in the adjacent cell population [1, 19]. 'Forward' signaling starts with the binding of an Ephrin ligand to a receptor dimer. This leads to trans-phosphorylation of the intracellular domain of the receptor, and results in a conformational change that can activate the kinase domain. 'Reverse' signaling occurs when the conserved tyrosine residues of the cytoplasmic domain of the Ephrin ligand are phosphorylated upon contact with the Eph receptor ectodomain, or by an Eph receptor-independent mechanism. This causes the recruitment of an SH2 (Src-homology-2) domain-containing adaptor protein and SH3 binding partners (reviewed by Kullander and Klein [41]). The exact contribution of forward and/or reverse signaling in arteriovenous specification is still unclear.

In zebrafish, *efnb2* expression is restricted to the arterial endothelial cells (ECs). Expression of *efnb2* is initiated at approximately the 20-somite stage, when the angioblasts have migrated to the midline but circulation has not yet commenced. *ephb4* receptor messenger RNA (mRNA) is also expressed in the vasculature but the expression is restricted to the venous ECs [87]. These clear expression patterns in arterial versus venous ECs make *EphB4* and *Efnb2* suitable markers for arterial and venous differentiation and, accordingly, their expression changes upon altered arteriovenous specification. For example, inhibition of Notch signaling results in a decrease of arterial fates, which can be appreciated by reduced expression of *efnb2* in the DA [44]. Herbert et al. suggested that Efnb2a limits the ventral migration of arterial angioblasts, whereas Ephb4a promotes it, based on results obtained upon transplanting *efnb2a or ephB4a* morpholino (MO) donor cells into wild-type host embryos. In hosts that received *efnb2a* MO donor cells, the donor cells ectopically localized to the vein. In host embryos that contain *ephb4* MO donor cells, fewer cells contribute to the vein compared with their controls, again suggesting a role for Efnb2 and EphB4 in arteriovenous specification [26]. This is largely consistent with other systems, and mice mutants for *Efnb2* have been shown to present the same phenotype as mutants for *EphB4*, characterized by defective morphogenesis of the vasculature. Mutant vasculature suffers from a lack of distinct boundaries between the arteries and the veins, again stressing the importance of Efnb2 and EphB4 in arteriovenous specification [20].

2.3.2 Vascular Endothelial Growth Factor and Sonic Hedgehog

In zebrafish, as in all other non-eutherian vertebrates, four VEGF receptors are present, namely *vegfr1 (flt1), vegfr2 (flk1/kdr), vegfr3 (flt4), and vegfr4 (kdrl)* [6, 14, 75]. The expression pattern of these receptors has been examined in detail [6, 50, 68], and transgenic reporter lines exist for most of them. Neuropilins, non-tyrosine kinase transmembrane molecules, have been shown to be needed for VEGF signaling in other systems, but in zebrafish, morpholino-based data [47, 51] are not entirely consistent with recently provided mutant data [40], and generating mutant lines for the duplicated *nrp1a/b and nrp2a/b* genes is required to shed light on a requirement for these co-receptors.

Morpholino-mediated knockdown of *vegf-Aa* was reported to result in deficiency of ISV sprouting, with no major deficiencies in the DA or PCV [53]. However, analyzing the dependence of vasculogenesis on Vegf-A is confounded by the duplication of zebrafish *vegf-A* genes [3], and by maternal expression of the respective mRNAs. Stable mutant lines or maternal zygotic mutants have not been generated, precluding a final assessment on the role of Vegf-A during the early stages of vasculogenesis; current evidence based on double knockdowns suggests diverged functions of Vegf-Aa and Vegf-Ab, but no major effects on vasculogenesis [3]. Moreoever, mutants for *vegfr4/kdrl* [23] and *vegfr3/flt4* [30] have been reported and clearly demonstrate that zygotic expression of these receptors is not essential for vasculogenesis to occur; mutants in either gene have an apparently normal DA and PCV. These genes have distinct functions at the later stages of vascular development, during arterial and venous ISV sprouting, but their role during earlier stages of vasculogenesis remains somewhat enigmatic; more than one VEGF receptor might be required to be mutated in order for a phenotype to be appreciated.

A key regulator in the early steps of angioblast migration, and in later events of arterial and venous specification, has been suggested in Shh, a member of the Hedgehog family which can act as a ligand for the transmembrane receptors Patched and Smoothened [45]. *Smoothened* mutants, which are devoid of Shh signaling, show comparatively normal angioblast migration [85], and the receptors for Shh, the duplicated *patched* genes, appear not to be expressed within the posterior LPM [48]. Shh regulates expression of *semaphorin 3a1*, which has been shown to have an effect on angioblast migration [69]. In addition, Shh signaling from the midline is essential for normal *vegf-A* expression in the medial aspects of the somites [45]. While more work needs to be carried out to clearly carve out the role for Shh and Vegf-A during the very first events of vasculogenesis, the requirement for both genes in the later steps of arteriovenous specification is better understood. *Shh* mutants show strongly reduced arterial marker gene expression, but this phenotype can be partially overcome by forced expression of *vegf-A* mRNA. Furthermore, overexpression of *shh* results in ectopic expression of arterial markers in venous ECs, again suggesting a specific role of Shh in arterial specification [21]. Interestingly, Shh has been demonstrated to positively influence the expression of *calcitonin receptor-like receptor*-a (*calcrla*) [54], which ultimately results in *vegf-A* expression upstream of Notch. This is particularly significant in light of a recent finding by Wilkinson et al. [85], where Hedgehog signaling was found to induce somitic *vegf-A* expression independent of Calcrla, while Hedgehog can also signal through Calcrla to induce arterial differentiation in angioblasts independent of Vegf-A function. Hence, at least during the later stages, there is room for both signaling pathways in parallel, which in turn might help to guide our thinking about a possibly redundant function for Shh and Vegf-A during the first steps of vasculogenesis. Indeed, it has been suggested that neither pathway is absolutely required for angioblast migration, and that one pathway can compensate for (partial) loss of the other. However, the requirement at the later stages is supported by a number of observations, and a picture has emerged where VEGF signaling in presumptive arteries induces Plc-γ1. Zebrafish mutants for *plc-γ1* show a marked defect in the formation of arteries and strongly reduced expression of *efnb2* [46]. *plc-γ1* mutants cannot be rescued with *vegf-A* overexpression, suggesting Plc-γ1 to act downstream of VEGF receptor function in arterial signaling [44, 46].

2.3.3 Notch and Hey2

The zebrafish Notch family members consist of four Notch receptors (Notch1a, 1b, 2, and 3) and several Notch ligands (DeltaA-D, Dll4, Jagged1a, 1b, and 2) in zebrafish, which are all membrane-bound proteins. Upon binding of these ligands to the Notch transmembrane receptor, a series of proteolytic cleavages release the Notch intercellular domain (NICD) of the receptor into the cytoplasm, after which it translocates to the nucleus. The NICD can then bind to Suppressor of Hairless [Su(H)], which in turn can cause activation of several transcription factors, such as the basic helix-loop-helix (bHLH) proteins, Hairy/Enhancer of Split (Hes) and Hes-related proteins (Hey/HRT/HERP). The promoter regions of *HRT* genes have a binding site for Su(H) [36, 52]. The Notch signaling pathway has long been recognized as a key driver of arterial identity, and in recent years has been extensively evaluated in this respect, and also in its involvement during tip cell/stalk cell formation [59]. The latter aspect has been reviewed in detail elsewhere [84], therefore we will focus only on the arteriovenous specification role of Notch–Delta signaling.

The Notch–Delta signaling pathway appears to be restricted to the arterial endothelium in zebrafish [79], and both loss-of-function as well as gain-of-function studies of Notch family members revealed a disrupted vasculature, with loss of Notch signaling, such as in *mindbomb* mutant embryos, resulting in decreased arterial marker expression and arterial-venous shunts [44]. Similarly, mutants in *hey2/grl*, a factor required downstream of Notch signaling, display altered arterial gene expression and develop a distinct shunt phenotype at the level of the cranial vasculature [87]. Gain-of-function of Notch family members causes a reduction of venous fate [44]. There is a tight link between Vegf signaling function and Notch–Delta activity; Vegf-A can induce the expression of *notch*, and Notch can rescue the arterial specification defect in *vegf-A* knockdown studies, suggesting that Notch acts downstream of Vegf in arterial specification [44, 79]

One of the Notch target proteins is the hairy/enhancer-of-split-related bHLH family member Hey2. The zebrafish ortholog of the mammalian *Hey2* gene is *gridlock*. Gridlock functions as a transcriptional repressor, and is already expressed in early development, in the angioblasts that are localized within the medial aspect of the LPM, whereafter *gridlock* expression continues to be restricted to arteries [87]. The loss of Gridlock function in early zebrafish development results in defective proliferation of angioblasts at the level of the LPM [10]. Later on, loss of *gridlock* results in a circulatory shortcut through a disrupted DA, with concomitant increase of the venous maker EphB4 and a decrease in the arterial marker *efnb2* [87]. More specifically, the point of fusion of the LDA to the DA is affected, which represents a remarkably specific and locally restricted phenotype [83, 88] which has recently been shown to be mimicked by the *sox7* and *efnb2a/b* mutants [27]. Furthermore, overexpression of *gridlock* causes suppression of venous markers [88]. Both *in vitro* and *in vivo* experiments showed that induction of Notch–ICD induces *gridlock* expression, again suggesting that Gridlock is acting downstream of Notch [52, 88]. Furthermore, inhibiting Hh or VEGF signaling results in loss of *gridlock* expression in the angioblasts that will form the DA, while stimulating *vegf* expression in *gridlock* morphants rescues the *gridlock* phenotype. These results show that gridlock functions downstream of the Hh–VEGF–Notch signaling pathway in arterial specification [57, 62]. As in other cases, the defect observable in mutants appears to be more pronounced in arteriovenous specification during angiogenesis and less during vasculogenesis.

2.3.4 SoxF Family Members

SRY-related high-mobility-group box (Sox) genes form a family of transcription factors involved in diverse developmental processes, including vascular development (reviewed by Chew and Gallow [9]). The Sox proteins contain two main domains—the high-mobility-group (HMG) domain, which can bind target DNA motifs, and the transactivation domain, which mediates the transcriptional response [4]. The *Sox* family is divided into several subfamilies, with the *Sox-F* family members comprised of the Sox7, Sox17, and Sox18 genes. Sox17 is involved in hematopoietic stem cell regulation and formation of endoderm [2, 11, 35, 37], but has also recently been shown to play a role in arterial specification in the mouse [12]. Sox7 and Sox18 have long been recognized as being involved in vascular development [8, 28, 55, 82]. The human syndrome hypotrichosis-lymphedema-telangiectasia (HLT) is linked to mutations in *Sox18*, with patients presenting disrupted blood and lymphatic vessels [31]. Mice with truncated Sox18 protein ('Ragged' mice) resemble HLT and display defects in blood and lymphatic vasculature development [15, 33, 56]. In mice, Sox18 starts to be expressed at approximately E9.0 in a subpopulation of venous endothelial cells, which induces Prox1 expression in these cells. Subsequently, these Prox1-expressing cells migrate away from the veins under the influence of VEGF-C and become lymphatic endothelial cells (LECs) which will later form the lymphatic vascular network [18, 24, 71] (see also Chap. 5). In contrast, zebrafish *Prox1* and *Sox18* are dispensable for lymphatic development, revealed by *Prox1* and *Sox18* mutants developing a lymphatic vasculature [77, 78].

In zebrafish, Sox18 and Sox7 appear to play redundant roles in vascular development. *Sox7* and *sox18* are expressed in the early pre-migratory angioblasts at the LPM, then in the migrating angioblast population and later in the specified vasculature. Double morpholino knockdown showed defective blood circulation, and the DA and PCV are fused together in the trunk of the embryo, which results in arteriovenous shunt formation at a relatively late stage of development. Venous markers are upregulated in the DA, whereas the arterial markers are downregulated, suggesting a role for Sox7/Sox18 in arteriovenous specification [28, 55]. This is further substantiated by the phenotype of *sox7* mutants, which show phenotypes identical to *gridlock* and *efnb2a/b* mutants [27]. The exact mechanism of Sox7 and Sox18 function in arteriovenous regulation is not yet known; however, recent evidence showed that the enhancer for the Notch ligand Dll4 contains a binding site for SoxF factors. Both SoxF and RBPJ transcription factors can bind and regulate Dll4 enhancer activity, suggesting an important role for SoxF family members in regulating Dll4 activity and subsequent arteriovenous specification [64].

2.3.5 Fox and ETS

Forkhead (fox) transcription factors are helix-turn-helix proteins. Foxc transcription factors are expressed in the vasculature and *Foxc*-null mice die during embryonic development with severe vascular defects, including arteriovenous malformations and loss of arterial markers [42, 67]. Similarly, in zebrafish, combined knockdown of *foxc1a* and *foxc1b* results in severe disruption of the vascular system [13]. *In vitro* studies showed that VEGF signaling can induce the transcriptional activity of Foxc proteins [67]. Overexpression of *foxc* genes induces expression of arterial markers, such as *notch1* and *dll4*. Foxc can bind and activate the *Dll4* promoter, suggesting that Foxc acts upstream of Notch signaling in arteriovenous specification [67]. Furthermore, Foxc2 can interact with the Su(H)/NICD complex to induce *Hey2* promoter activity [25]. Foxc, together with the Ets factor Etsrp, bind to a FOX:ETS motif-inducing enhancer activation. This FOX:ETS domain is present in many endothelial-specific enhancers, suggesting the importance of the Foxc and Ets transcription factors in vascular development [13].

2.4 Summary

Within the last few years, we have witnessed a considerable number of studies that have significantly advanced our understanding of vasculogenesis and angiogenesis. Careful lineage analysis and meticulous comparison of expression data of various marker genes have shed new light on the nature of early angioblasts. The later events of arteriovenous specification and differentiation have, in turn, benefited from genetic interference studies (summarized in Fig. 2.3) here, it will be necessary to repeat some of the work with stable mutant lines rather than relying on morpholino data [40, 66]. This notwithstanding, we will continue to gain more insight into the early events of cardiovascular development from additional work in zebrafish embryos, which are so well suited for the combined application of genetics and *in vivo* imaging.

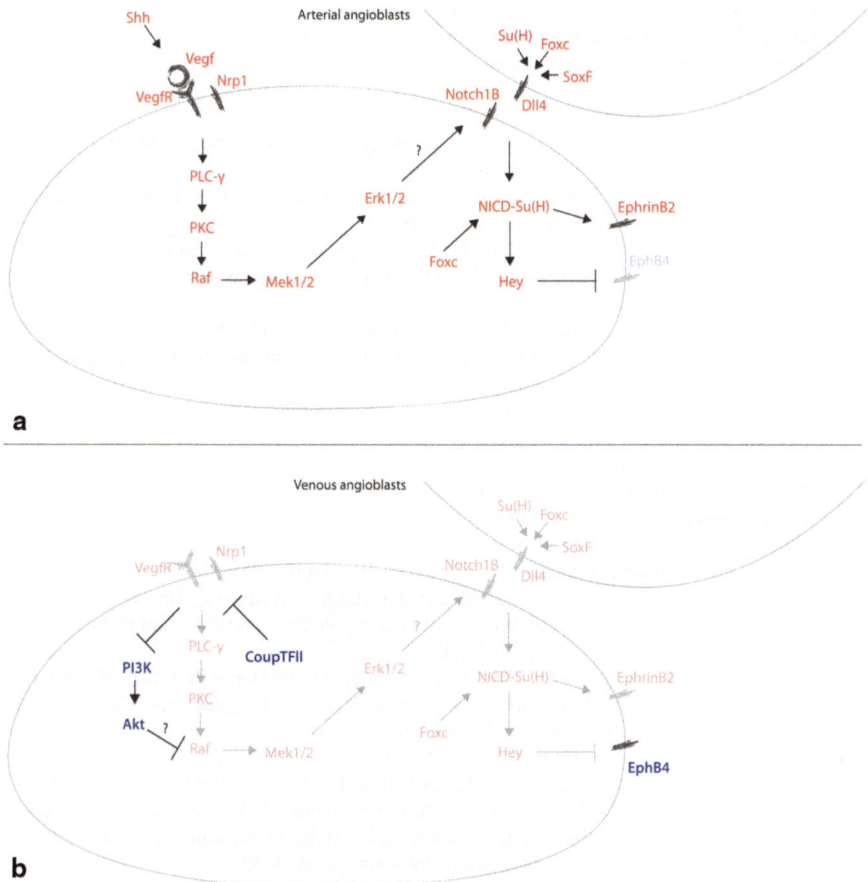

Fig. 2.3 A model of the molecular pathways in arterial- and venous-fated angioblasts. Molecular pathways and key factors discussed in this review are depicted. Arterial-specific genes are depicted in *red*, and venous-specific genes are depicted in *blue*. **a** The VEGF/NRP1 pathway activates the PLC-γ/Mek/ERK pathway in arterial cells [38], which can induce the Notch pathway. Su(H)/Foxc/SoxF can induce Notch signaling by binding to the Dll4 enhancer, while Notch/Hey2 signaling will result in the expression of EphrinB2 on arterial membranes and the inhibition of EphB4 function. **b** In venous-fated angioblasts, these pathways are inhibited by the PI3K/Akt pathway and, in mice, by COUPTF2. In venous cells, EphB4 expression is no longer inhibited and gets to be expressed on the membrane. Question marks indicate interactions that are unknown.

References

1. Adams RH, Wilkinson GA, Weiss C, Diella F, Gale NW, Deutsch U, Risau W, Klein R. Roles of ephrinB ligands and EphB receptors in cardiovascular development: demarcation of arterial/venous domains, vascular morphogenesis, and sprouting angiogenesis. Genes Dev. 1999;13(3):295–306.
2. Alexander J, Stainier DY. A molecular pathway leading to endoderm formation in zebrafish. Curr Biol. 1999;9(20):1147–57.

3. Bahary N, Goishi K, Stuckenholz C, Weber G, Leblanc J, Schafer CA, Berman SS, Klagsbrun M, Zon LI. Duplicate VegfA genes and orthologues of the KDR receptor tyrosine kinase family mediate vascular development in the zebrafish. Blood. 2007;110(10):3627–36.
4. Bowles J, Schepers G, Koopman P. Phylogeny of the SOX family of developmental transcription factors based on sequence and structural indicators. Dev Biol. 2000;227(2):239–55.
5. Bruckner K, Pasquale EB, Klein R. Tyrosine phosphorylation of transmembrane ligands for Eph receptors. Science. 1997;275(5306):1640–3.
6. Bussmann J, Lawson N, Zon L, Schulte-Merker S. Zebrafish VEGF receptors: a guideline to nomenclature. PLoS Genet. 2008;4(5):e1000064.
7. Bussmann J, Wolfe SA, Siekmann AF. Arterial-venous network formation during brain vascularization involves hemodynamic regulation of chemokine signaling. Development. 2011;138(9):1717–26.
8. Cermenati S, Moleri S, Cimbro S, Corti P, Del Giacco L, Amodeo R, Dejana E, Koopman P, Cotelli F, Beltrame M. Sox18 and Sox7 play redundant roles in vascular development. Blood. 2008;111(5):2657–66.
9. Chew LJ, Gallo V. The Yin and Yang of Sox proteins: activation and repression in development and disease. J Neurosci Res. 2009;87(15):3277–87.
10. Chun CZ, Remadevi I, Schupp MO, Samant GV, Pramanik K, Wilkinson GA, Ramchandran R. Fli+ etsrp+ hemato-vascular progenitor cells proliferate at the lateral plate mesoderm during vasculogenesis in zebrafish. PLoS One. 2011;6(2):e14732.
11. Chung MI, Ma AC, Fung TK, Leung AY. Characterization of Sry-related HMG box group F genes in zebrafish hematopoiesis. Exp Hematol. 2011;39(10):986–998.e5.
12. Corada M, Orsenigo F, Morini MF, Pitulescu ME, Bhat G, Nyqvist D, Breviario F, Conti V, Briot A, Iruela-Arispe ML, et al. Sox17 is indispensable for acquisition and maintenance of arterial identity. Nat Commun. 2013;4:2609.
13. De Val S, Chi NC, Meadows SM, Minovitsky S, Anderson JP, Harris IS, Ehlers ML, Agarwal P, Visel A, Xu SM, et al. Combinatorial regulation of endothelial gene expression by ets and forkhead transcription factors. Cell. 2008;135(6):1053–64.
14. de Vries C, Escobedo JA, Ueno H, Houck K, Ferrara N, Williams LT. The fms-like tyrosine kinase, a receptor for vascular endothelial growth factor. Science. 1992;255(5047):989–91.
15. Downes M, Francois M, Ferguson C, Parton RG, Koopman P. Vascular defects in a mouse model of hypotrichosis-lymphedema-telangiectasia syndrome indicate a role for SOX18 in blood vessel maturation. Hum Mol Genet. 2009;18(15):2839–50.
16. Ekker SC, Ungar AR, Greenstein P, von Kessler DP, Porter JA, Moon RT, Beachy PA. Patterning activities of vertebrate hedgehog proteins in the developing eye and brain. Curr Biol. 1995;5(8):944–55.
17. Fouquet B, Weinstein BM, Serluca FC, Fishman MC. Vessel patterning in the embryo of the zebrafish: guidance by notochord. Dev Biol. 1997;183(1):37–48.
18. Francois M, Caprini A, Hosking B, Orsenigo F, Wilhelm D, Browne C, Paavonen K, Karnezis T, Shayan R, Downes M, et al. Sox18 induces development of the lymphatic vasculature in mice. Nature. 2008;456(7222):643–7.
19. Gale NW, Holland SJ, Valenzuela DM, Flenniken A, Pan L, Ryan TE, Henkemeyer M, Strebhardt K, Hirai H, Wilkinson DG, et al. Eph receptors and ligands comprise two major specificity subclasses and are reciprocally compartmentalized during embryogenesis. Neuron. 1996;17(1):9–19.
20. Gerety SS, Wang HU, Chen ZF, Anderson DJ. Symmetrical mutant phenotypes of the receptor EphB4 and its specific transmembrane ligand ephrin-B2 in cardiovascular development. Mol Cell. 1999;4(3):403–14.
21. Gering M., Patient, R. Hedgehog signaling is required for adult blood stem cell formation in zebrafish embryos. Dev Cell. 2005;8(3):389–400.
22. Gonzalez-Crussi F. Vasculogenesis in the chick embryo. An ultrastructural study. Am J Anat. 1971;130(4):441–60.
23. Habeck H, Odenthal J, Walderich B, Maischein H, Schulte-Merker S. Analysis of a zebrafish VEGF receptor mutant reveals specific disruption of angiogenesis. Curr Biol. 2002;12(16):1405–12.

24. Hagerling R, Pollmann C, Andreas M, Schmidt C, Nurmi H, Adams RH, Alitalo K, Andresen V, Schulte-Merker S, Kiefer F. A novel multistep mechanism for initial lymphangiogenesis in mouse embryos based on ultramicroscopy. EMBO J. 2013;32(5):629–44.
25. Hayashi H, Kume T. Foxc transcription factors directly regulate Dll4 and Hey2 expression by interacting with the VEGF-Notch signaling pathways in endothelial cells. PLoS One. 2008;3(6):e2401.
26. Herbert SP, Huisken J, Kim TN, Feldman ME, Houseman BT, Wang RA, Shokat KM, Stainier DY. Arterial-venous segregation by selective cell sprouting: an alternative mode of blood vessel formation. Science. 2009;326(5950):294–8..
27. Hermkens DMA, Impel A, Urasaki A, Bussmann J, Duckers HJ, Schulte-Merker S. Sox7 controls arterial specification in conjunction with hey2 and efnb2 function. Development. 2015;142(9):1695–704.
28. Herpers R, van de Kamp E, Duckers HJ, Schulte-Merker S. Redundant roles for sox7 and sox18 in arteriovenous specification in zebrafish. Circ Res. 2008;102(1):12–5.
29. Hogan BM, Bos FL, Bussmann J, Witte M, Chi NC, Duckers HJ, Schulte-Merker S. Ccbe1 is required for embryonic lymphangiogenesis and venous sprouting. Nat Genet. 2009a;41(4):396–8.
30. Hogan BM, Herpers R, Witte M, Helotera H, Alitalo K, Duckers HJ, Schulte-Merker S. Vegfc/Flt4 signalling is suppressed by Dll4 in developing zebrafish intersegmental arteries. Development. 2009;136(23):4001–9.
31. Irrthum A, Devriendt K, Chitayat D, Matthijs G, Glade C, Steijlen PM, Fryns JP, Van Steensel MA, Vikkula M. Mutations in the transcription factor gene SOX18 underlie recessive and dominant forms of hypotrichosis-lymphedema-telangiectasia. Am J Hum Genet. 2003;72(6):1470–8.
32. Isogai S, Lawson ND, Torrealday S, Horiguchi M, Weinstein BM. Angiogenic network formation in the developing vertebrate trunk. Development. 2003;130(21):5281–90.
33. James K, Hosking B, Gardner J, Muscat GE, Koopman P. Sox18 mutations in the ragged mouse alleles ragged-like and opossum. Genesis. 2003;36(1):1–6.
34. Jin SW, Beis D, Mitchell T, Chen JN, Stainier DY. Cellular and molecular analyses of vascular tube and lumen formation in zebrafish. Development. 2005;132(23):5199–209.
35. Kanai-Azuma M, Kanai Y, Gad JM, Tajima Y, Taya C, Kurohmaru M, Sanai Y, Yonekawa H, Yazaki K, Tam PP, et al. Depletion of definitive gut endoderm in Sox17-null mutant mice. Development. 2002;129(10):2367–79.
36. Kao HY, Ordentlich P, Koyano-Nakagawa N, Tang Z, Downes M, Kintner CR, Evans RM, Kadesch T. A histone deacetylase corepressor complex regulates the Notch signal transduction pathway. Genes Dev. 1998;12(15):2269–77.
37. Kim I, Saunders TL, Morrison SJ. Sox17 dependence distinguishes the transcriptional regulation of fetal from adult hematopoietic stem cells. Cell. 2007;130(3):470–83.
38. Kim SH, Schmitt CE, Woolls MJ, Holland MB, Kim JD, Jin SW. Vascular endothelial growth factor signaling regulates the segregation of artery and vein via ERK activity during vascular development. Biochem Biophys Res Commun. 2013;430(4):1212–6.
39. Kohli V, Schumacher JA, Desai SP, Rehn K, Sumanas S. Arterial and venous progenitors of the major axial vessels originate at distinct locations. Dev Cell. 2013;25(2):196–206.
40. Kok FO, Shin M, Ni CW, Gupta A, Grosse AS, van Impel A, Kirchmaier BC, Peterson-Maduro J, Kourkoulis G, Male I, et al. Reverse genetic screening reveals poor correlation between morpholino-induced and mutant phenotypes in zebrafish. Dev Cell. 2015;32(1):97–108.
41. Kullander K, Klein R. Mechanisms and functions of Eph and ephrin signalling. Nat Rev Mol Cell Biol. 2002;3(7):475–86.
42. Kume T, Jiang H, Topczewska JM, Hogan BL. The murine winged helix transcription factors, Foxc1 and Foxc2, are both required for cardiovascular development and somitogenesis. Genes Dev. 2001;15(18):2470–82.
43. Lawson ND, Weinstein BM. In vivo imaging of embryonic vascular development using transgenic zebrafish. Dev Biol. 2002;248(2):307–18.
44. Lawson ND, Scheer N, Pham VN, Kim CH, Chitnis AB, Campos-Ortega JA, Weinstein BM. Notch signaling is required for arterial-venous differentiation during embryonic vascular development. Development. 2001;128(19):3675–83.

45. Lawson ND, Vogel AM, Weinstein BM. Sonic hedgehog and vascular endothelial growth factor act upstream of the Notch pathway during arterial endothelial differentiation. Dev Cell. 2002;3(1):127–36.
46. Lawson ND, Mugford JW, Diamond BA, Weinstein BM. Phospholipase C gamma-1 is required downstream of vascular endothelial growth factor during arterial development. Genes Dev. 2003;17(11):1346–51.
47. Lee P, Goishi K, Davidson AJ, Mannix R, Zon L, Klagsbrun M. Neuropilin-1 is required for vascular development and is a mediator of VEGF-dependent angiogenesis in zebrafish. Proc Natl Acad Sci U S A. 2002;99(16):10470–5.
48. Lewis KE, Concordet JP, Ingham PW. Characterisation of a second patched gene in the zebrafish Danio rerio and the differential response of patched genes to Hedgehog signalling. Dev Biol. 1999;208(1):14–29.
49. Liang D, Xu X, Chin AJ, Balasubramaniyan NV, Teo MA, Lam TJ, Weinberg ES, Ge R. Cloning and characterization of vascular endothelial growth factor (VEGF) from zebrafish, Danio rerio. Biochim Biophys Acta. 1998;1397(1):14–20.
50. Lohela M, Bry M, Tammela T, Alitalo K. VEGFs and receptors involved in angiogenesis versus lymphangiogenesis. Curr Opin Cell Biol. 2009;21(2):154–65.
51. Martyn U, Schulte-Merker S. The ventralized ogon mutant phenotype is caused by a mutation in the zebrafish homologue of Sizzled, a secreted Frizzled-related protein. Dev Biol. 2003;260(1):58–67.
52. Nakagawa O, McFadden DG, Nakagawa M, Yanagisawa H, Hu T, Srivastava D, Olson EN. Members of the HRT family of basic helix-loop-helix proteins act as transcriptional repressors downstream of Notch signaling. Proc Natl Acad Sci U S A. 2000;97(25):13655–60.
53. Nasevicius A, Larson J, Ekker SC. Distinct requirements for zebrafish angiogenesis revealed by a VEGF-A morphant. Yeast. 2000;17(4):294–301.
54. Nicoli S, Tobia C, Gualandi L, De Sena G, Presta M. Calcitonin receptor-like receptor guides arterial differentiation in zebrafish. Blood. 2008;111(10):4965–72.
55. Pendeville H, Winandy M, Manfroid I, Nivelles O, Motte P, Pasque V, Peers B, Struman I, Martial JA, Voz ML. Zebrafish Sox7 and Sox18 function together to control arterial-venous identity. Dev Biol. 2008;317(2):405–16.
56. Pennisi D, Gardner J, Chambers D, Hosking B, Peters J, Muscat G, Abbott C, Koopman P. Mutations in Sox18 underlie cardiovascular and hair follicle defects in ragged mice. Nat Genet. 2000;24(4):434–7.
57. Peterson RT, Shaw SY, Peterson TA, Milan DJ, Zhong TP, Schreiber SL, MacRae CA, Fishman MC. Chemical suppression of a genetic mutation in a zebrafish model of aortic coarctation. Nat Biotechnol. 2004;22(5):595–9.
58. Pham VN, Lawson ND, Mugford JW, Dye L, Castranova D, Lo B, Weinstein BM. Combinatorial function of ETS transcription factors in the developing vasculature. Dev Biol. 2007;303(2):772–83.
59. Phng LK, Gerhardt H. Angiogenesis: a team effort coordinated by notch. Dev Cell. 2009;16(2):196–208.
60. Poole TJ, Coffin JD. Vasculogenesis and angiogenesis: two distinct morphogenetic mechanisms establish embryonic vascular pattern. J Exp Zool. 1989;251(2):224–31.
61. Risau W, Flamme I. Vasculogenesis. Annu Rev Cell Dev Biol. 1995;11:73–91.
62. Rowlinson JM, Gering M. Hey2 acts upstream of Notch in hematopoietic stem cell specification in zebrafish embryos. Blood. 2010;116(12):2046–56.
63. Sabin FR. Preliminary note on the differentiation of angioblasts and the method by which they produce blood-vessels, blood-plasma and red blood-cells as seen in the living chick. 1917. J Hematother Stem Cell Res. 2002;11(1):5–7.
64. Sacilotto N, Monteiro R, Fritzsche M, Becker PW, Sanchez-Del-Campo L, Liu K, Pinheiro P, Ratnayaka I, Davies B, Goding CR, et al. Analysis of Dll4 regulation reveals a combinatorial role for Sox and Notch in arterial development. Proc Natl Acad Sci U S A. 2013;110(29):11893–8.

65. Schuermann A, Helker CS, Herzog W. Angiogenesis in zebrafish. Semin Cell Dev Biol. 2014;31C:106–14.
66. Schulte-Merker S, Stainier DY. Out with the old, in with the new: reassessing morpholino knockdowns in light of genome editing technology. Development. 2014;141(16):3103–4.
67. Seo S, Fujita H, Nakano A, Kang M, Duarte A, Kume T. The forkhead transcription factors, Foxc1 and Foxc2, are required for arterial specification and lymphatic sprouting during vascular development. Dev Biol. 2006;294(2):458–70.
68. Shibuya M. Vascular endothelial growth factor and its receptor system: physiological functions in angiogenesis and pathological roles in various diseases. J Biochem. 2013;153(1):13–9.
69. Shoji W, Isogai S, Sato-Maeda M, Obinata M, Kuwada JY. Semaphorin3a1 regulates angioblast migration and vascular development in zebrafish embryos. Development. 2003;130(14):3227–36.
70. Siekmann AF, Standley C, Fogarty KE, Wolfe SA, Lawson ND. Chemokine signaling guides regional patterning of the first embryonic artery. Genes Dev. 2009;23(19):2272–7.
71. Srinivasan RS, Geng X, Yang Y, Wang Y, Mukatira S, Studer M, Porto MP, Lagutin O, Oliver G. The nuclear hormone receptor Coup-TFII is required for the initiation and early maintenance of Prox1 expression in lymphatic endothelial cells. Genes Dev. 2010;24(7):696–707.
72. Stainier DY, Weinstein BM, Detrich HW 3rd, Zon LI, Fishman MC. Cloche, an early acting zebrafish gene, is required by both the endothelial and hematopoietic lineages. Development. 1995;121(10):3141–50.
73. Stainier DY, Fouquet B, Chen JN, Warren KS, Weinstein BM, Meiler SE, Mohideen MA, Neuhauss SC, Solnica-Krezel L, Schier AF, et al. Mutations affecting the formation and function of the cardiovascular system in the zebrafish embryo. Development. 1996;123:285–92.
74. Sumanas S, Lin S. Ets1-related protein is a key regulator of vasculogenesis in zebrafish. PLoS Biol. 2006;4(1):e10.
75. Terman BI, Carrion ME, Kovacs E, Rasmussen BA, Eddy RL, Shows TB. Identification of a new endothelial cell growth factor receptor tyrosine kinase. Oncogene. 1991;6(9):1677–83.
76. Thompson MA, Ransom DG, Pratt SJ, MacLennan H, Kieran MW, Detrich HW 3rd, Vail B, Huber TL, Paw B, Brownlie AJ, et al. The cloche and spadetail genes differentially affect hematopoiesis and vasculogenesis. Dev Biol. 1998;197(2):248–69.
77. van Impel A, Schulte-Merker S. A fisheye view on lymphangiogenesis. Adv Anat Embryol Cell Biol. 2014;214:153–65.
78. van Impel A, Zhao Z, Hermkens DM, Roukens MG, Fischer JC, Peterson-Maduro J, Duckers H, Ober EA, Ingham PW, Schulte-Merker S. Divergence of zebrafish and mouse lymphatic cell fate specification pathways. Development. 2014;141(6):1228–38.
79. Villa N, Walker L, Lindsell CE, Gasson J, Iruela-Arispe ML, Weinmaster G. Vascular expression of Notch pathway receptors and ligands is restricted to arterial vessels. Mech Dev. 2001;108(1–2):161–4.
80. Vogeli KM, Jin SW, Martin GR, Stainier DY. A common progenitor for haematopoietic and endothelial lineages in the zebrafish gastrula. Nature. 2006;443(7109):337–9.
81. Wang HU, Chen ZF, Anderson DJ. Molecular distinction and angiogenic interaction between embryonic arteries and veins revealed by ephrin-B2 and its receptor Eph-B4. Cell. 1998;93(5):741–53.
82. Wat MJ, Beck TF, Hernandez-Garcia A, Yu Z, Veenma D, Garcia M, Holder AM, Wat JJ, Chen Y, Mohila CA, et al. Mouse model reveals the role of SOX7 in the development of congenital diaphragmatic hernia associated with recurrent deletions of 8p23.1. Hum Mol Genet. 2012;21(18):4115–25.
83. Weinstein BM, Stemple DL, Driever W, Fishman MC. Gridlock, a localized heritable vascular patterning defect in the zebrafish. Nat Med. 1995;1(11):1143–7.
84. Wilkinson RN, van Eeden FJ. The zebrafish as a model of vascular development and disease. Prog Mol Biol Transl Sci. 2014;124:93–122.
85. Wilkinson RN, Koudijs MJ, Patient RK, Ingham PW, Schulte-Merker S, van Eeden FJ. Hedgehog signaling via a calcitonin receptor-like receptor can induce arterial differentiation independently of VEGF signaling in zebrafish. Blood. 2012;120(2):477–88.

86. Xiong JW, Yu Q, Zhang J, Mably JD. An acyltransferase controls the generation of hematopoietic and endothelial lineages in zebrafish. Circ Res. 2008;102(9):1057–64.
87. Zhong TP, Rosenberg M, Mohideen MA, Weinstein B, Fishman MC. Gridlock, an HLH gene required for assembly of the aorta in zebrafish. Science. 2000;287(5459):1820–4.
88. Zhong TP, Childs S, Leu JP, Fishman MC. Gridlock signalling pathway fashions the first embryonic artery. Nature. 2001;414(6860):216–20.
89. Zon LI. Developmental biology of hematopoiesis. Blood. 1995;86(8):2876–91.

Chapter 3
Pericytes in Vascular Development and Function

Richard Daneman and Annika Keller

Abbreviations

ALK	Activin receptor-like kinase
Ang1	Angiopoietin 1
BBB	Blood–brain barrier
BMP	Bone morphogenetic protein
CADASIL	Cerebral autosomal dominant arteriopathy with subcortical infarcts and leukoencephalopathy
CNS	Central nervous system
EM	Electron microscopy
HB-EGF	Heparin-binding epidermal growth factor
HHT	Hereditary hemorrhagic telangiectasia
IBGC	Idiopathic basal ganglia calcifications
MMP	Matrix metalloproteinase
PDGF-B	Platelet-derived growth factor-B
PDGFRβ	Platelet-derived growth factor receptor-β
SDF-1α	Stromal-derived growth factor-α
Shh	Sonic hedgehog
SMA	Smooth muscle actin
S1P	Sphingosine-1-phosphate
S1PR	Sphingosine-1-phosphate receptor
TAAD	Thoracic aortic aneurisms
TGFβ	Transforming growth factor-β
TIMP	Tissue inhibitors of metalloproteinases
vSMCs	Vascular smooth muscle cells

R. Daneman (✉)
Departments of Neuroscience and Pharmacology, University of California, San Diego, BSB 3092, 9500 Gilman Drive, La Jolla, CA 92093-0636, USA
e-mail: rdaneman@ucsd.edu

A. Keller (✉)
Division of Neurosurgery, University Hospital Zürich, Frauenklinikstrasse 10, CH-8091 Zürich, Switzerland
e-mail: Annika.Keller@usz.ch

M. HH. Schmidt, S. Liebner (eds.), *Endothelial Signaling in Development and Disease*, DOI 10.1007/978-1-4939-2907-8_3

3.1 Introduction

Pericytes are embedded in the vascular basement membrane lining the abluminal surface of microvasculature. These cells belong to a larger lineage of cells termed mural cells, which include vascular smooth muscle cells (vSMCs) that completely surround large vessels (arteries, arterioles, and veins), and pericytes that incompletely cover microvessels (capillaries and postcapillary venules) (Fig. 3.1) [1]. These cells were first described by Eberth in 1871 and Rouget in 1873, who described contractile cells that surround small blood vessels [2, 3]. In 1923, Zimmerman first coined the term 'pericyte' by renaming 'Rouget cells' based on their proximity to endothelial cells [4].

Although much has been learned about the cell biology, development, and function of pericytes in the last 140 years, considerable mystery persists regarding the role of these cells in regulating tissue development, vascular function, homeostasis, and response to injury and disease. Much of the difficulty in studying pericytes comes from the fact that pericytes lack clearly defined histological and molecular criteria. While many researchers identify pericytes based on their proximity to microvascular endothelial cells, many other cell types in different tissues are

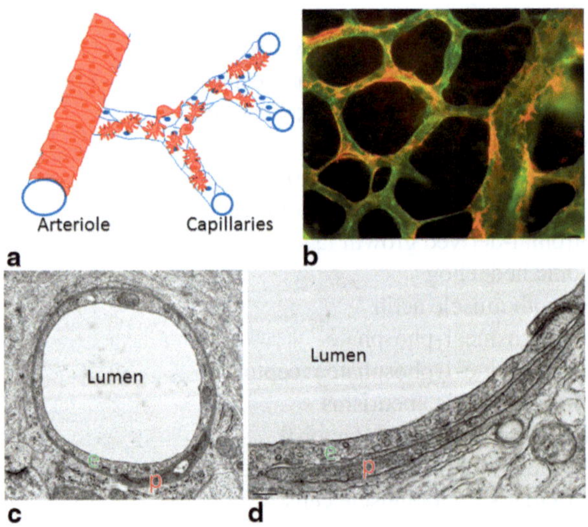

Fig. 3.1 Association of pericytes and endothelial cells. **a** Schematic representation of endothelial cells (*blue*) and mural cells (*pink*). Vascular smooth muscle cells cover larger vessels such as arterioles. In capillaries, pericytes are found on the abluminal surface of endothelial cells embedded in the vascular basement membrane. **b** Immunofluoresent micrograph of mouse retinal vasculature with endothelial cells stained with BSL-I (*green*) and pericytes stained with an anti-NG2 antibody. **c** Electron micrograph showing a cross-section of a central nervous system capillary. The endothelial cell (*e*) forms a tube for the passage of blood, and the pericyte (*p*) is situated on the abluminal surface of the endothelial cell. **d** Electron micrograph demonstrating the interaction of pericytes (*p*) with endothelial cells (*e*) in a mouse cerebral capillary. Pericytes are embedded in the vascular basement membrane with intermittent junctional contacts with the endothelial cell

associated with the abluminal surface of endothelial cells, including vSMCs, macrophages, fibroblasts, and epithelial cells. The lack of distinct molecular markers for pericytes has hindered the precise identification of each cell type, and thus there is often ambiguity regarding the cell population being studied. The most stringent definition of a pericyte requires that a cell is embedded within the vascular basement membrane [5]. However, studies suggest that pericytes exhibit dramatic plasticity, including cell proliferation, morphogenesis, and migration [1]. These processes that occur in response to injury and disease include breakdown of the extracellular matrix (ECM) and movement away from the vascular basement membrane. Therefore, defining pericytes strictly based on their localization in the basement membrane is too limited and fails to reflect these dynamic properties. Furthermore, there is evidence suggesting the heterogeneity of pericyte populations in different organs, and even within the same organ, further obscuring the identification of this cell type.

In the upcoming years, the identification of novel pericyte-specific molecular markers, and the generation of genetic tools to label pericytes *in vivo*, as well as new imaging techniques to recognize pericyte cell morphology and localization, may provide a better definition for this mysterious cell type.

3.2 Cell Biology of Pericytes

3.2.1 Cell Morphology

Pericytes are associated with the abluminal surface of microvessels, including capillaries and postcapillary venules throughout the body. These cells adhere to the abluminal surface of endothelial cell tubes, and are embedded in the vascular basement membrane [1, 6, 7]. Pericytes commonly extend long, branching cytoplasmic processes along the abluminal surface of the blood vessels that can stretch across multiple endothelial cell bodies (Fig. 3.1). The cell body of pericytes is often found at branch points of the microvascular network.

On capillaries, pericytes have a rounded cell body that contains a few primary cellular processes branching into perpendicular secondary processes that connect with endothelial cells. In postcapillary venules, pericytes display a more stellate pattern, with a flattened cell body and many thin, branching cellular processes [7].

Electron microscopy (EM) has identified that pericytes possess a discoid nucleus, few cytoplasmic organelles, and abundant plasmalemmal vesicles. Serial section EM has demonstrated that these vesicles interconnect to form a continuous network associated with the parenchymal-facing membrane of the pericyte [8]. The pericyte cell body is rich in cytoskeletal elements, such as intermediate filaments composed of desmin and vimentin extending into the primary extensions, and microtubules extending into both the primary and secondary cellular processes [1]. In addition, bands of actin, myosin, and tropomyosin are situated beneath the plasma membrane, indicating these cells are capable of contraction [9]. Although most of the cell body of the pericyte is separated from the endothelial cell by the vascular basement membrane, pericyte processes are observed to contact the endothelial cells in so-called peg-and-socket junctions [10–12].

As noted, one of the major challenges in studying pericytes comes from the lack of defining molecular markers. There is no molecular marker that is unique to pericytes. Recent findings demonstrating molecular heterogeneity of pericytes in different tissues and the ability of pericytes to change their morphology and molecular composition in response to injury, disease, and cell culture, further complicate the identification of pericytes. One commonly used molecular marker is platelet-derived growth factor receptor (PDGFR)-β, a receptor tyrosine kinase that is required for the recruitment of pericytes to the vascular tubes. However, PDGFRβ is also expressed in other cell types, including neural cells, cardiomyocytes, and fibroblasts [13, 14]. In addition, markers such as NG2, Anpep (CD13), desmin, RGS5, Abcc9, Kcnj8, Dlk, and Zic1 have all been utilized to identify pericytes, yet all lack specificity [1, 15–19].

3.2.2 *Heterogeneity of Pericytes*

Recent evidence suggests that there is remarkable heterogeneity of pericytes at different levels of the vascular tree, especially notable when comparing pericytes on postcapillary venules with pericytes on capillaries. Postcapillary venules have a greater number of flattened pericytes compared with capillaries, with thinner cellular processes and a greater extent of overlapping branches [20, 21].

In addition, there are clear differences between pericytes in various organs, as well as distinct subpopulations within an organ. These differences may indicate the importance for the vasculature to meet the unique requirements of each tissue and organ in the body.

Pericyte cellular density and extent of coverage of the vascular tube differ in various tissues. For example, central nervous system (CNS) capillaries have the largest number of pericytes, with an endothelial to pericyte ratio between 1:1 and 3:1, whereas skeletal muscle shows a pericyte:endothelial cell ratio of 100:1 [22]. However, it should be noted that such metrics are imprecise since there are no reliable markers for counting the exact number of pericytes in each tissue. The extent of coverage by pericytes of the external surface area of the vascular tube also varies and has been estimated to be 11 % in cardiac muscle, 21 % in skeletal muscle, 22–32 % in the cerebrum, and 41 % in the retina [23]. The density and vascular coverage of pericytes inversely correlates with the permeability of the vessel as well as the rate of endothelial cell turnover. These observations have led to the hypothesis that pericytes may be critical for endothelial stability and limiting permeability. Indeed, experiments using genetically engineered mice that lack pericytes have confirmed a role for pericytes in these processes [17, 24]. Pericytes in different tissues show variable levels of granularity, an indication of the abundance of cytoplasmic lysosomes. All cerebral pericytes are granular, a property that is increased following disruption of the blood–brain barrier (BBB) [5].

Pericyte populations within specific tissues have evolved unique morphological and physiological properties, many of which were identified initially by different names. For instance, stellate cells represent hepatic pericytes [25]. These cells are localized between endothelial cells of the sinusoidal capillaries and the parenchyma in the space of Disse [26] and are not embedded within the vascular basement membrane as sinusoidal capillaries have incomplete basement membranes and thus do not comply with classical definition of pericytes. Stellate cells contain cellular protrusions that adhere both to the endothelium and hepatocytes, and express desmin, glial fibrillary acidic protein, neural cell adhesion molecule, synaptophysin, nestin, and smooth muscle actin (SMA) when activated. These cells possess functions unique to the liver as they are involved in vitamin A storage, liver development, regeneration, and fibrosis (discussed in the 'Regulation of Tissue-Specific Properties' section) [27].

Evidence suggests that even within a given organ, there are different pericyte populations based on their localization. For instance, in the kidney there are two types of pericytes: the tubulointerstitial capillaries are covered by what are thought to be classical pericytes, whereas the glomerulus has specialized pericytes termed mesangial cells [28, 29]. Mesangial cells make up 30 % of all glomerular cells and play an important role in capillary tuft morphogenesis and the regulation of glomerular filtration (discussed in the 'Regulation of Tissue-Specific Properties' section). These cells express many classical pericyte markers, such as PDGFRβ and CD90, and are thus thought to be locally specialized pericytes [29]. In the cochlea, pericytes on vessels in different regions have different morphologies and express different molecular markers, depending on their localization on the capillary bed. Pericytes on vessels in the spiral ligament express α-SMA, desmin, and tropomyosin, whereas pericytes on vessels of the stria vascularis express only desmin [30].

In addition, different subpopulations of pericytes appear to be interspersed along the vascular tree. In the CNS, EM has been used to identify four subclasses of pericytes based on morphology of cellular processes and location of the nucleus. These different morphologic categories include (i) pericytes with broad cellular processes that are continuous with the surface of the endothelial cells; (ii) pericytes with thin, finger-like cellular projections confined to specific regions of the endothelial tube; (iii) pericytes that extend processes longitudinally along the axis of the capillary; and (iv) pericytes in which the cell body and processes are retracted from the endothelial tube [6]. These features may define specific subsets of pericytes, or may indicate unique pericyte characteristics at specific moments, including migration along the vessel during angiogenesis or migration away from the vessel in response to injury and disease. Furthermore, several studies have identified subtypes of pericytes in the lung, skin, kidney, and CNS that differ in their capacity to proliferate and form a scar tissue after tissue injury (see 'Regulation of Injury/Disease' section).

Furthermore, the location of pericytes along the vascular tube varies. For instance, in the choriocapillaris of the eye, pericytes are positioned distal to localized oxygen transport [31] and, in bovine, lung pericytes reside near endothelial cell junctions

[32]. In muscle, pericytes have been observed to cover histamine-induced gaps in the vasculature [33]. These data suggest that the location of pericytes within the vasculature may be important for specific vascular function within a tissue or organ.

3.2.3 Cell Adhesion to Endothelial Cells

Although pericytes line the outer surface of the endothelial cell tube, most of the pericyte cell body is not in contact with endothelial cells but is separated by the vascular basement membrane. However, there are discrete points where pericyte cellular processes contact endothelial cells to form cellular adhesions; it is estimated that a single pericyte can form up to 1000 adhesion contacts with endothelial cells [1].

The main type of cellular junction formed between pericytes and endothelial cells are described as 'peg-and-socket' junctions. In these junctions, cellular protrusions from the pericyte (pegs) adhere to invaginations (sockets) in the endothelial cells. These junctions are thought to be mediated by N-cadherin interaction based on the observation that N-cadherin expression has been visualized at points of pericyte–endothelial cell interactions, and the addition of N-cadherin blocking antibodies leads to defective pericyte adhesion to the vascular tube [34]. Further evidence suggests that transforming growth factor-β (TGFβ) signaling is required for adhesion since disruption of *Smad4* leads to loss of pericyte–endothelial cell interactions due to diminished N-cadherin expression [35] (discussed in the 'Vascular Recruitment' section).

In addition, several other types of cellular adhesions have been visualized between pericytes and endothelial cells. Adhesion plaques have been identified and are characterized as membrane contacts in which microfilament bundles are adjacent to the pericyte membrane, and an electron dense endothelial cytoplasm is adjacent to the endothelial membrane [36]. In brain microvessels, fibronectin has been observed at the adhesion plaques, suggesting that this could be a critical mechanical linkage, perhaps involved in vessel contractility. Several studies have suggested that gap junctions allow for electrical coupling of endothelial cells and pericytes. Functional gap junctions have been observed between endothelial cells and pericytes *in vitro* using a co-culture paradigm [37], and gap junctions have been visualized between endothelial cells and pericytes *in vivo* [38]; however, it has yet to be determined whether there is electrical coupling of endothelial cells and pericytes by these gap junctions in vivo. Although not fully characterized, tight junctions have been observed between endothelial cells and pericytes in vivo [12, 32].

3.2.4 Interaction with the Extracellular Matrix

Pericytes are embedded in the vascular basement membrane and thus share an ECM with CNS endothelial cells. Several studies using cell purification and microarray analysis have identified that pericytes express many different components of the EMC, including collagen subunits, laminin subunits, vitronectin, and asporin. Time course studies of vascular development suggest that pericyte recruitment to the endothelial tube correlates with the onset of vascular basement membrane deposition. Endothelial cell–pericyte co-cultures suggest that pericytes can also induce the expression of ECM components, such as nidogen and collagen, by endothelial cells [39]. Furthermore, disruption of platelet-derived growth factor-B (PDGF-B)/ PDGFRβ signaling, which inhibits the recruitment of pericytes to endothelial cells, causes marked disruption in the vascular basement membrane and leads to protrusions of endothelial cells into the CNS parenchyma [15]. Taken together, these data suggest that recruitment of pericytes to endothelial cells is critical for the deposition of EMC, which is produced by both cell types.

Additionally, it appears that the vascular basement membrane is critical to the recruitment and adherence of pericytes to the endothelial cells. RNA interference (RNAi)-mediated inhibition of tissue inhibitor of metalloproteinase 3 (TIMP-3) leads to increased matrix metalloproteinase disruption of the ECM and causes an increased vessel diameter and loss of endothelial cell–pericyte interactions [39]. One of the critical functions of the ECM is to localize the PDGF-B signal that is required for the recruitment of pericytes to the endothelial cells. Mutation of the ECM retention motif of PDGF-B leads to a poor pericyte recruitment to the endothelial tube [40] (discussed in the 'Vascular Recruitment' section).

While the ECM is critical to mediate endothelial-pericyte interactions during angiogenesis and adult tissue homeostasis, alterations in the basement membrane can lead to loss of pericyte–endothelial cell interactions during injury and disease. For instance, following a stroke there is a degradation of the vascular basement membrane mediated by matrix metalloproteinases (MMP)-2 and MMP-9, which coincides with the migration of the pericytes away from the endothelium [41, 42]. This can also be observed during tumor angiogenesis where the formation of new blood vessels involves the degradation of the vascular basement membrane and often results in endothelial cells poorly covered in pericytes.

A recent study suggests that pericyte–ECM interaction is an important regulator of the differentiation state of pericytes. In the CNS, loss of astrocyte-derived laminin was shown to drive expression of contractile proteins in pericytes, and pericyte interaction with laminin-111 via α2 integrins was required to maintain the noncontractile phenotype [43].

3.3 Cell Development

3.3.1 Cell Lineage

The few studies that have investigated the developmental origin of the mural cells covering microvasculature (i.e. pericytes) suggest a common origin for pericytes and vSMCs (reviewed by Armulik et al. [1]). However, mural cells have different developmental origins depending on which vascular bed they are located. The first blood vessels to develop in the embryo proper are the paired dorsal aortae that arise from the trunk. Initially, mural cells in the floor of the dorsal aorta are derived from splanchnic mesoderm, and mural cells on the roof are derived from sclerotome, similar to endothelial cells (reviewed by Sato [44]). Later during development, splanchnic-derived mural cells are replaced by sclerotome-derived cells [45]. Mural cells of the ascending and arch portions of the aorta, ductus arteriosus, the brachiocephalic, right subclavian arteries, and common carotid arteries arise from neural crest (reviewed by Majesky [46]). Similarly, mural cells in the heart exhibit several distinct developmental origins. Specifically, vSMCs of the proximal and major anterior coronary arteries arise from neural crest, whereas the rest derive from the proepicardial organ and epicardium. Coronary vein vSMCs have an atrial cardiomyocyte origin (reviewed by Riley and Smart [47]). Neural crest gives rise to vascular mural cells in the head region, CNS, and thymus (reviewed by Armulik et al. [1]). Mural cells in the gut, liver, and lung are derived from mesothelium (reviewed by Armulik et al. [1]) in contrast to those in the limbs and peritoneum, which are derived from the sclerotome [48]. It is unknown whether the different developmental origins of the mural cells in different vascular beds give rise to tissue-specific functions of these cells.

3.3.2 Differentiation

The transcriptional regulation of vSMC differentiation is relatively well understood, with vSMC-selective gene expression achieved by a unique combination of multiple ubiquitously expressed or selective factors (reviewed by Alexander and Owens [49]). Several signaling pathways (Notch, TGFβ [discussed in section 3.3.3], Wnt) have been shown to be important for initial vSMC differentiation surrounding the aorta [50–52].

The transcriptional mechanisms that control pericyte differentiation remain elusive. However, a recent report by Siegenthaler et al. demonstrates that genetic ablation of Foxc1 in pericytes leads to pericyte and endothelial cell hyperplasia, thus underscoring its role as an important regulator of pericyte maturation [53]. During development, several pericyte markers (e.g. RGS5, endosialin, NG2, desmin) show dynamic expression. The expression of RGS5, endosialin, and NG2 in pericytes is selectively decreased as the vasculature matures, whereas desmin expression increases [54–57]. According to a recent study, Dll4 and PDGF-B signaling induces the differentiation of pericytes from myoblasts [58]. It is not known to what extent PDGF-B/PDGFRβ and Notch signaling regulate pericyte differentiation during angiogenesis; however, Notch signaling has been shown to regulate PDGFRβ expression on vSMCs [59]. Additionally, Notch signaling in mesangial cells (specialized pericytes in kidney glomeruli) was shown to be necessary for specification of mesangial cell precursors that express a high level of PDGFRβ [60]. Of note, increased PDGFRβ signaling in brain pericytes results in higher pericyte proliferation and a less mature phenotype [61]. This suggests that the PDGFRβ activity that is required for pericyte recruitment may be needed to be switched off for pericyte maturation. It is interesting that brain pericytes have been suggested to differentiate within the mesenchyme that surrounds the telencephalon before entering the CNS [62].

3.3.3 Vascular Recruitment

Several well-characterized signaling pathways are critical for pericyte recruitment during angiogenesis, including the PDGF-B/PDGFRβ and TGFβ signaling axes (Fig. 3.2). In addition, other pathways that regulate pericyte recruitment are recognized, although largely in the context of pathological conditions (Fig. 3.2). It should be underlined that, based on current knowledge, no universal signaling pathway seems to regulate pericyte recruitment to all organ vascular beds during development, injury, or pathological angiogenesis. Most likely, pericyte recruitment to the developing vasculature in a given organ is guided by a combination of signaling pathways that is specific to each organ.

Signaling Axis: PDGF-B/PDGFRβ

Many studies have confirmed the importance of the PDGF-B/PDGFRβ signaling axis for pericyte recruitment during development and pathological angiogenesis. The importance of PDGFB in vascular development and pericyte recruitment was somewhat unexpected. In fact, *PDGFB*, the first oncogene to be cloned and sequenced, was recognized as a growth factor for mesenchymal cells. However, the critical role of PDGF-B in pericyte recruitment to the developing microvasculature was demonstrated by the genetic ablation of *Pdgfb*. Mice lacking *Pdgfb,* or its receptor *Pdgfrb,* showed almost identical phenotypes, with late embryonic lethality caused by dysfunctional vasculature [17, 63]. During angiogenesis, outgrowing endothelial cells express high levels of PDGF-B. Subsequently, secreted PDGF-BB is mobilized to the ECM adjacent to the endothelium. Pericytes, which express PDGFRβ, respond to the PDGF-BB gradient by proliferation and migration along the developing vasculature (reviewed by Armulik et al. [1]). The dependency of pericyte recruitment on PDGF-B/PDGFRβ signaling is organ-specific. *Pdgfb* or *Pdgfrb* knockout embryos lack pericytes in organs such as the brain, kidney, skin, and muscle, but not in the liver and thymus.

The binding of PDGF-BB to the ECM is critical for the bioavailability of this signaling molecule. The importance of localized presentation of PDGF-BB in pericyte recruitment was demonstrated in genetically-modified mice that express PDGF-B protein lacking the so called 'retention motif', thus preventing mobilization to the ECM [40]. Although these mice are viable, they possess a reduced number of pericytes along the capillary bed, and thus represent a tool to investigate the role of pericytes in the adult organism. Analysis of adult pericyte-deficient mice showed that pericytes are important regulators of CNS vascular permeability (discussed in section 3.4.4).

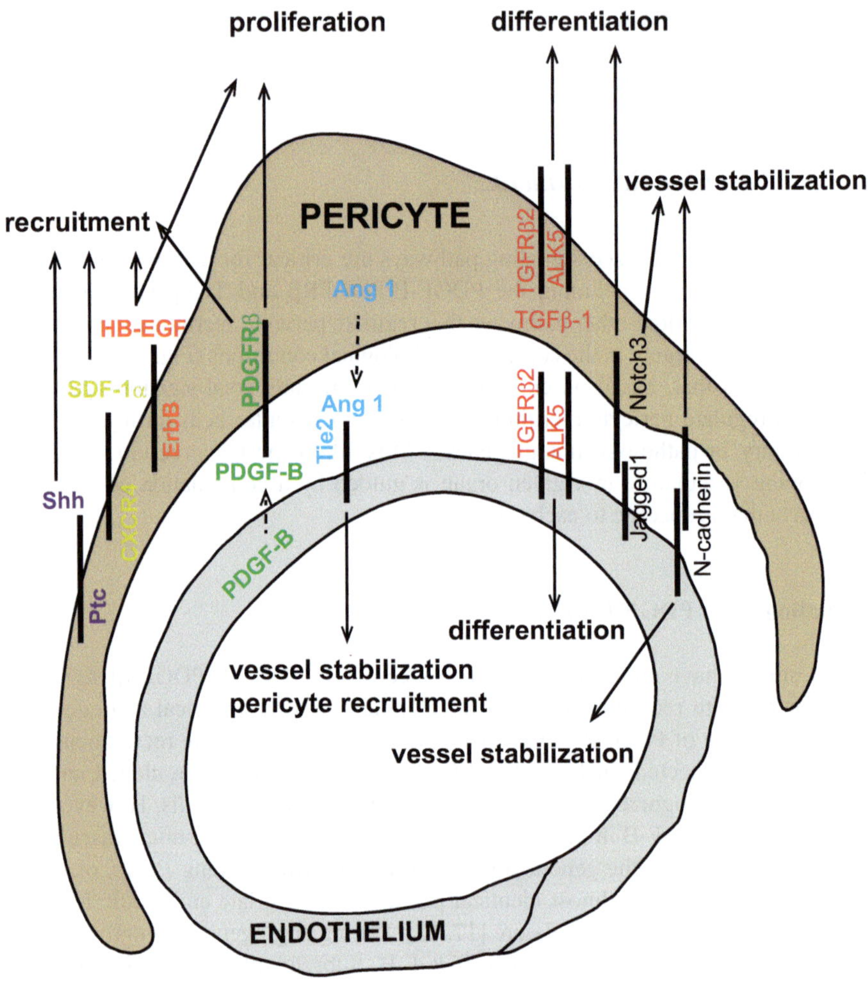

Fig. 3.2 Signaling pathways important for pericyte recruitment, differentiation, and vessel stabilization. Multiple ligand receptor complexes mediate pericyte recruitment to the endothelium during angiogenesis. Some of these pathways (SDF1α/CXC4R, HB-EGF/ErbB) appear relevant only in pathological angiogenesis. A ligand–receptor pair is indicated by the same color. Modified from Armulik et al. [1]

Furthermore, pericyte-deficient animals develop brain calcifications at anatomical locations similar to patients suffering from idiopathic basal ganglia calcifications (IBGC) [64]. IBGC is a neurodegenerative disease with an autosomal dominant inheritance pattern. It was recently demonstrated that approximately 25 % of IBGC patients have deleterious mutations in *PDGFB* or *PDGRRB* genes [64, 65]. Interestingly, analysis of mice with varying degrees of pericyte deficiency has demonstrated that the extent of brain calcifications correlates inversely with pericyte coverage and BBB defect [64].

Transforming Growth Factor-β

Mouse knockout studies of various genes encoding for components of the TGFβ signaling pathway have demonstrated the importance of the TGFβ signaling pathway in vascular development. TGFβ, together with activins and bone morphogenetic proteins (BMP), form a large superfamily of pleiotropic growth factors that play a role in many different developmental processes. The canonical TGFβ signaling pathway emanates from a ligand binding to a heteromeric receptor complex composed of type 1 (activin receptor-like kinases [ALKs]) and type 2 (e.g. TGFβRII, BMPRII) serine/threonine kinase receptors. Type 1 receptors phosphorylate the transcription factors SMAD2/3, which, upon binding to SMAD4, translocate into nucleus and activate TGFβ target genes. The specific outcome of the cellular response (proliferation, differentiation, migration, etc.) to TGF signaling depends on the repertoire of the type1 receptors expressed by a given cell and the presence of TGFβ co-receptors (e.g. endoglin) (reviewed by ten Dijke and Arthur [66]).

In humans, haploinsufficiency of ligands, receptors, or intracellular effectors causes thoracic aortic aneurysms and dissections (TAADs) [*TGFB2, FBN1, TGFBR1, TGFBR2, SMAD3*], and hereditary hemorrhagic telangiectasia (HHT) [*ENG, ACVRL1, SMAD4*]. TAAD patients have aortic dilatations that may develop into aneurysms or aortic dissections, potentially causing sudden aorta rupture and massive bleeding. HHT results in abnormal patterning of the vascular tree, characterized by the formation of arteriovenous anastomoses that predispose HHT patients to frequent hemorrhages. Since TGFβ receptors are expressed both in endothelial cells and mural cells, it has been difficult to assess the precise molecular mechanisms by which TGFβ regulate pericyte recruitment. ALK5 and TGFβRII are expressed by endothelium and pericytes, whereas ALK1 and endoglin are expressed by endothelial cells only. ALK1 induces phosphorylation of SMAD1/5, which promotes angiogenesis, whereas ALK5 signaling in endothelial cells inhibits proliferation acting via the SMAD2/3 complex (reviewed by ten Dijke and Arthur [66]). In addition, TGFβ signaling induces mural cell differentiation, and activation of TGFβ from its latent form requires endothelial–mural cell contact [67]. Thus, endothelial cells and mural are interdependent, with changes in one cell type leading to changes in the other.

The severity of the vascular phenotype of the modified TGFβ signaling pathway is dependent on the developmental stage. TGFβ signaling is not only required for formation of the vascular bed during development but also for vessel maintenance in the adult organism. Most studies that address the role of the TGFβ pathway in mouse vasculature have focused on endothelial-specific constitutive or conditional deletion of TGFβ pathway components (reviewed by Jakobsson and van Meeteren [68]). Generally, genetic ablation of either type 1 (ALKs) or type 2 receptors (TGFβRII, BMPRII) results in hemorrhage, a phenotype often accompanied by vascular smooth muscle defects (reviewed by Jakobsson and van Meeteren [68]). It is often challenging to precisely delineate the role of a single pathway since multiple TGFβ pathways are interdependent and simultaneously active in the cell. For example, ALK1 requires ALK5 kinase expression for its activity [69], and thus genetic ablation of ALK5 affects both pathways. The deletion of genes of the TGFβ/BMP pathways results in a relatively early vascular phenotype and *in utero* death (around E10) due to the presence of only big trunk vessels such as the dorsal aorta [66]. For example, the conditional of ablation of *Bmbr1a*, a type 1 receptor that mediates BMP signaling (ALK3), results in embryonic death at E11.5 due to extensive hemorrhage in the trunk [70]. Detailed analysis of the aorta demonstrated reduced mural cell coverage and the poor association between vSMCs and endothelium [70]. Whether this mutation also leads to defective pericyte coverage has not been reported; however, a recent report on ALK5/SMAD2/3 signaling in the endothelium reported poor coverage of mural cells on all caliber vessels, from the aorta to capillaries, which could be due to reduced PDGF-B expression in the endothelium [71]. Increased expression of PDGF-B by thalidomide stimulates mural cell recruitment in endoglin heterozygote animals, an experimental model for HHT [72]. In brain vasculature, the targeted disruption of *Smad4*, the central intracellular mediator of TGFβ/BMP signaling, also results in poor pericyte coverage along microvessels [35]. In support of this observation, it was demonstrated that TGFβ/BMP and Notch signaling pathways cooperate in the upregulatation of N-cadherin. As discussed above, N-cadherin mediated endothelial/pericyte interaction has been suggested to be important for vessel stabilization [34].

Based on the above studies, it is clear that aberrant TGFβ/BMP signaling in endothelial cells results in poor mural cell coverage. To date, data are not available to explain how pericyte-specific TGFβ signaling affects vessel development or maintenance. However, few studies on the cell-specific elimination of TGFβRII or ALK3 in vSMCs have proven the crucial role of TGFβ/BMP signaling in mural cell proliferation and differentiation along big trunk vessels [73, 74]. Thus, emerging data suggest that TGF signaling in endothelial cells promotes mural cell recruitment, at least partly by inducing PDGF-B expression and also by modulating expression of cell adhesion molecules required for endothelial/pericyte interaction.

Other Pathways

Notch signaling is critical for regulating angiogenesis in both endothelial and mural cells. As noted, Notch signaling is required for mural cell maturation (reviewed by Morrow et al. [75]). In humans, *NOTCH3* mutations cause stroke and a syndrome known as cerebral autosomal dominant arteriopathy with subcortical infarcts and leukoencephalopathy (CADASIL), which is associated with dementia. The pathogenic mechanism is not known but may be related to the degeneration of vSMCs caused by accumulated, non-functional Notch3 protein that leads to vessel occlusion and ischemic stroke. *Notch3* knockout mice show impaired maturation and reduced coverage of arterial vSMCs [76], but an effect on the pericyte population has not been reported. However, the recent analysis of microvasculature of CADASIL patients has demonstrated the involvement of proper NOTCH3 signaling in pericyte survival [77]. A critical Notch3 ligand on endothelial cells seems to be Jagged-1, which is required for Notch3 expression on vSMCs [78]. It is of note that the endothelial expression of Jagged-1 has been shown to be critical for vSMC differentiation *in vivo* [79].

The angiopoietin-1 (Ang1)/Tie-2 (Tek) signaling axis has long been suggested to be important for pericyte recruitment. This view is based on the analysis of *Angpt1* and *Tek* knockout embryos that lack mural cell coverage (reviewed by Gaengel et al. [80]). Ang1 is expressed by pericytes and Tie2 is expressed by the endothelium (reviewed Gaengel et al. [80]), yet it remains unclear how the loss of Ang1 or Tie2 results in a defect in pericyte coverage. A recent study has demonstrated that the vascular phenotype observed in Ang1 knockouts is due to abnormal cardiac development [81]. The genetic ablation Ang1 after E13.5 did not alter pericyte recruitment. Consistent with the proposed role of Ang1/Tie2 signaling in regulating vascular stability, absent *Angpt1* leads to pathological angiogenesis and tissue fibrosis in the setting of tissue injury or stress [81].

Under *in vivo* conditions, the heparin-binding epidermal growth factor (HB-EGF)/ErbB signaling pathway in vasculature is important for cardiac development [82]. In addition, *in vitro* studies have demonstrated the importance of endothelial HB-EGF in mural cell recruitment via ErbB1 and ErbB2 in mural cells [83]. Furthermore, the HB-EGF/Erb pathway has been shown, together with the PDGF-B/PDGFRβ pathway, to regulate pericyte recruitment [84]. Similar to Ang1, HB-EGF seems to play a role in maintaining vascular stability by promoting pericyte survival and proliferation during tissue injury [85].

In tumor vasculature, stromal-derived growth factor-1α (SDF1α) has been shown to promote PDGF-B-dependent pericyte migration via CXCR4 [86, 87]. Interestingly, glioblastoma stem cell recruitment to the endothelium has been shown to be dependent in SDF1α/CXCR4, and their subsequent differentiation into pericytes has been induced by TGFβ [88]. Also, semaphorin 4d-plexin-B1 signaling in endothelial cells was shown to promote pericyte recruitment to tumor vessels in a PDGF-B-dependent manner [89].

Sonic hedgehog (Shh) signaling regulates pericyte recruitment to the choroid plexus [90]. Shh signaling has been also shown to regulate mural cell recruitment during neoangiogenesis [91]. Ephrin-B2 expression in mural cells is needed for proper association between endothelial cells and pericytes [92]. Recently, it was demonstrated that Ephrin-B2 regulates PDGFRβ internalization and thus is important for controlling PDGFRβ signaling activity [93].

3.4 Regulation of Vascular Function

Recent advances using cell purification and culture, genomic analysis and genetic mouse models have identified many different roles for pericytes in development, tissue homeostasis, and disease. These functions include interactions with endothelial cells to regulate angiogenesis, vascular tone and permeability, and interactions with tissue-specific cells to regulate organ development and homeostasis. In addition, pericytes respond to injury and disease by regulating leukocyte trafficking, tissue regeneration, and fibrosis.

3.4.1 Regulation of Angiogenesis

Due to the close proximity of pericytes to endothelial cells, both during development and adulthood, it has long been thought that pericytes play a key role in regulating angiogenesis and vascular stability. Indeed, work with mice with mutations in genes required for pericyte recruitment has demonstrated that these cells play a role in regulating developmental angiogenesis, vascular stabilization, and homeostasis, as well as neoangiogenesis in the adult [80].

During development, many studies have identified the localization of pericytes along the nascent vascular tube and their interactions with endothelial tip cells and stalk cells. Endothelial sprouts have been observed to enter tissues in the absence of pericytes; however, pericytes invade and are recruited to the vascular tube shortly after initial invasion. In fact, pericytes can often be found at the leading edge of the endothelial sprouts interacting with tip cells [18, 94, 95]. It is thought that the endothelial tube plays a role as a migration signal for the pericytes, likely through the secretion of PDGF-BB. Pericytes are recruited either from adjacent vessels or from surrounding mesenchymal precursors.

The timing of pericyte recruitment to the nascent vasculature coincides with vessel stabilization. Indeed, work with mouse mutants that affect pericyte recruitment, such as PDGF-B/PDGFRβ or Ang1/Tie2, have demonstrated that pericytes are critical for vessel stabilization. In these mutants, endothelial cell sprouts are able to enter the developing tissues; however, the vascular tubes often have tortured patterning, increased vessel diameter, and the increased likelihood of hemorrhage [17, 24, 96, 97]. These data suggest that pericytes are not required for the initial invasion of endothelial cells into a tissue, or even the formation of the vascular tube, but they are important for the appropriate patterning of the vasculature as well as stabilizing the vascular tube. Corroborating these data are *in vitro* co-culture models that have demonstrated that pericytes can supress endothelial cell growth and migration [98, 99]. Furthermore, pericyte recruitment to the nascent vessels has been implicated in resistance to regression of the newly formed sprouts [100].

Pericytes have also been implicated in neoangiogenesis in adults in response to alterations in oxygen levels, injury, or disease. In the initial phase of angiogenic processes, pericyte cell bodies bulge and increase in volume, their processes shorten, and they proliferate [101]. In addition, they increase their expression of MMPs, such as MMP-9, which corresponds to the disruption of the basement membrane required for the migration of new angiogenic sprouts [102]. This process may be guided by the same signals that regulate developmental angiogenesis. Vascular endothelial growth factor (VEGF) production has been shown to induce proliferation and migration of pericytes during hypoxia via an indirect mechanism through the production of nitric oxide (NO) by the endothelium [103, 104]. In addition, VEGF can be produced by the pericytes to direct this process themselves [105].

Pericytes have also been implicated as important regulators of tumor angiogenesis and thus are potential targets in antitumor therapy [106]. Studies have identified pericytes on blood vessels along the leading edge of tumor growth with incomplete pericyte coverage implicated in vascular abnormalities, and increased vascular permeability has been observed in different tumors [107–109]. Anti-VEGF treatment, which targets the endothelium, has been shown to remove tumor vessels while sparing pericytes [110]. Alternatively, it also leads to vessel normalization with increased pericyte coverage [111, 112]. Although targeting of pericytes through inhibition of PDGFRβ showed no loss of tumor vasculature [113, 114], inhibition of PDGFRβ combined with anti-endothelial therapy has provided synergistic success in reducing tumor vasculature and size in multiple tumor models [115]. Therefore, targeting of pericytes to decrease tumor vasculature is an active area of research.

3.4.2 Regulation of Blood Flow

Almost 100 years ago, Krogh and co-workers first reported remote vasodilation of capillaries in response to neural stimuli (reviewed by Bagher and Segal [116]). Focal capillary stimulation has also been shown to affect local blood flow. Specifically, fluctuations in capillary blood flow have been explained by conduction of vasomotor responses by electrical signals via endothelium that leads to relaxation of vSMCs at distinct locations within the resistance network (reviewed by Bagher and Segal [116]). Pericytes, similarly to vSMCs, express contractile proteins and have the capacity to contract in response to vasodilative and vasoconstrictive agents *in vitro* (reviewed by Rucker et al. [117, 118]). Although the regulation of blood flow by pericytes seems plausible, (in fact, it was the first function suggested for pericytes), it has been difficult to demonstrate *in vivo*. Changes in the diameter of arteries and arterioles (i.e. vessels that have vSMC coverage) are thought to regulate blood flow in response to different metabolic needs. For example, in the brain, the regulation of functional hyperemia (i.e. blood flow increases in response to increased local neuronal activity) is initiated by neurotransmitters released by neurons, and occurs at the level of the arterioles (reviewed by Attwell et al. [119]). Based on the demonstration of pericyte-mediated constriction of vessels in *ex vivo* cerebellar slices and retina preparations [120, 121], it seems plausible that in vivo pericyte-mediation is possible at the level of capillaries in response to neuronal-stimuli. Although pericytes were shown to constrict in vivo, a role for pericytes in neurovascular coupling was difficult to identify [122]. However, a recent study reported that vasodilation and blood flow changes in the somatosensory cortex were initiated in capillaries after whisker pad stimulation [121]. Vasodilation of capillaries occurred, on average, 1.4 s before vasodilation was seen in the arterioles, thus implicating that vasodilation in capillaries is not a passive response to pressure changes occurring at the level of arterioles [121]. In addition, in pathological conditions, pericyte-mediated contraction could potentially disturb the flow. Pericyte-mediated contraction in response to ischemia was shown to constrict capillaries *in vivo*, suggesting that pericytes contribute to the 'no-reflow' phenomenon after cerebral ischemia [123]. Neural stem cell progenitor proliferation in the neurogenic niche in the subventricular zone is accompanied by an increase in local capillary bed blood flow. It was recently shown that specialized subventricular zone astrocytes (B cells) and neural progenitors regulate flow in the capillary bed by purinergic receptor activation on pericytes [124].

The kidney is another organ where pericytes are purported to regulate capillary blood flow. The renal medulla is vascularized by vasa recta capillaries, and blood flow in these capillaries requires tight control to facilitate the concentration of urine and avoid severe ischemia in the medulla. Similar to brain microvasculature, regulation of renal medullary blood flow may occur at the level of the capillary bed (vasa recta), independent of arterial perfusion pressure (reviewed by Kennedy-Lydon et al. [118]).

3.4.3 Regulation of Leukocyte Trafficking

For leukocytes to enter tissue, a tightly orchestrated series of events takes place on blood vessel walls. The leukocyte adhesion cascade involves rolling, activation, adhesion strengthening, intraluminal crawling, firm adhesion, and paracellular and transcellular migration. Leukocyte entry is crucial for immune surveillance and the control of tissue inflammation. Although there is extensive knowledge on the cellular and molecular mechanisms by which leukocytes cross the endothelium, relatively little attention has been paid to the role of pericytes in these processes.

Transmigration of neutrophils into the surrounding tissue takes place through venules at permissive sites that lack mural cell coverage and have different basement membrane components [125–128]. Real-time imaging to monitor neutrophil transmigration in living inflamed tissue has demonstrated that, after transendothelial migration, neutrophils crawl along pericytes, which are mediated via intercellular adhesion molecule-1 (ICAM1)–Macrophage-1 antigen (Mac1) interaction [127]. This interaction seems to be functionally important, since blocking this interaction reduces neutrophil extravasation into tissue or transmigration *in vitro* [127, 129, 130]. Mural cells also react to inflammatory stimuli (e.g. tumor necrosis factor) by the expansion of pericyte gaps, thus enabling neutrophil to exit these structures [127, 128]. Interestingly, the expansion of gaps between pericytes was induced by neutrophil–pericyte interaction that supressed actomyosin-based contractility in pericytes [128].

A recent study by Stark et al. specifically investigated the role of capillary pericytes in leukocyte transmigration in inflamed skin [130]. After neutrophils exit venules, they are attracted by capillary pericytes, which secrete chemoattractants in response to inflammatory stimuli [130]. More importantly, pericytes have been shown to control the activation status and survival of extravasated leukocytes [130]. Thymus pericytes are reported to produce sphingosine-1-phosphate (S1P) and promote T-cell exit from the thymus via S1P receptor (S1PR)-1 expressed in immature thymocytes [131].

In several tissues (i.e. CNS, testis), an active mechanism facilitates immune tolerance and ignorance, with these tissues considered to be immune-privileged. In addition, restrictive endothelial barriers physically block leukocyte entry into these tissues. Analysis of CNS endothelial cells from pericyte-deficient embryos have demonstrated that, in the absence of pericytes, endothelial cells express elevated levels of leukocyte adhesion molecules (e.g. ICAM1), accompanied by an increased number of subsets of Gr1-positive leukocytes within the CNS [15]. Interestingly, PDGFRβ activation in pericytes modulates expression of immune response genes in pericytes [61].

Therefore, different studies suggest opposing roles for pericytes in regulating leukocyte migration. For instance, in peripheral tissues such as muscle, pericytes are emerging as positive regulators of leukocyte transmigration, whereas in the CNS, pericytes are thought to inhibit leukocyte migration as pericyte-deficiency in the CNS is accompanied with neuroinflammation. These conflicting results may reside in the tissue-specific roles of pericytes, or may indicate a more complex regulation of leukocyte trafficking by pericytes that is context-dependent.

3.4.4 Regulation of Vascular Permeability

The permeability of the vascular bed to plasma molecules and proteins appears to be organ-specific, and inversely correlates with pericyte coverage of the capillaries. For instance, CNS vasculature, which has the highest pericyte coverage, is characterized by extreme tightness, achieved by specific endothelial characteristics, and collectively referred to as the BBB. The principal components of the BBB include closed endothelial cell–cell junctions, expression of solute carrier (SLC) and ATP-binding cassette (ABC) transporters, and a low level of transcytosis. *In vivo* analysis of various mouse mutants defective for CNS pericyte coverage indicate that pericytes are critical for regulating CNS vascular permeability during development, adulthood, and in aging. Analysis of *Pdgfrb* −/− embryos, which almost completely lack pericyte coverage of CNS vasculature, has demonstrated that pericytes regulate the endothelial permeability during embryogenesis [15]. Similarly, BBB permeability analysis of viable pericyte-deficient animals has shown that pericytes also regulate BBB permeability in the adult organism [132]. Whereas, early *in vitro* studies suggested that pericytes regulate the BBB at the level of endothelial cell–cell junctions [133], these *in vivo* studies have demonstrated that, while endothelial junctions are broader and more convoluted in the absence of pericytes, junctions are continuous and do not show accumulation of intravenously injected tracers [15, 132]. Intravenously administered tracers are detected in large vesicles in the endothelial cells of pericyte-deficient mice, as well as at the endothelial basement membrane, indicative of a transcellular route of passage [15, 132]. These data suggest that pericytes limit CNS vascular permeability by inhibiting the rate of transcytosis; however, the underlying mechanism by which this is achieved is not fully understood.

Although pericytes regulate the BBB in embryos, there seems to be subtle differences in endothelial cell differentiation in adults, which could contribute to increased permeability. Pericyte-deficient embryonic brain vessels show ectopic expression of PLVAP, a protein associated with endothelial cell vesicles and fenestrae [15]. Interestingly, expression of PLVAP in cerebral vasculature is not detected in adult pericyte-deficient mice [132]. Thus, adult vasculature may possess compensatory mechanisms to suppress PLVAP expression, which could be mediated by another cell type at the neurovascular unit (e.g. astrocyte). Accordingly, pericyte-deficiency leads to increased vesicular transport in the endothelium and pericytes regulate the BBB at the level of endothelial transport.

Less well explored is the pericyte-mediated regulation of vascular permeability in peripheral tissues. Analysis of pericyte-deficient adult mice (*Pdgfb*$^{ret/ret}$) has shown that, in addition to increased BBB permeability, these mice also show increased vascular permeability in the liver [134]. PDGF-B/PDGFRβ signaling is not required for pericyte (hepatic stellate cells) recruitment to liver vasculature [17], which suggests that the mechanism of increased liver permeability must be different from the brain. However, although the PDGF-B/PDGFRβ signaling is not important for pericyte recruitment, it may be necessary for organ-specific maturation of pericytes in the liver.

3.4.5 Regulation of Injury/Disease

Recently, a dual role has emerged for pericytes in regulating tissue response to injury and disease—on the one hand, as a potential source for tissue-specific stem cells, and on the other hand, as the scapegoat for fibrotic conditions that give rise to myofibroblasts, the scar-forming cells, in the periphery and in the CNS (reviewed by Armulik et al. [1] and Greenhalgh et al. [135]). While these observations have generated excitement due to their enormous translational potential, the lack of specific cellular markers and genetic tools, such as Cre drivers for identifying and fate-mapping pericytes, represents a current barrier to our understanding of these cells. Therefore, it is not surprising that conflicting results point to divergent origin of myofibroblasts in various fibrotic conditions.

Regardless of the etiology of the chronic renal disease, a common pathological manifestation is the development of fibrosis. In recent years many studies have revealed that pericytes represent the major source of myofibroblasts in kidney fibrosis (reviewed by Ren and Duffield [136]). However, a recent study using lineage tracing to identify myofibroblasts forming cells in renal fibrosis has identified a multicellular origin of these cells (e.g. tissue-resident fibroblasts, bone marrow-derived cells) and indicated that pericytes do not contribute significantly to this process [137].

It has also been reported that multiple cell types contribute to the formation of myofibroblasts in the lung. Of note, fate-tracing experiments using NG2ER-Cre have demonstrated that pericytes in the lung proliferate in response to injury but do not form myofibroblasts (i.e. SMA-positive cells) [138]. However, a recent study by Hung et al. showed that not all pericytes in the lung express NG2 [139]. Furthermore, careful fate mapping of lung mesenchymal cells has shown the presence of multiple pericyte populations that share the expression of certain pericyte markers (e.g. PDGFRβ), but with a distinct transcriptional profile [139]. They also found that in lung, similar to kidney, multiple mesenchymal cell populations (e.g. pericytes and perivascular fibroblasts) contribute to lung myofibroblasts [139]. Recently, spinal cord scar tissue was reported to originate from pericytes [140]. Göritz et al. used glutamate-aspartate transporter (GLAST)-Cre driver to lineage trace scar-forming cells after spinal cord injury. Although GLAST promoter is thought to be active in astrocytes, Göritz et al. noticed that scar-forming cells originated from GLAST and PDGFRα-positive perivascular cells (type A pericytes) embedded in the basement membrane of arterioles. However, another recent study identified perivascular fibroblasts as the culprit in spinal cord scar tissue [141]. In this study, genetic lineage tracing revealed that collagen 1α-producing cells (i.e. myofibroblasts) originate from perivascular fibroblasts adjacent to arterioles and not from capillary bed pericytes [141]. This raises the question whether type A pericytes in the CNS could actually represent PDGFRα-positive perivascular fibroblasts. In addition, CNS pericytes have also been identified as cells that contain multipotential stem cell activity. Purified pericytes have been shown to be capable of self-renewal and generate neuronal and glial cells in response to basic fibroblast growth factor. It is not clear whether these cells actually produce neural cells *in vivo* during development or in response to injury and disease [142, 143].

Dulauroy et al. have shed more light on the molecular phenotype of profibrotic cells [144]. They demonstrated that elimination of membrane-anchored metalloprotease ADAM12+ cells in skin and muscle decreases tissue fibrosis [144]. ADAM12+ cells were found to reside along the vascular wall and to have a distinct developmental origin (neural crest, mesenchymal), depending on the organ. Interestingly, perivascular ADAM12+ cells in normal tissue also express pericyte markers (PDGFRβ, NG2) and are embedded in the vascular basement membrane [144]. After injury, ADAM12+ cells are not embedded in the vascular basement membrane and it remains to be demonstrated whether myofibroblasts in skin and muscle are derived from ADAM12+ pericytes or from a population of ADAM12+ perivascular cells distinct from pericytes [144]. In muscle, pericytes have emerged as potential stem cells giving rise to new muscle cells after injury [145]. Interestingly, ADAM12+ profibrotic cells were not seen to contribute to muscle repair [144] and likely represent a pericyte population distinct from those activated upon tissue injury, and contribute to the formation of new myocytes.

Thus, various studies lend evidence to the heterogeneous origin of myofibroblasts in various organs. To what extent pericytes contribute to the formation of fibrotic tissue in different organs remains an open question. Investigations to answer this question are complicated by the lack of specific markers and Cre drivers for fate mapping. In addition, limited knowledge regarding the heterogeneity of pericytes within a specific organ makes data interpretation difficult and impedes progress on this front.

3.4.6 Regulation of Tissue-Specific Properties

In addition to regulating system-wide aspects of vascular function, pericytes in different organs have developed tissue-specific properties that allow them to regulate the development, function, and response to stress of these organs. In this section we will describe a few of these unique tissue-specific pericyte properties.

Hepatic Stellate Cells

Stellate cells are the pericytes of the liver and comprise 5–8% of all cells in this organ. These cells are vital for the storage and transport of Vitamin A within the liver. Hepatic stellate cells store 80% of the total body retinol as retinyl esters, mostly retinyl palmitate, in lipid droplets within the cell body, and release these retinols in response to systemic retinol levels [146]. These vitamin A droplets exhibit a unique autofluorescence that allows for imaging of these cells. Hepatic droplets exist in both smaller membrane-bound unites and larger membrane-free droplets, and contain triglyceride, cholesteryl ester, cholesterol, and phospholipids, in addition to retinols [26]. Lecithin:retinol acyltransferase, which catalyses the synthesis of retinoyl esters, is required for synthesis of these lipid droplets, and mice deficient for this enzyme lack these lip droplets. During injury and disease, activated hepatic stellate cells release their retinols, which are thought to be involved in liver regeneration and fibrosis [26].

Stellate cells are also critical for the liver's response to injury and disease. The liver can fully regenerate within a matter of weeks if up to two-thirds of the hepatic mass is removed. Stellate cells are thought to play important roles at multiple stages of liver regeneration, including positive regulators of early-phase cell proliferation and negative regulators of the later phase [147]. This is based on the observation that inhibiting activated stellate cells either with gliotoxin or L-cysteine prevents normal regeneration [148, 149]. They produce angiogenic factors that regulate both endothelial and hepatic cell proliferation, as well as producing modulators of the ECM that allow for regeneration. Further evidence suggests that they are capable of generating hepatocytes, furthering the notion that pericytes may function as multipotent progenitors in multiple organs [150].

In addition, stellate cells in the liver are thought to play a key role in regulating liver fibrosis. Following injury, hepatic stellate cells become highly proliferative, with enlargement of the rough endoplasmic reticulum, loss of cellular processes and lipid droplets, and gain of contractile filaments and α-SMA [151]. These activated stellate cells secrete ECM and TIMPs, which inhibit MMPs, and thus have limited ECM degradation, a hallmark of liver fibrosis [152]. Accordingly, activation of stellate cells appears to play a key role in the initiation and progression of liver fibrosis. This may not be a role specific to liver pericytes as pericytes in other organs, such as the kidney, have been suggested to contribute to fibrosis in these tissues.

Thymus Pericytes

While it is clear that pericytes regulate leukocyte trafficking from blood vessels into many tissues, recent data suggest that pericytes regulate the egress of T cells from the thymus into the blood stream, a process important for the adaptive immune response. T cells exit the thymus at the corticomedullary junction, and overexpression of the S1PR-1 receptor in thymocytes leads to the perivascular accumulation of thymocytes adjacent to pericytes and premature egress. Deletion of sphingosine kinases from neural crest-derived pericytes leads to the faulty egress of the thymocytes, demonstrating a key role for pericyte-secreted S1P in this process [131].

Renal Pericytes

In the kidney, there are two described pericyte subtypes: renal pericytes that are associated with tubulointersitial microvessels, and mesangial cells that are associated with the glomerulus. Renal pericytes are involved in different aspects of the kidney's response to injury and disease. For instance, migration of renal pericytes from the microvasculature may lead to capillary loss and subsequent hypoxia during chronic kidney disease [28]. Moreover, renal pericytes are thought to play an important role in the development of kidney fibrosis, since activation of renal pericytes leads to the deposition of ECM and the onset of fibrosis [153].

Mesangial cells in the kidney are important for the regulation of glomerulogenesis, as well as regulation of glomerular filtration. Mutation in PDGF-B signaling in mice results in lack of mesangial cells, and the glomerulus appears malformed [63, 154]. The contractile nature of mesangial cells allows them to regulate the filtration rate of the glomerulus [29].

3.5 Summary

Pericyte research is currently advancing rapidly in various areas of developmental biology, vascular homeostasis, and pathology. Once neglected by scientists, these 'dull' perivascular cells are now in the limelight of BBB research and tissue fibrosis. In addition, recent studies implicate pericytes in regulating blood flow, immune and inflammatory responses, and as a source of multipotent stem cells in adult tissues. There has been considerable progress in our understanding of the molecular mechanisms by which pericytes are recruited to the developing vasculature. However, the identification of pericytes *in vivo* remains a significant challenge, especially given the heterogeneity of pericytes within a given organ. Importantly, increasing evidence supports the concept of heterogeneity of pericytes where different subsets of pericytes are associated with the vasculature in every tissue. Thus, research that will define the molecular attributes of pericytes will facilitate the characterization of their role in regulating a large number of physiological processes and pathologies.

It will also help to determine whether all of the functions attributed to pericytes are carried out by all cells, or whether individual functions are carried out by subsets of pericytes (e.g. blood flow regulation). Although our current understanding regarding the biology of pericytes is only fragmentary, even today, pericytes are emerging as therapeutic targets in various pathological conditions, such as ischemia, fibrosis, and cancer. In addition, studies on the organ-specific functions of pericytes have the potential to guide the development of efficient drug delivery platforms for brain and to control tissue repair after injury.

References

1. Armulik A, Genove G, Betsholtz C. Pericytes: developmental, physiological, and pathological perspectives, problems, and promises. Dev Cell. 2011;21(2):193–215.
2. Eberth CJ. Von den Blutgefäßen. In Strickers Handbuch der Lehre von den Geweben des Menschen und der Tiere. Bd. 1. Leipzig; 1871.
3. Rouget C. Memoire sur le developpement, la structures et les proprietes des capillaires sanguins et lymphatiques. Archs Physiol Norm Pathol. 1873;5:603–33.
4. Zimmermann KW. Der feinere bau der blutcapillares. Z. Anat. Entwicklungsgesch. 1923;68:3–109.
5. Sims DE. The pericyte–a review. Tissue Cell. 1986;18(2):153–74.
6. Dore-Duffy P, Cleary K. Morphology and properties of pericytes. Methods Mol Biol. 2011;686:49–68.
7. Fujiwara T, Uehara Y. The cytoarchitecture of the wall and the innervation pattern of the microvessels in the rat mammary gland: a scanning electron microscopic observation. Am J Anat. 1984;170(1):39–54.
8. Frokjaer-Jensen J. The plasmalemmal vesicular system in striated muscle capillaries and in pericytes. Tissue Cell. 1984;16(1):31–42.
9. Joyce NC, Haire MF, Palade GE. Contractile proteins in pericytes: II. Immunocytochemical evidence for the presence of two isomyosins in graded concentrations. J Cell Biol. 1985;100(5):1387–95.
10. Matsusaka T. Tridimensional views of the relationship of pericytes to endothelial cells of capillaries in the human choroid and retina. J Electron Microsc (Tokyo). 1975;24(1):13–8.
11. Leeson T. Rat retinal blood vessels. Can J Ophthalmol. 1979;14(1):21–8.
12. Tilton R, Kilo C, Williamson J. Pericyte-endothelial relationships in cardiac and skeletal muscle capillaries. Microvasc Res. 1979;18(3):325–35.
13. Chintalgattu V, et al. Cardiomyocyte PDGFR-beta signaling is an essential component of the mouse cardiac response to load-induced stress. J Clin Invest. 2010;120(2):472–84.
14. Shen J, et al. PDGFR-beta as a positive regulator of tissue repair in a mouse model of focal cerebral ischemia. J Cereb Blood Flow Metab. 2012;32(2):353–67.
15. Daneman R, et al. Pericytes are required for blood-brain barrier integrity during embryogenesis. Nature. 2010;468(7323):562–6.
16. Bondjers C, et al. Transcription profiling of platelet-derived growth factor-B-deficient mouse embryos identifies RGS5 as a novel marker for pericytes and vascular smooth muscle cells. Am J Pathol. 2003;162(3):721–9.
17. Lindahl P, et al. Pericyte loss and microaneurysm formation in PDGF-B-deficient mice. Science. 1997;277(5323):242–5.
18. Nehls V, Denzer K, Drenckhahn D. Pericyte involvement in capillary sprouting during angiogenesis in situ. Cell Tissue Res. 1992;270(3):469–74.

19. Ruiter D, et al. Angiogenesis in wound healing and tumor metastasis. Behring Inst Mitt. 1993;(92):258–72.
20. Simionescu M, Simionescu N, Palade G. Segmental differentiations of cell junctions in the vascular endothelium. The microvasculature. J Cell Biol. 1975;67(3):863–85.
21. Rhodin J. Ultrastructure of mammalian venous capillaries, venules, and small collecting veins. J Ultrastruct Res. 1968;25(5):452–500.
22. Shepro D, Morel N, Pericyte physiology. FASEB J. 1993;7(11):1031–8.
23. Sims D. Recent advances in pericyte biology–implications for health and disease. Can J Cardiol. 1991;7(10):431–43.
24. Hellstrom M, et al. Lack of pericytes leads to endothelial hyperplasia and abnormal vascular morphogenesis. J Cell Biol. 2001;153(3):543–53.
25. Hellerbrand C. Hepatic stellate cells–the pericytes in the liver. Pflugers Arch. 2013;465(6):775–8.
26. Yin C, et al. Hepatic stellate cells in liver development, regeneration, and cancer. J Clin Invest. 2013;123(5):1902–10.
27. Geerts A. History, heterogeneity, developmental biology, and functions of quiescent hepatic stellate cells. Semin Liver Dis. 2001;21(3):311–35.
28. Smith SW, Chand S, Savage CO. Biology of the renal pericyte. Nephrol Dial Transplant. 2012;27(6):2149–55.
29. Abboud HE. Mesangial cell biology. Exp Cell Res. 2012;318(9):979–85.
30. Shi X, et al. The cochlear pericytes. Microcirculation. 2008;15(6):515–29.
31. Matsusaka T. [Ultrastructural differences between the choriocapillaries and retinal capillaries on the human eye]. Nihon Ganka Gakkai Zasshi. 1969;73(9):1603–5.
32. Sims D, Westfall J. Analysis of relationships between pericytes and gas exchange capillaries in neonatal and mature bovine lungs. Microvasc Res. 1983;25(3):333–42.
33. Sims D, et al. Ultrastructure of pericytes in early stages of histamine-induced inflammation. J Morphol. 1990;206(3):333–42.
34. Gerhardt H, Wolburg H, Redies C. N-cadherin mediates pericytic-endothelial interaction during brain angiogenesis in the chicken. Dev Dyn. 2000;218(3):472–9.
35. Li F, et al. Endothelial Smad4 maintains cerebrovascular integrity by activating N-cadherin through cooperation with Notch. Dev Cell. 2011;20(3):291–302.
36. Courtoy P, Boyles J. Fibronectin in the microvasculature: localization in the pericyte-endothelial interstitium. J Ultrastruct Res. 1983;83(3):258–73.
37. Larson DM, Carson MP, Haudenschild CC. Junctional transfer of small molecules in cultured bovine brain microvascular endothelial cells and pericytes. Microvasc Res. 1987;34(2):184–99.
38. Cuevas P, et al. Pericyte endothelial gap junctions in human cerebral capillaries. Anat Embryol (Berl). 1984;170(2):155–9.
39. Stratman AN, et al. Pericyte recruitment during vasculogenic tube assembly stimulates endothelial basement membrane matrix formation. Blood. 2009;114(24):5091–101.
40. Lindblom P, et al. Endothelial PDGF-B retention is required for proper investment of pericytes in the microvessel wall. Genes Dev. 2003;17(15):1835–40.
41. Clark A, et al. Increased gelatinase A (MMP-2) and gelatinase B (MMP-9) activities in human brain after focal ischemia. Neurosci Lett. 1997;238(1–2):53–6.
42. Fernandez-Klett F, et al. Early loss of pericytes and perivascular stromal cell-induced scar formation after stroke. J Cereb Blood Flow Metab. 2013;33(3):428–39.
43. Yao Y, et al. Astrocytic laminin regulates pericyte differentiation and maintains blood brain barrier integrity. Nat Commun. 2014;5:3413.
44. Sato Y. Dorsal aorta formation: separate origins, lateral-to-medial migration, and remodeling. Dev Growth Differ. 2013;55(1):113–29.
45. Wasteson P, et al. Developmental origin of smooth muscle cells in the descending aorta in mice. Development. 2008;135(10):1823–32.
46. Majesky MW. Developmental basis of vascular smooth muscle diversity. Arterioscler Thromb Vasc Biol. 2007;27(6):1248–58.
47. Riley PR, Smart N. Vascularizing the heart. Cardiovasc Res. 2011;91(2):260–8.

48. Pouget C, Pottin K, Jaffredo T. Sclerotomal origin of vascular smooth muscle cells and pericytes in the embryo. Dev Biol. 2008;315(2):437–47.
49. Alexander MR, Owens GK. Epigenetic control of smooth muscle cell differentiation and phenotypic switching in vascular development and disease. Annu Rev Physiol. 2012;74:13–40.
50. Wang Q, et al. Notch2 and Notch3 function together to regulate vascular smooth muscle development. PloS One. 2012;7(5):e37365.
51. Chang L, et al. Differentiation of vascular smooth muscle cells from local precursors during embryonic and adult arteriogenesis requires Notch signaling. Proc Natl Acad Sci U S A. 2012;109(18):6993–8.
52. Cohen ED, et al. Wnt signaling regulates smooth muscle precursor development in the mouse lung via a tenascin C/PDGFR pathway. J Clin Invest. 2009;119(9):2538–49.
53. Siegenthaler JA, et al. Foxc1 is required by pericytes during fetal brain angiogenesis. Biol Open. 2013;2(7):647–59.
54. Zhang H, et al. Origin-specific epigenetic program correlates with vascular bed-specific differences in Rgs5 expression. FASEB J. 2012;26(1):181–91.
55. Hughes S, Chan-Ling T, Characterization of smooth muscle cell and pericyte differentiation in the rat retina in vivo. Invest Ophthalmol Vis Sci. 2004;45(8):2795–806.
56. Bergers G, Song S. The role of pericytes in blood-vessel formation and maintenance. Neuro Oncol. 2005;7(4):452–64.
57. MacFadyen J, et al. Endosialin is expressed on stromal fibroblasts and CNS pericytes in mouse embryos and is downregulated during development. Gene Expr Patterns. 2007;7(3):363–9.
58. Cappellari O, et al. Dll4 and PDGF-BB convert committed skeletal myoblasts to pericytes without erasing their myogenic memory. Dev Cell. 2013;24(6):586–99.
59. Jin S, et al. Notch signaling regulates platelet-derived growth factor receptor-beta expression in vascular smooth muscle cells. Circ Res. 2008;102(12):1483–91.
60. Boyle SC, Liu Z, Kopan R. Notch signaling is required for the formation of mesangial cells from a stromal mesenchyme precursor during kidney development. Development. 2014;141(2):346–54.
61. Olson LE, Soriano P. PDGFRbeta signaling regulates mural cell plasticity and inhibits fat development. Dev Cell. 2011;20(6):815–26.
62. Yamanishi E, et al. Penetration and differentiation of cephalic neural crest-derived cells in the developing mouse telencephalon. Dev Growth Differ. 2012;54(9):785–800.
63. Soriano P. Abnormal kidney development and hematological disorders in PDGF beta-receptor mutant mice. Genes Dev. 1994;8(16):1888–96.
64. Keller A, et al. Mutations in the gene encoding PDGF-B cause brain calcifications in humans and mice. Nat Genet. 2013;45(9):1077–82.
65. Nicolas G, et al. Mutation of the PDGFRB gene as a cause of idiopathic basal ganglia calcification. Neurology. 2013;80(2):181–7.
66. ten Dijke P, Arthur HM. Extracellular control of TGFbeta signalling in vascular development and disease. Nat Rev Mol Cell Biol. 2007;8(11):857–69.
67. Sato Y, et al. Characterization of the activation of latent TGF-beta by co-cultures of endothelial cells and pericytes or smooth muscle cells: a self-regulating system. J Cell Biol. 1990;111(2):757–63.
68. Jakobsson L, van Meeteren LA. Transforming growth factor beta family members in regulation of vascular function: in the light of vascular conditional knockouts. Exp Cell Res. 2013;319(9):1264–70.
69. Goumans MJ, et al. Activin receptor-like kinase (ALK)1 is an antagonistic mediator of lateral TGFbeta/ALK5 signaling. Mol Cell. 2003;12(4):817–28.
70. Park C, et al. Bone morphogenetic protein receptor 1A signaling is dispensable for hematopoietic development but essential for vessel and atrioventricular endocardial cushion formation. Development. 2006;133(17):3473–84.
71. Itoh F, et al. Smad2/Smad3 in endothelium is indispensable for vascular stability via S1PR1 and N-cadherin expressions. Blood. 2012;119(22):5320–8.

72. Lebrin F, et al. Thalidomide stimulates vessel maturation and reduces epistaxis in individuals with hereditary hemorrhagic telangiectasia. Nat Med. 2010;16(4):420–8.
73. Langlois D, et al. Conditional inactivation of TGF-beta type II receptor in smooth muscle cells and epicardium causes lethal aortic and cardiac defects. Transgenic Res. 2010;19(6):1069–82.
74. El-Bizri N, et al. SM22alpha-targeted deletion of bone morphogenetic protein receptor 1A in mice impairs cardiac and vascular development, and influences organogenesis. Development. 2008;135(17):2981–91.
75. Morrow D, et al. Notch and vascular smooth muscle cell phenotype. Circ Res. 2008;103(12):1370–82.
76. Krebs LT, et al. Characterization of Notch3-deficient mice: normal embryonic development and absence of genetic interactions with a Notch1 mutation. Genesis. 2003;37(3):139–43.
77. Dziewulska D, Lewandowska E. Pericytes as a new target for pathological processes in CADASIL. Neuropathology. 2012;32(5):515–21.
78. Liu H, et al. Notch3 is critical for proper angiogenesis and mural cell investment. Circ Res. 2010;107(7):860–70.
79. High FA, et al. Endothelial expression of the Notch ligand Jagged1 is required for vascular smooth muscle development. Proc Natl Acad Sci U S A. 2008;105(6):1955–9.
80. Gaengel K, et al. Endothelial-mural cell signaling in vascular development and angiogenesis. Arterioscler Thromb Vasc Biol. 2009;29(5):630–8.
81. Jeansson M, et al. Angiopoietin-1 is essential in mouse vasculature during development and in response to injury. J Clin Invest. 2011;121(6):2278–89.
82. Nanba D, et al. Loss of HB-EGF in smooth muscle or endothelial cell lineages causes heart malformation. Biochem Biophys Res Commun. 2006;350(2):315–21.
83. Iivanainen E, et al. Angiopoietin-regulated recruitment of vascular smooth muscle cells by endothelial-derived heparin binding EGF-like growth factor. FASEB J. 2003;17(12):1609–21.
84. Stratman AN, et al. Endothelial-derived PDGF-BB and HB-EGF coordinately regulate pericyte recruitment during vasculogenic tube assembly and stabilization. Blood. 2010;116(22):4720–30.
85. Yu X, et al. Heparin-binding EGF-like growth factor protects pericytes from injury. J Surg Res. 2012;172(1):165–76.
86. Song N, et al. Overexpression of platelet-derived growth factor-BB increases tumor pericyte content via stromal-derived factor-1alpha/CXCR4 axis. Cancer Res. 2009;69(15):6057–64.
87. Hamdan R, Zhou Z, Kleinerman ES. SDF-1alpha induces PDGF-B expression and the differentiation of bone marrow cells into pericytes. Mol Cancer Res. 2011;9(11):1462–70.
88. Cheng L, et al. Glioblastoma stem cells generate vascular pericytes to support vessel function and tumor growth. Cell. 2013;153(1):139–52.
89. Zhou H, Yang YH, Basile JR. The Semaphorin 4D-Plexin-B1-RhoA signaling axis recruits pericytes and regulates vascular permeability through endothelial production of PDGF-B and ANGPTL4. Angiogenesis. 2014;17(1):261–74.
90. Nielsen CM, Dymecki SM. Sonic hedgehog is required for vascular outgrowth in the hindbrain choroid plexus. Dev Biol. 2010;340(2):430–7.
91. Yao Q, et al. Sonic hedgehog mediates a novel pathway of PDGF-BB-dependent vessel maturation. Blood. 2014;123(15):2429–37.
92. Foo SS, et al. Ephrin-B2 controls cell motility and adhesion during blood-vessel-wall assembly. Cell. 2006;124(1):161–73.
93. Nakayama A, et al. Ephrin-B2 controls PDGFRbeta internalization and signaling. Genes Dev. 2013;27(23):2576–89.
94. Gerhardt H, et al. VEGF guides angiogenic sprouting utilizing endothelial tip cell filopodia. J Cell Biol. 2003;161(6):1163–77.
95. Amselgruber W, Schäfer M, Sinowatz F. Angiogenesis in the bovine corpus luteum: an immunocytochemical and ultrastructural study. Anat Histol Embryol. 1999;28(3):157–66.
96. Jones N, et al. Tie receptors: new modulators of angiogenic and lymphangiogenic responses. Nature Rev Mol Cell Biol. 2001;2(4):257–67.
97. Patan S. TIE1 and TIE2 receptor tyrosine kinases inversely regulate embryonic angiogenesis by the mechanism of intussusceptive microvascular growth. Microvasc Res. 1998;56(1):1–21.

98. Orlidge A, D'Amore P. Inhibition of capillary endothelial cell growth by pericytes and smooth muscle cells. J Cell Biol. 1987;105(3):1455–62.
99. Sato Y, Rifkin D. Inhibition of endothelial cell movement by pericytes and smooth muscle cells: activation of a latent transforming growth factor-beta 1-like molecule by plasmin during co-culture. J Cell Biol. 1989;109(1):309–15.
100. Benjamin L, Hemo I, Keshet E. A plasticity window for blood vessel remodelling is defined by pericyte coverage of the preformed endothelial network and is regulated by PDGF-B and VEGF. Development. 1998;125(9):1591–8.
101. Diaz-Flores L, Gutierrez R, Varela H. Behavior of postcapillary venule pericytes during postnatal angiogenesis. J Morphol. 1992;213(1):33–45.
102. Yana I, et al. Crosstalk between neovessels and mural cells directs the site-specific expression of MT1-MMP to endothelial tip cells. J Cell Sci. 2007;120(Pt 9):1607–14.
103. Kashiwagi S, et al. NO mediates mural cell recruitment and vessel morphogenesis in murine melanomas and tissue-engineered blood vessels. J Clin Invest. 2005;115(7):1816–27.
104. Yamagishi S, et al. Vascular endothelial growth factor acts as a pericyte mitogen under hypoxic conditions. Lab Invest. 1999;79(4):501–9.
105. Darland D, et al. Pericyte production of cell-associated VEGF is differentiation-dependent and is associated with endothelial survival. Dev Biol. 2003;264(1):275–88.
106. Ribatti D, Nico B, Crivellato E. The role of pericytes in angiogenesis. Int J Dev Biol. 2011;55(3):261–8.
107. Schlingemann R, et al. Expression of the high molecular weight melanoma-associated antigen by pericytes during angiogenesis in tumors and in healing wounds. Am J Pathol. 1990;136(6):1393–405.
108. Wesseling P, et al. Early and extensive contribution of pericytes/vascular smooth muscle cells to microvascular proliferation in glioblastoma multiforme: an immuno-light and immuno-electron microscopic study. J Neuropathol Exp Neurol. 1995;54(3):304–10.
109. Gerhardt H, Semb H. Pericytes: gatekeepers in tumour cell metastasis? J Mol Med (Berl). 2008;86(2):135–44.
110. Morikawa S, et al. Abnormalities in pericytes on blood vessels and endothelial sprouts in tumors. Am J Pathol. 2002;160(3):985–1000.
111. Baluk P, Hashizume H, McDonald DM, Cellular abnormalities of blood vessels as targets in cancer. Curr Opin Genet Dev. 2005;15(1):102–11.
112. Jain RK. Normalization of tumor vasculature: an emerging concept in antiangiogenic therapy. Science. 2005;307(5706):58–62.
113. Abramsson A, Lindblom P, Betsholtz C. Endothelial and nonendothelial sources of PDGF-B regulate pericyte recruitment and influence vascular pattern formation in tumors. J Clin Invest. 2003;112(8):1142–51.
114. Song S, et al. *PDGFR*beta + perivascular progenitor cells in tumours regulate pericyte differentiation and vascular survival. Nat Cell Biol. 2005;7(9):870–9.
115. Bergers G, et al. Benefits of targeting both pericytes and endothelial cells in the tumor vasculature with kinase inhibitors. J Clin Invest. 2003;111(9):1287–95.
116. Bagher P, Segal S. Regulation of blood flow in the microcirculation: role of conducted vasodilation. Acta Physiol (Oxf). 2011;202(3):271–84.
117. Rucker H, Wynder H, Thomas W. Cellular mechanisms of CNS pericytes. Brain Res Bull. 2000;51(5):363–9.
118. Kennedy-Lydon T, et al. Renal pericytes: regulators of medullary blood flow. Acta Physiol (Oxf). 2013;207(2):212–25.
119. Attwell D, et al. Glial and neuronal control of brain blood flow. Nature. 2010;468(7321):232–43.
120. Peppiatt CM, et al. Bidirectional control of CNS capillary diameter by pericytes. Nature. 2006;443(7112):700–4.
121. Hall CN, et al. Capillary pericytes regulate cerebral blood flow in health and disease. Nature. 2014;508(7494):55–60.

122. Fernandez-Klett F, et al. Pericytes in capillaries are contractile in vivo, but arterioles mediate functional hyperemia in the mouse brain. Proc Natl Acad Sci U S A. 2010;107(51):22290–5.
123. Yemisci M, et al. Pericyte contraction induced by oxidative-nitrative stress impairs capillary reflow despite successful opening of an occluded cerebral artery. Nat Med. 2009;15(9):1031–7.
124. Lacar B, et al. Neural progenitor cells regulate capillary blood flow in the postnatal subventricular zone. J Neurosci. 2012;32(46):16435–48.
125. Wang S, et al. Venular basement membranes contain specific matrix protein low expression regions that act as exit points for emigrating neutrophils. J Exp Med. 2006;203(6):1519–32.
126. Voisin MB, Probstl D, Nourshargh S. Venular basement membranes ubiquitously express matrix protein low-expression regions: characterization in multiple tissues and remodeling during inflammation. Am J Pathol. 2010;176(1):482–95.
127. Proebstl D, et al. Pericytes support neutrophil subendothelial cell crawling and breaching of venular walls in vivo. J Exp Med. 2012;209(6):1219–34.
128. Wang S, et al. Pericytes regulate vascular basement membrane remodeling and govern neutrophil extravasation during inflammation. PloS ONE. 2012;7(9):e45499.
129. Ayres-Sander CE, et al. Transendothelial migration enables subsequent transmigration of neutrophils through underlying pericytes. PloS ONE. 2013;8(3):e60025.
130. Stark K, et al. Capillary and arteriolar pericytes attract innate leukocytes exiting through venules and 'instruct' them with pattern-recognition and motility programs. Nat Immunol. 2013;14(1):41–51.
131. Zachariah MA, Cyster JG. Neural crest-derived pericytes promote egress of mature thymocytes at the corticomedullary junction. Science. 2010;328(5982):1129–35.
132. Armulik A, et al. Pericytes regulate the blood-brain barrier. Nature. 2010;468(7323):557–61.
133. Dohgu S, et al. Brain pericytes contribute to the induction and up-regulation of blood-brain barrier functions through transforming growth factor-beta production. Brain Res. 2005;1038(2):208–15.
134. Raines SM, et al. Loss of PDGF-B activity increases hepatic vascular permeability and enhances insulin sensitivity. Am J Physiol Endocrinol Metab. 2011;301(3):E517–26.
135. Greenhalgh SN, Iredale JP, Henderson NC. Origins of fibrosis: pericytes take centre stage. F1000Prime Rep. 2013;5:37.
136. Ren S, Duffield JS. Pericytes in kidney fibrosis. Curr Opin Nephrol Hypertens. 2013;22(4):471–80.
137. LeBleu VS, et al. Origin and function of myofibroblasts in kidney fibrosis. Nat Med. 2013;19(8):1047–53.
138. Rock JR, et al. Multiple stromal populations contribute to pulmonary fibrosis without evidence for epithelial to mesenchymal transition. Proc Natl Acad Sci U S A. 2011;108(52):E1475–83.
139. Hung C, et al. Role of lung pericytes and resident fibroblasts in the pathogenesis of pulmonary fibrosis. Am J Respir Crit Care Med. 2013;188(7):820–30.
140. Goritz C, et al. A pericyte origin of spinal cord scar tissue. Science. 2011;333(6039):238–42.
141. Soderblom C, et al. Perivascular fibroblasts form the fibrotic scar after contusive spinal cord injury. J Neurosci. 2013;33(34):13882–7.
142. Dore-Duffy P, et al. CNS microvascular pericytes exhibit multipotential stem cell activity. J Cereb Blood Flow Metab. 2006;26(5):613–24.
143. Dore-Duffy P. Pericytes: pluripotent cells of the blood brain barrier. Curr Pharm Des. 2008;14(16):1581–93.
144. Dulauroy S, et al. Lineage tracing and genetic ablation of ADAM12(+) perivascular cells identify a major source of profibrotic cells during acute tissue injury. Nat Med. 2012;18(8):1262–70.
145. Dellavalle A, et al. Pericytes of human skeletal muscle are myogenic precursors distinct from satellite cells. Nat Cell Biol. 2007;9(3):255–67.
146. Blomhoff R, Blomhoff HK. Overview of retinoid metabolism and function. J Neurobiol. 2006;66(7):606–30.

147. Chen L, et al. HSCs play a distinct role in different phases of oval cell-mediated liver regeneration. Cell Biochem Funct. 2012;30(7):588–96.
148. Shen K, et al. Depletion of activated hepatic stellate cell correlates with severe liver damage and abnormal liver regeneration in acetaminophen-induced liver injury. Acta Biochim Biophys Sin (Shanghai). 2011;43(4):307–15.
149. Pintilie DG, et al. Hepatic stellate cells' involvement in progenitor-mediated liver regeneration. Lab Invest. 2010;90(8):1199–208.
150. Yang L, et al. Fate-mapping evidence that hepatic stellate cells are epithelial progenitors in adult mouse livers. Stem cells. 2008;26(8):2104–13.
151. Roskams T. Relationships among stellate cell activation, progenitor cells, and hepatic regeneration. Clin Liver Dis. 2008;**12**(4):853–60, ix.
152. Mormone E, George J, Nieto N. Molecular pathogenesis of hepatic fibrosis and current therapeutic approaches. Chem Biol Interact. 2011;193(3):225–31.
153. Fligny C, Duffield JS. Activation of pericytes: recent insights into kidney fibrosis and microvascular rarefaction. Curr Opin Rheumatol. 2013;25(1):78–86.
154. Leveen P, et al. Mice deficient for PDGF B show renal, cardiovascular, and hematological abnormalities. Genes Dev. 1994;8(16):1875–87.

Chapter 4
Fibroblast Growth Factor Signaling in Vascular Development

Igor Kovacevic, Meike Hoffmeister and Stefanie Oess

4.1 Overview of the Fibroblast Growth Factor (FGF) Family

The family of fibroblast growth factors (FGFs) in humans is composed of 22 different members [4]; they are thought to exert their biological effects through autocrine, paracrine, endocrine, or intracrine mechanisms [4, 5]. These properties are correlated with the extra- or intracellular localization of certain FGFs and their interaction with extracellular matrix (ECM) or other regulatory components.

The majority of members of the FGF protein family contains a signal peptide sequence and is secreted through the secretory pathway. However, extracellular FGF-1 (acidic FGF) and FGF-2 (basic FGF) do not contain such a sequence and can be directly translocated through the cell membrane in an unconventional protein secretion pathway [6, 7]. Structurally, FGFs are characterized by the presence of a core region, comprising 12 antiparallel β-strands [8]. This core region contains the heparan sulfate glycosaminoglycan-binding site, and the affinity for heparan sulfate is important for determining the range of action of a certain FGF protein. The endocrine FGFs (FGF-19, -21, and -23[1]) display reduced heparan sulfate affinity, which confers their long-range endocrine function [10]. The intracrine FGFs (FGF-11,

[1] In humans the term FGF-15 is not assigned, explaining the existence of FGF-23, although the FGF family has only 22 members. In mouse and rat, FGF-19 has not been identified. *Fgf15* and *Fgf19* are considered orthologous genes in vertebrates and are often referred to as *Fgf15/19* [9].

S. Oess (✉) · M. Hoffmeister
Institute for Biochemistry II, Goethe University Frankfurt, Medical School, Theodor Stern-Kai 7, 60590 Frankfurt/Main, Germany
e-mail: oess@biochem2.de

I. Kovacevic
Department of Molecular Cell Biology, Sanquin Research and Landsteiner Laboratory, Academic Medical Center, University of Amsterdam, Plesmanlaan 125, 1066 CX Amsterdam, The Netherland
e-mail: i.kovacevic@sanquin.nl

M. HH. Schmidt, S. Liebner (eds.), *Endothelial Signaling in Development and Disease,* DOI 10.1007/978-1-4939-2907-8_4

-12, -13, and -14) are not secreted and their biological activity is independent of the FGF transmembrane receptor [4, 11]. The remaining members of the FGF family function in an autocrine or paracrine manner [4, 5].

The complexity of the FGF family is further increased by the expression of several isoforms, generated through alternative initiation of translation and proteolytic processing [12]. For instance, the human *Fgf2* gene codes for several low and high molecular weight isoforms, ranging from 18 to 24 kDa. High molecular weight isoforms of FGF-2 are localized in the nucleus due to the presence of a nuclear localization signal (NLS), while low molecular weight isoforms are mainly cytosolic [5, 13]. Both high and low molecular weight isoforms have angiogenic potential [14], but they may employ distinct signaling mechanisms and induce different cellular phenotypes [5].

Endothelial cells express several different FGFs. Antoine et al. demonstrated the expression of FGF-1, -2, -4, -7, -16, and -18 in cultured human umbilical vein endothelial cells (HUVECs) and human aorta [15]. Among these, FGF-1 and FGF-2 belong to the first proangiogenic factors described and were shown to induce endothelial cell proliferation and migration *in vitro* and *in vivo* [1–3, 16, 17]. Therefore, in the following sections we will focus mainly on FGF-1- and FGF-2-dependent signaling mechanisms.

4.2 The Family of FGF Receptors (FGFRs)

FGF receptors (FGFRs) are transmembrane tyrosine kinase receptors. FGFRs generally comprise a ligand-binding extracellular domain, a single transmembrane domain, and the cytoplasmic tyrosine kinase domain [18] (Fig. 4.1). The FGFR protein family consists of four typical members (FGFR1, FGFR2, FGFR3, and FGFR4), as well as a recently discovered FGFR5 (or FGFR1L), which lacks the tyrosine kinase domain [4, 19]. The extracellular portion of the FGFR is composed of two to three immunoglobulin (Ig)-like domains (designated D1–D3), the acid box of seven to eight acidic residues, located between D1 and D2, and the heparin-binding site within the D2 domain. The juxtamembrane region in the cytoplasmic part of the FGFR is important for the recruitment of several adaptor proteins, which are involved in the signal transduction from the activated FGFR. Finally, a split tyrosine kinase domain and a C-terminal cytoplasmic tail mediate the phosphorylation events and the regulation of the receptor trafficking, respectively [4, 9].

4.3 Alternative Splicing Yields Structurally and Functionally Diverse FGFRs

The complexity of the FGF/FGFR signaling network is additionally enhanced by alternative splicing, which results in the expression of multiple isoforms of a single FGFR family member. There is considerable structural diversity between the

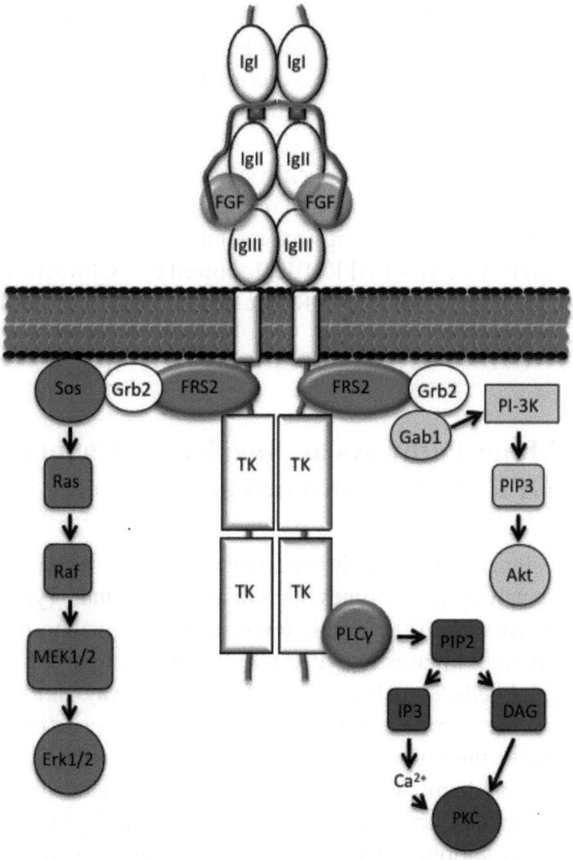

Fig. 4.1 Fibroblast growth factor receptor 1 (*FGFR1*) signal transduction. The binding of FGF-1, -2, or -4 at the interface of the IgII and IgIII domains, together with the binding of heparin or heparan sulfate to the acid box, induces dimerization and subsequent activation of FGFR1. Autophosphorylation of seven tyrosine residues in the intracellular domain of the receptor triggers the activation of three major signaling cascades. The mitogen-activated protein kinase (*MAPK*) pathway is activated through the binding of the FGFR substrate (FRS)-2 adaptor protein and sequential activation of Ras-Raf-MEK1/2 kinases. The Akt pathway is induced via FRS2-mediated recruitment of the adaptor protein Gab1 and activation of PI3K. The protein kinase C (*PKC*) pathway is activated by direct binding of phospholipase C (*PLC*)-γ to Tyr766 of FGFR1. Active PLC-γ hydrolyses phosphatidylinositol-4,5-diphosphate (*PIP2*) to diacylglycerol (*DAG*) and inositol-1,4,5-triphosphate (*IP3*); DAG subsequently induces activation of PKC, and IP3 triggers the release of Ca^{2+} from intracellular stores

different FGFR splicing isoforms; for example, soluble, secreted FGFRs, FGFRs lacking the kinase domain, or receptors containing two or three Ig-like domains resulting in altered ligand binding specificity are produced [20, 21]. One of the most carefully studied splicing events in the FGFR protein family is the splicing of the messenger RNA (mRNA) coding for the C-terminal part of the D3 (IgIII) domain. This domain can be encoded in two different ways: isoform IgIIIb is encoded by exons 7/8, and isoform IgIIIc is encoded by exons 7/9. The alternative splicing of FGFR1, FGFR2, and FGFR3 produces both IgIIIb and IgIIIc variants, while FGFR4 splicing yields only the IgIIIc isoform. These isoforms confer different ligand-binding specificity, e.g. FGFR3IIIb binds to FGF-1 and -9, and the FGFR3IIIc isoform binds to FGF-1, -2, -4, -8, -9, -17, -18, and -23 [22]. Moreover, different FGFR isoforms display a cell-type-specific expression pattern, e.g. IgIIIb isoforms are predominantly expressed in epithelial tissues, and IgIIIc isoforms in mesenchymal tissues [23]. The most prominently expressed FGFR isoform in cultured human endothelial cells (HUVECs) is FGFR1IIIc, followed by FGFR3IIIc and FGFR2IIIc [15]. Moreover, the expression of FGFR1IIIc and FGFR2IIIc isoforms were shown in the human aorta [15].

4.4 Three Major Pathways are Involved in Canonical FGF/FGFR Signal Transduction

The FGF/FGFR signaling network is important in numerous cell types and has been studied more extensively in cell types other than endothelium. Typically, the signal transduction downstream of the FGFR involves three signaling mechanisms: mitogen-activated protein kinases (MAPKs), PI3K/Akt, and phospholipase C (PLC) [24]. In endothelial cells, signaling via the MAPK pathway and PI3K/Akt is well established, while the significance of FGFR-induced PLC-γ activation is still controversial [25]. Since the activation process is studied in greatest detail for FGFR1, and FGFR1 is the predominant FGFR isoform in endothelial cells, we will focus here on the description of activation and signaling elicited by FGFR1 (Fig. 4.1).

At the basal auto-inhibited state of FGFR1, a proline residue in the C-terminal part of the kinase activation loop prevents binding of the kinase substrate. Interestingly, ATP can bind to the nucleotide-binding site under these conditions, occuring even without ligand stimulation [26]. Binding of FGF-1, -2, or -4 to the interface of the IgII and IgIII domains, accompanied by binding of heparin to the acid box in the IgII domain, induces conformational changes in the FGFR1 structure. Heparin or heparan sulfate are required to stabilize the complex between the FGF ligand and the FGFR [27]. Subsequent dimerization of the FGFR results in autophosphorylation of seven cytoplasmic tyrosine residues [27, 28]. In addition, adaptor proteins of the FGFR substrate (FRS)-2 family—FRS2α and FRS2β—that are associated with the juxtamembrane domain of FGFR1 already in its inactivated state, become tyrosine-phosphorylated by the activated receptor [29]. The phosphorylated tyrosine residues now serve as docking sites for the recruitment of adaptor proteins and enzymes containing either a phosphotyrosine-binding domain (PTB) or an Src homology 2 (SH2) domain.

As mentioned above, FRS2α and FRS2β bind to the juxtamembrane domain of FGFR1 via their PTB domains, independently of the receptor activation status [30]. Upon activation of the FGFR1, FRS2α recruits other adaptor proteins, namely Shp2 and Grb2. Grb2 forms a complex with the guanidine nucleotide exchange factor

(GEF) Sos, which in turn activates the small GTPase Ras. This event induces a whole cascade of protein kinases, finally leading to phosphorylation and activation of the MAPK ERK1/2, p38, and JNK. Among those, Erk1/2 kinases phosphorylate a wide range of cytoplasmic proteins (such as cytoskeleton-associated proteins) and transcription factors (e.g. c-Fos, Ets, Elk-1, etc.) and thereby regulate processes such as cell proliferation, differentiation and migration [31].

Activation of the PI3K/Akt pathway downstream of FGFR1 also depends on FRS2. In this pathway, upon FGFR1 activation, FRS2 recruits the adaptor protein Gab1, which leads to activation of PI3K and, subsequently, Akt. Active Akt kinase induces prosurvival signaling pathways, reflecting once more the diversity of cellular signaling events initiated by activation of FGFR1 [24].

Finally, PLC-γ binds to the phosphorylated tyrosine 766 in the C-terminal tail of FGFR1 via its SH2 domain [28]. In the next step, activated PLC-γ hydrolyses phosphatidylinositol-4,5-diphosphate (PIP2) to diacylglycerol (DAG) and inositol-1,4,5-triphosphate (IP3). Finally, DAG induces activation of the protein kinase C (PKC) and IP3 triggers the release of Ca^{2+} from intracellular stores, resulting in activation of Ca^{2+}-dependent signaling events [24].

Beyond this canonical FGF signal transduction via the FGFRs, the exogenous FGF-1 and FGF-2 molecules have been shown to be able to reach the cell cytosol and nucleus; therefore it has been proposed that they have a dual role in signal transduction, partly independent of the FGFRs [32–34]. This has particularly been shown in the context of cell growth regulation and rRNA synthesis [35–37]; however, the molecular details of this signaling process and its importance for endothelial cell biology are not completely understood.

4.5 Negative Regulation of FGFR1 Signaling

In order to keep the FGFR1 signaling under tight control, a mechanism of negative regulation of signal transduction is required. This is achieved via recruitment of proteins such as Cbl, Sprouty (Spry), Sef, and MAPK phosphatases.

Upon activation of the FGFR1, the ubiquitin ligase Cbl is recruited via FRS2 and Grb2 adaptor proteins. Cbl subsequently ubiquitinates the FGFR1 as well as FRS2, leading to attenuation of FGFR1 signaling [38], possibly by regulating the trafficking of FGFR1 to the lysosome [39]. However, the process of endocytosis and sorting of activated FGFRs is not thoroughly investigated in endothelial cells. In epithelial cells, the activated FGFR1 is co-internalized with E-cadherin, and co-localization in EEA1- and Rab5-positive endosomes shows that both proteins share the initial endocytosis and sorting mechanism in this cell type [40]. Furthermore, *in vivo* studies in *Xenopus* indicated that extended-synaptotagmin-2 mediates the endocytosis of the FGFR1, most probably through clathrin-dependent endocytosis [41]. Finally, a series of studies by Wiedlocha and Sorensen have demonstrated that FGFR endocytosis can occur through both clathrin- and caveolin-dependent pathways [34].

An additional level of negative regulation of the activated MAPK pathway downstream of the FGFR is achieved through the action of the four members of the Spry protein family [42]. Spry proteins form homo- or hetero-oligomers, which can inhibit the MAPK pathway via interaction with the FRS2-Grb2-Sos complex or directly via interaction with Raf kinase [43]. Overexpression of Spry-1 or Spry-2 in endothelial cells has been shown to inhibit FGF-2-dependent MAPK activation [44].

The transmembrane protein Sef (similar expression to FGF) is another negative regulator of FGFR signaling. Sef is a direct inhibitor of tyrosine phosphorylation of the FGFR1, as well as several kinases in the MAPK pathway [45, 46].

Finally, MAPK phosphatases, also known as dual-specificity phosphatases (DUSP), are important for the attenuation of FGFR1 signaling [47–49]. Recently, it has been shown that ERK-mediated phosphorylation of Ser777 in the C-terminal tail of FGFR1 itself leads to inhibition of FGFR1 signaling, proliferation, and migration, and constitutes a negative feedback mechanism in the control of FGF signaling [50].

4.6 FGF Co-Receptors and Extracellular Matrix Components Contribute to FGF-Dependent Signal Transduction

The FGF co-receptors Klotho (one of the three fates in Greek mythology), Cfr (cysteine-rich FGFR) and Flrt (fibronectin leucine-rich transmembrane protein), localized on the extracellular face of the plasma membrane, can support the formation of certain FGF/FGFR complexes and promote the FGFR signaling activity [51–54]. The endocrine FGFs (FGF-19, -21, and -23), which have a low affinity to heparan sulfate, rely on Klotho as essential tissue-specific co-receptors [53]. Furthermore, proteins involved in cell-adhesion processes, such as integrins, neural cell adhesion molecule (NCAM), and N-cadherin, are involved in the regulation of FGFR signal transduction [55].

As mentioned above, components of the ECM, such as heparan sulfate proteoglycans (HSPGs), also perform an important regulatory function in FGFR signaling. HSPGs regulate the distribution of FGF ligands in the extracellular environment, and thereby the accessibility of FGFs for activation of its receptor [56]. One of the HSPGs, syndecan-4, has been shown to play a role in the regulation of FGFR signaling in endothelial cells. In their study, Elfenbein and Simons have shown that syndecan-4 promotes the internalization of the FGFR1/FGF-2 complex in endothelial cells. The authors described a mechanism whereby syndecan-4 mediates macropinocytosis of the FGF-2/FGFR1 complex and modulates the kinetics of FGFR1-dependent MAPK activation [57, 58].

Moreover, a study addressing a potential FGFR-independent role of syndecans in FGF signaling has been performed. In this study, overexpression of full-size syndecan-4 in human endothelial cells enhanced FGF-2-dependent signal transduction. In contrast, overexpression of the extracellular domain of syndecan-4 linked to the glycosylphosphatidylinositol (GPI) anchor sequence from glypican-1 failed to induce FGF-2 signaling [59]. Since the intracellular domain of syndecan-4 binds to several phospholipids and proteins, e.g. PIP2 and PKCα, the authors argued that the overexpression effects were due to the intrinsic FGF-dependent signaling properties of the full-size syndecan-4 protein [59], independent of FGFR1 [57, 58]. Additional experiments in rat fat pad endothelial cells (RFPECs) by the same group demonstrated that only the expression of full-size syndecan-4 was able to enhance the FGF-2-dependent signaling response [60]. Furthermore, it has been proposed by the same group that syndecan-4 is involved in the activation of the small GTPase Rac1 in endothelial cells [61, 62]. However, additional studies in alternative experimental settings are required in order to confirm the role of syndecans as FGFR-independent FGF signal transducers.

In addition to the co-receptors and ECM components mentioned, the biological activity of FGFs might be regulated by binding to gangliosides, thrombospondin-1, the fibronectin fragment fibstatin, or FGF-binding protein, as well as to a number of serum components such as soluble FGFR1, fibrinogen, α2-macroglobulin (α2 M), pentraxin-3, and several cytokines [5].

4.7 Current Knowledge About FGFs and Their Receptors in Vascular Development is Limited

In comparison to other ligand receptor pairs, such as vascular endothelial growth factor/vascular endothelial growth factor receptor (VEGF/VEGFR), Eph/EphR, or Ang/Tie, the function of FGFs and their receptors in the development of the vascular system is poorly understood [63, 64]. Loss-of-function studies in mice have not been informative because FGF-1 and FGF-2 knockout mice (both single and double knockouts) do not show a strong phenotype, both in respect to the vascular system as well as development in general [65]. This might be explained by the existence of various FGFs and their functional redundancy [63]. However, it has been reported more recently that FGF-2 knockout mice show a decreased number of lymphatic vessels in the eye cornea [66] (see 4.15), indicating that more specific defects in the vascular system might be observed upon detailed analysis. In addition, complete FGFR1 and FGFR2 knockout mice are developmentally arrested before the onset of vascularization [67–69], making the analysis of the function of FGFR1 and FGFR2 during vascular development impossible.

The Ornitz Laboratory has recently generated tissue-specific FGFR1/FGFR2 double knockout mice with deletion of both genes in endothelial/hematopoietic cells. These mice are viable, indicating that deletion of FGFR1 and FGFR2 in these cell types does not result in lethal developmental defects. An initial analysis of vascular developmental in these mice suggests that endothelial FGFR1/2 signaling is not strictly required for vascular development or homeostasis, but is important for neovascularization in response to injury (Ornitz D, personal communication). These mice represent a highly valuable tool for the future detailed study of FGFR1/2 signaling in the various aspects of vascular development.

To date, evidence for the function of FGFs in angiogenesis stems from a variety of *in vitro* and *in vivo* models. These will be discussed in the following section, including information regarding the underlying signaling mechanisms, where available.

4.8 The Vascular System is Built Through Vasculogenesis and Angiogenesis

The vascular system forms as one of the first functional structures during embryonic development. It is built through two interlinked, but mechanistically distinct, processes, referred to as vasculogenesis and angiogenesis. Vasculogenesis describes an assembly process of precursor-derived endothelial cells into a primitive vascular plexus, while angiogenesis is defined as the formation of new blood vessels from a pre-existing network [70]. The mesoderm is the origin of endothelial cell and, in amniotes, the differentiation occurs in several distinct regions: (i) in the embryo proper, where endothelial precursors are first detected within paraxial and lateral plate mesoderm around embryonic day (E)7.0 of mouse development; and (ii) in the extraembryonic mesoderm of the yolk sac, allantois, and placenta [71–73]. Cells from the extraembryonic splanchnopleuric mesoderm are induced to develop into hemangioblasts, which are common precursors of blood and endothelial cells and form the blood islands on the yolk sac around E7 [74]. In the following developmental process, the inner cells of the blood islands differentiate into hematopoietic stem cells surrounded by endothelial precursors, so-called angioblasts. From these angioblasts, the first blood vessels are formed *de novo* in the process of vasculogenesis. The intraembryonic vascular networks in amniotes, but also in fish and amphibians, arise from single angioblast progenitors situated in the mesoderm that surrounds a developing organ and these angioblasts, e.g. give rise to the dorsal aorta [74–76].

4.9 FGF is an Important Inducer of Angioblasts during Vasculogenesis

The endoderm provides the growth factors for the induction of the mesoderm; it was shown that one of the key players in the process of mesoderm induction is FGF-2 [77, 78], together with TGFβ [79] early studies perfomed *in vivo* on *Xenopus* embryos and *in vitro* on dissociated pre-gastrulation avian embryos identified FGFs as potent inducers of mesoderm and subsequent angioblast formation [74, 80]. Furthermore, the expression of dominant negative FGFR (*Xenopus* homolog of FGFR1, containing only the extracellular and transmembrane domains) in *Xenopus* disrupted mesoderm formation [81]. Cox and Poole employed vasculogenesis assays using quail/chick chimeras, where somite-derived mesodermal precursors were transplanted from a quail donor and gave rise to angioblasts in a chick recipient. In this experimental setting, the injection of FGF-2-blocking antibodies, along with the quail somite, significantly reduced the number of angioblasts originating from the somite. By implantation of beads soaked with FGF-2 into quail/chick chimeras, the authors further demonstrated that FGF-2 not only regulates the angioblast induction from uncommitted mesoderm but is also required to establish an initial vascular pattern in the developing embryos [82]. The importance of FGF-2 for vasculogenesis was also confirmed in cultures of quail blastodisc-derived cells. These cells efficiently differentiated into blood vessels only in the presence of FGF-2 [83].

In contrast to the embryonic stem-cell-derived embryoid bodies from avian and *Xenopus* embryos, mouse embryoid bodies are less suitable for studying the involvement of single growth factors in angioblast induction because they have a tendency to spontaneously undergo vasculogenesis. However, the addition of a growth factor cocktail, including FGF-2, significantly increases the development of primitive vascular-like structures from mouse embryoid bodies [84]. The important function of FGF-2 in inducing vasculogenesis has recently been confirmed in a study using mouse embryonic stem cells or embryoid bodies. Treatment with α2 M induced both vasculogenesis as well as angiogenesis via induction of FGF-2 expression. Incubation with the FGFR1 inhibitor SU5402 prevented the α2 M-dependent activation of both vasculogenesis and angiogenesis [85]. Furthermore, FGF-2 in combination with leukemia inhibitory factor (LIF) has been shown to induce vasculogenesis in endothelial cells derived from mouse embryonic stem (ES) cells *in vitro* [86].

Interdependence of the FGF and VEGF signaling network in the regulation of vasculogenesis (and angiogenesis) was shown in experiments performed on explanted mouse embryonic hearts. In this setting, FGF-1, -2, -4, -8, -9, and -18 induced coronary vasculogenesis as well as angiogenesis. The provasculogenic/angiogenic effects of all tested FGFs in this assay were dependent on VEGF signaling and vice versa [87], indicating an extensive crosstalk and synergism between FGFs and members of the VEGF family, which we will continue to discuss below (4.16).

4.10 Developmental Angiogenesis is a Multistep Process Producing Stereotypical Vascular Patterns

Developmental angiogenesis is a highly stereotypical process, which leads to the establishment of organ-specific vascular patterns with reproducible anatomy [88]. The process of angiogenesis can be broken down in a series of events, namely destabilization of an existing vessel and degradation of the ECM, proliferation and directed migration of endothelial cells, the formation of new endothelial cell/cell contacts and vascular tubes followed by coverage with pericytes or vascular smooth muscle cells and vessel stabilization [89]. The functional unit in the process of establishing new vascular branches is the angiogenic sprout, which consists of several types of specialized endothelial cells. The leading position of the vascular sprout is taken by the so-called tip cell. Tip cells are highly polarized and form numerous cellular protrusions, referred to as filopodia [90]. Through these filopodia, the tip cells constantly sense the microenvironment for guidance cues in order to navigate the growing vessel. Tip cells also regulate capillary branching by detecting and connecting to neighboring sprouts. Sensing of the directional cues through filopodia, as well as translation into directed migration, strongly depend on coordination of the tip cell cytoskeleton with membrane dynamics and the activity of small GTPases is thought to be important for tip cell function [91–95]. Stalk cells follow the leading tip cell; they proliferate and thereby elongate the growing branch [88, 90, 91]. However, the assignment of these specialized functions is only transient, and endothelial cells dynamically shuffle their relative positioning in the angiogenic sprout, probably due to continuous competition for the tip cell function [96]. Finally, quiescent phalanx cells build the inner lining of the new vessel after its outline has been set [88, 90, 91].

4.11 The Role of FGF in Vessel Destabilization and Matrix Remodeling

During the first phase of angiogenesis, degradation of the ECM is an important step, and matrix metalloproteinases (MMPs), as well as the plasmin/plasminogen system, have been implicated in this process [5, 97, 98]. Urokinase plasminogen activator (uPA) converts the inactive proenzyme plasminogen into the active serine protease plasmin, which degrades fibrin and other ECM components [99]. FGF-1, FGF-2, and FGF-4 induce uPA expression in endothelial cells and, in addition, enhance the cell surface expression of the uPA receptor, which might mediate localization of the proteolytic activity to the site of active migration [100]. In line with these findings, FGF-2-induced vessel formation was decreased on the cornea of uPA and plasminogen knockout mice [101], suggesting an important role of the plasmin/plasminogen system in mediating the proangiogenic FGF signal.

The expression of several MMPs is also positively regulated by FGFs [5, 98]. Importantly, FGF-2 stimulation leads to increased shedding of pro-MMP-2 and -9 as well as MT1-MMP-containing vesicles from endothelial cells, thereby enhancing proteolytic activity in the pericellular compartment [102]. Notably, MT1-MMP is highly expressed in tip cells [103]. These MMPs degrade capillary basement membrane components such as type IV collagen or HSPGs, but also serve to reveal cryptic proangiogenic binding sites, e.g. for integrins [98]. In addition, MMPs, such as MMP-3 and -13, as well as plasmin have been shown to liberate HSPG-tethered FGF-2 from the matrix, demonstrating the intricate interplay between growth factors, proteinases, and the ECM [98, 101, 104].

4.12 FGFs in Angiogenic Sprout Formation and Function

As described above, the angiogenic sprouting process requires the complex action of a number of different specialized endothelial cells, which need to coordinate a variety of distinct cellular functions, such as matrix degradation, proliferation, migration, and regulation of cell/cell contacts. Therefore, it is important to analyse the role of proangiogenic factors, such as FGFs, in a system that allows for these complex cell/cell interactions in a setting most closely related to the *in vivo* situation.

FGFs promote sprouting angiogenesis from embryoid bodies [105]. FGF-2 has been shown to induce angiogenesis from pre-existing vessels in the chorioallantoic membrane (CAM) and the yolk sac membrane of the developing chick [106, 107], and FGF-2 administration on the quail CAM induces vessel growth from small vessels in the arterial tree [108]. Interestingly, FGF-2 application induced a strong angiogenic response at E7, when FGFR expression was high, but FGF-2 response was much less pronounced at later embryonic stages, when FGFR expression was decreased [108]. Likewise, angiogenesis is induced in endothelial corneal cells after implantation of FGF-2-containing hydrogel into the rabbit mid-stroma corneal pocket [109]. It has also been demonstrated that endogenous FGF-2 induces vessel outgrowth from heart explants [110]. Moreover, the administration of FGF-2 into the perivitelline space of developing zebrafish embryos induced dose-dependent vascular sprouting from adjacent subintestinal vein vessels [111].

In addition to these model systems, a number of mouse models have been developed that allow us to study the role of the FGF/FGFR system in vascular development. General overexpression of FGF-2 in transgenic mice leads to a so-called 'latent angiogenic phenotype' characterized by a predisposition to angiogenic reactions and amplified angiogenesis demonstrated in the matrigel plug assay [112]. In line with these experiments, overexpression of FGF-2 in the retina of transgenic mice has been reported to cause a similar proangiogenic phenotype in the context of cell injury [113]. In accordance with this proangiogenic role of FGF-2, the adenovirus-mediated delivery of FGF-2-antisense RNA to endothelial cells of mouse embryos cultured *ex utero* disrupts vascular development. Major phenotypes observed after FGF-2 knockdown at E7.5 included abnormal development of the yolk sac vasculature and growth cessation, which were reversed by application of FGF-2 complementary DNA (cDNA) [114]. The same group has circumvented the early lethality of FGFR1 knockout mice, and used adenoviral delivery of a dominant negative FGFR1 mutant to endothelial cells of E9 mouse embryos cultured *ex utero*. The disruption of FGFR1 function caused defects in developmental angiogenesis, including incomplete branching of the yolk sac vasculature and defects in intersomitic vessels and brain vascularization [115], demonstrating an important role of FGFR1 in these developmental processes. More recently, a role for FGFR2 has been suggested (in the mouse heart) since endothelial-cell-targeted overexpression of constitutively active FGFR2 induces increased migration and tube formation in isolated endothelial cells, and conveys cardioprotection and enhanced angiogenesis after myocardial infarction [116].

4.13 Function of FGFs in Proliferation and Migration of Endothelial Cells

The complex nature of the angiogenic process entails that the role of FGFs in its discrete steps is difficult to analyse in the physiologically relevant *in vivo* situation. Therefore, *in vitro* cell culture systems for endothelial cells are employed, which allow single aspects of cell behavior to be specifically addressed. Although there is a certain caveat concerning the significance of the results obtained in these *in vitro* settings, they have allowed valuable insight into the involvement of FGFs in the key processes of angiogenesis.

In 1984, FGF-2 purified from chondrosarcoma ECM has been shown to stimulate the proliferation of capillary endothelial cells *in vitro*, and the proproliferative action of FGF on endothelial cells has subsequently been confirmed by a large number of laboratories [2, 3, 16]. FGF-2-induced proliferation of immortalized capillary endothelial cells is accompanied by tyrosine phosphorylation of the adaptor protein FRS2, which in turn interacts with Grb2, leading to strong and sustained activation of the MAPK ERK2. Accordingly, the addition of the MEK inhibitor PD98059 suppressed MAPK activity, which inhibited the proliferative response to FGF-2 [117]. Moreover, proliferation of choriocapillary endothelial cells has been shown to be stimulated by FGF-2 *in vitro* and required both ERK1/2 and PI3Kinase/Akt pathways, demonstrating the importance of these signaling pathways for FGF-2-induced proliferation in endothelial cells [118]. In contrast, FGF-2-induced activation of the p38 pathway in endothelial cells is involved in the negative regulation of cell survival, proliferation, and differentiation [119]. Moreover, FGF-2-dependent Akt signaling has been shown to be an important survival signal in endothelial cells [120].

The coordinated movement of tip-cell-directed angiogenic sprouts is crucial for the establishment of the vascular branching pattern; however, the role of FGFRs in the specification of endothelial tip versus stalk cells and/or the guidance of vascular sprouts is not entirely clear. The function of FGF/FGFRs in directed migration and branching is most extensively studied in the *Drosophila* airway system, which is highly reminiscent of tip-cell-guided developmental angiogenesis. Here, FGFs determine the specification of the epithelial tip cell, which directly responds to the FGF ligand and leads airway branch outgrowth [91]. Accordingly, in a mixed endothelial cell population, in which FGFR1 was silenced in half of the cells, FGF-2-responsive cells lead migration, whereas nonresponsive cells adopted a subsidiary trailing position in the context of sheet migration [121]. Interestingly, an FGFR1 mutant lacking the C-terminally located 63 amino acids of the cytoplasmic domain failed to mediate chemotaxis in a modified Boyden chamber assay, but efficiently mediated MAPK activation and FRS2 phosphorylation. In addition, the migration defect was independent of PLC-γ1 or phospholipase A_2 activation, but sensitive to PI3K inhibition, suggesting a possible involvement of PI3K/Akt in mediating endothelial cell migration in this experimental setting [122].

The role of the FGF/FGFR system in endothelial cell migration is further highlighted by the fact that endothelial cells isolated from FGF-2 knockout mice have a defect in migration in response to mechanical damage in the so-called scratch assay *in vitro*. The migration defect is accompanied by reduced MAPK activation, and can be rescued by the addition of exogenous FGF-2 [123]. In addition, FGF-2 (alongside other FGFs) stimulates migration of endothelial cells from heart explants in the context of coronary angiogenesis. Importantly, coronary tubulogenesis of embryonic epicardium is responsive to many FGF family members and requires both FGF and VEGF-A [124]. Finally, expression of the dominant negative FGFR1 mutant (used by Lee et al. [115] in embryos *ex utero* as described above) in HUVECs leads to an impairment of MAPK signaling, reduction of endothelial cell number, induction of apoptosis and inhibition of migration in the modified Boyden chamber assay, confirming the role of FGF-2/FGFR1-initiated MAPK activation in these endothelial key properties [115].

In accordance with the well established role of small GTPases in the control of cell movement, FGF-2-induced angiogenesis in the *in vivo* matrigel plug assay depends on the presence of the small GTPase Rac [125]. Our laboratory has recently shown that FGF-2-induced activation of Rac1 in primary mouse lung endothelial cells depends on the presence of the F-BAR protein NOSTRIN. NOSTRIN serves as a membrane-bound multivalent adaptor to assemble a signaling complex containing FGFR1, Rac1, and its activating exchange factor Sos-1. The loss of NOSTRIN in NOSTRIN knockout mice, or after morpholino-mediated knockdown in zebrafish embryos, leads to reduced endothelial cell proliferation and directed migration, with a pronounced defect in tip cell filopodia formation, suggesting an important role for the FGF-2/FGFR1/Rac1 axis in regulating angiogenic sprouts *in vivo* [92]. Moreover, PI3K has been shown to regulate the activity of the small GTPases Cdc42 and Rac via activation of the GEF proteins [126], which are thought to be vital for mediating the directed migration of endothelial cells [91].

4.14 Function of FGFs in Capillary Branching and Vessel Maturation

To date, there has been no general scheme established describing the function of FGF signaling in branching morphogenesis; however, there is ample evidence for an important role of FGF during this process [127]. Upon overexpression of FGF-1 in the myocardium of transgenic mice, coronary artery density and branching are increased [128]. Likewise, overexpression of FGF-2 in transgenic mice leads to increased vessel density and vessel arborescence in the heart [129], illustrating the properties of these FGFs to positively regulate vascular branching. This has been confirmed by studies indicating that FGF-2 is a key determinant for vascular branching in luteal angiogenesis [130]. FGF-dependent signaling involves FGFR1, since a dominant negative version of FGFR1 expressed in retinal pigmented epithelial cells leads to a poorly branched vascular bed in the choroid and an avascular neonatal retina [131]. In accordance with these findings, embryoid bodies derived from FGFR1 knockout mice are characterized by abundant, but morphologically distinct, elongated vessels [132–134], pointing to the fact that FGFs are important for the establishment of vascular network morphology.

After completion of the early, invasive phase of angiogenesis, the late phase of angiogenesis serves the maturation of the newly formed vessel, and requires the formation of tight cell/cell contacts, deposition of ECM, and recruitment of pericytes or smooth muscle cells. FGFs have been suggested to contribute to these processes, e.g. through the regulation of integrin and cadherin expression and ECM deposition [5, 89]. FGFs play an important role in the formation and, more explicitly, the maintenance of cell/cell contacts. Inhibition of FGF signaling through the introduction of soluble FGF traps or dominant negative FGFR mutants *in vitro* and *in vivo* leads to dissociation of adherence and tight junctions followed by loss of endothelial barrier function and vascular integrity in adult mice [135]. The loss of FGF signaling disrupts the VE-cadherin-catenin complex at adherence junctions, and FGF-dependent regulation of expression and stability of the protein tyrosine phosphatase Shp2 have been implicated in this process [136]. However, the role of FGFs in maturation of newly formed vessels during developmental angiogenesis has not been investigated in detail.

4.15 FGFs and Lymphangiogenesis

While angiogenesis is a term used to describe development of the novel blood vessels from the pre-existing vasculature, lymphangiogenesis is a process of formation of novel lymphatic vessels. FGF-2, together with VEGF-C and VEGF-D is a potent lymphangiogenic factor. Analysis of FGF-2 knockout mice showed a decreased number of lymphatic vessels in the eye cornea compared with wild-type mice [66]. Application of FGF-2 was shown to induce lymphangiogenesis in mouse cornea in a dose-dependent manner. At lower doses, FGF-2 promoted formation of lymph vessels, whereas the endothelial angiogenic response was neglectable. The mechanism of FGF-2-induced lymphangiogenesis in this assay was dependent on the increase in expression of VEGF-C and VEGF-D [137, 138]. Moreover, VEGFR3 was shown to be important for FGF-2-dependent lymphangiogenesis in the mouse cornea [139]. A study by Matsuo et al. applying chemical inhibitors, suggested that FGF-2-induced lymphangiogenesis in temperature-sensitive rat endothelial lymphatic cells (TR-LE cells) was mediated via the Akt/mTOR/p70S6 kinase pathway [140]. More detailed experiments on FGF-2-mediated signaling in lymphangiogenesis were performed in primary lymphatic endothelial cells (LECs) isolated from embryonic mouse skin. These data show that FGFR1 is the primary receptor for FGF-2 in LECs, and that the FGF-2/FGFR1 signaling axis mediates proliferation and migration of LECs, while VEGF signaling is important for LEC sprouting and elongation [141]. In addition to FGFR1, additional FGFR isoforms might be important for FGF-2 signal transduction in LECs [142]. Several recent studies confirmed the crucial role of FGF-2/FGFR1 signaling in tumor lymphangiogenesis and proposed a potential treatment strategy for the inhibition of cancer metastasis [143, 144].

4.16 Crosstalk between FGF and Vascular Endothelial Growth Factor Signaling during Vascular Development

There is an intimate crosstalk between FGFs and members of the VEGF family during vasculogenesis and angiogenesis, as well as lymphangiogensis [5, 145, 146] (for a recent review on the role of VEGF see Chap. 1 of this book). FGF-2 has been referred to as 'the master regulator of angiogenesis' due to its strict requirement for mesoderm induction and angioblast formation, as well as its potential to induce the expression of VEGF and its receptor [135, 146]. However, the crosstalk between these two important proangiogenic signaling pathways is far more complex. In some experimental settings, FGF-2 requires activation of the VEGF/VEGFR system to induce neovascularization. On the other hand, VEGF has been shown to rely on the action of FGFs to promote angiogenesis. Despite this synergism, FGF and VEGF display distinct biological properties, resulting in different endothelial cell behavior during angiogenesis [5, 127, 145], highlighted by the distinct morphology of vessels induced in the absence or presence of the different growth factors [132–134].

4.17 Summary

FGFs are pleotropic growth factors with important functions in a variety of different cell types. The broad specificity of FGFs and the redundancy in respect of biological functions render it difficult to assign precise roles for FGF/FGFR signaling modules in vascular development. The combined information from various *in vitro*, *ex vivo*, and *in vivo* models collectively describes an important role of FGFs and, foremost, FGF-2 in the regulation and fine-tuning of vasculogenesis, angiogenesis, and lymphangiogenesis. Endothelial cell-specific knockout mouse models for different FGFRs combinations will allow the involvement of FGF/FGFR signaling in various aspects of vascular development and neovascularization to be studied in more detail.

FGFs and their cognate FGFRs have recently gained great attention as promising therapeutic targets in cancer therapy. Oncogenic mutations and overexpression of FGFs or FGFRs have frequently been observed in a variety of different cancers, and FGFs/FGFRs have a key role in promoting tumor angiogenesis. This has lead to the development of a number of agents designed to disrupt FGF/FGFR signaling on the basis of small molecule tyrosine kinase inhibitors, monoclonal antibodies, FGF-ligand traps, and allosteric inhibitors [147–149]. Solid tumors co-opt developmental signaling programs in order to promote tumor vascularization, and not least, therefore, it is of great importance to understand the role of FGFs and their receptors in vascular development in molecular detail.

References

1. Maciag T, Mehlman T, Friesel R, Schreiber AB. Heparin binds endothelial cell growth factor, the principal endothelial cell mitogen in bovine brain. Science. 1984;225:932–5.
2. Shing Y, Folkman J, Sullivan R, Butterfield C, Murray J, Klagsbrun M. Heparin affinity: purification of a tumor-derived capillary endothelial cell growth factor. Science. 1984;223:1296–9.
3. Ribatti D, Vacca A, Presta M. The discovery of angiogenic factors: a historical review. Gen Pharmacol. 2000;35:227–31.
4. Itoh N. The Fgf families in humans, mice, and zebrafish: their evolutional processes and roles in development, metabolism, and disease. Biol Pharm Bull. 2007;30:1819–25.
5. Presta M, Dell'Era P, Mitola S, Moroni E, Ronca R, Rusnati M. Fibroblast growth factor/fibroblast growth factor receptor system in angiogenesis. Cytokine Growth Factor Rev. 2005;16:159–78.
6. Schafer T, Zentgraf H, Zehe C, Brugger B, Bernhagen J, Nickel W. Unconventional secretion of fibroblast growth factor 2 is mediated by direct translocation across the plasma membrane of mammalian cells. J Biol Chem. 2004;279:6244–51.
7. Nickel W. Pathways of unconventional protein secretion. Curr Opin Biotechnol. 2010;21:621–6.
8. Beenken A, Mohammadi M. The FGF family: biology, pathophysiology and therapy. Nat Rev Drug Discov. 2009;8:235–53.
9. Itoh N, Ornitz DM. Fibroblast growth factors: from molecular evolution to roles in development, metabolism and disease. J Biochem (Tokyo). 2011;149:121–30.
10. Goetz R, et al. Molecular insights into the klotho-dependent, endocrine mode of action of fibroblast growth factor 19 subfamily members. Mol Cell Biol. 2007;27:3417–28.
11. Olsen SK, Garbi M, Zampieri N, Eliseenkova AV, Ornitz DM, Goldfarb M, Mohammadi M. Fibroblast growth factor (FGF) homologous factors share structural but not functional homology with FGFs. J Biol Chem. 2003;278:34226–36.
12. Bikfalvi A, Klein S, Pintucci G, Rifkin DB. Biological roles of fibroblast growth factor-2. Endocr Rev. 1997;18:26–45.
13. Florkiewicz RZ, Baird A, Gonzalez AM. Multiple forms of bFGF: differential nuclear and cell surface localization. Growth Factors. 1991;4:265–75.
14. Gualandris A, Urbinati C, Rusnati M, Ziche M, Presta M. Interaction of high-molecular-weight basic fibroblast growth factor with endothelium: biological activity and intracellular fate of human recombinant M(r) 24,000 bFGF. J Cell Physiol. 1994;161:149–59.
15. Antoine M, Wirz W, Tag CG, Mavituna M, Emans N, Korff T, Stoldt V, Gressner AM, Kiefer P. Expression pattern of fibroblast growth factors (FGFs), their receptors and antagonists in primary endothelial cells and vascular smooth muscle cells. Growth Factors. 2005;23:87–95.
16. Gospodarowicz D, Abraham JA, Schilling J. Isolation and characterization of a vascular endothelial cell mitogen produced by pituitary-derived folliculo stellate cells. Proc Natl Acad Sci U S A. 1989;86:7311–5.
17. Yan W, Bentley B, Shao R. Distinct angiogenic mediators are required for basic fibroblast growth factor- and vascular endothelial growth factor-induced angiogenesis: the role of cytoplasmic tyrosine kinase c-Abl in tumor angiogenesis. Mol Biol Cell. 2008;19:2278–88.
18. Schlessinger J. Cell signaling by receptor tyrosine kinases. Cell. 2000;103:211–25.
19. Trueb B. Biology of FGFRL1, the fifth fibroblast growth factor receptor. Cell Mol Life Sci. 2011;68:951–64.
20. Tiong KH, Mah LY, Leong CO. Functional roles of fibroblast growth factor receptors (FGFRs) signaling in human cancers. Apoptosis. 2013;18(12):1447–68.
21. Miki T, Bottaro DP, Fleming TP, Smith CL, Burgess WH, Chan AM, Aaronson SA. Determination of ligand-binding specificity by alternative splicing: two distinct growth factor receptors encoded by a single gene. Proc Natl Acad Sci U S A. 1992;89:246–50.
22. Ornitz DM, Xu J, Colvin JS, McEwen DG, MacArthur CA, Coulier F, Gao G, Goldfarb M. Receptor specificity of the fibroblast growth factor family. J Biol Chem. 1996;271:15292–7.

23. Orr-Urtreger A, Bedford MT, Burakova T, Arman E, Zimmer Y, Yayon A, Givol D, Lonai P. Developmental localization of the splicing alternatives of fibroblast growth factor receptor-2 (FGFR2). Dev Biol. 1993;158:475–86.
24. Eswarakumar VP, Lax I, Schlessinger J. Cellular signaling by fibroblast growth factor receptors. Cytokine Growth Factor Rev. 2005;16:139–49.
25. McLaughlin AP, De Vries GW. Role of PLCgamma and Ca(2+) in VEGF- and FGF-induced choroidal endothelial cell proliferation. Am J Physiol Cell Physiol. 2001;281:C1448–C56.
26. Mohammadi M, Schlessinger J, Hubbard SR. Structure of the FGF receptor tyrosine kinase domain reveals a novel autoinhibitory mechanism. Cell. 1996;86:577–87.
27. Schlessinger J, Plotnikov AN, Ibrahimi OA, Eliseenkova AV, Yeh BK, Yayon A, Linhardt RJ, Mohammadi M. Crystal structure of a ternary FGF-FGFR-heparin complex reveals a dual role for heparin in FGFR binding and dimerization. Mol Cell. 2000;6:743–50.
28. Mohammadi M, Dikic I, Sorokin A, Burgess WH, Jaye M, Schlessinger J. Identification of six novel autophosphorylation sites on fibroblast growth factor receptor 1 and elucidation of their importance in receptor activation and signal transduction. Mol Cell Biol. 1996;16:977–89.
29. Gotoh N. Regulation of growth factor signaling by FRS2 family docking/scaffold adaptor proteins. Cancer Sci. 2008;99:1319–25.
30. Groth C, Lardelli M. The structure and function of vertebrate fibroblast growth factor receptor 1. Int J Dev Biol. 2002;46:393–400.
31. Rubinfeld H, Seger R. The ERK cascade: a prototype of MAPK signaling. Mol Biotechnol. 2005;31:151–74.
32. Olsnes S, Klingenberg O, Wiedlocha A. Transport of exogenous growth factors and cytokines to the cytosol and to the nucleus. Physiol Rev. 2003;83:163–82.
33. Sorensen V, Nilsen T, Wiedlocha A. Functional diversity of FGF-2 isoforms by intracellular sorting. Bioessays. 2006;28:504–14.
34. Wiedlocha A, Sorensen V. Signaling, internalization, and intracellular activity of fibroblast growth factor. Curr Top Microbiol Immunol. 2004;286:45–79.
35. Baldin V, Roman AM, Bosc-Bierne I, Amalric F, Bouche G. Translocation of bFGF to the nucleus is G1 phase cell cycle specific in bovine aortic endothelial cells. EMBO J. 1990;9:1511–7.
36. Bouche G, Gas N, Prats H, Baldin V, Tauber JP, Teissie J, Amalric F. Basic fibroblast growth factor enters the nucleolus and stimulates the transcription of ribosomal genes in ABAE cells undergoing G0—G1 transition. Proc Natl Acad Sci U S A. 1987;84:6770–4.
37. Wiedlocha A, Falnes PO, Madshus IH, Sandvig K, Olsnes S. Dual mode of signal transduction by externally added acidic fibroblast growth factor. Cell. 1994;76:1039–51.
38. Wong A, Lamothe B, Lee A, Schlessinger J, Lax I. FRS2 alpha attenuates FGF receptor signaling by Grb2-mediated recruitment of the ubiquitin ligase Cbl. Proc Natl Acad Sci U S A. 2002;99:6684–9.
39. Haugsten EM, Malecki J, Bjorklund SM, Olsnes S, Wesche J. Ubiquitination of fibroblast growth factor receptor 1 is required for its intracellular sorting but not for its endocytosis. Mol Biol Cell. 2008;19:3390–403.
40. Bryant DM, Wylie FG, Stow JL. Regulation of endocytosis, nuclear translocation, and signaling of fibroblast growth factor receptor 1 by E-cadherin. Mol Biol Cell. 2005;16:14–23.
41. Jean S, Mikryukov A, Tremblay MG, Baril J, Guillou F, Bellenfant S, Moss T. Extended-synaptotagmin-2 mediates FGF receptor endocytosis and ERK activation in vivo. Dev Cell. 2010;19:426–39.
42. Guy GR, Jackson RA, Yusoff P, Chow SY. Sprouty proteins: modified modulators, matchmakers or missing links? J Endocrinol. 2009;203:191–202.
43. Yusoff P, Lao DH, Ong SH, Wong ES, Lim J, Lo TL, Leong HF, Fong CW, Guy GR. Sprouty2 inhibits the Ras/MAP kinase pathway by inhibiting the activation of Raf. J Biol Chem. 2002;277:3195–201.

44. Impagnatiello MA, Weitzer S, Gannon G, Compagni A, Cotten M, Christofori G. Mammalian sprouty-1 and -2 are membrane-anchored phosphoprotein inhibitors of growth factor signaling in endothelial cells. J Cell Biol. 2001;152:1087–98.
45. Kovalenko D, Yang X, Nadeau RJ, Harkins LK, Friesel R. Sef inhibits fibroblast growth factor signaling by inhibiting FGFR1 tyrosine phosphorylation and subsequent ERK activation. J Biol Chem. 2003;278:14087–91.
46. Kovalenko D, Yang X, Chen PY, Nadeau RJ, Zubanova O, Pigeon K, Friesel R. A role for extracellular and transmembrane domains of Sef in Sef-mediated inhibition of FGF signaling. Cell Signal. 2006;18:1958–66.
47. Muda M, Boschert U, Dickinson R, Martinou JC, Martinou I, Camps M, Schlegel W, Arkinstall S. MKP-3, a novel cytosolic protein-tyrosine phosphatase that exemplifies a new class of mitogen-activated protein kinase phosphatase. J Biol Chem. 1996;271:4319–26.
48. Muda M, Theodosiou A, Rodrigues N, Boschert U, Camps M, Gillieron C, Davies K, Ashworth A, Arkinstall S. The dual specificity phosphatases M3/6 and MKP-3 are highly selective for inactivation of distinct mitogen-activated protein kinases. J Biol Chem. 1996;271:27205–8.
49. Slack DN, Seternes OM, Gabrielsen M, Keyse SM. Distinct binding determinants for ERK2/p38alpha and JNK map kinases mediate catalytic activation and substrate selectivity of map kinase phosphatase-1. J Biol Chem. 2001;276:16491–500.
50. Zakrzewska M, Haugsten EM, Nadratowska-Wesolowska B, Oppelt A, Hausott B, Jin Y, Otlewski J, Wesche J, Wiedlocha A. ERK-mediated phosphorylation of fibroblast growth factor receptor 1 on Ser777 inhibits signaling. Sci Signal. 2013;6: ra11.
51. Bottcher RT, Pollet N, Delius H, Niehrs C. The transmembrane protein XFLRT3 forms a complex with FGF receptors and promotes FGF signalling. Nat Cell Biol. 2004;6:38–44.
52. Goetz R, et al. Klotho coreceptors inhibit signaling by paracrine fibroblast growth factor 8 subfamily ligands. Mol Cell Biol. 2012;32:1944–54.
53. Kuro-o M. Endocrine FGFs and Klothos: emerging concepts. Trends Endocrinol Metab. 2008;19:239–45.
54. Miyaoka Y, Tanaka M, Imamura T, Takada S, Miyajima A. A novel regulatory mechanism for Fgf18 signaling involving cysteine-rich FGF receptor (Cfr) and delta-like protein (Dlk). Development. 2010;137:159–67.
55. Murakami M, Elfenbein A, Simons M. Non-canonical fibroblast growth factor signalling in angiogenesis. Cardiovasc Res. 2008;78:223–31.
56. Matsuo I, Kimura-Yoshida C. Extracellular modulation of fibroblast growth factor signaling through heparan sulfate proteoglycans in mammalian development. Curr Opin Genet Dev. 2013;23:399–407.
57. Elfenbein A, Lanahan A, Zhou TX, Yamasaki A, Tkachenko E, Matsuda M, Simons M. (2012) Syndecan 4 regulates FGFR1 signaling in endothelial cells by directing macropinocytosis. Sci Signal. 2012;5: ra36.
58. Elfenbein A, Simons M. Syndecan-4 signaling at a glance. J Cell Sci. 2013;126:3799–804.
59. Volk R, Schwartz JJ, Li J, Rosenberg RD, Simons M. The role of syndecan cytoplasmic domain in basic fibroblast growth factor-dependent signal transduction. J Biol Chem. 1999;274:24417–24.
60. Horowitz A, Tkachenko E, Simons M. Fibroblast growth factor-specific modulation of cellular response by syndecan-4. J Cell Biol. 2002;157:715–25.
61. Tkachenko E, Elfenbein A, Tirziu D, Simons M. Syndecan-4 clustering induces cell migration in a PDZ-dependent manner. Circ Res. 2006;98:1398–404.
62. Elfenbein A, Rhodes JM, Meller J, Schwartz MA, Matsuda M, Simons M. Suppression of RhoG activity is mediated by a syndecan 4-synectin-RhoGDI1 complex and is reversed by PKCalpha in a Rac1 activation pathway. J Cell Biol. 2009;186:75–83.
63. Auguste P, Javerzat S, Bikfalvi A. Regulation of vascular development by fibroblast growth factors. Cell Tissue Res. 2003;314:157–66.
64. Javerzat S, Auguste P, Bikfalvi A. The role of fibroblast growth factors in vascular development. Trends Mol Med. 2002;8:483–9.

65. Miller DL, Ortega S, Bashayan O, Basch R, Basilico C. Compensation by fibroblast growth factor 1 (FGF1) does not account for the mild phenotypic defects observed in FGF2 null mice. Mol Cell Biol. 2000;20:2260–8.
66. Nakao S, Maruyama K, Zandi S, Melhorn MI, Taher M, Noda K, Nusayr E, Doetschman T, Hafezi-Moghadam A. Lymphangiogenesis and angiogenesis: concurrence and/or dependence? Studies in inbred mouse strains. FASEB J. 2010;24:504–13.
67. Arman E, Haffner-Krausz R, Chen Y, Heath JK, Lonai P. Targeted disruption of fibroblast growth factor (FGF) receptor 2 suggests a role for FGF signaling in pregastrulation mammalian development. Proc Natl Acad Sci U S A. 1998;95:5082–7.
68. Deng CX, Wynshaw-Boris A, Shen MM, Daugherty C, Ornitz DM, Leder P. Murine FGFR-1 is required for early postimplantation growth and axial organization. Genes Dev. 1994;8:3045–57.
69. Yamaguchi TP, Harpal K, Henkemeyer M, Rossant J. Fgfr-1 is required for embryonic growth and mesodermal patterning during mouse gastrulation. Genes Dev. 1994;8:3032–44.
70. Risau W. Mechanisms of angiogenesis. Nature. 1997;386:671–4.
71. Arora R, Papaioannou VE. The murine allantois: a model system for the study of blood vessel formation. Blood. 2012;120:2562–72.
72. Caprioli A, Minko K, Drevon C, Eichmann A, Dieterlen-Lievre F, Jaffredo T. Hemangioblast commitment in the avian allantois: cellular and molecular aspects. Dev Biol. 2001;238:64–78.
73. Patel-Hett S, D'Amore PA. Signal transduction in vasculogenesis and developmental angiogenesis. Int J Dev Biol. 2011;55:353–63.
74. Risau W, Flamme I. Vasculogenesis. Annu Rev Cell Dev Biol. 1995;11:73–91.
75. Coffin JD, Harrison J, Schwartz S, Heimark R. Angioblast differentiation and morphogenesis of the vascular endothelium in the mouse embryo. Dev Biol. 1991;148:51–62.
76. Risau W. Embryonic angiogenesis factors. Pharmacol Ther. 1991;51:371–6.
77. Poole TJ, Finkelstein EB, Cox CM. The role of FGF and VEGF in angioblast induction and migration during vascular development. Dev Dyn. 2001;220:1–17.
78. Riese J, Zeller R, Dono R. Nucleo-cytoplasmic translocation and secretion of fibroblast growth factor-2 during avian gastrulation. Mech Dev. 1995;49:13–22.
79. Smith JC, Cooke J, Green JB, Howes G, Symes K. (1989) Inducing factors and the control of mesodermal pattern in Xenopus laevis. Development. 1989;107 (Suppl):149–159.
80. Slack JM, Darlington BG, Heath JK, Godsave SF. Mesoderm induction in early Xenopus embryos by heparin-binding growth factors. Nature. 1987;326:197–200.
81. Amaya E, Musci TJ, Kirschner MW. Expression of a dominant negative mutant of the FGF receptor disrupts mesoderm formation in Xenopus embryos. Cell. 1991;66:257–70.
82. Cox CM, Poole TJ. Angioblast differentiation is influenced by the local environment: FGF-2 induces angioblasts and patterns vessel formation in the quail embryo. Dev Dyn. 2000;218:371–82.
83. Krah K, Mironov V, Risau W, Flamme I. Induction of vasculogenesis in quail blastodisc-derived embryoid bodies. Dev Biol. 1994;164:123–32.
84. Vittet D, Prandini MH, Berthier R, Schweitzer A, Martin-Sisteron H, Uzan G, Dejana E. Embryonic stem cells differentiate in vitro to endothelial cells through successive maturation steps. Blood. 1996;88:3424–31.
85. Sauer H, Ravindran F, Beldoch M, Sharifpanah F, Jedelska J, Strehlow B, Wartenberg M. α2-Macroglobulin enhances vasculogenesis/angiogenesis of mouse embryonic stem cells by stimulation of nitric oxide generation and induction of fibroblast growth factor-2 expression. Stem Cells Dev. 2013;22:1443–54.
86. Gendron RL, Tsai FY, Paradis H, Arceci RJ. Induction of embryonic vasculogenesis by bFGF and LIF in vitro and in vivo. Dev Biol. 1996;177:332–46.
87. Tomanek RJ, Zheng W. Role of growth factors in coronary morphogenesis. Tex Heart Inst J. 2002;29(4):250–4.
88. Larrivee B, Freitas C, Suchting S, Brunet I, Eichmann A. Guidance of vascular development: lessons from the nervous system. Circ Res. 2009;104:428–41.

89. Papetti M, Herman IM. Mechanisms of normal and tumor-derived angiogenesis. Am J Physiol Cell Physiol. 2002;282:C947–70.
90. Gerhardt H, et al. VEGF guides angiogenic sprouting utilizing endothelial tip cell filopodia. J Cell Biol. 2003;161:1163–77.
91. De Smet F, Segura I, De Bock K, Hohensinner PJ, Carmeliet P. Mechanisms of vessel branching: filopodia on endothelial tip cells lead the way. Arterioscler Thromb Vasc Biol. 2009;29:639–49.
92. Kovacevic I, et al. The F-BAR protein NOSTRIN participates in FGF signal transduction and vascular development. EMBO J. 2012;31:3309–22.
93. Suchting S, Bicknell R, Eichmann A. Neuronal clues to vascular guidance. Exp Cell Res. 2006;312:668–75.
94. Tan W, Palmby TR, Gavard J, Amornphimoltham P, Zheng Y, Gutkind JS. An essential role for Rac1 in endothelial cell function and vascular development. FASEB J. 2008;22:1829–38.
95. Wang Y, et al. Ephrin-B2 controls VEGF-induced angiogenesis and lymphangiogenesis. Nature. 2010;465:483–6.
96. Jakobsson L, et al. Endothelial cells dynamically compete for the tip cell position during angiogenic sprouting. Nat Cell Biol. 2010;12:943–53.
97. Breuss JM, Uhrin P. VEGF-initiated angiogenesis and the uPA/uPAR system. Cell Adh Migr. 2012;6:535–615.
98. Ghajar CM, George SC, Putnam AJ. Matrix metalloproteinase control of capillary morphogenesis. Crit Rev Eukaryot Gene Expr. 2008;18:251–78.
99. Lijnen HR. Elements of the fibrinolytic system. Ann NY Acad Sci. 2001;936:226–36.
100. Mignatti P, Rifkin DB. Nonenzymatic interactions between proteinases and the cell surface: novel roles in normal and malignant cell physiology. Adv Cancer Res. 2000;78:103–57.
101. Oh CW, Hoover-Plow J, Plow EF. The role of plasminogen in angiogenesis in vivo. J Thromb Haemost. 2003;1:1683–7.
102. Taraboletti G, D'Ascenzo D'Ascenzo S, Borsotti P, Giavazzi R, Pavan A, Dolo V. Shedding of the matrix metalloproteinases MMP-2, MMP-9, and MT1-MMP as membrane vesicle-associated components by endothelial cells. Am J Pathol. 2002;160:673–80.
103. Yana I, et al. Crosstalk between neovessels and mural cells directs the site-specific expression of MT1-MMP to endothelial tip cells. J Cell Sci. 2007;120:1607–14.
104. Folkman J. Antiangiogenic activity of a matrix protein. Cancer Biol Ther. 2003;2:53–4.
105. Feraud O, Cao Y, Vittet D. Embryonic stem cell-derived embryoid bodies development in collagen gels recapitulates sprouting angiogenesis. Lab Invest. 2001;81:1669–81.
106. Shing Y, Folkman J, Haudenschild C, Lund D, Crum R, Klagsbrun M. Angiogenesis is stimulated by a tumor-derived endothelial cell growth factor. J Cell Biochem. 1985;29:275–87.
107. Ribatti D, Urbinati C, Nico B, Rusnati M, Roncali L, Presta M. Endogenous basic fibroblast growth factor is implicated in the vascularization of the chick embryo chorioallantoic membrane. Dev Biol. 1995;170:39–49.
108. Parsons-Wingerter P, Elliott KE, Clark JI, Farr AG. Fibroblast growth factor-2 selectively stimulates angiogenesis of small vessels in arterial tree. Arterioscler Thromb Vasc Biol. 2000;20:1250–6.
109. Gaudric A, N'Guyen N'Guyen T, Moenner M, Glacet-Bernard A, Barritault D. Quantification of angiogenesis due to basic fibroblast growth factor in a modified rabbit corneal model. Ophthalmic Res. 1992;24:181–8.
110. Tomanek RJ, Sandra A, Zheng W, Brock T, Bjercke RJ, Holifield JS. Vascular endothelial growth factor and basic fibroblast growth factor differentially modulate early postnatal coronary angiogenesis. Circ Res. 2001;88:1135–41.
111. Nicoli S, De Sena G, Presta M. Fibroblast growth factor 2-induced angiogenesis in zebrafish: the zebrafish yolk membrane (ZFYM) angiogenesis assay. J Cell Mol Med. 2009;13:2061–8.
112. Fulgham DL, Widhalm SR, Martin S, Coffin JD. FGF-2 dependent angiogenesis is a latent phenotype in basic fibroblast growth factor transgenic mice. Endothelium. 1999;6:185–95.

113. Yamada H, Yamada E, Kwak N, Ando A, Suzuki A, Esumi N, Zack DJ, Campochiaro PA. Cell injury unmasks a latent proangiogenic phenotype in mice with increased expression of FGF2 in the retina. J Cell Physiol. 2000;185:135–42.
114. Leconte I, Fox JC, Baldwin HS, Buck CA, Swain JL. Adenoviral-mediated expression of antisense RNA to fibroblast growth factors disrupts murine vascular development. Dev Dyn. 1998;213:421–30.
115. Lee SH, Schloss DJ, Swain JL. Maintenance of vascular integrity in the embryo requires signaling through the fibroblast growth factor receptor. J Biol Chem. 2000;275:33679–87.
116. Matsunaga S, et al. Endothelium-targeted overexpression of constitutively active FGF receptor induces cardioprotection in mice myocardial infarction. J Mol Cell Cardiol. 2009;46:663–73.
117. Klint P, Kanda S, Kloog Y, Claesson-Welsh L. Contribution of Src and Ras pathways in FGF-2 induced endothelial cell differentiation. Oncogene. 1999;18:3354–64.
118. Zubilewicz A, Hecquet C, Jeanny JC, Soubrane G, Courtois Y, Mascarelli F. Two distinct signalling pathways are involved in FGF2-stimulated proliferation of choriocapillary endothelial cells: a comparative study with VEGF. Oncogene. 2001;20:1403–13.
119. Matsumoto T, Turesson I, Book M, Gerwins P, Claesson-Welsh L. p38 MAP kinase negatively regulates endothelial cell survival, proliferation, and differentiation in FGF-2-stimulated angiogenesis. J Cell Biol. 2002;156:149–60.
120. Partovian C, Simons M. Regulation of protein kinase B/Akt activity and Ser473 phosphorylation by protein kinase Calpha in endothelial cells. Cell Signal. 2004;16:951–7.
121. Vitorino P, Meyer T. Modular control of endothelial sheet migration. Genes Dev. 2008;22:3268–81.
122. Landgren E, Klint P, Yokote K, Claesson-Welsh L. Fibroblast growth factor receptor-1 mediates chemotaxis independently of direct SH2-domain protein binding. Oncogene. 1998;17:283–91.
123. Pintucci G, Moscatelli D, Saponara F, Biernacki PR, Baumann FG, Bizekis C, Galloway AC, Basilico C, Mignatti P. Lack of ERK activation and cell migration in FGF-2-deficient endothelial cells. FASEB J. 2002;16:598–600.
124. Tomanek RJ, Christensen LP, Simons M, Murakami M, Zheng W, Schatteman GC. Embryonic coronary vasculogenesis and angiogenesis are regulated by interactions between multiple FGFs and VEGF and are influenced by mesenchymal stem cells. Dev Dyn. 2010;239:3182–91.
125. Dormond O, Foletti A, Paroz C, Ruegg C. NSAIDs inhibit alpha V beta 3 integrin-mediated and Cdc42/Rac-dependent endothelial-cell spreading, migration and angiogenesis. Nat Med. 2001;7:1041–7.
126. Welch HC, Coadwell WJ, Stephens LR, Hawkins PT. Phosphoinositide 3-kinase-dependent activation of Rac. FEBS Lett. 2003;546:93–7.
127. Horowitz A, Simons M. Branching morphogenesis. Circ Res. 2008;103:784–95.
128. Fernandez B, et al. Transgenic myocardial overexpression of fibroblast growth factor-1 increases coronary artery density and branching. Circ Res. 2000;87:207–13.
129. Sheikh F, Sontag DP, Fandrich RR, Kardami E, Cattini PA. Overexpression of FGF-2 increases cardiac myocyte viability after injury in isolated mouse hearts. Am J Physiol Heart Circ Physiol. 2001;280:H1039–H50.
130. Woad KJ, Hunter MG, Mann GE, Laird M, Hammond AJ, Robinson RS. Fibroblast growth factor 2 is a key determinant of vascular sprouting during bovine luteal angiogenesis. Reproduction. 2012;143:35–43.
131. Rousseau B, Larrieu-Lahargue F, Bikfalvi A, Javerzat S. Involvement of fibroblast growth factors in choroidal angiogenesis and retinal vascularization. Exp Eye Res. 2003;77:147–56.
132. Magnusson P, Rolny C, Jakobsson L, Wikner C, Wu Y, Hicklin DJ, Claesson-Welsh L. Deregulation of Flk-1/vascular endothelial growth factor receptor-2 in fibroblast growth factor receptor-1-deficient vascular stem cell development. J Cell Sci. 2004;117:1513–23.
133. Magnusson PU, Dimberg A, Mellberg S, Lukinius A, Claesson-Welsh L. FGFR-1 regulates angiogenesis through cytokines interleukin-4 and pleiotrophin. Blood. 2007;110:4214–22.

134. Magnusson PU, Ronca R, Dell'Era P, Carlstedt P, Jakobsson L, Partanen J, Dimberg A, Claesson-Welsh L. Fibroblast growth factor receptor-1 expression is required for hematopoietic but not endothelial cell development. Arterioscler Thromb Vasc Biol. 2005;25:944–9.
135. Murakami M, Nguyen LT, Zhuang ZW, Moodie KL, Carmeliet P, Stan RV, Simons M. The FGF system has a key role in regulating vascular integrity. J Clin Invest. 2008;118:3355–66.
136. Hatanaka K, Lanahan AA, Murakami M, Simons M. Fibroblast growth factor signaling potentiates VE-cadherin stability at adherens junctions by regulating SHP2. PLoS ONE. 2012;7:e37600.
137. Chang LK, et al. Dose-dependent response of FGF-2 for lymphangiogenesis. Proc Natl Acad Sci U S A. 2004;101:11658–63.
138. Hajrasouliha AR, Sadrai Z, Chauhan SK, Dana R. b-FGF induces corneal blood and lymphatic vessel growth in a spatially distinct pattern. Cornea. 2012;31:804–9.
139. Kubo H, Cao R, Brakenhielm E, Makinen T, Cao Y, Alitalo K. Blockade of vascular endothelial growth factor receptor-3 signaling inhibits fibroblast growth factor-2-induced lymphangiogenesis in mouse cornea. Proc Natl Acad Sci U S A. 2002;99:8868–73.
140. Matsuo M, Yamada S, Koizumi K, Sakurai H, Saiki I. Tumour-derived fibroblast growth factor-2 exerts lymphangiogenic effects through Akt/mTOR/p70S6kinase pathway in rat lymphatic endothelial cells. Eur J Cancer. 2007;43:1748–54.
141. Kazenwadel J, Secker GA, Betterman KL, Harvey NL. In vitro assays using primary embryonic mouse lymphatic endothelial cells uncover key roles for FGFR1 signalling in lymphangiogenesis. PLoS ONE. 2012;7:e40497.
142. Shin JW, et al. Prox1 promotes lineage-specific expression of fibroblast growth factor (FGF) receptor-3 in lymphatic endothelium: a role for FGF signaling in lymphangiogenesis. Mol Biol Cell. 2006;17:576–84.
143. Cao R, et al. Collaborative interplay between FGF-2 and VEGF-C promotes lymphangiogenesis and metastasis. Proc Natl Acad Sci U S A. 2012;109:15894–9.
144. Larrieu-Lahargue F, Welm AL, Bouchecareilh M, Alitalo K, Li DY, Bikfalvi A, Auguste P. Blocking fibroblast growth factor receptor signaling inhibits tumor growth, lymphangiogenesis, and metastasis. PLoS ONE. 2012;7:e39540.
145. Murakami M, Simons M. Fibroblast growth factor regulation of neovascularization. Curr Opin Hematol. 2008;15:215–20.
146. Seghezzi G, et al. Fibroblast growth factor-2 (FGF-2) induces vascular endothelial growth factor (VEGF) expression in the endothelial cells of forming capillaries: an autocrine mechanism contributing to angiogenesis. J Cell Biol. 1998;141:1659–73.
147. Bono F, et al. Inhibition of tumor angiogenesis and growth by a small-molecule multi-FGF receptor blocker with allosteric properties. Cancer Cell. 2013;23:477–88.
148. Brooks AN, Kilgour E, Smith PD. Molecular pathways: fibroblast growth factor signaling: a new therapeutic opportunity in cancer. Clin Cancer Res. 2012;18:1855–62.
149. Liang G, Liu Z, Wu J, Cai Y, Li X. Anticancer molecules targeting fibroblast growth factor receptors. Trends Pharmacol Sci. 2012;33:531–41.

Chapter 5
Development and Differentiation of the Lymphatic Vascular System

Jeremiah Bernier-Latmani, Amélie Sabine and Tatiana V. Petrova

5.1 Introduction

The lymphatic vasculature is a highly specialized part of the cardiovascular system, present in virtually all tissues of the body. Lymphatic vessels are essential for homeostasis by promoting removal of extracellular fluid and various macromolecules from the tissues back into blood circulation, and by mediating immune cell trafficking from the periphery to lymph nodes. The importance of the lymphatic system can be observed in clinics around the world in patients with lymphedema. Lymphatic vessel disruption, either through genetic or mechanical means, in patients with lymphedema results in limb fluid build-up which causes, at best, discomfort, and, at worst, life-threatening disfiguration [1]. Furthermore, lymphatic vessels are one of the main conduits for metastasizing cancer cells, the main driver of terminal cancer [2]. Moreover, obesity is linked to lymphedema, suggesting that lymphatic dysfunction rates may rise with the current obesity epidemic in Western countries [3]. Lastly, lymphangiogenesis is commonly increased in inflammatory diseases [4]. However, targeted molecular approaches for modulating growth and function of the lymphatic vasculature are only emerging, making the study of the lymphatic system an important aspect of improving human health.

In this chapter, we review molecular mechanisms controlling developmental lymphangiogenesis and lymphatic vessel maturation. In learning about how the normal lymphatic network develops, and the factors regulating these processes, we hope to use this knowledge to provide treatments for human diseases associated with lymphatic vessels.

T. V. Petrova (✉) · J. Bernier-Latmani · A. Sabine
Department of Oncology, University of Lausanne, Epalinges, 1006 Lausanne, Switzerland
e-mail: Tatiana.petrova@unil.ch

T. V. Petrova
Swiss Institute for Cancer Research, Ecole Polytechnique Federale de Lausanne, Lausanne, Switzerland

M. HH. Schmidt, S. Liebner (eds.), *Endothelial Signaling in Development and Disease,* DOI 10.1007/978-1-4939-2907-8_5

5.2 Lymphatic Basics

The lymphatic network is comprised of capillaries, precollecting and collecting vessels. Immune cells, dietary fats, and interstitial fluid first pass into permeable lymphatic capillaries (Fig. 5.1). The permeability of the lymphatic vessels is due to the semicontinuous nature of adherens junctions, which are organized into specialized 'button-like' structures [5]. Lymphatic capillaries can be also distinguished by the high expression of the lymphatic vessel endothelial hyaluronan receptor (LYVE-1), sparse or absent basement membrane, and absence of mural cells. Collecting lymphatic vessels are the least permeable lymphatic vessel type, and they are lined with endothelial cells, connected by continuous 'zipper-like' cell–cell junctions [5]. Lymphatic collectors do not express LYVE-1 but have a basement membrane and are covered by smooth muscle cells (Fig. 5.1) (reviewed by Schulte-Merker et al. [6]). Lymphatic smooth muscle cell contractions, together with body movements and arterial pulsations, are important for propulsion of lymph. Intraluminal lymphatic valves, positioned at regular intervals in collecting lymphatic vessels, prevent lymph back flow towards lymphatic capillaries. Precollector vessels, which

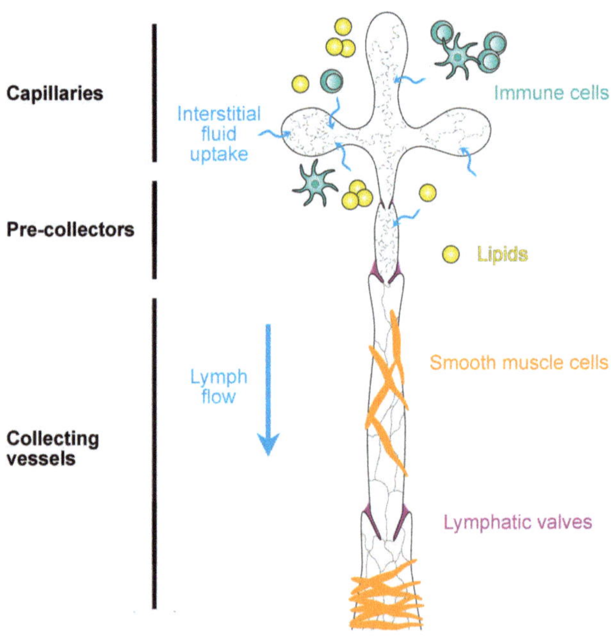

Fig. 5.1 Lymphatic vessel basics. Interstitial fluid and immune cells pass through discontinuous adherens junctions on lymphatic capillaries in a unidirectional manner. One of the main markers for lymphatic capillaries is LYVE-1. Precollecting vessels share features of both lymphatic capillaries (*LYVE-1 expression*) and collecting vessels (*lymphatic valves*). Lymphatic collecting vessels have valves that ensure unidirectional flow of lymph, continuous basement membrane and contractile smooth muscle cells

join capillaries and collecting lymphatic vessels, have features of both vessel types (Fig. 5.1). For example, they express LYVE-1 but also have intraluminal valves [6].

The diversity of lymphatic vessel types found in adults is a product of a complex interplay between lymphatic endothelial identity factors, growth factor-mediated signaling to lymphatic-specific receptors, and interaction with external factors, e.g. interstitial fluid pressure, and flow and shear stress. We summarize in this chapter our current understanding of the molecular mechanisms that produce the lymphatic vasculature.

5.3 Developmental Lymphangiogenesis

5.3.1 Lymphatic Network Formation

Florence Sabin first proposed the venous origin of lymphatic vessels [7], and recent studies, using lineage tracing and high-resolution imaging in thick sections and whole embryos, have confirmed this hypothesis [8–10]. Lymphatic endothelial cell (LEC) progenitors emerge from the cardinal veins and intersomitic vessels through a process of budding without disturbing the integrity of the venous wall. The streams of LECs, connected by continuous adherens junctions, migrate from the veins and form an initial lymphatic plexus. LECs then coalesce and form large lumenized lymphatic vessels, commonly called ‘lymph sacs’, one of which will become the thoracic duct [8, 10]. This process of lymphatic plexus formation, followed by lumen formation, continues in a caudal–dorsal manner to form an almost complete lymphatic network before birth (Fig. 5.2a, b, c, d).

Although many questions remain regarding the process of lymphangiogenesis during embryonic and postnatal growth, the molecular mechanisms controlling many steps of LEC identity, differentiation, migration, proliferation, and vessel formation have been identified.

5.3.2 Establishment of Lymphatic Endothelial Cell Identity

The first gene described as a regulator of LEC identity was PROX1 [11], a homeobox transcription factor that is first expressed in a subset of cells in the cardinal vein around E9.5 (Fig. 5.2a). Nascent LECs, which bud off into the mesenchyme, also express lymphatic endothelial-specific markers, including VEGFR-3, podoplanin, CCL21, neuropilin-2 (NRP2), and LYVE-1, as well as a specific integrin repertoire, such as ITGα6β1 (Fig. 5.2b) [8, 12]. PROX1 is essential for lymphatic vascular development as *Prox1*-null mice fail to develop a lymphatic vascular network because of the failure of LECs to migrate from veins [10, 11]. Furthermore, PROX1 is necessary for lymphatic identity maintenance after the formation of the lymphatic network, as shown by both *in vitro* and *in vivo* experiments [13, 14].

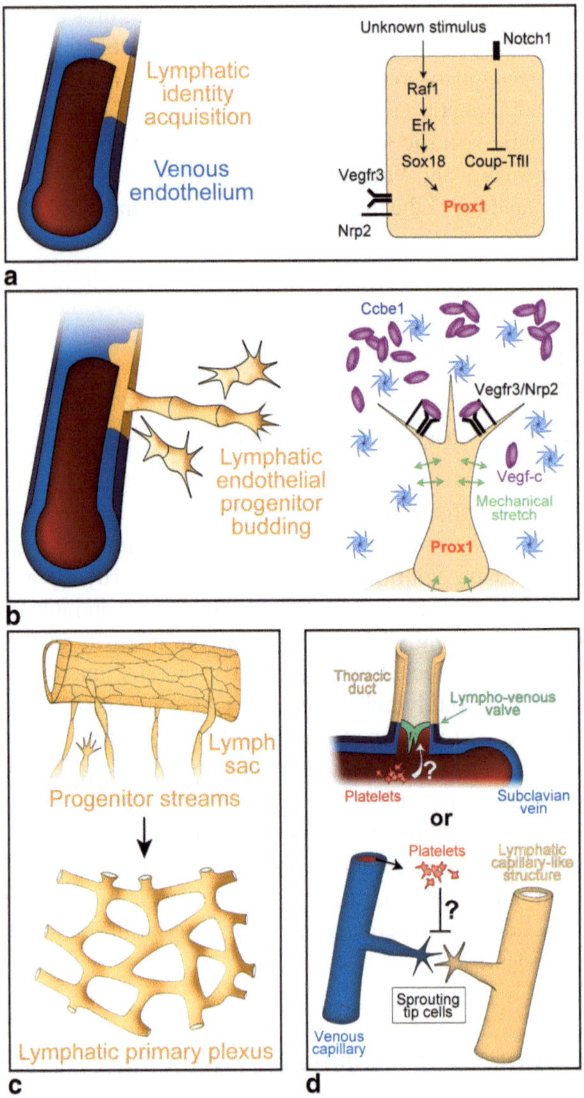

Fig. 5.2 Molecular mechanisms controlling lymphatic cell identity, sprouting, and segregation from blood vessels. (**a**) ERK signaling activation delineates a subset of venous cells as progenitor lymphatic endothelial cells (*LECs*), which, in turn, express SOX18, COUP-TFII, and PROX1. (**b**) LECs migrate from the vein in response to vascular endothelial growth factor (*VEGF*)-C/VEGF receptor (*VEGFR*)-3 signaling. The VEGFR-3 co-receptor neuropilin-2 (NRP2) also promotes VEGF-C/VEGFR-3 signaling. CCBE-1 is also necessary for LEC migration from the cardinal vein. LEC mechanical stretching contributes to activation of VEGFR-3 signaling and LEC proliferation. (**c**) LECs migrate in streams away from the cardinal vein to form the lymphatic primary plexus and primordial thoracic duct. (**d**) Normal platelet aggregation is important for keeping lymphatic and blood vasculatures segregated. Platelets may act either by modulating formation or function of lymphovenous valves via unknown mechanisms or by preventing ectopic fusion of lymphatic and blood vessels

The transcription factor SOX18 acts upstream of PROX1, and its expression in a subset of dorsolateral cardinal vein endothelial cells is necessary for the induction of PROX1 expression [15]. Mutations in *SOX18* were described in patients with hypotrichosis-lymphedema-telangiectasia [16], which underscores the role of this transcription factor in lymphatic vascular development. Recent work suggests that localized activation of the MEK/ERK signaling cascade regulates polarized SOX18 expression and subsequent development of lymphatic vasculature [17]. Endothelial-specific activation of the MEK/ERK pathway, through an activating mutation in Raf1, caused increased commitment of venous endothelial cells to the lymphatic fate, which led to expansion of lymphatic vessels and decreased size of veins. Strikingly, activated ERK was able to induce PROX1 expression in arterial endothelial cells, suggesting a fundamental role of this signaling cascade in the establishment of the LEC identity program [17].

Another factor necessary for lymphatic cell identity maintenance is COUP-TFII. In blood vessels, COUP-TFII maintains venous identity by suppressing Notch signaling [18]. COUP-TFII is critical for the early stages of lymphatic identity establishment and maintenance in LEC progenitors during migration away from the vein by promoting expression of PROX1, VEGFR-3, NRP2 and LYVE-1 [9, 19, 20]. However, COUP-TFII is not necessary to maintain PROX1 after initial LEC identity establishment as PROX1 expression is unchanged after inducible deletion of COUP-TFII at E13.5 [19]. The importance of COUP-TFII seems to be limited to growing lymphatics as COUP-TFII deletion in adults resulted in no detectable phenotype [19]. Nevertheless, COUP-TFII deletion suppressed tumor lymphangiogenesis and *in vitro* LEC sprouting, and these effects seem to be mediated by a direct effect of COUP-TFII on *NRP2* expression, a modulator of VEGFR-3 receptor tyrosine kinase (RTK) activity [19, 21].

Notch signaling plays a major role in arteriovenous differentiation (reviewed by Swift and Weinstein [22]), and was also recently shown to play a role in lymphatic endothelial differentiation [23]. Lymphatic-specific Notch1 deletion from E9.75 to E13.5 resulted in an increase of PROX1-positive cells, enlarged lymph sacs and dermal lymphatics, and LEC migration into veins [23]. Consistent with Notch as a negative regulator of LEC identity, LEC-specific Notch activation, by inducible expression of the Notch1 intracellular domain, during the same timeframe, caused an opposite phenotype: decreased expression of PROX1, podoplanin, VEGFR-2, VEGFR-3, LYVE-1, and hypoplastic, abnormal, blood-filled lymphatics near the cardinal vein [23]. Furthermore, LEC-specific Notch1 activation also decreased COUP-TFII expression, suggesting that Notch repression is necessary for COUP-TFII-mediated PROX1 activation. However, the effect of Notch signaling may not be so simple as endothelial cell-specific deletion of RBPJ, the main mediator of Notch target gene expression, could not rescue PROX1 expression in embryos with ablated COUP-TFII [20]. Whatever the exact mechanism, Notch plays an important role in constraining lymphatic cell differentiation during embryogenesis.

5.3.3 Lymphangiogenic Sprouting and Migration

Once LECs have differentiated in cardinal veins they migrate and proliferate to form the lymphatic vascular network (Fig. 5.2c). Vascular endothelial growth factors (VEGFs) are the primary stimuli driving developmental angiogenesis and lymphangiogenesis by binding VEGF receptors, which in turn activate intracellular signaling cascades promoting migration, proliferation, and survival (reviewed by Jeltsch et al. [24]). VEGF-A and VEGF-C are the main angiogenic and lymphangiogenic growth factors, respectively, and both are ligands for VEGFR-2 and VEGFR-3 (reviewed by Adams and Alitalo [25]). LECs express high levels of VEGFR-3 [26]. In addition, sprouting blood vessels, as well as some tumor vessels and fenestrated endothelial cells, also express low levels of VEGFR-3 [27–29]. VEGF-C expression is first detected at E12.5 near the cardinal vein where the first LECs are differentiating, corresponding with high levels of LEC VEGFR-3 expression [30]. *VEGF-C*-null embryos, which die before birth, have severe edema due to the lack of lymphatic vessels. Indeed, in the absence of VEGF-C, PROX1 is expressed in LEC progenitor cells; however, these cells do not migrate from veins to form 'lymphatic sacs', suggesting VEGF-C is one of the main stimuli promoting LEC migration [8, 31]. The importance of VEGF-C is highlighted by the fact that even *VEGF-C*$^{+/-}$ mice exhibit chylous ascites and defects in postnatal lymphangiogenesis with hypoplastic dermal lymphatics [31]. Furthermore, *VEGF-C* overexpression in skin results in hyperplastic and proliferating lymphatic vessels [32]. Mutations in *VEGFR-3* cause congenital lymphedema, and similar mutations in mice lead to severe lymphatic vascular hypoplasia [33]. In addition, expression of soluble VEGFR-3, which prevents VEGF-C-mediated activation of VEGFR-3, in the skin, caused complete regression of the dermal lymphatic network in mouse pups without affecting blood vessels [34]. Thus, the VEGF-C/VEGFR-3 signaling axis is likely the most important mechanism regulating lymphatic endothelial migration and survival.

5.3.4 Extracellular Signaling Pathways in Lymphangiogenesis

One accessory protein that promotes VEGFR-3 signaling in LECs is NRP2. NRP1 and 2 were first described as axon guidance factors belonging to the class-3 semaphorin subfamily, but were later shown to play an important role in the vasculature. NRP1 is a co-receptor of VEGFR-2 and is essential for embryonic angiogenesis and vascular maturation (reviewed by Koch and Claesson-Welsh [35]). NRP2 binds VEGF-C and VEGFR-3 and, after ligand binding, both NRP2 and VEGFR-3 are internalized in LECs [36]. NRP2 is highly expressed in actively growing lymphatic vessels, where it controls VEGF-C/VEGFR-3-dependent lymphatic vessel sprouting [21, 37, 38].

Another factor associated with VEGF receptors is claudin-like protein 24 (CLP24), a member of the claudin protein family of intercellular junction proteins

[39]. Frogs or fish with *CLP24* ablation failed to develop a lymphatic network, whereas *CLP24*-deficient mice display a milder phenotype with dilated lymphatic capillaries and ectopic smooth muscle cell coverage [40]. CLP24 associates with VEGFR-2 and VEGFR-3 and restricts activation of Ca^{2+}/cAMP/CREB, while other downstream signaling pathways, such as p38 mitogen-activated protein kinase (MAPK), are not affected [40].

Collagen- and calcium-binding EGF domains 1 (CCBE1) protein is an extracellular matrix (ECM) component critical for lymphangiogenesis in zebrafish, mice, and humans [41–44]. Hennekam syndrome is characterized by general lymphatic dysplasia with patients having limb lymphedema and lymphangiectasia of the lung, intestine, pericardium, thyroid, and kidney [41]. Some Hennekam syndrome patients have inherited point mutations in the calcium-binding EGF domain of *CCBE1* [41]. *CCBE1*-deficient mouse embryos have severe edema and die before birth because of a complete lack of lymphatic vessels [42]. Similar to the phenotypes of *VEGF-C*- and *PROX1*-deficient mice, the PROX1-positive LEC precursors cells are unable to bud from the cardinal vein, indicating that LEC migration cues are disturbed in these mice [42].

An elegant study recently shed light on how mechanosensory input influences developmental lymphangiogenesis [45]. LEC stretching was found to increase VEGF-C- and integrin-β1-dependent VEGFR-3 phosphorylation and LEC proliferation [45]. Therefore, interstitial fluid pressure, which increases at the onset of lymphatic vasculature expansion at E10.5–E12.5 and causes stretching of LECs, is one of the important mechanical factors that contributes and cooperates with tissue gradients of VEGF-C to induce growth of lymphatic vessels [45].

The angiopoietins (Ang1 and Ang2) are ligands for Tie receptors (Tie1 and Tie2), another class of RTKs. Ang1 and Ang2 are important in blood vessel maturation and sprouting, respectively (for review, see Augustin et al. [46]). $Ang2^{-/-}$ mice have chylous ascites, hypoplastic lymphatic capillaries, and ectopic smooth muscle cell coverage of lymphatic capillaries, suggesting that Ang2 plays an important role during developmental angiogenesis [47]. However, the lymphatic phenotype observed in $Ang2^{-/-}$ mice could be rescued by inserting *Ang1* into the *Ang2* locus, suggesting that the differential roles of Ang1 and Ang2 in lymphangiogenesis need to be resolved [46]. Tie1 is necessary for functional lymphatic vessels. Hypomorphic mutation or induced deletion of *Tie1* results in embryos with edema. This edema is likely caused by a loss of lymphatic vessel caliber regulation, suggesting a role for Tie1 in modulating pro-lymphangiogenic signaling [48, 49].

The Eph receptors and their ligands, ephrins, are a class of RTK. Ephrin ligands bind and stimulate phosphorylation of the cytoplasmic end of Eph receptors in 'forward' signaling. Eph/ephrin interactions can also result in 'reverse' signaling whereby the Eph receptors act as ligands and stimulate cytoplasmic phosphorylation of ephrin ligands (reviewed by Kullander and Klein [50]). EphB4 and ephrinB2 are expressed in lymphatic vessels [51]. The cytoplasmic domain of ephrinB2 is important for lymphangiogenic sprouting as sprouting defects were observed in the postnatal dermis in mutant animals [51]. The mechanism of reduced sprouting in eprhinB2 mutants may be through inhibition of VEGFR-3 internalization after

VEGF-C binding [52]. Therefore, EphB4/ephrinB2 signaling plays an important part in mediating the effects of the VEGF-C/VEGFR-3 signaling axis.

Notch signaling plays a fundamental role in the regulation of sprouting angiogenesis. The current view is that angiogenic tip cells induce canonical Notch signaling in stalk cells, which represses VEGF receptor signaling and thereby reduces sprouting potential [53, 54]. Such maintenance of differential sprouting potential is essential for efficient angiogenesis [55]. The role of Notch signaling in lymphatic vascular sprouting and remodeling has recently been investigated. Systematic analysis of Notch signaling components in zebrafish revealed that Dll4 and Notch1b are necessary for thoracic duct formation and LEC sprouting and migration [56]. In mice, the role of Notch signaling in sprouting and migration is less well-defined. A study by Zheng et al., using Dll4-Fc, showed that Notch blockade increased LEC spheroid sprouting *in vitro*, especially in response to VEGF-A [57]. Co-injection of Dll4-Fc and VEGF-A-expressing adenovirus caused an increase in dermal lymphatic vessels in adult mice compared with VEGF-A alone, suggesting that Notch signaling in lymphatic vessels restricts VEGFR-2-dependent signaling in LECs [57]. In contrast, a study by Niessen et al. found that postnatal Notch inhibition, using Notch1 and Dll4 blocking antibodies, decreased lymphatic vessel density and VEGFR-3 surface staining in dermal lymphatic vessels [58]. Therefore, the role of Notch signaling in lymphatic sprouting, and possible differences between adult and postnatal lymphangiogenesis, remains an open question.

ALK1 and ALK5 are endothelial type I receptors of the transforming growth factor (TGF)-β superfamily of ligands [59]. ALK1-Fc treatment in pups induced chylous ascites as well as a massive increase in retinal blood vessel density [60]. Dermal and intestinal lymphatic capillaries also fail to fully develop after ALK1 blocking; however, collecting vessels are unaffected [60]. Interestingly, genetic inactivation of one of the high-affinity ALK1 ligands, BMP9, has no effect on lymphatic capillary sprouting, while the formation of collecting vessels is compromised ([61]; see below). This suggests that other ALK1 ligands or ligand-independent signaling should be considered for explaining the effects of ALK1-Fc. Endothelial- or LEC-specific inactivation of TGFβr1, or its co-receptor TGFβr2, during embryonic development prevents LEC sprouting and leads to enlarged, hyperproliferative dermal lymphatic vessels [62]. Thus, signaling via ALK1, TGFβr1, and TGFβr2 is important to ensure the balance of sprouting versus proliferating LECs during lymphangiogenesis, and is necessary for the correct patterning of the vascular network [60, 62].

5.3.5 Intracellular Signaling Pathways in Lymphangiogenesis

Survival and proliferation of LECs *in vitro* is mediated by the PI3K/AKT and MEK/ERK signaling pathways [63]. Indirect evidence suggesting a role for the PI3K/AKT pathway was the observation that mice with germline deletions of the regulatory subunits of PI3K develop chylous ascites [64]. Intestinal submucosal and diaphragm lymphatic capillary density was decreased in the mice with germline deletion of PI3K regulatory subunits, suggesting that PI3K plays a role in LEC survival postnatally [65]. Furthermore, PI3K catalytic subunit interaction with Ras is necessary for LEC survival as mice with point mutations disturbing this interaction have chylous ascites and defects in lymphatic vasculature development [66].

AKT-deficient mice provided further evidence for the role of PI3K signaling in lymphangiogenesis. There are three isoforms of AKT (AKT1, 2, and 3), and AKT1 is the predominant form in BECs [67]. Dermal lymphatic capillaries in *AKT1*, but not *AKT2* or *AKT3*, germline knockout mice were hypoplastic due to a decreased number of LECs [68]. In addition, AKT1 was also important for the formation of precollector valves; however, valves in larger collecting vessels were unaffected. Nevertheless, AKT1 inactivation did not affect VEGF-C-induced LEC sprouting *in vivo*, suggesting that it may be more important for cell survival [68]. Interestingly, the development of blood vessels is normal in the absence of AKT1 [69], which suggests differential usage of this signaling pathway between blood and lymphatic vessels.

A modulator of the Ras/MEK/ERK pathway, RASA1, was also found to be involved in lymphatic function. *RASA1*-null mice have numerous blood vessel defects, including decreases in branching and stability [70]. Constitutive or lymphatic-specific deletion of *RASA1*-induced chylothorax and chylous ascites in both weaning pups and adults, presumably due to loss of VE-cadherin/LYVE1 button-like junctions [71]. Lymphatic-specific deletion of *RASA1* also caused increased LEC proliferation, coincident with an increase in Ras/ERK signaling in *RASA1*-deficient lymphatic vessels *in vivo*. Aberrant ERK signaling is activated via the VEGF-C/VEGFR-3 pathway, as *RASA1*-deficient mice treated with VEGFR-3 blocking antibodies did not have a lymphatic phenotype. Therefore, RASA1 provides a link between pro-lymphangiogenic growth factors and mechanisms of VEGFR-3-mediated transduction of these signals through Ras [71].

RAC1, a small GTPase in the Rho family, has been shown to play a role in endothelial cell migration *in vitro* [72]. Surprisingly, endothelial-specific deletion of RAC1 *in vivo* is not sufficient to impair sprouting angiogenesis, but is required for the migration of committed LECs from the cardinal vein [73]. *RAC1* inactivation reduces LEC migration in response to VEGF-C, which ultimately leads to close association of early lymphatic vessels ('lymph sacs') and the cardinal vein. *RAC1*-deficient embryos also display blood-filled lymphatic vessels [73]; however, complete interpretation of this phenotype needs to be evaluated in the context of the roles of lymphovenous valve status and/or platelet function in lymphatic-venous separation (see below).

5.3.6 Platelets and Lymphatics: More Complex than Initially Thought

The mature lymphatic vasculature is connected to blood circulation at only few specific regions, such as the junction between the thoracic duct and subclavian vein. The reflux of blood to the thoracic duct is prevented by a lymphovenous valve that forms early during embryonic development [74]. A number of genetic model phenotypes include blood-filled lymphatic vessels, indicative of additional patent connections between the blood and lymphatic vasculatures. Importantly, genes deleted in models that display the blood-filled lymphatic phenotype encode components of signaling pathways regulating platelet development or aggregation, such as MEIS1 [75], podoplanin (or the enzyme important for its post-translational modification) [76, 77], PLCγ2 [78], CLEC-2, SLP76, and Syk [79, 80]. Podoplanin, a cell surface glycoprotein, is highly expressed in LECs. Binding of podoplanin to CLEC-2 on platelets induces activation of the Syk/SLP76/PLCγ2 signaling cascade and platelet aggregation (reviewed by D'Amico and Alitalo [81]). Until recently, it was thought that platelet thrombi, induced by podoplanin on the surface of the nascent lumenized lymphatic vessel, 'seal off' new lymphatic vasculature from the cardinal vein [77, 80], in a mechanism somewhat analogous to closure of the ductus arteriosus [82]. However, more recent evidence shows that LECs migrate from veins as nonlumenized streams of cells, without disturbing venous wall integrity [8, 10]. Therefore, the role of platelet aggregation in the formation of abnormal connections between blood and lymphatic vessels is probably more complex, and may, at least in part, include malformations or dysfunction of the lymphovenous valve and cytokine-induced fusion of the lymphatic and blood vasculatures (Fig. 5.2d) [83].

5.4 Lymphatic Vessel Maturation

5.4.1 Maturation of Lymphatic Capillaries

Fully mature lymphatic capillaries contain LECs with discontinuous highly specialized 'button-like' junctions, allowing free passage of interstitial components and immune cells into the vessel lumen while maintaining vessel integrity [5]. In contrast, LECs in actively sprouting lymphatic vessels are connected by continuous adherens junctions (Fig. 5.3a). The transition from 'zipper-like' to 'button-like' junction phenotype begins at E17.5 and is completed by P28 [84]. The mature state of lymphatic intercellular junctions is lost during the lymphangiogenic response in inflammation but can be restored by dexamethasone treatment, which appears to act through the direct regulation of glucocorticoid receptors in LECs [84] (Fig. 5.3a). Lymphatic endothelial-derived sphingosine-1-phosphate is another factor contributing to the maintenance of mature junctions, although this effect may be, in part, indirect due to increased lymphatic endothelial sprouting [85]. In contrast, inhibition of lymphangiogenesis using VEGFR-3-blocking antibody did not increase junction maturation [84], suggesting that VEGF-C/VEGFR-3 signaling and formation of specialized lymphatic junctions are not directly coupled. In summary, intercellular junctions of capillary LECs are highly dynamic structures, which are regulated by a variety of pathological and physiological stimuli; this regulation may have important consequences for lymphatic vessel functions.

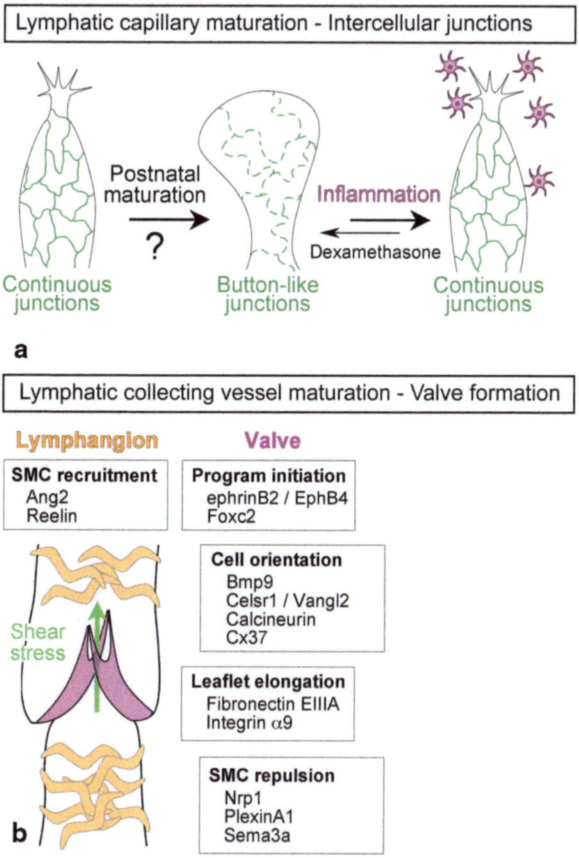

Fig. 5.3 Lymphatic vessel maturation. (**a**) Lymphatic endothelial cells in growing capillaries are connected by continuous adherens junctions, which are transformed into discontinuous 'button-like' junctions in mature lymphatic capillaries [84]. 'Button-like' junctions can revert to the 'zipper-like' state during inflammation. Dexamethasone promotes maturation of lymphatic endothelial cell (LEC) junctions. (**b**) Collecting lymphatic vessels develop from the capillary-like primary lymphatic plexus through the acquisition of lymphatic valves, deposition of basement membrane, and recruitment of smooth muscle cells. Lymphatic valve development is controlled by FOXC2/NFATC1/shear stress/connexin37 signaling, planar cell polarity proteins CELSR1 and VANGL2 regulate rotation and collective migration of lymphatic-valve-forming cells, and integrin-α9/fibronectin EIIIA promote extracellular matrix (*ECM*) fibril assembly and leaflet elongation. Ang2 and BMP9 are two growth factors important for formation of lymphatic valves. Recruitment of SMCs to collecting vessels is regulated by reelin and Ang2, whereas SEMA3A/NRP1/Plexin A1 signaling is important for keeping the valve area free of smooth muscle cells

5.4.2 Formation of Collecting Lymphatic Vessels

Lymphatic collecting vessel formation involves development of lymphatic valves, increased deposition of basement membrane components, and recruitment of smooth muscle cells. In the mouse embryo, mesenteric collecting lymphatic vessels begin to form around E15.5, and this process is accompanied by the downregulation of lymphatic capillary markers LYVE-1, VEGFR-3, and CCL21 [86]. Sites of forming lymphatic valves are marked by increased expression of PROX1 and the forkhead transcription factor FOXC2. Once specified, lymphatic valve-forming cells rotate and invaginate into the vessel lumen and further elongate to form bicuspid leaflets containing specialized ECM, covered on each side by endothelial cells [86, 87]. Lymphatic valve endothelial cells express high levels of transcription factors PROX1, FOXC2, and GATA2, adhesion receptor integrin-α9, the glycocalyx component podocalyxin, and ECM components laminin-α5 and fibronectin EIIIA [86–89].

Formation of lymphatic valves is a complex morphogenetic process that requires cooperation of multiple signaling pathways (Fig. 5.3b). Inactivation of FOXC2 completely prevents formation of lymphatic valves and further maturation of vessels [86, 90]. Heterozygous loss-of-function mutations of *FOXC2* are found in a human hereditary disease, lymphedema-distichiasis, characterized by hyperplastic initial lymphatic vessels and abnormal lymphatic drainage, likely due to absent or dysfunctional lymphatic valves [91–93]. Genome-wide analysis of FOXC2 DNA-binding sequences in LECs suggested cooperation between the FOXC2 and calcineurin/nuclear factor of activated T cells (NFAT) signaling pathways [86]. Indeed, activated nuclear NFATc1 is present in lymphatic valve-forming cells, and lymphatic endothelial-specific loss of the Ser/Thr phosphatase calcineurin, which controls NFATc1 activation, arrests the development of lymphatic valves [86, 89]. A number of gap-junction proteins are also implicated in the maturation of lymphatic vessels. Lymphatic valve development defects are documented in *Cx37*- and *Cx43*-deficient mice, as well as in mice double heterozygous for *Cx43* and *Cx37* [89, 94]. Mechanistically, loss of Cx37 does not affect lymphatic valve cell identity establishment but prevents formation of a defined lymphatic valve territory, and the number of valves is severely reduced in $Cx37^{-/-}$ animals [89, 94]. In humans, mutations in *GJC2* (connexin47) and *GJA1* (connexin43) are associated with lymphedema [95–97], further highlighting the important role of connexins in lymphatic vascular biology.

Lymphatic valve agenesis is observed in mice bearing a mutation in the PDZ-binding domain of ephrinB2 [51]. Thus, in addition to its role in the regulation of VEGFR-3 endocytosis and lymphatic vessel sprouting [52], ephrinB2 reverse signaling is important for lymphatic vascular remodeling. More recently, studies of adult corneal lymphatic vessels showed that EphB4, an ephrinB2 receptor, is highly expressed in the valves, and administration of EphB4-Fc fusion proteins prevented regeneration of valves after corneal injury [98].

Reorientation of lymphatic valve-forming cells requires planar cell polarity proteins CELSR1 and VANGL2 [99]. PCP proteins delay the recruitment of VE-cadherin and adherens junction stabilization to allow efficient collective migration and change of orientation of valve-forming cells, which is requisite for leaflet formation [99]. Further leaflet formation and elongation is controlled by the adhesion receptor integrin-α9, which forms a complex with integrin-β1 and regulates the assembly of the valve ECM component fibronectin EIIIA [87].

Inactivation of BMP9, a member of the TGFβ/BMP growth factor family and a ligand of ALK1, leads to a decreased number of mature valves and abnormal lymph flow [61]. Treatment of LECs *in vitro* with BMP9 induces the expression of *CX37, FOXC2, integrin-α9,* and *EPHRINB2* in an ALK1-dependent manner [61], suggesting that BMP9/ALK1 signaling contributes to many aspects of collecting vessel development.

Lymphatic valves are frequently formed at the sites of vessel branching, suggesting that disturbed lymph flow contributes to valve development [89]. Indeed, *in vitro* experiments demonstrated that both calcineurin activation and *CX37* expression are induced by shear stress, and this induction requires PROX1 and FOXC2. Moreover, Cx37 mediates uniform activation of Ca^{2+}/calcineurin/NFATc1 in response to shear stress. Thus, PROX1, FOXC2, and shear stress act upstream of Cx37 and calcineurin signaling to induce the lymphatic valve phenotype [89]. While the mechanisms of flow sensing remain elusive, intriguingly Clsr1-rich protrusions in lymphatic valve-forming cells are oriented away from lymph flow, suggesting that they may act as flow sensors during lymphatic valve development [99]. Both connexins and Ca^{2+}/calcineurin signaling are important for collective cell migration and the PCP pathway in other cellular systems; therefore, it may be interesting to explore a potential connection between CELSR1/VANGL2 and shear stress/connexin37/calcineurin signaling in lymphatic valve-forming cells.

To date, only a limited number of studies addressed the mechanism of interaction and recruitment of SMCs to collecting lymphatic vessels. Although commonly considered to be a subtype of vascular smooth muscle cells, lymphatic smooth muscle cells express contractile proteins from both smooth muscle and cardiac muscle, and they display functional properties consistent with this unusual composition [100–102]. Currently, the molecular mechanisms of such lymphatic smooth muscle specialization are not understood but it may be necessary to generate sufficient force for lymph propulsion (reviewed by von der Weid and Zawieja [103]).

Recruitment of smooth muscle cells to collecting lymphatic vessels is a relatively late event, occurring after the formation of lymphatic valves and initial deposition of the ECM component collagen IV [86, 104] (Fig. 5.3b). Lymphatic smooth muscle cells express α-smooth muscle actin and desmin, but not blood vascular pericyte marker NG2 [104]. Decreased smooth muscle coverage is observed in collecting lymphatic vessels o*f Ang2$^{-/-}$* mice [47], although the mechanism is not well understood. Reelin is a large extracellular protein, mostly studied for its role in nervous system development. In the vasculature, reelin is highly expressed in lymphatic vessels [104, 105], with intracellular localization in capillary LECs and extracellular localization in collecting lymphatic vessels [104]. Reelin-deficient mice have abnormal dermal collecting lymphatic vessels, with decreased smooth muscle cell coverage, ectopic expression of the capillary marker LYVE-1, and defective lymph flow. *In vitro,* contact of SMCs with LECs elicits secretion and processing of reelin, which in turn induces expression of the SMC chemotactic factor monocyte chemoattractant protein MCP-1 in LECs. Thus, reelin acts as an LEC-specific ECM component, important for the communication between SMCs and LECs [104]. Given that large collecting vessels are not affected by reelin deficiency, other mechanisms for the recruitment and stabilization of lymphatic SMCs should be further investigated.

Smooth muscle coverage of collecting vessels is discontinuous, with low or absent SMCs in the area of lymphatic valves. Such arrangement is important for lymphatic valve function since the valve region undergoes significant variations in diameter during valve opening and closing. Blockade or genetic inactivation of axon guidance molecule SEMA3A, or its receptors NRP1 and PlexinA1, leads to the uniform coverage of collecting vessels with SMCs, shortening of lymphatic valve leaflets, and abnormal lymph flow [106, 107]. Enhanced smooth muscle cell recruitment and ectopic coverage of the valve area with SMCs is also found in *CLP24*$^{-/-}$ mice. Whether this is linked to modulation of VEGFR-2 or VEGFR-3 signaling by CLP24 remains to be studied [40].

5.5 Conclusions

Over the past 25 years, understanding of lymphatic biology has increased immensely and many of the pathways controlling lymphatic vessels have been elucidated. However, organ-specific features of lymphatic vessels and roles of additional cell types, e.g. immune or smooth muscle cells, in the regulation of lymphatic vascular development and function should be further evaluated under physiological and pathological conditions. Furthermore, the role of intracellular signaling pathways in lymphatic growth and maturation is only beginning to be addressed. There are an increasing number of tools available for the study of the lymphatic system. Inducible lymphatic endothelial-specific gene deletion, reporter models, and advances in imaging technology should allow increased understanding of molecular mechanisms controlling lymphatic biology. Application of the knowledge obtained in preclinical mouse models should be applied to improving human health.

References

1. Alitalo K, Tammela T, Petrova TV. Lymphangiogenesis in development and human disease. Nature. 2005;438(7070):946–53.
2. Stacker SA, Achen MG, Jussila L, Baldwin ME, Alitalo K. Lymphangiogenesis and cancer metastasis. Nat Rev Cancer. 2002;2(8):573–83.
3. Greene AK, Grant FD, Slavin SA. Lower-extremity lymphedema and elevated body-mass index. N Engl J Med. 2012;366(22):2136–7.
4. Kim H, Kataru RP, Koh GY. Regulation and implications of inflammatory lymphangiogenesis. Trends Immunol. 2012;33(7):350–6.
5. Baluk P, Fuxe J, Hashizume H, Romano T, Lashnits E, Butz S, et al. Functionally specialized junctions between endothelial cells of lymphatic vessels. J Exp Med. 2007;204(10):2349–62.
6. Schulte-Merker S, Sabine A, Petrova TV. Lymphatic vascular morphogenesis in development, physiology, and disease. J Cell Biol. 2011;193(4):607–18.
7. Sabin FR. On the origin of the lymphatic system from the veins and the development of the lymph hearts and thoracic duct in the pig. Am J Anat. 1902;1:367–91.

8. Hagerling R, Pollmann C, Andreas M, Schmidt C, Nurmi H, Adams RH, et al. A novel multistep mechanism for initial lymphangiogenesis in mouse embryos based on ultramicroscopy. EMBO J. 2013;32(5):629–44.
9. Srinivasan RS, Dillard ME, Lagutin OV, Lin FJ, Tsai S, Tsai MJ, et al. Lineage tracing demonstrates the venous origin of the mammalian lymphatic vasculature. Genes Dev. 2007;21(19):2422–32.
10. Yang Y, Garcia-Verdugo JM, Soriano-Navarro M, Srinivasan RS, Scallan JP, Singh MK, et al. Lymphatic endothelial progenitors bud from the cardinal vein and intersomitic vessels in mammalian embryos. Blood. 2012;120(11):2340–8.
11. Wigle JT, Oliver G. Prox1 function is required for the development of the murine lymphatic system. Cell. 1999;98(6):769–78.
12. Wigle JT, Harvey N, Detmar M, Lagutina I, Grosveld G, Gunn MD, et al. An essential role for Prox1 in the induction of the lymphatic endothelial cell phenotype. EMBO J. 2002;21(7):1505–13.
13. Johnson NC, Dillard ME, Baluk P, McDonald DM, Harvey NL, Frase SL, et al. Lymphatic endothelial cell identity is reversible and its maintenance requires Prox1 activity. Genes Dev. 2008;22(23):3282–91.
14. Petrova TV, Makinen T, Makela TP, Saarela J, Virtanen I, Ferrell RE, et al. Lymphatic endothelial reprogramming of vascular endothelial cells by the Prox-1 homeobox transcription factor. EMBO J. 2002;21(17):4593–9.
15. Francois M, Caprini A, Hosking B, Orsenigo F, Wilhelm D, Browne C, et al. Sox18 induces development of the lymphatic vasculature in mice. Nature. 2008;456(7222):643–69.
16. Irrthum A, Devriendt K, Chitayat D, Matthijs G, Glade C, Steijlen PM, et al. Mutations in the transcription factor gene SOX18 underlie recessive and dominant forms of hypotrichosis-lymphedema-telangiectasia. Am J Hum Genet. 2003;72(6):1470–8.
17. Deng Y, Atri D, Eichmann A, Simons M. Endothelial ERK signaling controls lymphatic fate specification. J Clin Invest. 2013;123(3):1202–15.
18. You LR, Lin FJ, Lee CT, DeMayo FJ, Tsai MJ, Tsai SY. Suppression of Notch signalling by the COUP-TFII transcription factor regulates vein identity. Nature. 2005;435(7038):98–104.
19. Lin FJ, Chen XP, Qin J, Hong YK, Tsai MJ, Tsai SY. Direct transcriptional regulation of neuropilin-2 by COUP-TFII modulates multiple steps in murine lymphatic vessel development. J Clin Invest. 2010;120(5):1694–707.
20. Srinivasan RS, Geng X, Yang Y, Wang YD, Mukatira S, Studer M, et al. The nuclear hormone receptor Coup-TFII is required for the initiation and early maintenance of Prox1 expression in lymphatic endothelial cells. Genes Dev. 2010;24(7):696–707.
21. Xu YL, Yuan L, Mak J, Pardanaud L, Caunt M, Kasman I, et al. Neuropilin-2 mediates VEGF-C-induced lymphatic sprouting together with VEGFR3. J Cell Biol. 2010;188(1):115–30.
22. Swift MR, Weinstein BM. Arterial-venous specification during development. Circ Res. 2009;104(5):576–88.
23. Murtomaki A, Uh MK, Choi YK, Kitajewski C, Borisenko V, Kitajewski J, et al. Notch1 functions as a negative regulator of lymphatic endothelial cell differentiation in the venous endothelium. Development. 2013;140(11):2365–76.
24. Jeltsch M, Leppanen V-M, Saharinen P, Alitalo K. Receptor tyrosine kinase-mediated angiogenesis. Cold Spring Harb Perspect Biol. 2013;5(9):1–20.
25. Adams RH, Alitalo K. Molecular regulation of angiogenesis and lymphangiogenesis. Nat Rev Mol Cell Biol. 2007;8(6):464–78.
26. Kaipainen A, Korhonen J, Mustonen T, van Hinsbergh VW, Fang GH, Dumont D, et al. Expression of the fms-like tyrosine kinase 4 gene becomes restricted to lymphatic endothelium during development. Proc Natl Acad Sci U S A. 1995;92(8):3566–70.
27. Lymboussaki A, Partanen TA, Olofsson B, Thomas-Crusells J, Fletcher CDM, de Waal RMW, et al. Expression of the vascular endothelial growth factor C receptor VEGFR-3 in lymphatic endothelium of the skin and in vascular tumors. Am J Pathol. 1998;153(2):395–403.
28. Partanen TA, Arola J, Saaristo A, Jussila L, Ora A, Miettinen M, et al. VEGF-C and VEGF-D expression in neuroendocrine cells and their receptor, VEGFR-3, in fenestrated blood vessels in human tissues. FASEB J. 2000;14(13):2087–96.

29. Tammela T, Zarkada G, Wallgard E, Murtomaki A, Suchting S, Wirzenius M, et al. Blocking VEGFR-3 suppresses angiogenic sprouting and vascular network formation. Nature. 2008;454(7204):656–60.
30. Kukk E, Lymboussaki A, Taira S, Kaipainen A, Jeltsch M, Joukov V, et al. VEGF-C receptor binding and pattern of expression with VEGFR-3 suggests a role in lymphatic vascular development. Development. 1996;122(12):3829–37.
31. Karkkainen MJ, Haiko P, Sainio K, Partanen J, Taipale J, Petrova TV, et al. Vascular endothelial growth factor C is required for sprouting of the first lymphatic vessels from embryonic veins. Nat Immunol. 2004;5(1):74–80.
32. Jeltsch M, Kaipainen A, Joukov V, Meng XJ, Lakso M, Rauvala H, et al. Hyperplasia of lymphatic vessels in VEGF-C transgenic mice. Science. 1997;276(5317):1423–5.
33. Karkkainen MJ, Ferrell RE, Lawrence EC, Kimak MA, Levinson KL, McTigue MA, et al. Missense mutations interfere with VEGFR-3 signalling in primary lymphoedema. Nat Genet. 2000;25(2):153–9.
34. Makinen T, Jussila L, Veikkola T, Karpanen T, Kettunen MI, Pulkkanen KJ, et al. Inhibition of lymphangiogenesis with resulting lymphedema in transgenic mice expressing soluble VEGF receptor-3. Nat Med. 2001;7(2):199–205.
35. Koch S, Claesson-Welsh L. Signal transduction by vascular endothelial growth factor receptors. Cold Spring Harb Perspect Med. 2012;2(7):a006502.
36. Karpanen T, Heckman CA, Keskitalo S, Jeltsch M, Ollila H, Neufeld G, et al. Functional interaction of VEGF-C and VEGF-D with neuropilin receptors. FASEB J. 2006;20(9):1462–72.
37. Yuan L, Moyon D, Pardanaud L, Breant C, Karkkainen MJ, Alitalo K, et al. Abnormal lymphatic vessel development in neuropilin 2 mutant mice. Development. 2002;129(20):4797–806.
38. Caunt M, Mak J, Liang WC, Stawicki S, Pan Q, Tong RK, et al. Blocking neuropilin-2 function inhibits tumor cell metastasis. Cancer Cell. 2008;13(4):331–42.
39. Kearsey J, Petit S, De Oliveira C, Schweighoffer F. A novel four transmembrane spanning protein, CLP24. A hypoxically regulated cell junction protein. Eur J Biochem. 2004 Jul;271(13):2584–92.
40. Saharinen P, Helotera H, Miettinen J, Norrmen C, D'Amico G, Jeltsch M, et al. Claudin-like protein 24 interacts with the VEGFR-2 and VEGFR-3 pathways and regulates lymphatic vessel development. Genes Dev. 2010;24(9):875–80.
41. Alders M, Hogan BM, Gjini E, Salehi F, Al-Gazali L, Hennekam EA, et al. Mutations in CCBE1 cause generalized lymph vessel dysplasia in humans. Nat Genet. 2009;41(12):1272–4.
42. Bos FL, Caunt M, Peterson-Maduro J, Planas-Paz L, Kowalski J, Karpanen T, et al. CCBE1 is essential for mammalian lymphatic vascular development and enhances the lymphangiogenic effect of vascular endothelial growth factor-C in vivo. Circ Res. 2011;109(5):486–91.
43. Connell FC, Kalidas K, Ostergaard P, Brice G, Murday V, Mortimer PS, et al. CCBE1 mutations can cause a mild, atypical form of generalized lymphatic dysplasia but are not a common cause of non-immune hydrops fetalis. Clin Genet. 2012;81(2):191–7.
44. Hogan BM, Bos FL, Bussmann J, Witte M, Chi NC, Duckers HJ, et al. CCBE1 is required for embryonic lymphangiogenesis and venous sprouting. Nat Genet. 2009;41(4):396–8.
45. Planas-Paz L, Strilic B, Goedecke A, Breier G, Fassler R, Lammert E. Mechanoinduction of lymph vessel expansion. EMBO J. 2012;31(4):788–804.
46. Augustin HG, Koh GY, Thurston G, Alitalo K. Control of vascular morphogenesis and homeostasis through the angiopoietin-Tie system. Nat Rev Mol Cell Biol. 2009;10(3):165–77.
47. Gale NW, Thurston G, Hackett SF, Renard R, Wang Q, McClain J, et al. Angiopoietin-2 is required for postnatal angiogenesis and lymphatic patterning, and only the latter role is rescued by angiopoietin-1. Dev Cell. 2002;3(3):411–23.
48. D'Amico G, Korhonen EA, Waltari M, Saharinen P, Laakkonen P, Alitalo K. Loss of endothelial Tie1 receptor impairs lymphatic vessel development. Arterioscler Thromb Vasc Biol. 2010;30(2):207–9.
49. Qu XH, Tompkins K, Batts LE, Puri M, Baldwin S. Abnormal embryonic lymphatic vessel development in Tie1 hypomorphic mice. Development. 2010;137(8):1285–95.

50. Kullander K, Klein R. Mechanisms and functions of EPH and ephrin signalling. Nat Rev Mol Cell Biol. 2002;3(7):475–86.
51. Makinen T, Adams RH, Bailey J, Lu Q, Ziemiecki A, Alitalo K, et al. PDZ interaction site in ephrinB2 is required for the remodeling of lymphatic vasculature. Genes Dev. 2005;19(3):397–410.
52. Wang YD, Nakayama M, Pitulescu ME, Schmidt TS, Bochenek ML, Sakakibara A, et al. Ephrin-B2 controls VEGF-induced angiogenesis and lymphangiogenesis. Nature. 2010;465(7297):483–6.
53. Geudens I, Gerhardt H. Coordinating cell behaviour during blood vessel formation. Development. 2011;138(21):4569–83.
54. Benedito R, Rocha SF, Woeste M, Zamykal M, Radtke F, Casanovas O, et al. Notch-dependent VEGFR3 upregulation allows angiogenesis without VEGF-VEGFR2 signalling. Nature. 2012;484(7392):110–4.
55. Jakobsson L, Franco CA, Bentley K, Collins RT, Ponsioen B, Aspalter IM, et al. Endothelial cells dynamically compete for the tip cell position during angiogenic sprouting. Nat Cell Biol. 2010;12(10):943–53.
56. Geudens I, Herpers R, Hermans K, Segura I, Ruiz de Almodovar C, Bussmann J, et al. Role of Delta-like-4/Notch in the formation and wiring of the lymphatic network in zebrafish. Arterioscler Thromb Vasc Biol. 2010;30(9):1695–702.
57. Zheng W, Tammela T, Yamamoto M, Anisimov A, Holopainen T, Kaijalainen S, et al. Notch restricts lymphatic vessel sprouting induced by vascular endothelial growth factor. Blood. 2011;118(4):1154–62.
58. Niessen K, Zhang G, Ridgway JB, Chen H, Kolumam G, Siebel CW, et al. The Notch1-Dll4 signaling pathway regulates mouse postnatal lymphatic development. Blood. 2011;118(7):1989–97.
59. Lebrin F, Deckers M, Bertolino P, ten Dijke P. TGF-beta receptor function in the endothelium. Cardiovasc Res. 2005;65(3):599–608.
60. Niessen K, Zhang G, Ridgway JB, Chen H, Yan MH. ALK1 signaling regulates early postnatal lymphatic vessel development. Blood. 2010;115(8):1654–61.
61. Levet S, Ciais D, Merdzhanova G, Mallet C, Zimmers TA, Lee S-J, et al. Bone morphogenetic protein 9 (BMP9) controls lymphatic vessel maturation and valve formation. Blood. 2013;122(4):598–607.
62. James JM, Nalbandian A, Y-s M. TGF-beta signaling is required for sprouting lymphangiogenesis during lymphatic network development in the skin. Development. 2013;140(18):3903–14.
63. Makinen T, Veikkola T, Mustjoki S, Karpanen T, Catimel B, Nice EC, et al. Isolated lymphatic endothelial cells transduce growth, survival and migratory signals via the VEGF-C/D receptor VEGFR-3. EMBO J. 2001;20(17):4762–73.
64. Fruman DA, Mauvais-Jarvis F, Pollard DA, Yballe CM, Brazil D, Bronson RT, et al. Hypoglycaemia, liver necrosis and perinatal death in mice lacking all isoforms of phosphoinositide 3-kinase p85 alpha. Nat Genet. 2000;26(3):379–82.
65. Mouta-Bellum C, Kirov A, Miceli-Libby L, Mancini ML, Petrova TV, Liaw L, et al. Organ-specific lymphangiectasia, arrested lymphatic sprouting, and maturation defects resulting from gene-targeting of the PI3K regulatory isoforms p85 alpha, p55 alpha, and p50 alpha. Dev Dyn. 2009;238(10):2670–9.
66. Gupta S, Ramjaun AR, Haiko P, Wang YH, Warne PH, Nicke B, et al. Binding of Ras to phosphoinositide 3-kinase p110 alpha is required for Ras-driven tumorigenesis in mice. Cell. 2007;129(5):957–68.
67. Chen JH, Somanath PR, Razorenova O, Chen WS, Hay N, Bornstein P, et al. Akt1 regulates pathological angiogenesis, vascular maturation and permeability in vivo. Nat Med. 2005;11(11):1188–96.
68. Zhou F, Chang Z, Zhang LQ, Hong YK, Shen B, Wang B, et al. Akt/protein kinase B Is required for lymphatic network formation, remodeling, and valve development. Am J Pathol. 2010;177(4):2124–33.

69. Chen WS, Xu P-Z, Gottlob K, Chen M-L, Sokol K, Shiyanova T, et al. Growth retardation and increased apoptosis in mice with homozygous disruption of the akt1 gene. Genes Dev. 2001;15(17):2203–8.
70. Henkemeyer M, Rossi DJ, Holmyard DP, Puri MC, Mbamalu G, Harpal K, et al. Vascular system defects and neuronal apoptosis in mice lacking Ras GTPase-activating protein. Nature. 1995;377(6551):695–701.
71. Lapinski PE, Kwon S, Lubeck BA, Wilkinson JE, Srinivasan RS, Sevick-Muraca E, et al. RASA1 maintains the lymphatic vasculature in a quiescent functional state in mice. J Clin Invest. 2012;122(2):733–47.
72. Garrett TA, van Buula JD, Burridge K. VEGF-induced Rac1 activation in endothelial cells is regulated by the guanine nucleotide exchange factor Vav2. Exp Cell Res. 2007;313(15):3285–97.
73. D'Amico G, Jones DT, Nye E, Sapienza K, Ramjuan AR, Reynolds LE, et al. Regulation of lymphatic-blood vessel separation by endothelial Rac1. Development. 2009;136(23):4043–53.
74. Srinivasan RS, Oliver G. Prox1 dosage controls the number of lymphatic endothelial cell progenitors and the formation of the lymphovenous valves. Genes Dev. 2011;25(20):2187–97.
75. Carramolino L, Fuentes J, Garcia-Andres C, Azcoitia V, Riethmacher D, Torres M. Platelets play an essential role in separating the blood and lymphatic vasculatures during embryonic angiogenesis. Circ Res. 2010;106(7):1197–201.
76. Fu J, Gerhardt H, McDaniel JM, Xia B, Liu X, Ivanciu L, et al. Endothelial cell O-glycan deficiency causes blood/lymphatic misconnections and consequent fatty liver disease in mice. J Clin Invest. 2008;118(11):3725–37.
77. Uhrin P, Zaujec J, Breuss JM, Olcaydu D, Chrenek P, Stockinger H, et al. Novel function for blood platelets and podoplanin in developmental separation of blood and lymphatic circulation. Blood. 2010;115(19):3997–4005.
78. Ichise H, Ichise T, Ohtani O, Yoshida N. Phospholipase gamma2 is necessary for separation of blood and lymphatic vasculature in mice. Development. 2009;136(2):191–5.
79. Abtahian F, Guerriero A, Sebzda E, Lu M-M, Zhou R, Mocsai A, et al. Regulation of blood and lymphatic vascular separation by signaling proteins SLP-76 and Syk. Science. 2003;299(5604):247–51.
80. Bertozzi CC, Schmaier AA, Mericko P, Hess PR, Zou Z, Chen M, et al. Platelets regulate lymphatic vascular development through CLEC-2-SLP-76 signaling. Blood. 2010;116(4):661–70.
81. D'Amico G, Alitalo K. Inside bloody lymphatics. Blood. 2010;116(4):512–3.
82. Hamrick SEG, Hansmann G. Patent ductus arteriosus of the preterm infant. Pediatrics. 2010;125(5):1020–30.
83. Bohmer R, Neuhaus B, Buhren S, Zhang D, Stehling M, Bock B, et al. Regulation of developmental lymphangiogenesis by Syk+ leukocytes. Dev Cell. 2010;18(3):437–49.
84. Yao LC, Baluk P, Srinivasan RS, Oliver G, McDonald DM. Plasticity of button-like junctions in the endothelium of airway lymphatics in development and inflammation. Am J Pathol. 2012;180(6):2561–75.
85. Pham THM, Baluk P, Xu Y, Grigorova I, Bankovich AJ, Pappu R, et al. Lymphatic endothelial cell sphingosine kinase activity is required for lymphocyte egress and lymphatic patterning. J Exp Med. 2010;207(1):17–27.
86. Norrmen C, Ivanov KI, Cheng JP, Zangger N, Delorenzi M, Jaquet M, et al. FOXC2 controls formation and maturation of lymphatic collecting vessels through cooperation with NFATc1. J Cell Biol. 2009;185(3):439–57.
87. Bazigou E, Xie S, Chen C, Weston A, Miura N, Sorokin L, et al. Integrin-alpha 9 Is required for fibronectin matrix assembly during lymphatic valve morphogenesis. Dev Cell. 2009;17(2):175–86.
88. Kazenwadel J, Secker GA, Liu YJJ, Rosenfeld JA, Wildin RS, Cuellar-Rodriguez J, et al. Loss-of-function germline GATA2 mutations in patients with MDS/AML or MonoMAC syndrome and primary lymphedema reveal a key role for GATA2 in the lymphatic vasculature. Blood. 2012;119(5):1283–91.

89. Sabine A, Agalarov Y, Maby-El Hajjami H, Jaquet M, Hagerling R, Pollmann C, et al. Mechanotransduction, PROX1, and FOXC2 cooperate to control Connexin37 and Calcineurin during lymphatic-valve formation. Dev Cell. 2012;22(2):430–45.
90. Petrova TV, Karpanen T, Norrmen C, Mellor R, Tamakoshi T, Finegold D, et al. Defective valves and abnormal mural cell recruitment underlie lymphatic vascular failure in lymphedema distichiasis. Nat Med. 2004;10(9):974–81.
91. Fang JM, Dagenais SL, Erickson RP, Arlt MF, Glynn MW, Gorski JL, et al. Mutations in FOXC2 (MFH-1), a forkhead family transcription factor, are responsible for the hereditary lymphedema-distichiasis syndrome. Am J Hum Genet. 2000;67(6):1382–8.
92. Mellor RH, Brice G, Stanton AWB, French J, Smith A, Jeffery S, et al. Mutations in FOXC2 are strongly associated with primary valve failure in veins of the lower limb. Circulation. 2007;115(14):1912–20.
93. Sholto-Douglas-Vernon C, Bell R, Brice G, Mansour S, Sarfarazi M, Child AH, et al. Lymphoedema-distichiasis and FOXC2: unreported mutations, de novo mutation estimate, families without coding mutations. Hum Genet. 2005;117(2–3):238–42.
94. Kanady JD, Dellinger MT, Munger SJ, Witte MH, Simon AM. Connexin37 and Connexin43 deficiencies in mice disrupt lymphatic valve development and result in lymphatic disorders including lymphedema and chylothorax. Dev Biol. 2011;354(2):253–66.
95. Brice G, Ostergaard P, Jeffery S, Gordon K, Mortimer PS, Mansour S. A novel mutation in GJA1 causing oculodentodigital syndrome and primary lymphoedema in a three generation family. Clin Genet. 2013;84(4):378–81.
96. Ferrell RE, Baty CJ, Kimak MA, Karlsson JM, Lawrence EC, Franke-Snyder M, et al. GJC2 missense mutations cause human lymphedema. Am J Hum Genet. 2010;86(6):943–8.
97. Finegold DN, Baty CJ, Knickelbein KZ, Perschke S, Noon SE, Campbell D, et al. Connexin 47 mutations increase risk for secondary lymphedema following breast cancer treatment. Clin Cancer Res. 2012;18(8):2382–90.
98. Katsuta H, Fukushima Y, Maruyama K, Hirashima M, Nishida K, Nishikawa SI, et al. EphrinB2-EphB4 signals regulate formation and maintenance of funnel-shaped valves in corneal lymphatic capillaries. Invest Ophthalmol Vis Sci. 2013;54(6):4102–8.
99. Tatin F, Taddei A, Weston A, Fuchs E, Devenport D, Tissir F, et al. Planar cell polarity protein Celsr1 regulates endothelial adherens junctions and directed cell rearrangements during valve morphogenesis. Dev Cell. 2013;26(1):31–44.
100. Muthuchamy M, Gashev A, Boswell N, Dawson N, Zawieja D. Molecular and functional analyses of the contractile apparatus in lymphatic muscle. FASEB J. 2003;17(3):920–2.
101. Zhang RZ, Gashev AA, Zawieja DC, Davis MJ. Mechanical properties of rat mesenteric lymphatic and arterial smooth muscle. FASEB J. 2005;19(4):A165.
102. Zhang R, Taucer AI, Gashev AA, Muthuchamy M, Zawieja DC, Davis MJ. Maximum shortening velocity of lymphatic muscle approaches that of striated muscle. Am J Physiol Heart Circ Physiol. 2013;305:1494–507.
103. von der Weid P-Y, Zawieja DC. Lymphatic smooth muscle: the motor unit of lymph drainage. Int J Biochem Cell Biol. 2004;36(7):1147–53.
104. Lutter S, Xie S, Tatin F, Makinen T. Smooth muscle-endothelial cell communication activates Reelin signaling and regulates lymphatic vessel formation. J Cell Biol. 2012;197(6):837–49.
105. Samama B, Boehm N. Reelin immunoreactivity in lymphatics and liver during development and adult life. Anat Rec A Discov Mol Cell Evol Biol. 2005;285A(1):595–9.
106. Bouvree K, Brunet I, del Toro R, Gordon E, Prahst C, Cristofaro B, et al. Semaphorin3A, Neuropilin-1, and PlexinA1 are required for lymphatic valve formation. Circ Res. 2012;111(4):437–45.
107. Jurisic G, Maby-El Hajjami H, Karaman S, Ochsenbein AM, Alitalo A, Siddiqui SS, et al. An unexpected role of Semaphorin3A-Neuropilin-1 signaling in lymphatic vessel maturation and valve formation. Circ Res. 2012;111(4):426–36.

Part II
Homeostasis of the Vascular System

Chapter 6
Junctional Signaling in Endothelial Cells

Luca Bravi and Maria Grazia Lampugnani

6.1 The Architecture of Endothelial Cell-to-Cell Junctions: Focus on Adherens Junctions

Cell-to-cell junctions represent highly specialized domains of the plasma membrane that control fundamental activities of the endothelium, such as barrier properties, the growth-arrested phenotype predominant in adult vessels, and response to proliferation and differentiation factors during physiological angiogenesis. Consistent with their critical functional role, cell-to-cell junctions are often altered in vascular diseases and in general pathologies that involve vascular dysfunctions [22].

To sustain such diverse and crucial tasks, the molecular architecture of endothelial junctions is finely specialized (see Dejana et al. [22] for an extensive description). In this chapter, we will particularly focus on a specific subdomain of endothelial cell-to-cell junctions, the adherens junctions. Our laboratory has contributed to this field since the early 1990s with the identification and initial characterization of vascular endothelial (VE)-cadherin, the endothelial-selective transmembrane constituent of endothelial adherens junctions [47].

VE-cadherin does not act alone in adherens junctions; indeed, several different molecules have been reported to interact with VE-cadherin at the level of both extracellular and cytoplasmic domains (for an introductory list, see Lampugnani [46]). Considering the plethora of molecules that have been found to form complexes with VE-cadherin, it appears most likely that several distinct types of molecular clusters

M. G. Lampugnani (✉) · L. Bravi
FIRC Institute for Molecular Oncology Fondazione, Via Adamello 16, 20139 Milan, Italy
e-mail: mariagrazia.lampugnani@ifom.eu

M. G. Lampugnani
Mario Negri Institute for Pharmacological Research, Via La Masa 19, 20156 Milan, Italy

L. Bravi
e-mail: luca.bravi@ifom.eu

M. HH. Schmidt, S. Liebner (eds.), *Endothelial Signaling in Development and Disease,* DOI 10.1007/978-1-4939-2907-8_6

can organize around VE-cadherin and define specialized microdomains at cell-to-cell junctions to support specific cellular functions.

In addition, while endothelial adherens junctions exist in the organisms [5, 75], the definition of their molecular specialization in different organs and tissues is still initial (see below [75]).

Undoubtedly, junctional signaling acts locally to regulate the tightness of cell contacts and the permeability of the monolayer, as illustrated in the examples below. However, the consequences of such signaling extend beyond the junctions, determining deep modifications of the transcriptional profile of the cell. The molecular details of such transcriptional regulations are in large part not known. A few examples of the most explored regulatory pathways are reported below.

6.2 Local Signaling at Adherens Junctions: In Control of Vascular Structure and Permeability

VE-cadherin can be regarded as a signaling receptor that, upon recognition of an identical VE-cadherin molecule on an adjoining cell, acting as the ligand, becomes engaged in an adhesive trans-interaction [71]. Such molecular interaction has the consequence of inducing the clustering of VE-cadherin molecules at the sites of membrane contact between two adjoining cells. This can be considered the first signaling response. Molecules of the nectin family can facilitate the initial concentration of VE-cadherin at nascent junctions [66].

The junction then requires further stabilization which is achieved through interactions among VE-cadherin molecules present on the same cell (cis-interaction), especially through association of the VE-cadherin complex to the cortical actin cytoskeleton [71]. Such peripheral actin bundles can organize independently of VE-cadherin [73]. However, VE-cadherin can recruit several cytoplasmic molecules that regulate the local organization of the actin cytoskeleton. One of these is Tiam [48], a guanine nucleotide exchange factor (GEF) that activates Rac for the organization of peripheral actin [96]. The small GTPase, Rap1, which plays a crucial role in the stabilization of endothelial junctions ([32]; see below), can stimulate translocation to the membrane of Tiam [6]. Rap1 can also mediate the association of the Raf1/Rok-α complex to VE-cadherin to moderate phosphorylation of myosin light chain 2 (MLC2) and actomyosin contractility at junctions [111]. A provisional list of small GTPases and their regulators recruited to VE-cadherin complex and controlling actin polymerization and actomyosin contractility is presented in Table 6.1. For a general model of formation of cadherin-based junctions see Nelson [71].

How can the information developed through such molecular interactions at junctions be transmitted inside the cell to modulate its behavior? As anticipated above, this is primarily achieved through the recruitment at the cytoplasmic domain of VE-cadherin of molecules that possess either adaptor or enzymatic functions (β-catenin, cerebral cavernous malformation [CCM] proteins [see below], Rap1, and many others; see Lampugnani [46] for a more extensive list). VE-cadherin can also enroll transmembrane molecules at junctions, among which are both kinase receptors

Table 6.1 Small GTPases can both enhance and decrease adherens junction stability

Junction stabilization		
Small GTPase	*Effectors*	*Molecular function*
	GEF	
Rap1[a]	*Epac*[a] (cAMP-dependent) *PDZ-GEF 1*[a] (cAMP-independent) *C3G*[b]	Stabilization of VE-cadherin at junctions
Rac	*Tiam* (Rac specific)	Polimerization of cortical actin
Cdc42		
R-RAS[c]		Inhibition of VE-cadherin internalization
	GAP	
Rho	*P190Rho-GAP*[d]	Inhibition of Rho
Junction destabilization		
Rho		ROCK activation, phosphorylation of myosin light chain, actomyosin
	GEF	
Rac	*Vav2*	VE-cadherin endocytosis

cAMP cyclic adenosine monophosphate, *ROCK* Rho-associated coiled-coil-containing protein kinase, *VE* vascular endothelial

[a] Molecular complex with VE-cadherln reported: Rap1 bound to CCM1 [31]; Epac bound to PDE4D [83]; PDZ-GEF1 bound through MAGI/β-catenin [90]

[b] Src-dependent phosphorylation in response to thrombin [6]

[c] Sawada et al. [92]

[d] p120-mediated complex with N-cadherin reported, complex with VE-cadherin putative [110]

and phosphatase receptors [46]. In addition, post-transcriptional modifications of VE-cadherin can modulate such associations, as well as permanence at junctions of VE-cadherin itself (see below).

6.2.1 *β-Catenin: The Canonical Cytoplasmic Partner of Vascular Endothelial (VE)-Cadherin*

The classical cytoplasmic partner of VE-cadherin is β-catenin, a very versatile adaptor molecule that binds and recruits several regulatory and scaffold molecules to cell junctions (for a provisional list of molecules engaged to adherens junctions through β-catenin see Lampugnani [46]). Therefore, β-catenin provides a fundamental contribution to the molecular architecture and function of endothelial junctions. Indeed, truncated VE-cadherin, lacking the carboxy-terminal half of the cytoplasmic domain and unable to bind β-catenin, induces defects *in vivo* and *in vitro* superimposable to those observed after null mutation of VE-cadherin [11, 70].

It has been proposed that the phosphorylation of tyrosine residues of β-catenin can decrease its binding affinity for VE-cadherin. This has been reported in response to vascular endothelial growth factor (VEGF) [67]. In addition, focal adhesion kinase (FAK) could contribute to the weakening of endothelial junctions in response to VEGF, phosphorylating the tyrosine residue 142 of β-catenin and decreasing its association with VE-cadherin *in vivo* [13].

However, other reports are in contrast to the model in which VE-cadherin/β-catenin complex is modulated by tyrosine phosphorylation of β-catenin (or VE-cadherin itself; see below). Tyrosine phosphorylation of β-catenin could contribute to loss of barrier function, without requiring detachment from VE-cadherin. Indeed, Timmerman et al. [102] report tyrosine phosphorylation of VE-cadherin-associated β-catenin in response to thrombin, and no dissociation of the complex. Nonetheless, tyrosine phosphorylation of β-catenin is critical to sustain the permeability increase in response to thrombin. Indeed, depletion of the phosphatase SHP2, which dephosphorylates β-catenin, inhibits recovery of the barrier after thrombin challenge [102]. This may suggest that tyrosine phosphorylated β-catenin could change its association with partners other than VE-cadherin, and in this way could convey different signals. In summary, no general agreement on this issue has so far been reached. The specific mode of modulation of the VE-cadherin/β-catenin complex appears dependent on the stimulus and, possibly, on the origin of the endothelium. These variables could determine phosphorylation of specific tyrosine residues of β-catenin [56], resulting in apparently contrasting results (for example, see the opposite results obtained either with human umbilical vein endothelial cells [HUVEC] cultured on fibronectin, or human pulmonary microvascular endothelial cells cultured on collagen, reported by Timmerman et al. [102] and Monaghan-Benson and Burridge [67], respectively).

In some conditions of severely and constantly disorganized junctions, β-catenin can partially dissociate from VE-cadherin. For example, this has been reported in CCM1-silenced endothelial cells [31], and we have observed a similar event in *CCM3*-knockout endothelial cells [8]. A reasonable hypothesis is that, in endothelial cells, even a limited decrease in the association between VE-cadherin and β-catenin is sufficient to create subtle local heterogeneity in the molecular composition of cell-to-cell junctions that weaken the continuity of the monolayer and impair its barrier function.

In addition to its adaptor role at junction, β-catenin is a crucial regulator of transcription, particularly in association with transcription factors of the Tcf family [16]. This aspect is discussed below (see also comments in the legend for Fig. 6.1).

A particular example of indirect control of transcription by β-catenin localized to junctions is the regulation of nuclear localization of the transcription factor FoxO1 (see below).

6.2.2 Other Cytoplasmic Associates of VE-Cadherin with a Crucial Role in Endothelial Physiology: Rap1 and the Cerebral Cavernous Malformations (CCMs)

Among the several molecules that associate with VE-cadherin at cell-to-cell junctions and control vascular permeability and vascular structure, we report here on

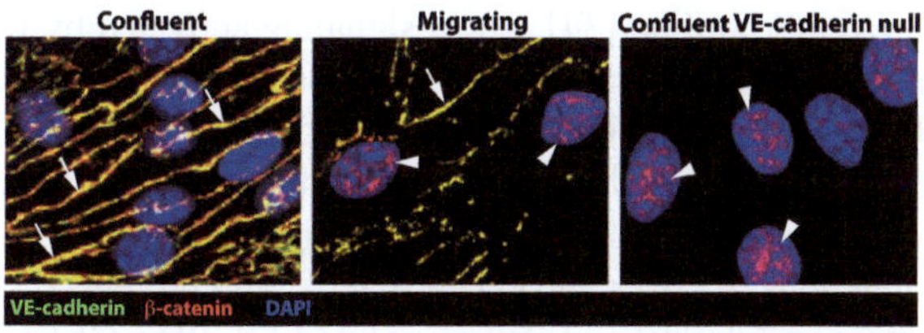

Fig. 6.1 Distribution of β-catenin to the nucleus inversely correlates to the stability of endothelial junctions. **a** In confluent endothelial layers in culture, β-catenin (*red*) is prevalently associated with cell-to-cell junctions where it co-distributes with VE-cadherin (*green*, co-distribution; *yellow, arrows*). **b** In migrating endothelial cells, which have dynamic adherens junctions in comparison to the stable monolayer, β-catenin can be observed in the nucleus (*red, arrowheads*), in addition to junctions, where it still co-localizes with VE-cadherin (*green*, co-distribution; *yellow, arrow*). Endothelial cells at the migrating front of a wounded monolayer are shown. **c** In VE-cadherin-null endothelial cells, β-catenin is also prevalently nuclear (*red, arrowheads*) in highly dense cultures. Endothelial cells in culture, and isolated from either wild-type or VE-cadherin-null mouse embryos, are shown. Active β-catenin (in the *three panels*), dephosphorylated on the N-terminal residues Ser37 or 41, is immune-stained with the mAb clone 8E7 (Merck Millipore, Darmstadt, Germany). Junctional β-catenin acts as a crucial scaffold for several signaling molecules (see Lampugnani [46] for a provisional list of such molecules). Nuclear β-catenin in the different situations shown is competent for regulation of transcription. This has been analyzed in detail for the transcription of claudin-5, a component of tight junctions with crucial activity in the control of permeability [100]. The mechanisms that direct β-catenin localization in the nucleus are not yet defined in molecular details. See the text for further discussion

the small GTPase Rap1 [32] and the CCM proteins (CCM1, 2, and 3) [85] as our laboratory also contributed to the definition of their function in the endothelium.

These molecules are discussed together as CCM1 has been demonstrated to recruit and localize Rap1 to cell-to-cell junctions [31]. CCM1 associates with adherens junctions, directly binding β-catenin through its FERM domain [31]. In addition, co-immunoprecipitation of CCM1 with the adherens junction components VE-cadherin, α-catenin, and afadin has been reported [31]. However, interaction with β-catenin appears to be nonexclusive for junctional localization of CCM1. Indeed, CCM1 can be recruited to cell junctions through interaction of its FERM domain (the groove between subdomain 1 and 3) with the C-terminal region of the transmembrane protein heart of glass (HEG1) [29]. In addition, CCM1 appears to sustain functions downstream of HEG1. Indeed HEG1, CCM1, or CCM2 murine and zebrafish mutants show convergent traits in their abnormal phenotypes [40].

The stabilizing activity of Rap1 on endothelial cell-to cell junctions has long been identified and analyzed in several models of endothelial cells cultured *in vitro* [32, 42]. The cyclic adenosine monophosphate (cAMP) analog 8-pCPT-2′-Me-cAMP (007), a synthetic cell-permeable activator of Epac1 (an Rap1 GEF), is a potent stabilizer of VE-cadherin-based endothelial cell-to-cell junctions [32, 42]. In addition, *Rap1b* isoform knockout mice show hemorrhagic phenotype due to unstable blood vessels [14]. When ablated in mice, Rap1a, the other isoform of Rap1, does not determine a vascular phenotype. Similar effects of vessel fragility that result predominantly in cerebral hemorrhages have been observed in zebrafish larvae treated with *Rap1b* morpholino [34]. Local signaling activated by Rap1 would determine polymerization of cortical actin through the activation of Rac [6, 32]. Rap1 can recruit an Raf1/Rok-α complex to junctions for local modulation of actomyosin contraction [111], as discussed above). However, the molecular mechanism of junction stabilization in response to Rap1 activation still remains largely unknown. Association with junctions of GEFs and GTPase-activating proteins (GAPs) can regulate the localization and duration of Rap signaling [32]. PDZ-GEF1, an Rap1 GEF, can be found to be associated with VE-cadherin through β-catenin and MAGI1 [90, 32]. Another Rap1 GEF, Epac1, can associate with VE-cadherin through the phosphodiesterase PDE4D [83]. Afadin (AF6) [60], which can recruit both Rap-GTP and RapGAPs, can associate with adherens junctions [6, 32].

While constitutive activation of Rap1 is required for maintaining stable junctions, and its activation further reinforces them, it is debated whether Rap1 deactivation or delocalization can destabilize established cell-to-cell junctions [32].

Interestingly, besides CCM1, which binds directly to Rap1 [94], CCM2 and CCM3, which can form a complex with CCM1 [97, 107, 114], also strongly contribute to the maintenance of stable endothelial cell-to-cell junctions in both *in vitro* and *in vivo* models [7, 8, 51, 58]. Indeed, as soon as they are ablated, adherens junctions (and, as a consequence, tight junctions; see below) become disorganized and permeability is increased.

Local activation of Rap1 at adherens junctions, as promoted by CCM1, is critical for stabilization of endothelial junctions. Indeed, stimulation of Epac1 with 8-pCPT-2′-Me-cAMP (007) in CCM1-ablated cells does not restore either endothelial barrier or localization of Rap1 to cell-to-cell junctions [31]. Consistently, activation of Epac1 does not restore junction and permeability in Rap1-depleted endothelial cells [77]. While it is established that CCM2 binds directly to CCM1, and CCM3 binds to CCM2 [97], it is not known how the absence of either CCM2 or CCM3 could influence the localization or activity of Rap1 bound to CCM1. In addition to affecting Rap1 GTPase, ablation of CCM1, 2, or 3 stimulate the activity of Rho, with consequent phosphorylation of MLC2, actomyosin contraction, and disorganization of cell-to-cell junctions [99, 109] (for Rho activation after CCM1 and 2 ablation); [8] (for Rho activation after CCM3 ablation). Inhibition of Rho activity can restore the structure and function of junctions in CCM-ablated endothelial cells, and inhibit formation of vascular lesions *in vivo* (see below). Rap1 could mediate the downregulation of Rho by CCM proteins recruiting Rac and inhibiting Rho activation at cell-to-cell junctions [6, 99]. Figure 6.2 summarizes the molecular interaction at cell-to-cell junctions between VE-cadherin, β-catenin, CCM proteins, and small GTPases.

As illustrated by the individual examples discussed above, signaling conveyed by stabilized junctions depends on the association of specific molecules with junctions, and on their interactions. This relies primarily on clustering of VE-cadherin to cell-to-cell contacts. Therefore, the presence of VE-cadherin at junctions is crucial and can be regulated as discussed in the following paragraph.

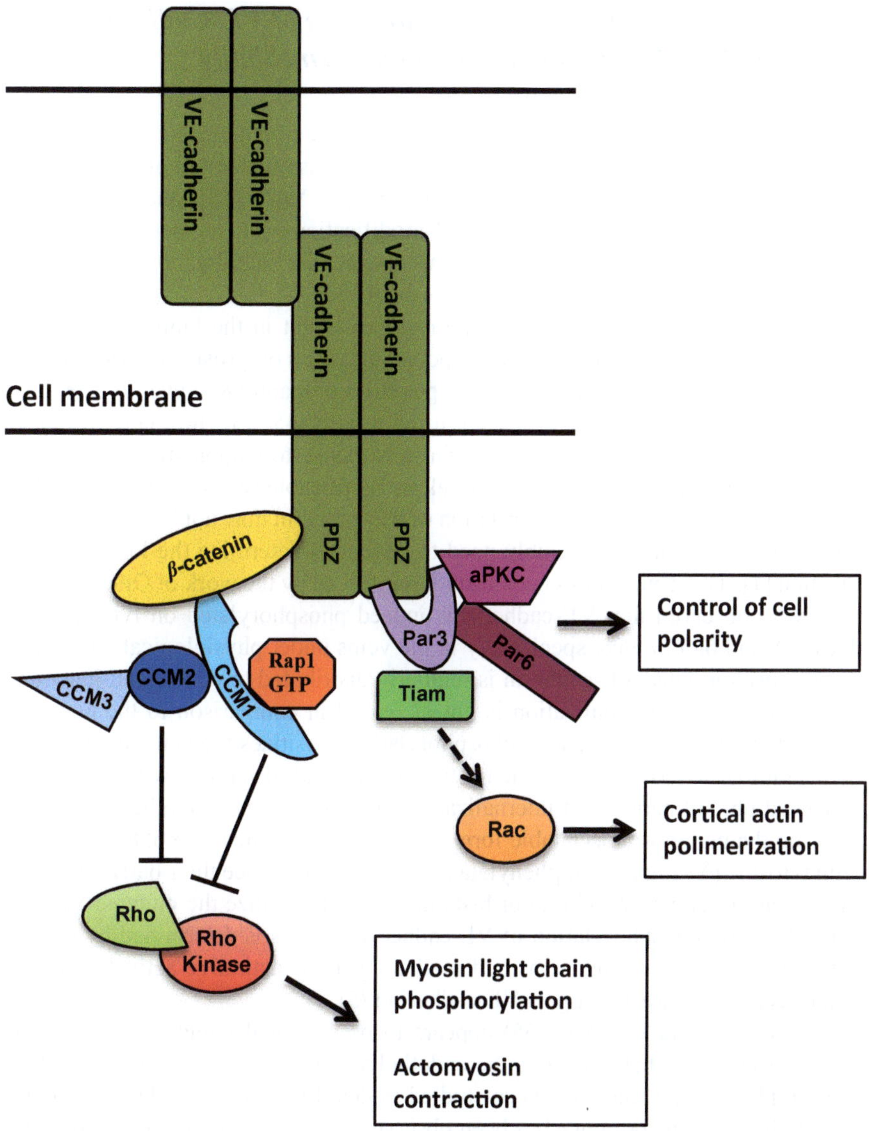

Fig. 6.2 Schematic representation of the molecular interactions at endothelial cell-to-cell junctions between VE-cadherin, cerebral cavernous malformation (*CCM*) proteins, small GTPases, and the polarity complex. CCM1 interacts with both β-catenin and Rap1 [31]. As a consequence, Rap1 is activated locally (GTP-bound) and stabilizes cell-to-cell junctions through a mechanism that is still poorly understood [32]. Active Rap1, possibly activating Rac and inhibiting Rho, induces stabilization of the cortical actin cytoskeleton.In addition, active Rap1, stabilizing VE-cadherin to cell-to-cell junctions [42], indirectly promotes localization to cell-to-cell contacts of the polarity complex, Par3/Par6/atypicalPKC (*aPKC*) together with Tiam [48]. This complex can associate with VE-cadherin through recognition, by the third PDZ domain of Par3, of the type II PDZ-domain-binding motif at the C-terminus of VE-cadherin [38], and directs apical-basal polarity and the formation of a regular and polarized lumen [51]. CCM2 and CCM3 can form a complex

6.2.3 Phosphorylation and Ubiquitination of VE-Cadherin and the Regulation of Vascular Permeability

The permanence of clustered VE-cadherin at cell-to-cell junctions can be modulated by post-transcriptional modification of the cytoplasmic domain of VE-cadherin, particularly phosphorylation on tyrosine (Y658 and Y685 [75]) and serine (S665 [26]) residues. Such post-transcriptional modifications can regulate the internalization rate of VE-cadherin and, as a consequence, the stability, composition, and signaling activity of adherens junctions.

VE-cadherin posses several tyrosine residues (eight in the human protein and nine in the mouse protein) that could be potential targets of tyrosine kinases [81]. Of these residues, only tyrosine residues at positions 658 and 685 have, to date, been reported to be phosphorylated *in vivo* in the mouse [45, 75]. In a current model, tyrosine phosphorylation of VE-cadherin in response to angiogenic or inflammatory stimuli has been associated with weak and permeable cell-to-cell junctions [81, 103]. However, tyrosine phosphorylation of VE-cadherin may not be sufficient, and other signaling pathways possibly need to concur to determine the loss-of-barrier function [1]. This concept has been further reinforced by the work of Orsenigo et al. [75], who observed that VE-cadherin is indeed phosphorylated on tyrosine residues 658 and 685 *in vivo*, specifically in the veins under physiological conditions. Such phosphorylated VE-cadherin is regularly distributed to cell-to-cell junctions, and the rate of its internalization is not increased in comparison to the nonphosphorylated form. However, such phosphorylation sensitizes the cells to respond to permeability-increasing agents such as bradykinin and histamine, with consequent enhanced ubiquitination and internalization of VE-cadherin. Endothelial cells expressing the nonphosphorylatable form of VE-cadherin, such as Y658F or Y685F, with tyrosine (Y) mutated in phenylalanine (F), do not reduce their barrier function in response to either bradykinin or histamine, nor internalize the mutated VE-cadherin. Specific phosphorylation of VE-cadherin in veins is determined by the low shear stress characterizing veins and mediating the constitutive activation of Src that directly or indirectly targets VE-cadherin [75].

The serine residue 665 (S665) appears to be a critical target in the action of VEGF as an agent that decreases the endothelial barrier [26]. Although the effects of such phospho-residue have been studied in detail in *in vitro* models of cultured endothelial cells, its presence has been observed as a short-term response to VEGF also *in vivo* in VE-cadherin from mouse skin vessels [27]. Once more, a key factor would be Src, which, activated by VEGF receptor 2 (VEGFR2), upon VEGF

[107] that might be recruited to cell-to-cell contacts in association with CCM1 [97], contributing to junction stability. Although localization to junctions of such ternary complex has to be definitely proved, the absence of any of the CCM proteins profoundly affects endothelial cell-to-cell junctions. Indeed, active Rap1 is delocalized from junctions (Glading et al. 2008; [34], junctions are dismantled, apical-basal polarity is lost [7, 51, 58] and lumen is abnormal in CCM-ablated (any *CCM* gene) endothelial cells cultured *in vitro* (Glading et al. 2008 [7, 51, 58], in murine models of CCM pathology [7, 58] and in vascular lesions of CCM patients [7, 58]

binding, would phosphorylate Vav2. This GEF starts a chain of activating events: activation of the small-GTPase Rac, which stimulates its downstream target p21-activated-kinase (PAK) to phosphorylate the residue S665 in VE-cadherin. B2-arrestin recognizes such phosphorylated residue and, upon binding to VE-cadherin, promotes clathrin-dependent internalization of VE-cadherin. The consequence of this process is weakening of cell-to-cell junctions. Interestingly, this sequence of events can be blocked by angiopoietin-1 (Ang1). Ang1, promoting the sequestering of Src through mDia, inhibits the phosphorylation of VE-cadherin on S665, its internalization, and the ensuing weakening of junctions.

Phosphatases are expected to modulate the phosphorylation of VE-cadherin. Until now, the tyrosine phosphatase SHP2, which is associated with VE-cadherin through β-catenin, has been reported to dephosphorylate β-catenin, γ-catenin, and p120 associated with VE-cadherin, but not VE-cadherin itself [102, 105].

The endothelial-specific vascular endothelial phosphatase (VE-PTP) is associated with VE-cadherin at junctions through interaction of the extracellular domains of the two proteins. VE-PTP can indeed dephosphorylate VE-cadherin, besides Tie2 and plakoglobin [24, 89]. It has recently been reported that VEGFR2, when associated with junctions, can also be a target of VE-PTP [35].

In addition, clathrin-dependent internalization of VE-cadherin can be inhibited by p120 catenin masking an endocytic signal [69].

6.2.4 VE-Cadherin can be Associated with, and Modulate the Activity of, Angiogenic Receptors

VE-cadherin clustered at junctions can also modulate the activity of receptors that act as crucial regulators of vessel organization, such as VEGFR2 [11, 50] and transforming growth factor (TGF)-βR2 [88]. Activated by VEGF, VEGFR2 becomes phosphorylated on several tyrosine residues [15]. These phosphorylated tyrosines become docking sites for different mediators that can activate locally-specific signaling pathways [15], depending on the stability of junctions. In5 particular, we have observed that when VE-cadherin is engaged in stable contacts, as in confluent endothelial monolayers, activated VEGFR2 associates with VE-cadherin through β-catenin and activates phosphatidylinositol3-kinase (PI3K) [also associated with VE-cadherin through β-catenin] to phosphorylate Akt and to signal cell survival and resistance to apoptosis [11, 100]. When VE-cadherin is engaged in dynamic junctions, as in cells undergoing angiogenic responses, VEGF-activated VEGFR2 preferentially stimulates the phospholipase C (PLC) and mitogen-activated protein kinase (MAPK) pathway to signal proliferation [49]. Such signaling can take place from endocytic compartments in which VEGFR2 is internalized in association with VE-cadherin through a clathrin-dependent process [50].

VE-cadherin clustered in stable contacts can modulate endothelial behavior, forming a complex with another regulator of the vascular phenotype, the TGFβ receptor TGFβR2 (as well as with Alk1 and Alk5 [88]). The functional consequence of such local regulation is the activation, by TGFβ, of antiproliferative and antimigratory responses. Such signaling would contribute to the stability of the mature endothelial layer. On the contrary, in endothelial cells with weakened junctions, as is the case after treatment with a VE-cadherin-blocking antibody [88] or with a reduced level of VE-cadherin [58], TGFβR2 is not associated, or is less associated, with VE-cadherin [88]. This deactivates the antiproliferative and antimigratory response elicited by TGFβ, and can even redirect TGFβ signaling towards dedifferentiation (endothelial-mesenchymal transition [EndMT]) of the endothelial cells, as described in more detail in the following paragraphs in CCM pathology.

VE-cadherin has also recently been shown to be associated with fibroblast growth factor receptor 1 (FGFR1) and to reduce its phosphorylation in response to FGF2 engaging the phosphatase Dep1 in a multiprotein complex [28].

6.3 Adherens Junctions and Nuclear Signaling: Mediators and Targets in the Establishment of Barrier Properties and Differentiated Phenotype of the Endothelium

In addition to regulating endothelial functions through the local organization of specific signaling complexes to adherens junctions, as described in the previous paragraphs, the 'degree of tightness' of VE-cadherin association with junctions appears to regulate the transcriptional profile of endothelial cells [100]. 'Degree of tightness' means distinct and specific molecular complexes recruited by VE-cadherin to adherens junctions as a function of the duration and stability of its engagement in an adhesive interaction. In general, the molecular details of the dynamic remodeling of such complexes are still poorly known. Some examples have been presented in the previous paragraphs. Among the interactors of VE-cadherin, β-catenin plays an important role in coordinating the state of the junction compartment to the quality and type of transcription in the nuclear compartment. Although such regulation is evident from a functional point of view, details of the underlying molecular mechanism are still virtually unknown.

Examples of loss and acquisition of differentiated phenotype through the crosstalk between adherens junctions and the nucleus mediated by β-catenin in endothelial cells will be discussed in more detail below; however, before then, we have to face the still unsolved problem of cell biology.

6.3.1 Nuclear β-Catenin: How Adherens Junctions Could Control It?

In addition to its scaffold role in adherens junctions, β-catenin is a crucial regulator of transcription in association with Tcf transcription factors [16]. At present, the relationship between the junctional and nuclear pool of β-catenin remains an unsolved general issue in cell biology, besides endothelial cells. It is not known whether β-catenin concentrated in the nucleus originates from junctions or whether it comes from a distinct pool. In Drosophila, the existence of two distinct pools, junctional and nuclear, of the β-catenin homolog armadillo, is ascertained [91], and is confirmed in *Caenorhabditis elegans* [43]. In contrast, in mammalian epithelial cells, photoactivatable GFP-tagged β-catenin has been observed to re-localize from E-cadherin to the nucleus upon dissociation of adherens junctions [39].

While it is reasonable to hypothesize that some communication exists between these two pools, the molecular details of such a connection are not defined. A possible mechanism through which cell-to-cell junctions could regulate the level of the cytoplasmic pool of β-catenin, and indirectly the level of nuclear β-catenin, has been reported in epithelial cells [59]. In these cells, crucial components of the β-catenin phosphodestruction complex, such as axin, adenomatous polyposis coli 2 (APC2), and glycogen synthase kinase (GSK)-3β, are localized to cell-to-cell contacts, and cadherins can promote N-terminal phosphorylation of β-catenin, which targets β-catenin to proteasomal degradation. As a consequence, when junctions are tightly organized, N-terminal phosphorylation of β-catenin and its turnover are enhanced. The reduced pool of cytoplasmic β-catenin would limit the nuclear distribution of β-catenin and its transcriptional activity. Whether the phosphodestruction complex acts on β-catenin released from cadherin or on an independent pool has not presently been clarified. N-terminal phosphorylated β-catenin can also be found at cell-to-cell junctions "in a complex which is molecularly distinct from the cadherin-catenin complex" [59]. However, such coordination between tightness of junctions and targeted degradation of β-catenin would regulate the cytoplasmic and nuclear amount of β-catenin and allow the cell to mount a transcriptional response mediated by β-catenin and appropriate to the state of junctions.

6.3.2 Nuclear β-Catenin in Endothelial Cells

In addition to the above general questions, endothelial cells present a peculiarity in respect to the classical model of a direct balance (proportion/ratio) between the cytoplasmic and nuclear level of β-catenin. In our experience, endothelial cells can concentrate β-catenin in the nucleus without apparently requiring accumulation of (stabilized) β-catenin in the cytoplasm, as generally observed after stimulation with Wnt in other cell types. Constant finding in endothelial cells is that when adherens junctions are poorly organized, β-catenin concentrates in the nucleus and regulates transcription. Examples of weak cell-to-cell contacts are: endothelial cells in non-confluent layers as during angiogenic response; models of pathological conditions as after KD of the CCM1 [31] and knockout of the *CCM3* gene [8]; and the extreme experimental situation of VE-cadherin-null endothelial cells [100]. Remarkably, in each of these circumstances, although the total level of β-catenin can decrease, even to an extremely low level, as in VE-cadherin-null endothelial cells, residual β-catenin concentrates into the nucleus where it is transcriptionally active. The mechanism of such nuclear localization of the low total level of β-catenin has not presently been defined. Notably, β-catenin lacks nuclear localization sequences [16, 23].

When junctions are poorly organized, β-catenin induces a transcriptional profile characterized by loss of differentiation markers and acquisition of EndMT markers [8, 58, 100], as discussed below.

On the other hand, endothelial cells can also use β-catenin to direct a program of differentiation. Indeed, endothelial cells of the central nervous system (CNS), which participate in the blood–brain barrier (BBB) when stimulated by either Wnt3a [55], Wnt7a, or Wnt7b [98] activate a canonical response and block proteasomal degradation of β-catenin that now concentrates into the nucleus where it induces a BBB transcriptional profile [55, 78]. In parallel, Wnt induces stabilization of adherens junctions, with increased localization to junctions of VE-cadherin and β-catenin, and tightening of tight junctions through upregulation of claudin transcription [55, 98] (see below). As far as adherens junctions are concerned, both VE-cadherin and β-catenin appear more regularly disposed to cell-to-cell contact in Wnt3a-treated endothelial cells from brain microvessels; however, this is not the consequence of transcriptional activation of these genes. The molecular mechanisms of strengthening of adherens junctions in response to Wnt is not yet defined in endothelial cells, although it might involve regulation of small GTPases [93].

All in all, the transcriptional responses elicited by nuclear β-catenin appear to be coordinated to the state of endothelial cell-to-cell adherens junctions. The molecular mechanisms that direct such alternative responses of endothelial cells are still mostly unknown (see also below).

6.3.3 Adherens Junctions and Transcriptional Regulation of Permeability: The Case of Claudin-5 and Claudin-3

The extreme example of fragile endothelial junctions is that determined by the absence of VE-cadherin. Until now, this situation has mostly been studied in experimental models of *in vitro* cultured endothelial cells. Indeed, in the organism, constitutive inactivation of the VE-cadherin gene is embryonically lethal in the mouse [11] and in zebrafish [68], and VE-cadherin-inactivating antibody produces an acute lethal phenotype in the adult mouse [18, 19]. VE-cadherin can be dispensable for the early phase of sprouting angiogenesis when tip cells interact through filopodial contacts [52]. However, VE-cadherin is required for coordinating the following steps and allowing cells to properly recognize each other and mature the initial contacts, as observed in the zebrafish embryo [52].

In the absence of VE-cadherin, molecular determinants of the endothelial mature phenotype, such as claudin-5, VE-PTP, and von Willebrand factor, are transcriptionally downregulated [100] (unpublished results). In particular, claudin-5, an endothelial-selective constituent of tight junctions, is crucially involved in the control of endothelial permeability in the CNS [72]. The inhibition of claudin-5 transcription involves the formation of a nuclear complex between β-catenin and FoxO1, which acts as a transcriptional repressor of the *Claudin-5* gene [100]. In addition, FoxO1 represents an interesting example of molecular cross-talk between junctional and nuclear compartments. Indeed, FoxO1 is phosphorylated through active Akt (phosphorylated on Thr308 and Ser473) localized to stable cell-to-cell contacts. Akt is activated locally at cell-to-cell junctions via PI3K, which associates with VE-cadherin through β-catenin only in stable contacts [11], as described above. Relevant for regulation of transcription is the fact that phosphorylation both promotes FoxO1 exclusion from the nucleus (phosphorylation on Thr24) and inhibits its binding to DNA (phosphorylation on Ser256) [9, 115]. As a consequence, transcription of claudin-5 (and other target genes such as VE-PTP and von Willebrand factor), is derepressed in endothelial cells with stable junctions, and the barrier properties are enhanced by the organization of claudin-based tight junctions.

An opposite example of transcriptional control of permeability by β-catenin is represented by the response of brain endothelial cells to Wnt3a and Wnt7a/b [55, 98]. As discussed above, in response to Wnt stimulus, β-catenin concentrates in the nucleus where it can activate a specific transcriptional program that drives differentiation of endothelial cells towards the CNS-specific phenotype (the BBB [21]). In particular, as far as control of permeability is concerned, the transcription of claudin-3 is upregulated both *in vivo* and *in vitro* [55, 98], and this has the functional effect of reinforcing the barrier properties of the endothelium [55, 98], as specifically needed at the BBB.

As discussed above, the molecular mechanism that determines very distinct β-catenin-driven transcriptional programs, in relation to different stability of adherens junctions, is still scarcely defined. As it has been shown for FoxO1, the cooperation between β-catenin and transcriptional co-regulators, the level or nuclear localization of which is regulated by the strength of adherens junctions, could be crucial. In addition, fine-tuning of the nuclear level of β-catenin could contribute to such distinct responses [33].

6.3.4 Adherens Junctions and Endothelial Differentiation: The Case of Endothelial-Mesenchymal Transition

EndMT represents a process of dedifferentiation driven by transcriptional reprogramming that also targets the composition and function of adherens junctions [64]. In the course of this process, the promigratory N-cadherin partially substitutes for VE-cadherin [28, 99]. The so-called cadherin switch, with N-cadherin substituting E-cadherin, is indeed a well-established trait of epithelial-mesenchymal transition (EMT) in epithelial cells [44]. Transcriptional repression of E-cadherin is well-defined in these cells, and similar mechanisms involving Twist, Snail and Snai2 also operate in endothelial cells [57]. However, reorganization of cell-to-cell junctions might also represent a priming event of EndMT. Although the precise relationship between junctional and nuclear β-catenin are poorly defined, as discussed above, β-catenin might represent one of the molecular link(s) in the reshaping of both junction and transcription profile that characterizes the EndMT. Indeed, β-catenin can drive the transcriptional upregulation of EndMT markers also cooperating with the TGFβ pathway [65, 116].

A physiological example of EndMT takes place during the formation of the heart cushion in the embryo [82]. In this case, not only is β-catenin transcriptionally active in the endothelial cells undergoing the transformation but it is also required for initiation of the process [54], which is subsequently reinforced by the activation of endothelial cells by TGFβ2 produced in the heart tissue. At the initial stages of transformation, endothelial cells co-express α-smooth muscle actin and the endothelial marker VE-cadherin [54]. Although the cells undergoing such transformation become highly motile, the organization of endothelial cell-to-cell junctions has not been examined in detail; in particular, it remains to be defined whether expression of the promigratory N-cadherin is increased [28].

EndMT has increasingly been studied for its crucial contribution to various pathologies characterized by fibrosis, as well as to cancer [44, 64]. As far as vascular pathology is concerned, EndMT takes place and has a causative role [58] in the CCM, which has already been introduced above. In this case, endothelial adherens junctions constitute an early target of the loss-of-function mutation of any of the three *CCM* genes. VE-cadherin and β-catenin become highly disorganized and the endothelium loses its barrier properties. Subsequently, N-cadherin substitutes VE-cadherin at junctions, which remain highly disorganized, as a consequence of delocalization of Rap1 and activation of Rho [31, 99, 109] (see above). Reshaping of adherens junctions is accompanied by induction of an array of EndMT and stem cell markers, among which are CD44, s100a4, Ly6a, Klf4, which contribute to the pathological vascular phenotype [58]. If and how disorganization of junctions has a causal role in transcriptional reshaping and initiation of EndMT, besides being a target of this process, remains to be defined.

However, also in this case, the dedifferentiation process requires β-catenin transcriptional signaling [8]. Association of β-catenin to junction-delocalized VE-cadherin is reduced ([8] for CCM3 knockout; [31] for CCM1). However, once more, it is not known whether the β-catenin released from VE-cadherin is directly responsible for the β-catenin-driven transcriptional signaling observed. In addition, EndMT is reinforced by activation of the TGFβ signaling that, in CCM1-knockout endothelial cells, is cell autonomous, through the production of BMP6 [58]. These observations confirm the cooperation between the β-catenin and TGFβ pathway in the control of EndMT.

6.4 Endothelial Adherens Junctions In Vivo: Learning from Vascular Dysfunctions in Animal Models and Human Pathologies

Endothelial adherens junctions play a crucial role *in vivo*, both for correct vascular morphogenesis and for establishment and maintenance of the barrier function of the endothelium. This has been observed both in genetic mouse mutants and zebrafish embryos, these last treated with specific morpholinos for the main component of adherens junctions and VE-cadherin, as well as for several molecules recruited to adherens junctions [46]. In humans, only a few pathologies linked to genetic defects of molecular components of endothelial adherens junctions have been reported (see [22], and below). In addition, endothelial cell-to-cell junctions are typically altered in several human diseases (see below).

6.4.1 Animal Models

Either ubiquitous or endothelial-restricted inactivation or mutation of genes for component of adherens junctions in transgenic mice mostly results in early embryonic lethality due to vascular disorganization and impaired control of permeability. We have reported such phenotypes in transgenic mice with homozygous null mutation for VE-cadherin or expressing a carboxy-terminal truncated VE-cadherin mutant unable to bind β-catenin [11]. Vascular defects, both in organization and barrier function, have also been observed in zebrafish larvae treated with morpholinos to VE-cadherin [52, 67]. In addition, endothelial-selective inactivation of β-catenin produces vascular hemorrhagic lacunae at bifurcations of cerebral, as well as vitelline, umbilical and placental vessels and embryonic lethality between 11.5 and 13.5 dpc [12].

Embryonic lethality with vascular phenotypes has been reported after inactivation in p120, N-cadherin (both endothelial-selective knockout), VE-PTP, Dep1, and Rap1b mice, among the others. Please refer to Lampugnani [46] for an extensive description of these and other related phenotypes.

In recent years, our laboratory has particularly focused on the role of CCMs in vascular pathologies [51, 58].

Although not endothelial-specific, CCM molecules show a specific activity in the endothelium. Indeed, constitutive and ubiquitous ablation in the mouse of any of the three genes determines generalized vascular disorganization and early embryonic lethality (around 10 dpc) [36, 108, 109]. CNS-specificity for vascular lesions appears if gene deactivation is induced after birth, using either endothelial-restricted knockout [7, 58] or heterozygosity for CCM on a mutation-prone genetic background (loss of tumor suppressor gene *Trp53* [95], and deletion of mismatch repair complex gene *Msh2* [62]). The vasculature of the CNS is particularly affected with the development of focal vascular malformations resembling human cavernomas in the brain, cerebellum, and retina [85]. The molecular mechanism of this typical organ specificity still remains to be explained. This might depend either on specific characteristics of the differentiated endothelium in the CNS or on the persistence of endothelial precursors that could be particularly affected by mutation in *CCM* genes [58]. This hypothesis may fit with the observation that, in the animal model, the number of cavernomas and the rate of their appearance is maximal when gene recombination is induced soon after birth, and decreases sharply when recombination is postponed by even a few days [7]. In addition, it has been reported that endothelial cells in the CNS acquire a fully differentiated phenotype early after birth [55].

6.4.2 Human Pathologies

Numerous human pathologies affecting disparate organs present morphologically abnormal and permeable vessels (see Cattelino et al. [12] for a comprehensive list).

In light of the phenotype of the animal mutants discussed above, endothelial adherens junctions are likely affected and also contribute to pathologies in humans. In various human diseases, the abnormality of endothelial junctions has been confirmed and molecular details on endothelial adherens junctions in pathological samples from different organs are accumulating. Endothelial junctions in tumors have represented a typical subject for this type of study [4]. It has also been shown that VE-cadherin in endothelial cells of tumor vessels expresses distinct antigenicity [61]. The monoclonal antibody that specifically recognizes VE-cadherin in tumor vasculature was also effective in inhibiting the neoplastic growth in experimental models of murine tumors without increasing permeability [53]. More recently, non-neoplastic diseases, such as inflammatory, infectious, and degenerative pathologies, in particular of brain and lung, have also been increasingly investigated under the perspective of endothelial adherens junctions [20, 37, 86, 112]. A website collecting and organizing the increasing mass of such reports could represent a useful tool for the scientific community.

Genetic defects in endothelial cells have been recognized to be the cause of various vascular pathologies in humans [104]. However, mutations in genes for components of adherens junctions and resulting in vascular defects are rare in humans. To our knowledge, *RASA1*/p120RasGAP and the three *CCM* genes represent the only examples of such a relationship.

The *RASA1* (p120-RASGAP) gene codifies for a Ras GTPase and is mutated in patients (heterozygous for loss-of-function mutation) with brain arteriovenous malformations [84]. Its possible that relationship with adherens junctions is suggested by the role of RASA1 in activating p190RhoGAP, which positively regulates adherens junction assembly in cooperation with catenin p120 [110]. However, this molecular pathway targeting adherens junctions has, until now, been proved in fibroblasts expressing N-cadherin.

Loss-of-function mutation in any of the *CCM* genes is the cause of the genetic form of CCM disease (20–30 % of all CCM patients [80]). Some molecular functions of CCM1, 2, and 3 in relation to adherens junctions have already been described above. The inheritance of the pathology is autosomal dominant, in the sense that patients are heterozygous for the CCM mutation. It has not been definitely established whether the formation of focal vascular lesions requires local loss of heterozygosity of the CCM genes or a second hit, either genetic or environmental [2, 25, 76]. Sporadic CCM patients diagnosed on the basis of magnetic resonance imaging (MRI) similarity with genetic patients do not present germline mutation, and the genetic situation in the brain lesion is extremely difficult to ascertain in a number of samples with statistical significance.

6.4.3 *Targeting Endothelial Adherens Junctions for Therapy: Pharmacological Tools to Stabilize/Destabilize Adherens Junctions*

As discussed above, adherens junctions can significantly contribute to vascular dysfunctions and pathologies; therefore, they could also represent targets for therapeutic intervention.

A huge number of biological substances and chemicals have been shown to affect the organization and functions of endothelial adherens junctions. These effects have been demonstrated in cells in culture and in animal models, while, for most, specific clinical trials in patients are still lacking. However, some of the substances that regularize adherens junctions in experimental models are already in clinical use, with different therapeutic indications than the targeting of adherens junctions in the endothelium. We will briefly discuss two such drugs—simvastatin and fasudil.

Simvastatin and fasudil both interfere with signaling from small GTPases, inhibiting the Rho-family of small GTPases and the Rho-associated coiled-coil-containing protein kinase (ROCK), a crucial effector of Rho, respectively.

Simvastatin is a statin that has long been used in clinics to lower serum cholesterol. It has pleiotropic effects and inhibits isoprenylation of Rho GTPases, targeting the 3-hydroxy-3-methylglutaryl coenzyme A (HMG-CoA) reductase and depleting the cell of the geranylgeranyl pyrophosphate intermediate necessary for post-translational lipidation of Rho GTPases. This results in the inhibition of Rho localization to the plasmatic membrane and downregulation of Rho activity [17, 79].

Fasudil has been used in patients to protect against neuronal damage induced by ischemia [113]. It inhibits ROCK, competing for ATP binding. The result is downregulation of the organization of actin stress fibers induced by phosphorylation and activation of MLC2.

Simvastatin and fasudil have been used in experimental murine models of CCM2 and CCM1 mutation *in vivo*, as well as in endothelial cells in culture. As extensively described above, CCM proteins are required for the control of permeability in endothelial cells, in addition to their role in vessel organization [99, 109]. It has been reported that simvastatin can correct the defect in barrier functions of CCM2 heterozygous mice elicited by VEGF and induce organization of cortical actin in cultured HUVEC after knockdown of CCM2 by small interfering RNA (siRNA) [109].

Similarly, fasudil inhibits vascular leakage, both basal and induced by Lipopolysaccharide (LPS) in the brain and lung of both CCM1 and CCM2 heterozygous mice [99]. It also reduces the enhanced permeability of endothelial cells with impaired expression of CCM1 or CCM2 (endothelial cells either isolated from brain and lung of CCM1 heterozygous mice or treated with siRNA to downregulate CCM1 and CCM2 expression [99]). In addition, it reduces the number and dimensions of brain vascular malformations in a murine model of CCM1 pathology [63].

The drugs discussed above all intervene in stabilizing cell-to-cell contacts. It appears that drugs aimed at opening endothelial junctions are much rarer. In particular,

Table 6.2 Agonists targeting adherens junctions can regulate endothelial stability and permeability

Molecule	Type of molecule	Mediator
Destabilization		
Thrombin	Inflammatory	Rho
Histamine		
Bradykinin		
IL1/TNFα/IFN-γ		
VEGF	Angiogenic	VEGFR2 (several downstream mediators)
Stabilization		
Adrenomedullin	Endogenous mediator	cAMP
Prostacyclin		
PGE2		
β-adrenergic agonists		
S1P		S1P receptor/Rac activation
Angiopoietin 1		Tie2 receptor
8-pCPT-2′-Me-cAMP (007)[a]	Experimental drug	Epac (Rap1 GEF) Tie2
Angiocomp		Tie2
Simvastatin	Drug in clinical use	Inhibition of RhoGTPase isoprenylation[b]
Fasudil		Inhibition of ROCK

cAMP cyclic adenosine monophosphate, *HMG-CoA* 3-hydroxy-3-methylglutaryl-CoA, *IFN* interferon, *IL* interleukin, *PGE2* prostaglandin E2, *ROCK* Rho-associated coiled-coil-containing protein kinase 1, *S1P* sphingosine1-phosphate, *TNF* tumor necrosis factor, *VEGF* vascular endothelial growth factor, *VEGFR2* vascular endothelial growth factor receptor 2 [a] Soluble and cell permeable cAMP analog specific for EPAC

[b] Inhibition of HMG-CoA reductase

reversible opening of adherens junctions could be important for focal delivery of drugs to tissues difficult to reach (as after ischemia or as the CNS which is protected by the BBB). Delivery of drugs to the CNS has also been attempted using physical systems for focal drug delivery, such as MRI-guided focused ultrasound to activate local drug release from nanoparticles [101]. Table 6.2 reports a more extensive list of drugs and substances modulating the stability of adherens junctions.

6.5 Conclusions

Adherens junctions in endothelial cells represent an extremely complex signaling center, the molecular details of which are now starting to be described. Endothelial adherens junctions play a central role in coordinating the state of cell-to-cell interaction with the transcriptional responses. It remains to be defined which molecular specificities characterize adherens junctions in the endothelium of different types of vessels (artery versus vein, large vessels versus microvessels) and organs. This knowledge will be useful to envisage treatments that, targeting the function of specific components of endothelial adherens junction, could show efficacy and selectivity in the therapy of several human pathologies.

In general, Rac activation mostly results in stabilization of cell-to-cell contacts through inhibition of Rho and stabilization of cortical actin cytoskeleton [96]. However, Rac activation can also induce increased vascular permeability through the activation of nicotinamide adenine dinucleotide phosphate (NADPH) oxidase and production of reactive oxygen species [106].

On the contrary, Rho activation means decreased barrier function as a consequence of ROCK activation, actin stress fiber organization, phosphorylation of myosin light chain, and actomyosin contraction. This mechanism has been demonstrated to play an important role in CCM pathology in which loss of CCM expression induces activation of Rho and destabilization of adherens junctions [99, 109]. However, limited Rho activation by the angiopoietin-activated Tie2 receptor can stimulate mDia to bind Src. Sequestration of Src blocks the sequence of signaling steps that ensues in VE-cadherin phosphorylation on S665, VE-cadherin internalization, and enhanced permeability [27]. In addition, transient activation of Rho at junctions may be required in the early stages of cell-to cell adhesion [3, 111].

Furthermore, catenin p120, in addition to stabilizing VE-cadherin at the plasma membrane inhibiting its clathrin-dependent internalization [69], can enhance the adhesive contact area of endothelial cells activating Rac-dependent cell spreading [74]. p120 has also been shown to inhibit Rho in other cell types [3].

Localization to adherens junctions of small GTPases and their regulators, GEFs and GAPs, is crucial to locally control the activity of small GTPases and modulate specific targets. This has been demonstrated, for example, for Rap1 [32]. Molecular complexes with VE-cadherin have been shown for some GTPases, GEF and GAP, as reported in the footnote to Table 6.1. Small GTPases and respective GEFs and GAPs are aligned in parallel columns.

A selection of representative agonists has been reported. Destabilization of the adherens junction is induced by inflammatory and angiogenic factors. Endogenous mediators, among which cAMP-increasing agents (such as adrenomedullin, prostacyclin, prostaglandin E2 (PGE2), and β-adrenergic agonists) can induce stabilization. Interesting experimental compounds include 8-pCPT-2′-Me-cAMP (007), a synthetic, cell-membrane permeable cAMP analog that specifically activates Epac, the Rap1 GEF [32], and Comp-Ang1, the soluble and stable variant of Ang1 in which the coiled-coil domain (45 amino acid) of the cartilage oligomeric matrix protein COMP replaces the N-terminal portion of Ang1 (245 amino acid) [41]. At least two drugs in clinical use have been demonstrated to effectively restore destabilized endothelial junctions, i.e. simvastasin and fasudil, which repair adherens junctions of endothelial cells affected *in vitro* and *in vivo* by either loss-of-function mutation (heterozygous) of *CCM1* and *CCM2* genes or by siRNA downregulated expression of CCM1 and CCM2. In addition, we have observed that the metabolites of sulindac (another drug in clinical use), sulindac sulfide, and sulfone effectively regularize adherens junctions of CCM3-ablated endothelial cells, both *in vitro* and *in vivo* [8].

References

1. Adam AP, Sharenko AL, Pumiglia K, Vincent PA. Src-induced tyrosine phosphorylation of VE-cadherin is not sufficient to decrease barrier function of endothelial monolayers. J Biol Chem. 2010;285:7045–55.
2. Akers AL, Johnson E, Steinberg GK, Zabramski JM, Marchuk DA. Biallelic somatic and germline mutations in cerebral cavernous malformations (CCMs): evidence for a two-hit mechanism of CCM pathogenesis. Hum Mol Genet. 2009;18:919–30.
3. Anastasiadis PZ. p120-ctn: a nexus for contextual signaling via Rho GTPases. Biochim Biophys Acta. 2007;1773:34–46.
4. Baluk P, Hashizume H, McDonald DM. Cellular abnormalities of blood vessels as targets in cancer. Curr Opin Genet Dev. 2005;15:102–11.
5. Baluk P, et al. Functionally specialized junctions between endothelial cells of lymphatic vessels. J Exp Med. 2007;204:2349–62.
6. Birukova AA, Tian X, Tian Y, Higginbotham K, Birukov KG. Rap-afadin axis in control of Rho signaling and endothelial barrier recovery. Mol Biol Cell. 2013;24:2678–88.
7. Boulday G, et al. Developmental timing of CCM2 loss influences cerebral cavernous malformations in mice. J Exp Med. 2011;208:1835–47.
8. Bravi L, et al. Sulindac metabolites decrease cerebrovascular malformations in CCM3-knockout mice. Proc Natl Acad Sci U S A. 2015;112(27):8421–6.
9. Brunet A, et al. Akt promotes cell survival by phosphorylating and inhibiting a Forkhead transcription factor. Cell. 1999;96:857–68.
10. Carmeliet P. Angiogenesis in health and disease. Nat Med. 2003;9:653–60.
11. Carmeliet P, et al. Targeted deficiency or cytosolic truncation of the VE-cadherin gene in mice impairs VEGF-mediated endothelial survival and angiogenesis. Cell. 1999;98:147–57.
12. Cattelino A, et al. The conditional inactivation of the beta-catenin gene in endothelial cells causes a defective vascular pattern and increased vascular fragility. J Cell Biol. 2003;162:1111–22.
13. Chen XL, et al. VEGF-induced vascular permeability is mediated by FAK. Dev Cell. 2012;22:146–57.
14. Chrzanowska-Wodnicka M, Kraus AE, Gale D, White GC 2nd & Vansluys J. Defective angiogenesis, endothelial migration, proliferation, and MAPK signaling in Rap1b-deficient mice. Blood. 2008;111:2647–56.
15. Claesson-Welsh L, Welsh M. VEGFA and tumour angiogenesis. J Intern Med. 2013;273:114–27.
16. Clevers H, Nusse R. Wnt/beta-catenin signaling and disease. Cell. 2012;149:1192–205.
17. Collisson EA, Carranza DC, Chen IY, Kolodney MS. Isoprenylation is necessary for the full invasive potential of RhoA overexpression in human melanoma cells. J Invest Dermatol. 2002;119:1172–6.
18. Corada M, et al. Vascular endothelial-cadherin is an important determinant of microvascular integrity in vivo. Proc Natl Acad Sci U S A. 1999;96:9815–20.
19. Corada M, et al. Monoclonal antibodies directed to different regions of vascular endothelial cadherin extracellular domain affect adhesion and clustering of the protein and modulate endothelial permeability. Blood. 2001;97:1679–84.
20. Cristante E, et al. Identification of an essential endogenous regulator of blood-brain barrier integrity, and its pathological and therapeutic implications. Proc Natl Acad Sci U S A. 2013;110:832–41.
21. Daneman R, et al. Wnt/beta-catenin signaling is required for CNS, but not non-CNS, angiogenesis. Proc Natl Acad Sci U S A. 2009;106:641–6.
22. Dejana E, Tournier-Lasserve E, Weinstein BM. The control of vascular integrity by endothelial cell junctions: molecular basis and pathological implications. Dev Cell. 2009;16:209–21.
23. Fagotto F. Looking beyond the Wnt pathway for the deep nature of beta-catenin. EMBO Rep. 2013;14:422–33.

24. Fukuhara S, et al. Differential function of Tie2 at cell-cell contacts and cell-substratum contacts regulated by angiopoietin-1. Nat Cell Biol. 2008;10:513–26.
25. Gault J, et al. Cerebral cavernous malformations: somatic mutations in vascular endothelial cells. Neurosurgery. 2009;65:138–44; discussion 144–5.
26. Gavard J, Gutkind JS. VEGF controls endothelial-cell permeability by promoting the beta-arrestin-dependent endocytosis of VE-cadherin. Nat Cell Biol. 2006;8:1223–34.
27. Gavard J, Patel V, Gutkind JS. Angiopoietin-1 prevents VEGF-induced endothelial permeability by sequestering Src through mDia. Dev Cell. 2008;14:25–36.
28. Giampietro C, et al. Overlapping and divergent signaling pathways of N-cadherin and VE-cadherin in endothelial cells. Blood. 2012;119:2159–70.
29. Gingras AR, Liu JJ, Ginsberg MH. Structural basis of the junctional anchorage of the cerebral cavernous malformations complex. J Cell Biol. 2012;199:39–48.
30. Glading AJ, Ginsberg MH. Rap1 and its effector KRIT1/CCM1 regulate beta-catenin signaling. Dis Model Mech. 2010;3:73–83.
31. Glading A, Han J, Stockton RA, Ginsberg MH. KRIT-1/CCM1 is a Rap1 effector that regulates endothelial cell cell junctions. J Cell Biol. 2007;179:247–54.
32. Gloerich M, Bos JL. Regulating Rap small G-proteins in time and space. Trends Cell Biol. 2011;21:615–23.
33. Goentoro L, Kirschner MW. Evidence that fold-change, and not absolute level, of beta-catenin dictates Wnt signaling. Mol Cell. 2009;36:872–84.
34. Gore AV, Lampugnani MG, Dye L, Dejana E, Weinstein BM. Combinatorial interaction between CCM pathway genes precipitates hemorrhagic stroke. Dis Model Mech. 2008;1:275–81.
35. Hayashi M, et al. VE-PTP regulates VEGFR2 activity in stalk cells to establish endothelial cell polarity and lumen formation. Nat Commun. 2013;4:1672.
36. He Y, et al. Stabilization of VEGFR2 signaling by cerebral cavernous malformation 3 is critical for vascular development. Sci Signal. 2010;3:ra26.
37. Herwig MC, Tsokos M, Hermanns MI, Kirkpatrick CJ, Muller AM. Vascular endothelial cadherin expression in lung specimens of patients with sepsis-induced acute respiratory distress syndrome and endothelial cell cultures. Pathobiology. 2013;80:245–51.
38. Iden S, et al. A distinct PAR complex associates physically with VE-cadherin in vertebrate endothelial cells. EMBO Rep. 2006;7:1239–46.
39. Kam Y, Quaranta V. Cadherin-bound beta-catenin feeds into the Wnt pathway upon adherens junctions dissociation: evidence for an intersection between beta-catenin pools. PloS One. 2009;4:e4580.
40. Kleaveland B, et al. Regulation of cardiovascular development and integrity by the heart of glass-cerebral cavernous malformation protein pathway. Nat Med. 2009;15:169–76.
41. Koh GY. Orchestral actions of angiopoietin-1 in vascular regeneration. Trends Mol Med. 2013;19:31–9.
42. Kooistra MR, Corada M, Dejana E, Bos JL. Epac1 regulates integrity of endothelial cell junctions through VE-cadherin. FEBS Lett. 2005;579:4966–72.
43. Korswagen HC, Herman MA, Clevers HC. Distinct beta-catenins mediate adhesion and signalling functions in C. elegans. Nature. 2000;406:527–32.
44. Kovacic JC, Mercader N, Torres M, Boehm M, Fuster V. Epithelial-to-mesenchymal and endothelial-to-mesenchymal transition: from cardiovascular development to disease. Circulation. 2012;125:1795–808.
45. Lambeng, N. et al. Vascular endothelial-cadherin tyrosine phosphorylation in angiogenic and quiescent adult tissues. Circ Res. 2005;96:384–91.
46. Lampugnani MG. Endothelial cell-to-cell junctions: adhesion and signaling in physiology and pathology. Cold Spring Harb Perspect Med 2012;2:pii: a006528.
47. Lampugnani MG, et al. A novel endothelial-specific membrane protein is a marker of cell-cell contacts. J Cell Biol. 1992;118:1511–22.
48. Lampugnani MG, et al. VE-cadherin regulates endothelial actin activating Rac and increasing membrane association of Tiam. Mol Biol Cell. 2002;13:1175–89.

49. Lampugnani MG, et al. Contact inhibition of VEGF-induced proliferation requires vascular endothelial cadherin, beta-catenin, and the phosphatase DEP-1/CD148. J Cell Biol. 2003;161:793–804.
50. Lampugnani MG, Orsenigo F, Gagliani MC, Tacchetti C, Dejana E. Vascular endothelial cadherin controls VEGFR-2 internalization and signaling from intracellular compartments. J Cell Biol. 2006;174:593–604.
51. Lampugnani MG, et al. CCM1 regulates vascular-lumen organization by inducing endothelial polarity. J Cell Sci. 2010;123:1073–80.
52. Lenard A, et al. In vivo analysis reveals a highly stereotypic morphogenetic pathway of vascular anastomosis. Dev Cell. 2013;25:492–506.
53. Liao F, et al. Selective targeting of angiogenic tumor vasculature by vascular endothelial-cadherin antibody inhibits tumor growth without affecting vascular permeability. Cancer Res. 2002;62:2567–75.
54. Liebner S, et al. Beta-catenin is required for endothelial-mesenchymal transformation during heart cushion development in the mouse. J Cell Biol. 2004;166:359–67.
55. Liebner S, et al. Wnt/beta-catenin signaling controls development of the blood-brain barrier. J Cell Biol. 2008;183:409–17.
56. Lilien J, Balsamo J. The regulation of cadherin-mediated adhesion by tyrosine phosphorylation/dephosphorylation of beta-catenin. Curr Opin Cell Biol. 2005;17:459–65.
57. Lopez D, Niu G, Huber P, Carter WB. Tumor-induced upregulation of Twist, Snail, and Slug represses the activity of the human VE-cadherin promoter. Arch Biochem Biophys. 2009;482:77–82.
58. Maddaluno L, et al. EndMT contributes to the onset and progression of cerebral cavernous malformations. Nature. 2013;498:492–6.
59. Maher MT, Flozak AS, Stocker AM, Chenn A, Gottardi CJ. Activity of the beta-catenin phosphodestruction complex at cell-cell contacts is enhanced by cadherin-based adhesion. J Cell Biol. 2009;186:219–28.
60. Mandai K, Rikitake Y, Shimono Y, Takai Y. Afadin/AF-6 and canoe: roles in cell adhesion and beyond. Prog Mol Biol Translational Sci. 2013;116:433–54.
61. May C, et al. Identification of a transiently exposed VE-cadherin epitope that allows for specific targeting of an antibody to the tumor neovasculature. Blood. 2005;105:4337–44.
62. McDonald DA, et al. A novel mouse model of cerebral cavernous malformations based on the two-hit mutation hypothesis recapitulates the human disease. Hum Mol Genet. 2011;20:211–22.
63. McDonald DA, et al. Fasudil decreases lesion burden in a murine model of cerebral cavernous malformation disease. Stroke. 2012;43:571–4.
64. Medici D, Kalluri R. Endothelial-mesenchymal transition and its contribution to the emergence of stem cell phenotype. Semin Cancer Biol. 2012;22:379–84.
65. Medici D, Hay ED, Olsen BR. Snail and Slug promote epithelial-mesenchymal transition through beta-catenin-T-cell factor-4-dependent expression of transforming growth factor-beta3. Mol Biol Cell. 2008;19:4875–87.
66. Miyoshi J, Takai Y. Nectin and nectin-like molecules: biology and pathology. Am J Nephrol. 2007;27:590–604.
67. Monaghan-Benson E, Burridge K. The regulation of vascular endothelial growth factor-induced microvascular permeability requires Rac and reactive oxygen species. J Biol Chem. 2009;284:25602–11.
68. Montero-Balaguer M, et al. Stable vascular connections and remodeling require full expression of VE-cadherin in zebrafish embryos. PloS ONE. 2009;4:e5772.
69. Nanes BA, et al. p120-catenin binding masks an endocytic signal conserved in classical cadherins. J Cell Biol. 2012;199:365–80.
70. Navarro P, et al. Catenin-dependent and -independent functions of vascular endothelial cadherin. J Biol Chem. 1995;270:30965–72.
71. Nelson WJ. Regulation of cell-cell adhesion by the cadherin-catenin complex. Biochem Soc Trans. 2008;36:149–55.

72. Nitta T, et al. Size-selective loosening of the blood-brain barrier in claudin-5-deficient mice. J Cell Biol. 2003;161:653–60.
73. Noda K, et al. Vascular endothelial-cadherin stabilizes at cell-cell junctions by anchoring to circumferential actin bundles through alpha- and beta-catenins in cyclic AMP-Epac-Rap1 signal-activated endothelial cells. Mol Biol Cell. 2010;21:584–96.
74. Oas RG, et al. p120-catenin and beta-catenin differentially regulate cadherin adhesive function. Mol Biol Cell. 2013;24:704–14.
75. Orsenigo F, et al. Phosphorylation of VE-cadherin is modulated by haemodynamic forces and contributes to the regulation of vascular permeability in vivo. Nat Commun. 2012;3:1208.
76. Pagenstecher A, Stahl S, Sure U, Felbor U. A two-hit mechanism causes cerebral cavernous malformations: complete inactivation of CCM1, CCM2 or CCM3 in affected endothelial cells. Hum Mol Genet. 2009;18:911–8.
77. Pannekoek WJ, et al. Epac1 and PDZ-GEF cooperate in Rap1 mediated endothelial junction control. Cell Signal. 2011;23:2056–64.
78. Paolinelli R, et al. Wnt activation of immortalized brain endothelial cells as a tool for generating a standardized model of the blood brain barrier in vitro. PloS ONE. 2013;8:e70233.
79. Park HJ, et al. 3-hydroxy-3-methylglutaryl coenzyme A reductase inhibitors interfere with angiogenesis by inhibiting the geranylgeranylation of RhoA. Circ Res. 2002;91:143–50.
80. Petit N, Blecon A, Denier C, Tournier-Lasserve E. Patterns of expression of the three cerebral cavernous malformation (CCM) genes during embryonic and postnatal brain development. Gene Expr Patterns. 2006;6:495–503.
81. Potter MD, Barbero S, Cheresh DA. Tyrosine phosphorylation of VE-cadherin prevents binding of p120- and beta-catenin and maintains the cellular mesenchymal state. J Biol Chem. 2005;280:31906–12.
82. Potts JD, Dagle JM, Walder JA, Weeks DL, Runyan RB. Epithelial-mesenchymal transformation of embryonic cardiac endothelial cells is inhibited by a modified antisense oligodeoxynucleotide to transforming growth factor beta 3. Proc Natl Acad Sci U S A. 1991;88:1516–20.
83. Rampersad SN, et al. Cyclic AMP phosphodiesterase 4D (PDE4D) Tethers EPAC1 in a vascular endothelial cadherin (VE-Cad)-based signaling complex and controls cAMP-mediated vascular permeability. J Biol Chem. 2010;285:33614–22.
84. Revencu N, et al. Parkes Weber syndrome, vein of Galen aneurysmal malformation, and other fast-flow vascular anomalies are caused by RASA1 mutations. Hum Mutat. 2008;29:959–65.
85. Riant F, Bergametti F, Ayrignac X, Boulday G, Tournier-Lasserve E. Recent insights into cerebral cavernous malformations: the molecular genetics of CCM. FEBS J. 2010;277:1070–5.
86. Roberts TK, et al. CCL2 disrupts the adherens junction: implications for neuroinflammation. Lab Invest. 2012;92:1213–33.
87. Roura S, Miravet S, Piedra J, Garcia de Herreros A, Dunach M. Regulation of E-cadherin/Catenin association by tyrosine phosphorylation. J Biol Chem. 1999;274:36734–40.
88. Rudini N, et al. VE-cadherin is a critical endothelial regulator of TGF-beta signalling. EMBO J. 2008;27:993–1004.
89. Saharinen P, et al. Angiopoietins assemble distinct Tie2 signalling complexes in endothelial cell-cell and cell-matrix contacts. Nat Cell Biol. 2008;10:527–37.
90. Sakurai A, et al. MAGI-1 is required for Rap1 activation upon cell-cell contact and for enhancement of vascular endothelial cadherin-mediated cell adhesion. Mol Biol Cell. 2006;17:966–76.
91. Sanson B, White P, Vincent JP. Uncoupling cadherin-based adhesion from wingless signalling in Drosophila. Nature. 1996;383:627–30.
92. Sawada J, et al. Small GTPase R-Ras regulates integrity and functionality of tumor blood vessels. Cancer Cell. 2012;22:235–49.
93. Schlessinger K, Hall A, Tolwinski N. Wnt signaling pathways meet Rho GTPases. Genes Dev. 2009;23:265–77.
94. Serebriiskii I, Estojak J, Sonoda G, Testa JR, Golemis EA. Association of Krev-1/rap1a with Krit1, a novel ankyrin repeat-containing protein encoded by a gene mapping to 7q21–22. Oncogene. 1997;15:1043–9.

95. Shenkar R, et al. Advanced magnetic resonance imaging of cerebral cavernous malformations: part II. Imaging of lesions in murine models. Neurosurgery. 2008;63:790–7 (discussion 797–8).
96. Spindler V, Schlegel N, Waschke J. Role of GTPases in control of microvascular permeability. Cardiovasc Res. 2010;87:243–53.
97. Stahl S, et al. Novel CCM1, CCM2, and CCM3 mutations in patients with cerebral cavernous malformations: in-frame deletion in CCM2 prevents formation of a CCM1/CCM2/CCM3 protein complex. Hum Mutat. 2008;29:709–17.
98. Stenman JM, et al. Canonical Wnt signaling regulates organ-specific assembly and differentiation of CNS vasculature. Science. 2008;322:1247–50.
99. Stockton RA, Shenkar R, Awad IA, Ginsberg MH. Cerebral cavernous malformations proteins inhibit Rho kinase to stabilize vascular integrity. J Exp Med. 2010;207:881–96.
100. Taddei A, et al. Endothelial adherens junctions control tight junctions by VE-cadherin-mediated upregulation of claudin-5. Nat Cell Biol. 2008;10:923–34.
101. Thanou M, Gedroyc W. MRI-guided focused ultrasound as a new method of drug delivery. J Drug Deliv. 2013;2013:616197.
102. Timmerman I, et al. The tyrosine phosphatase SHP2 regulates recovery of endothelial adherens junctions through control of beta-catenin phosphorylation. Mol Biol Cell. 2012;23:4212–25.
103. Turowski P, et al. Phosphorylation of vascular endothelial cadherin controls lymphocyte emigration. J Cell Sci. 2008;121:29–37.
104. Uebelhoer M, Boon LM, Vikkula M. Vascular anomalies: from genetics toward models for therapeutic trials. Cold Spring Harbor Perspect Med. 2012;2:pii: a009688.
105. Ukropec JA, Hollinger MK, Salva SM, Woolkalis MJ. SHP2 association with VE-cadherin complexes in human endothelial cells is regulated by thrombin. J Biol Chem. 2000;275:5983–6.
106. van Wetering S, et al. Reactive oxygen species mediate Rac-induced loss of cell-cell adhesion in primary human endothelial cells. J Cell Sci. 2002;115:1837–46.
107. Voss K, et al. CCM3 interacts with CCM2 indicating common pathogenesis for cerebral cavernous malformations. Neurogenetics. 2007;8:249–56.
108. Whitehead KJ, Plummer NW, Adams JA, Marchuk DA, Li DY. Ccm1 is required for arterial morphogenesis: implications for the etiology of human cavernous malformations. Development. 2004;131:1437–48.
109. Whitehead KJ, et al. The cerebral cavernous malformation signaling pathway promotes vascular integrity via Rho GTPases. Nat Med. 2009;15:177–84.
110. Wildenberg GA, et al. p120-catenin and p190RhoGAP regulate cell-cell adhesion by coordinating antagonism between Rac and Rho. Cell. 2006;127:1027–39.
111. Wimmer R, Cseh B, Maier B, Scherrer K, Baccarini M. Angiogenic sprouting requires the fine tuning of endothelial cell cohesion by the Raf-1/Rok-alpha complex. Dev Cell. 2012;22:158–71.
112. Xie Y, et al. Genomic characteristics of adhesion molecules in patients with symptomatic pulmonary embolism. Mol Med Rep. 2012;6:585–90.
113. Yamashita K, et al. Fasudil, a Rho kinase (ROCK) inhibitor, protects against ischemic neuronal damage in vitro and in vivo by acting directly on neurons. Brain Res. 2007;1154:215–24.
114. Zawistowski JS, et al. CCM1 and CCM2 protein interactions in cell signaling: implications for cerebral cavernous malformations pathogenesis. Hum Mol Genet. 2005;14:2521–31.
115. Zhang X, et al. Phosphorylation of serine 256 suppresses transactivation by FKHR (FOXO1) by multiple mechanisms. Direct and indirect effects on nuclear/cytoplasmic shuttling and DNA binding. J Biol Chem. 2002;277:45276–84.
116. Zhou B, et al. Interactions between beta-catenin and transforming growth factor-beta signaling pathways mediate epithelial-mesenchymal transition and are dependent on the transcriptional co-activator cAMP-response element-binding protein (CREB)-binding protein (CBP). J Biol Chem. 2012;287:7026–38.

Chapter 7
More than a Scaffold: Extracellular Matrix in Vascular Signaling

Iva Nikolic

7.1 Extracellular Matrix (ECM): Ligands and Receptors

7.1.1 ECM Proteins

Extracellular matrix (ECM) proteins comprise a diverse network of molecules that can be broadly classified into two main classes of macromolecules: proteoglycans (PGs) and fibrous proteins such as collagens and laminins. Nonfibrous PGs are extremely hydrophilic and they fill the majority of interstitial space, providing mechanical buffering and hydration to the matrices. In contrast, fibrous proteins form highly enmeshed fibrils, which not only act predominantly as structural elements of the ECM but also as regulators of different cellular functions. In addition, matricellular proteins and matrikines have recently been recognized as novel nonstructural ECM components that are able to further modify cell–matrix interactions and perform a broad spectrum of functions in a tissue-specific manner.

Proteoglycans

PGs encompass a group of over 30 different species of macromolecules consisting of glycosaminoglycan chains (GAG) that are covalently bound to a specific protein core; based on differences in these components, they can be classified into small leucine-rich, modular, and cell-surface PGs [1] (Fig. 7.1a, e, Table 7.1) Although PGs constitute a minor component of vascular tissue, these macromolecules regulate vessel properties such as elasticity, permeability, hemostasis, and thrombosis [2]. In addition, a subset of modular PGs is found in basement membranes, where

I. Nikolic (✉)
Cancer Research Division, The Garvan Institute of Medical Research, 384 Victoria Street, Sydney, NSW 2010, Australia
e-mail: i.nikolic@garvan.org.au

M. HH. Schmidt, S. Liebner (eds.), *Endothelial Signaling in Development and Disease,* DOI 10.1007/978-1-4939-2907-8_7

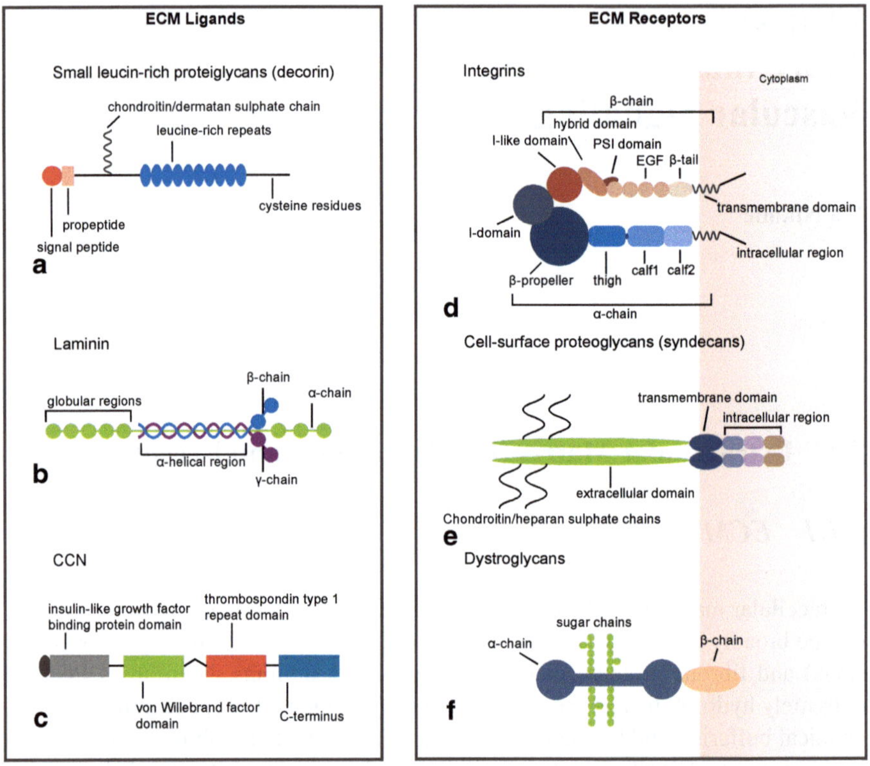

Fig. 7.1 Extracellular matrix (*ECM*) ligands and receptors. Both ECM ligands and receptors typically possess modular structure, containing domains that allow them to bind to each other as well as a broad range of different molecules. Depicted here are the representatives of (**a**) proteoglycans (*decorin*), (**b**) structural ECM proteins (*laminin*), and (**c**) matricellular proteins (*CCNs*). In addition, different types of ECM receptors, such as (**d**) integrins, (**e**) syndecans, and (**f**) dystroglycans, are included

they display both pro- and anti-angiogenic effects by binding growth factors and regulating their availability [3]. For a comprehensive review of proteoglycan biology see Couchman and Pataki [4].

Structural ECM Proteins

Collagen constitutes the most abundant component of the ECM, where it provides tensil strength, regulates cell adhesion and migration, and directs tissue development [5]. Its basic building blocks consist of three alpha chains intertwined to form a triple helix, which further assemble into extended fibrilar structures with tissue-specific alignment and distribution. A dramatic diversity of collagen molecules—at least 28 different vertebrate collagens have been reported to date—derive from distinct α-chain composition as well as diverse supramolecular structures [6]. For

Table 7.1 Representative extracellular matrix ligands and receptors in the endothelium

	Receptor	Function	Source
Collagens			
I	$\alpha_1\beta_1$; $\alpha_2\beta_1$	*EC morphogenesis; tensile strength of the vessel wall*	*Whelan and Senger [75]*
III	$\alpha_1\beta_1$; $\alpha_2\beta_1$	*Elasticity of the vessel wall*	*Heino [158]*
IV	$\alpha_1\beta_1$; $\alpha_2\beta_1$	*Structural component of vascular basement membrane*	*Kern et al. [159]*
V	$\alpha_1\beta_1$; $\alpha_2\beta_1$	*Inhibition of EC adhesion and proliferation*	*Fukuda et al. [160]*
VI	β_1 *integrins*	*Adhesion of smooth muscle cells*	*Kielty et al. [161]*
Elastin and microfibrillar proteins			
Elastin		*Elasticity of the vessel wall*	*Midwood and Schwarzbauer [162]*
Fibrillins (1–2)	$\alpha_v\beta_3$; $\alpha_5\beta_1$; $\alpha_v\beta_6$	*Structural components of microfibrils*	*Muiznieks and Keeley [6]; Jovanovic et al. [163]*
Fibulins (1–7)	$\alpha_v\beta_3$; $\alpha_v\beta_5$; $\alpha_9\beta_1$	*Structural components of microfibrils*	*Muiznieks and Keeley [6]; Timpl et al. [164]*
Laminins			
Laminin-411/511	$\alpha_3\beta_1$; $\alpha_6\beta_1$; *syndecans; dystroglycan*	*Structural component of vascular basement membrane*	*Kostourou and Papalazarou [9]*
Fibronectin	$\alpha_v\beta_3$; $\alpha_5\beta_1$	*EC migration; vessel lumen formation*	*Zou et al. [165], Wang and Milner [166]*
Matricellular proteins			
Thrombospondins (1–5)	*CD36; IAP;* β_3 *integrins*	*Inhibition of EC adhesion, migration, survival, and tube formation*	*Lawler and Lawler [167]*
CCNs (1–2)	$\alpha_v\beta_3$; $\alpha_6\beta_1$; $\alpha_{II}b\beta_3$; $\alpha_M\beta_3$	*EC adhesion, migration, differentiation, and survival*	*Chaqour [168]*
Tenascins (-C, -R, -X, -W)	$\alpha_v\beta_3$; *syndecan-4*	*EC adhesion, migration*	*van Obberghen-Schilling et al. [169]*
Small leucine-rich proteoglycans			
Decorin		*Structural component of vessel wall*	*Williams [170]*
Biglycan		*Structural component of vessel wall*	*Williams [170]*
Basement membrane proteoglycans			
Perlecan	$\alpha_2\beta_1$	*Structural component of vascular basement membrane*	*Iozzo [3]*

EC endothelial cell

instance, fibril-forming collagens pack together side-by-side to form thick fibrils, whereas network-forming collagens form open mesh-like structures with various geometries. Vessel walls, especially arterial, can contain up to 17 different collagen types, with collagens I, III, IV, V, and VI having the highest expression levels [7] (Table 7.1).

Another major fibrillar ECM protein is elastin, which imparts the property of elasticity to tissues that undergo repeated stretch, such as the lung, skin, and blood vessels [8]. In fact, elastin is one of the earliest structural matrix proteins to be expressed by vascular smooth muscle cells (SMCs) in large vessels. Similar to collagen, secreted tropoelastin monomers (precursors of elastin) assemble into elastin fibers through cross-linking of their lysine residues and activity of the lysyl oxidase (LOX) enzyme family. In addition, elastin strongly associates with a scaffold of fibrillin-rich microfibrils, which contribute to the integrity of elastin fibers [6].

Laminins are one of the main constituents of basement membranes, where they act as scaffolds necessary for the initial assembly of the membrane. Laminin molecules are heterotrimeric glycoproteins consisting of one α, one β, and one γ chain held together by disulfide bonds [9] (Fig. 7.1b). In vertebrates, different types of each of the chains have been found (α1–5, β1–3, γ1–3), and their expression varies between the cell types and during development; together, they constitute 18 different laminin trimers. The most important laminins found in endothelial basement membranes are LN 411 and LN 511, which contain α4β1γ1 and α5β1γ1 chains, respectively.

Another protein that acts as a scaffold for the assembly of other matrix components is fibronectin, a large dimer molecule that exists in different isoforms and different conformations. In addition to its role in directing the organization of the ECM, fibronectin is essential for cell attachment and migration during both physiological and pathological conditions [10]. Its ability to stretch and expose cryptic sites within its molecule enables fibronectin to elicit pleiotropic effects on cellular behavior and to modulate cellular responses [11].

Matricellular Proteins and Matrikines

Although structural ECM proteins can affect cellular functions through mechanotransduction and through binding to many cell-surface receptors, tissues employ an additional mechanism to further modulate these cell–matrix interactions—matricellular proteins. This fast-growing group of proteins appears not to contribute directly to the organization or physical properties of the ECM, but it rather modulates a broad range of cell regulatory functions through various mechanisms [12]. Typical representatives are thrombospondin (TSP), SPARC (secreted protein, acidic and rich in cysteine), tenascin, osteopontin, and CCN (Cyr61), but numerous other novel proteins such as EGFL7 [13] and MAGP-1 [14] seem to share many of their structural and functional properties. Confirming their nonstructural roles, phenotypes of mice lacking a matricellular protein are either normal or very mild, and typically exacerbated upon mechanical stress and wound healing [15, 16]. Similarly, these proteins are only expressed during embryonic development and in response to wound injury, tissue remodeling, inflammation, cancer, and other chronic diseases [17, 18]. Their structure is typically modular, containing domains that are able to bind a variety of other ECM proteins, cell-surface receptors, growth factors, cytokines, and proteases, enabling them to control an array of cellular functions in a context-dependent manner (Fig. 7.1c). In the endothelium, matricellular proteins exert either pro- or antiangiogenic effects, affecting virtually every phase of blood vessel formation (see the 'ECM and Blood Vessel Growth' section).

Finally, both structural and nonstructural ECM proteins undergo partial proteolytic cleavage, generating fragments—known as matrikines—that also exert effects on cellular functions. Typical examples include angiostatin derived from plasminogen, endostatin from collagen XVIII, restin from collagen XV, and anastellin from fibronectin, as well as SPARC- and TSP-derived fragments. These bioactive fragments play an active role in angiogenesis by affecting different steps of the angiogenic process and by employing different mechanisms of action [19].

7.1.2 ECM Receptors

Integrins

Structurally and functionally diverse ECM proteins exert their effects through different cell-surface receptors, the most important of which are integrins. This is a group of large heterodimeric transmembrane glycoproteins that is highly conserved throughout evolution, ranging from sponges to humans. Integrin receptors consist of noncovalently associated α and β subunits, and each subunit contains an extracellular domain, a single transmembrane region, and a short cytoplasmic region [20] (Fig. 7.1d). Together, 18α and 8β subunits can form 24 different integrin heterodimers, each specific for a unique set of ECM ligands.

Rather than just simple adhesion molecules, integrins act as complex relay points that transmit bidirectional signals across the plasma membrane. Upon ligation, these receptors change their conformation from inactive to active, and cluster together to form large signaling hubs, which further mediate the effects of the extracellular environment on cell proliferation, motility, survival, and other processes. In contrast, integrins are also able to respond to signals coming from inside the cell and engage in so-called inside-out signaling; this further allows the cells to control the affinity and avidity of integrin receptors for their ligands (for extensive review of integrin signaling see Hood and Cheresh [21]).

The most important vascular integrins that are expressed on the quiescent endothelium include $\alpha_1\beta_1$, $\alpha_2\beta_1$, $\alpha_3\beta_1$, $\alpha_6\beta_1$, and $\alpha_6\beta_4$, which predominantly bind to collagens and laminins, but also $\alpha_5\beta_1$ and $\alpha_v\beta_5$, which bind to fibronectin and vitronectin, respectively. In remodeling endothelium, in contrast, $\alpha_5\beta_1$ and $\alpha_v\beta_3$ undergo a dramatic increase in expression and exert diverse functions through binding an assortment of ECM ligands [22].

Other ECM Receptors

In addition to integrins, ECM proteins can signal through previously mentioned cell-surface PG syndecans. These are single transmembrane proteoglycan molecules that carry heparan sulfate and chondroitin sulfate chains, which facilitate interactions with a diverse group of proteins including different growth factors and ECM proteins such as fibronectin and laminin [23] (Fig. 7.1e). In addition, syndecans act as co-receptors that are able to further modify the activity of integrin receptors.

Finally, dystroglycans have been characterized as novel nonintegrin receptors able to bind laminins and mediate their effects [24]. They consist of a highly glycosylated α chain noncovalently anchored to a transmembrane β chain (Fig. 7.1f). Although their function still remains elusive, dystroglycans are upregulated in endothelial cells (ECs) during both physiological and pathological angiogenesis.

7.2 ECM and Blood Vessel Growth

7.2.1 Introduction

Growth of new blood vessels proceeds through two main mechanisms: vasculogenesis, which entails differentiation of mesodermal cells into angioblasts and their subsequent fusion into a primitive vascular plexus, and angiogenesis, or formation of new blood vessels from the pre-existing blood vessels [25, 26]. The former predominantly occurs during embryonic development, although certain pathological processes such as cancer involve incorporation of circulating progenitors into the vascular network; the latter process defines vessel growth not only in the adult but also during the later stages of vascular plexus remodeling.

Angiogenic sprouting ensues through a well-defined cascade of events, and its underlying molecular machinery has been extensively studied over the last decade. Angiogenic stimuli activate quiescent ECs, which then assume a hierarchical organization into highly motile tip cells and trailing stalk cells. Tip cells extend numerous filopodia, sense the environment, and lead the sprout into a certain direction while stalk cells actively proliferate and concurrently form vascular lumens. When tip cells from opposing sprouts meet, they lose their invasive phenotype and generate tight junctions, allowing formation of continuous, patent vessels and promoting their subsequent maturation. Such complex and tightly regulated events are under control of an intricate network of different molecular players spearheaded largely by the vascular endothelial growth factor (VEGF)/Notch signaling axis, which is extensively reviewed elsewhere [27]. However, angiogenic sprouting does not occur in isolation but within an elaborate fabric of ECM cues that guide cellular interactions with the surrounding environment (Fig. 7.2). Therefore, this chapter will focus on how ECM components regulate different phases of vessel growth.

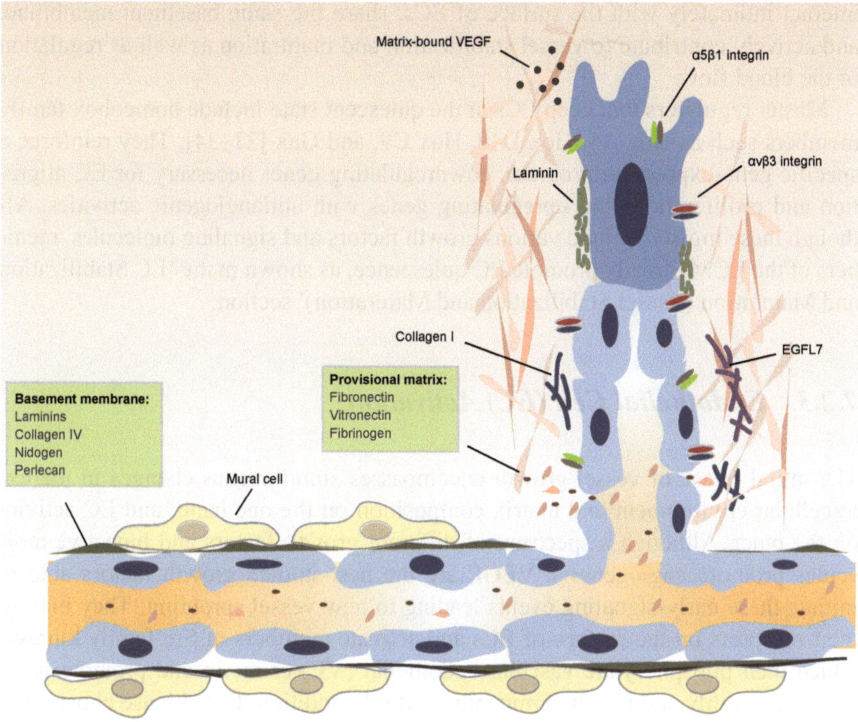

Fig. 7.2 Extracelluar matrix (*ECM*) drives physiological angiogenesis. At the onset of angiogenic sprouting, blood-borne proteins such as fibronectin, vitronectin, and fibrinogen extravasate through leaky vasculature into the perivascular space and interact with the pre-existing ECM to form a provisional matrix. In addition, the basement membrane gets degraded, exposing the endothelial cells (*ECs*) to the provisional matrix and proteins such as collagen I, which is abundantly present in the interstitial space. The activity of the ECs matches the changes in the extracellular environment, resulting in the upregulation of proangiogenic factors such as $\alpha_v\beta_3$ and $\alpha_5\beta_1$ integrins. Furthermore, ECs secrete various ECM proteins that support different phases of the angiogenic process

7.2.2 Resting Vasculature

ECs that line the walls of blood vessels in an organism are typically quiescent and packed tightly enough to facilitate the integrity of the vessel but permeable enough to enable flux of material between blood and interstitium; different combinations of tight junctions and adherens junctions, through which ECs interact with each other, regulate this permeability in a tissue-specific manner [28, 29]. However, much of the EC surface is in contact with a vascular basal membrane comprising laminins, collagen IV, nidogen, perlecan, and other PGs, which together assemble into a matrix that insulates ECs from the surrounding interstitial space. Another type of vascular cells that further regulate EC function are mural cells—smooth muscle cells in arteries, arterioles, and veins; and pericytes in capillaries and venules [30]. They interact intimately with the surface of ECs, share the same basement membrane, and actively contribute to vessel stabilization and maturation as well as regulation of the blood flow.

Master regulators that keep ECs in the quiescent state include homeobox family members such as Hox A5, Hox D10, Hox C9, and Gax [31–34]. They reinforce a specific gene-expression program, downregulating genes necessary for EC migration and proliferation, and upregulating genes with antiangiogenic activities. Although these mostly include various growth factors and signaling molecules, members of the ECM directly promote EC quiescence, as shown in the 'EC Stabilization and Maturation (Vessel Stabilization and Maturation)' section.

7.2.3 Endothelial Cell (EC) Activation

The initial phase of vessel growth encompasses simultaneous changes in the extracellular environment and matrix composition on the one hand, and EC activity on the other. Although a spectrum of different growth factors and bioactive molecules promote angiogenesis, VEGF are the best-studied growth factors able to initiate these early-signaling events leading to new vessel sprouting. They bind to their receptors on the surface of ECs and activate members of Src family kinases, which then phosphorylate vascular endothelial (VE)-cadherin and promote junction disassembly and vessel permeability [35]. In addition, FAK kinase phosphorylates β-catenin and facilitates VE-cadherin–β-catenin dissociation [36]. This allows blood-borne ECM proteins, such as fibronectin, vitronectin, and fibrinogen, to extravasate into the perivascular space, interact with the pre-existing ECM, and form a 'provisional ECM', which facilitates all the subsequent phases of vessel growth.

These events are perfectly orchestrated with the changes in EC morphology and behavior. Namely, in addition to affecting EC junctions and EC communication, angiogenic signals coordinate simultaneous deactivation of quiescence-maintaining Hox genes and induction of proangiogenic members of the Hox gene family, such as HoxB3, HoxD3, and HoxA9 [37–39], which promote invasive EC phenotype [39]. For instance, activated ECs rapidly change the repertoire of integrin receptors on their surface, dramatically upregulating $\alpha_v\beta_5$, $\alpha_5\beta_1$ and, in particular, $\alpha_v\beta_3$ integrin, which preferentially bind components of the provisional ECM [22]. In addition, Hox genes control the expression of different proteases as well as Eph receptors, which further regulate sprout formation. Taken together, changes in EC phenotype upon angiogenic signaling match the dynamics of remodeling ECM, enabling the cells to sense and respond to the new matrix topology.

7.2.4 Tip/Stalk Cell Selection

As mentioned in the introduction, upon activation ECs establish a hierarchy of leading tip cells and trailing stalk cells. Although it is well established that VEGF-induced Dll4 upregulation in tip cells and activation of the Notch pathway in the stalk cells define tip/stalk cell phenotypes, there has been evidence that ECM plays a role in tip cell selection. Namely, Estrach et al. demonstrated that ligation of $\alpha_2\beta_1$ and $\alpha_6\beta_1$ integrins by laminin-111 induces FoxC2 signaling and high Dll4 expression [40]. Moreover, Lama4 mutant mice display excessive filopodial branching and tip cell formation as a result of decreased Dll4 signaling [41]. Finally, gene expression profiling of ECs enriched for tip cells revealed laminin β_1 as one of the upregulated genes, reinforcing the hypothesis that basement membrane components such as laminins could have additional functions in addition to their role in vessel maturation and stabilization [42]. In addition, other extracellular-associated proteins were shown to directly affect Notch signaling. Namely, both EGFL7 and MAGP-2 act as Notch antagonists in ECs, suggesting they may actively participate in regulating the balance between tip and stalk cells within the angiogenic sprout [43–45].

7.2.5 EC Invasion and Migration

To lead the sprout in the right direction, tip cells interact with and modify the surrounding matrix. Therefore, a group of factors specifically enriched in these cells involves ECM degrading enzymes such as cathepsin S, a disintegrin and metalloproteinase with TSP motifs (ADAMTS), and urokinase-plasminogen-activated receptor (uPAR) [42]. These proteases release growth factors which are stably deposited in the ECM through binding other proteins and PGs. For instance, both VEGF and basic fibroblast growth factor (bFGF) associate with heparan-like glycosaminoglycans and require haparinases and proteases to become fully active [46]. In addition, plasminogen activators release transforming growth factor (TGF)-β, which then further promotes expression of angiogenic signals and ECM-degrading proteases [47].

Furthermore, proteases participate in the degradation of the basement membrane and remodeling of the surrounding matrix, thus regulating EC migration on multiple levels [48]. Namely, disassembly of the basement membrane exposes ECs to high concentrations of interstitial collagen I, which drives morphogenesis of new vessel sprouts. This abundant ECM protein supports VEGF-induced EC migration through β_1 integrin signaling and Erk1/Erk2 activation, but it also drives directed cell migration in the absence of angiogenic factors [49]. Similarly, components of the provisional ECM, especially fibronectin, play a crucial role in this phase of angiogenic sprouting, and highly motile tip cells, for instance, show enrichment for β_1 integrin, a fibronectin receptor [42]. These ECM proteins and their integrin receptors modulate EC migration by controlling cellular adhesion and by allowing functional connection between focal adhesions and actin cytoskeleton [50]. Ultimately, reorganization of the cytoskeleton enables ECs to complete their migration cycle through activation of small Rho GTPases, including CDC42, Rac1, and RhoA [51, 52]. However, the control of EC migration through cellular adhesion seems to be slightly more complex. Namely, fibronectin plays an additional role of orchestrating ECM matrix assembly by acting as a scaffold for a range of proteins produced by activated ECs [53]. The most intriguing subset includes matricellular proteins, which display unique effects on EC adhesion. When plated on tenascin-C and TSP-1, ECs do not undergo actin cytoskeleton remodeling and stress-fiber formation, although they engage integrin receptors and spread to a certain extent [54, 55]. In addition, soluble matricellular proteins induce focal adhesion restructuring and alterations in the stress fibers in strongly adherent cells without affecting integrin clustering and cell shape [56]. Such intermediate state of adhesion appears to favor cellular motility, enabling maximal migration [57]. In contrast, strong adhesion prevents the turnover of ECM-cellular contacts, and weak adhesion does not generate contractile force necessary for directed cellular migration [58]. Interestingly, although numerous members of the matricellular protein family, as well as some novel ECM-associated proteins such as EGFL7, appear to promote EC migration in such a manner, they all employ unique receptors as well as distinct signaling pathways to induce the state of intermediate adhesion [59, 60]. This strongly argues against their functional redundancy and suggests that their roles are highly contextual.

7.2.6 EC Proliferation and Tube Formation

While tip cells explore the surrounding environment and guide the sprout in the right direction, stalk cells actively proliferate and form lumens to extend and stabilize the sprout. Although the current model of vascular branching explains how the balance of Notch and Wnt signaling in the stalk cells maintains their active proliferation [61, 62], components of the ECM are indispensable for the regulation of both cell cycle and cell survival. In particular, the Ras-mitogen-activated protein (MAP) pathway seems to act as a master regulator of these cellular functions. For instance, fibronectin and vitronectin engage a subset of β_1 integrins and $\alpha_v\beta_3$ integrin to activate Shc adaptor protein and Erk2 kinase, which in turn regulates cell-cycle progression [63]. In addition, integrins seem to cooperate with receptor tyrosine kinases (RTKs) to regulate the activity of cyclin-dependent kinases [64]. On the other hand, EC survival is largely dependent on cell adhesion and spreading, as well as activation of FAK and PI3K kinase signaling [65, 66]. Indeed, disruption of ECM-integrin interactions promotes apoptotic cell death [67]. In this context, matricellular proteins that maintain intermediate cell adhesion could be particularly important for the advancing vascular sprout as they would keep apoptosis at bay in migrating cells.

As previously mentioned, stalk cells following the invading tip cells eventually undergo complex molecular changes to form vascular lumens, although it still remains unclear whether this happens concurrently or subsequently to the sprout invasion [68, 69]. Numerous studies, using elegant *in vitro* and *in vivo* models, have demonstrated that this process requires a precise sequence of events and that it occurs through different mechanisms, depending on the size of the vessel or type of the vascular bed [70]. The first mechanism, also known as cell hollowing, has been demonstrated both *in vitro* and *in vivo*, and it involves formation of intracellular vacuoles that subsequently fuse and enable the cells to form continuous lumens [71, 72]. In contrast, cord hollowing occurs when lumen appears between the adjacent cells following complex junctional rearrangements, and changes in cell polarity and cell shape [73, 74]. However, a common feature of these different mechanisms is that they critically depend on numerous integrin–ECM interactions. For instance, *in vitro* studies have shown that collagen and fibrin/fibronectin matrices induce EC tubular morphogenesis through $\alpha_2\beta_1$, $\alpha_1\beta_2$, and $\alpha_v\beta_3$, $\alpha_5\beta_1$ integrins, respectively [75, 76]. Indeed, inhibition or ablation of β_1 integrin prevents lumen formation in chicken embryos [77] as well as mouse embryos through disruption of EC polarity [78]. These ECM–integrin interactions activate Src and FAK kinases as well as Rho GTPases to control intracellular vacuole formation and their coalescence [79, 80]. In contrast, ECM displays additional functions during cord hollowing. For example, it has recently been shown that a matricellular-like protein EGFL7 regulates EC adhesion and cell shape through activation of the RhoA pathway, enabling the formation of central lumen between the neighboring ECs [81, 82]. Finally, several recent reports revealed a surprising discovery that members of the vascular basement membrane do not only play a role in tube stabilization but also act earlier during the phase of lumen formation. Deletion of laminin γ_1 in stem cells increases lumen diameter in angiogenic sprouts, and inhibition of basement membrane deposition in three-dimensional EC culture leads to lumen enlargement [83]. Interestingly, it seems that this is intricately connected with the sprout invasion and the tip cells themselves, as they show high expression of both nidogen-1 and nidogen-2, as well as laminin β_1 [42].

7.2.7 Vessel Stabilization and Maturation

Finally, to end the angiogenic cycle and assume their resting phenotype, newly formed angiogenic branches recruit mural cells through activation of signaling pathways such as PDGF-(PDGF)-β, Ang1-Tie2, TGF-β, and SIP1-EDG1 [84]. One of the major functions that mural cells exert during this phase is production and deposition of vascular basement membrane matrix [85], which directly contributes to EC quiescence. For instance, both laminin and collagen XVIII knockout mice show increased angiogenic sprouting following application of angiogenic factors [86, 87], and these basement membrane proteins may directly supress EC proliferation and tube formation [88, 89]. To reinforce this process, a matricellular protein, CCN2, promotes pericyte recruitment by potentiating platelet-derived growth factor (PDGF) signaling; it also directs basement membrane assembly [90]. Another way through which basement membrane proteins stabilize angiogenic branches is the generation of matrikines. For instance, collagen XVIII undergoes partial proteolysis to form endostatin, which induces cell-cycle arrest and inhibits EC migration [91, 92]. Likewise, numerous fragments derived from collagen IV, including arresten, canstatin, and turmstatin, exert antiangiogenic activity through α_v and β_1 integrins, and through inhibition of matrix metalloproteinase (MMP) activity [19, 93]. Finally, concurrent with these dynamic changes in the extracellular environment, ECs and mural cells shift their gene expression program towards a more mature phenotype. ECs upregulate α_3 and α_6 integrin messenger RNA (mRNA) while pericytes dramatically increase the expression of α_1, α_3, and α_6 integrin—receptors that preferentially bind laminins, nidogens, and collagen IV in the basement membrane [85]. In addition, activation of Hox genes that maintain EC quiescence (see the 'Resting EC' section) dramatically upregulate TSP-2, which inhibits EC adhesion, migration, survival, and tube formation [94, 95].

7.3 ECM and Tumor Vascularization

7.3.1 Introduction

One of the major steps during tumor growth and expansion is the development of a tumor vascular network, a process also known as angiogenic switch, which allows tumors to maintain their oxygen and nutrient supply and to remove the waste products [96]. Beyond such function, which is closely coupled with the blood supply, it has been well-documented that the angiogenic switch occurs at different stages of carcinogenesis, suggesting that tumor endothelium provides additional instructive cues necessary for tumor progression [97]. The main mechanism shown to contribute to tumor vascularization is angiogenic sprouting, which exploits much of the same molecular circuitry that drives physiological angiogenesis (see the 'ECM and Blood Vessel Growth' section). However, tumor-derived endothelium displays dramatic differences compared with its normal counterpart, reflected in irregular morphology, loss-of-function-specific properties, increased leakiness, defective basement membrane, and absence of close contact with mural cells [98]. This in turn causes activation of mechanisms that release angiogenic and permeability factors, which perpetuate the angiogenic cycle, maintaining the vessels in an immature state and continuously promoting the growth of new vessels [99].

7.3.2 Tumor Angiogenesis

A major component of reactive tumor stroma includes activated fibroblasts, which aberrantly deposit ECM proteins and enzymes, and therefore remodel local matrix topology and structure to form permissive grounds for tumor angiogenesis. Similar to what has been described in a physiological setting, ECM components regulate both initiation and maintenance of vessel growth. For instance, MMPs play a crucial role in releasing and activating matrix-bound proangiogenic factors [100] that initiate angiogenic sprouting. Indeed, mice lacking MMP-1 and MMP-9 show reduced tumor growth and angiogenesis [101, 102]. In addition, ECM ligands support all phases of vessel extension by regulating different functions of tumor ECs. The matricellular protein CCN1 promotes EC migration and aberrant neovascularization in pancreatic cancer [103]. EC survival during hypoxic conditions is mediated by a matrix-associated protein EGFL7 [104], while different collagens, which are excessively deposited within the stroma, may control EC proliferation through regulation of ECM stiffness [105, 106]. Finally, tenascin-C, which is highly expressed in tumor-derived endothelium [107], regulates multiple aspects of vascular sprouting by instructing EC migration, proliferation, and VEGF expression [108, 109].

On the other hand, integrins have also been implicated in angiogenic sprouting during tumor development. Both $\alpha_v\beta_3$ and $\alpha_5\beta_1$ integrin, which are strongly upregulated in activated endothelium, seem to be critical for this process; their antagonists effectively block tumor angiogenesis and induce tumor regression [110, 111]. In addition, collagen and laminin-binding receptors have been shown to play a role; function-directed antibodies against $\alpha_1\beta_1$ and $\alpha_2\beta_1$ integrin reduce tumor growth and angiogenesis [112]. Interestingly, however, genetic ablation models of these integrins in mice do not corroborate this data and the outcomes differ dramatically depending on the tumor type. Namely, mice lacking $\alpha_2\beta_1$ integrin exhibit increased melanoma growth and angiogenesis, while in Lewis lung carcinoma these processes remain unchanged [113]. Furthermore, mice lacking β_3 and β_5 integrins show enhanced tumor growth and angiogenesis or, alternatively, they induce only transient inhibition of angiogenesis and tumor progression, without an effect on already-established tumors [114, 115]. This confirms that their role in tumor angiogenesis is highly context-dependent.

7.3.3 Additional Modes of Tumor Vascularization

Although angiogenic sprouting has been traditionally recognized as the main source of new vessels in developing tumors, extensive research within the last several years revealed unexpected and novel mechanisms of neovascularization. Not surprisingly, ECM plays a critical role in driving and facilitating these processes (Fig. 7.3).

Endothelial Progenitor Cells

Contribution of endothelial progenitor cells (EPCs) to tumor vascularization has been intensively studied ever since they were isolated from the bone marrow [116] and peripheral blood [117], and shown to home into sites of active neovascularization. In sites such as tumors, there has been evidence that EPCs directly incorporate into already-existing vessels, although the extent of the contribution remains the subject of controversy [118]. Alternatively, EPCs appear to indirectly support tumor angiogenesis through paracrine mechanisms [119, 120]. Importantly, different com-

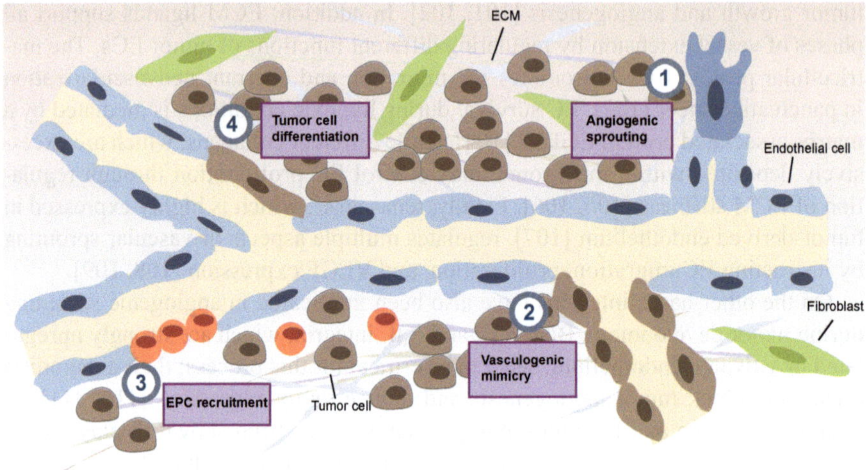

Fig. 7.3 Extracellular matrix (*ECM*) supports different modes of tumor vascularization. Similar to their role in physiological processes, ECM proteins promote angiogenic sprouting within tumors. However, recent findings uncovered alternative ways of how tumors increase their vascular supply and promote oncogenic signaling, and ECM components seem to be critical for these processes. The figure depicts some of these newly discovered modes of tumor vascularization

ponents of the ECM promote and regulate each of the steps that EPCs need to complete to accomplish their function: mobilization, invasion of the tumor site, differentiation into mature ECs, and/or regulation of the pre-existing function of the ECs. For instance, integrins $\alpha_4\beta_1$ and $\alpha_4\beta_7$ are essential for mobilization of EPCs from the bone marrow microenvironment, while $\alpha_6\beta_1$ integrin allows EPCs to respond to laminins, which act as a homing signal within the vascular basement membrane [121]. On the other hand, integrin $\alpha_5\beta_1$ facilitates EPC homing to vascular injury sites where it binds fibronectin, which is abundantly present within the remodeling matrix [122, 123]. Additionally, EPCs adhere directly to the surface of the activated ECs by employing $\alpha_v\beta_3$ and $\alpha_v\beta_5$ integrins [124, 125]. During EPC differentiation into mature ECs, several ECM proteins play a crucial role, the most important of which is fibronectin. Namely, EPCs show higher adhesion and differentiation when plated on fibronectin compared with collagen, and they display continuous expression of fibronectin-binding integrins α_4, α_5, and α_v [117, 126]. Fibronectin consequently promotes VEGF-induced EPC differentiation through ligation of $\alpha_5\beta_1$ integrin [127]. In addition, a matricellular protein, CCN1, promotes EPC differentiation through regulation of negative transcriptional regulator Id1 [128]. Finally, in contrast to the process of differentiation, EPCs affect the angiogenic process through production of paracrine factors, which are dynamically regulated by different ECM substrates. For instance, gelatin, fibronectin, and fibrin proteolytic fragment E promote expression of VEGF, TGF-β1, stromal-cell derived factor (SDF)-1, and interleukin (IL)-8, which in turn facilitate EC tube formation and wound healing [129].

Tumor-Derived Endothelium and Vasculogenic Mimicry

Several recent studies made a surprising discovery that stem-like population of tumor cells in glioblastoma has the potential to differentiate along endothelial lineage, giving rise to EPCs and, subsequently, mature ECs [130, 131]. These tumor-derived ECs progressively acquire more pronounced EC-like properties and a propensity to incorporate into the tumor vasculature [132]. Related to this observation, aggressive melanoma cells have been shown to dedifferentiate *in vivo* and organize in vasculogenic-like matrix-embedded networks that contain plasma and erythrocytes [133]. This process has been termed vasculogenic mimicry and has been observed in many different cancers [134–136].

Although distinct, these phenomena describe an important feature of aggressive cancer cells—the capacity to cycle between different cellular states and under specific conditions promote cancer progression, for example, by amplifying tumor vascularization. Importantly, the local microenvironment and instructive cues from the ECM appear to play an outstanding role in cancer cell plasticity. Namely, numerous studies have shown that differentiation of stem cells along different cellular lineages is tightly regulated by the stiffness, as well as the composition, of the ECM [137]. In the context of cancer, neuroblastoma cells expressing tenascin-C differentiate into endothelial-like cells in the presence of VEGF, while deletion of tenascin-C completely abolishes this effect [138]. On the other hand, cancer cells that undergo vasculogenic mimicry upregulate numerous angiogenesis- and vasculogenesis-related genes, including laminin-5, MMP-2, and MT1-MMP, which appear to be necessary for the network formation. Knockdown of galectin-3, an ECM-associated protein, abolishes the capacity of melanoma cells to form tubular networks in type I collagen gel [139]. Finally, matrices conditioned by aggressive melanoma cells induce poorly aggressive melanoma cells, as well as normal melanocytes, to undergo reprogramming and assume vasculogenic phenotype, undeniably showing that instructive information is encoded within the ECM [140, 141].

Taken together, it has become increasingly clear that tumors represent incredibly complex entities that undergo accelerated evolution and adaptation to even the most extreme conditions. However, this evolution requires cooperation between cancer cells themselves and stromal elements such as tumor endothelium, fibroblasts, and infiltrating immune cells. Considering that ECM with its diverse components mediates much of the communication between these different elements, its targeting could be an important direction in the area of cancer therapy, a topic discussed in the following section.

7.4 Conclusions and Perspectives

Throughout this chapter, we learned that ECM proteins and their receptors cooperate with growth factors and other classes of molecules to regulate EC biology during formation of blood vessels. Recapitulating their physiological roles, ECM components also drive neovascularization in tumors, although in this setting they display additional functions that can further fuel cancer progression. Not surprisingly, the last several years saw an expansion of ECM-based therapeutics that are currently tested in clinical trials for the treatment of different types of hematopoietic as well as solid cancers.

Ever since integrin $\alpha_v\beta_3$ and $\alpha_5\beta_1$ were shown to be driving tumor angiogenesis and tumor growth, they have been considered the prime targets for cancer therapy and diagnostics. Several therapeutics targeting these molecules have been developed, including both peptides and antibodies: ATN-661 peptide and volociximab antibody were designed to specifically block $\alpha_5\beta_1$ integrin signaling, while cilengitide and etaracizumab inhibit $\alpha_v\beta_3$ and $\alpha_v\beta_5$ integrin or, specifically, $\alpha_v\beta_3$ integrin, respectively [142]. In addition, because $\alpha_v\beta_3$ integrin almost selectively labels neoangiogenic sprouts, therapeutics such as maraciclatide-Tc-99m, which recognizes this integrin, were developed for noninvasive tumor imaging [143]. However, the idea of targeting tumor vasculature through these receptors seems to be less straightforward than originally thought as several late-stage clinical trials failed to report survival benefit for patients; for instance, a phase III clinical trial that tested the efficacy of cilengitide in glioblastoma patients showed no impact on overall survival. Moreover, several recent findings are starting to offer hints as to why this may be the case. Namely, the discrepancy between a phenotype in $\alpha_v\beta_3$ knockout mice and the effects of $\alpha_v\beta_3$-blocking antibodies suggested that this integrin might not be indispensable for blood vessel formation (see the 'ECM and Tumor Vascularization' section). However, Steri et al. recently showed that $\alpha_v\beta_3$ integrin is indeed required for angiogenesis, but only transiently; over the longer-term, ECs undergo rewiring and use other molecular mechanisms to execute their basic functions [115, 144]. Furthermore, doses of inhibitors used to treat the tumors seem to be of outstanding importance, as low doses of cilengitide promote tumor growth and angiogenesis [145]. Finally, this particular integrin has been shown to bind a wide spectrum of ECM ligands and growth factors that elicit both pro- and antiangiogenic effects, suggesting that its functions are highly contextual, and further complicating its use as an anticancer agent.

One way to circumvent the fact that integrin receptors act through different ligands, and therefore display differential effects, would be to target the ligands themselves. Indeed, several ECM protein-based therapeutics have been showing promising results in preclinical and clinical studies. The matricellular protein CCN2 elicits potent angiogenic activity, and the CCN2-specific humanized antibody FG-3019 inhibits tumor growth and metastasis in pancreatic cancer [146]. Antibodies targeting the ECM-associated protein EGFL7 improve the efficacy of anti-VEGF therapy through induction of EC apoptosis and inhibition of tumor progression in nonsmall cell lung cancer [104]. Finally, different derivatives of antibodies recognizing fibronectin extra-domain B (EDB), as well as large tenascin-C isoforms, have been in use for radioimmunotherapy in different cancers [147]. Apart from blocking the proangiogenic ligands, another strategy would be to mimic the effects of the antiangiogenic ligands, such as TSP. Indeed, several ongoing clinical trials are testing the efficacy of the TSP-1 analog, ABT-510, in both hematopoietic and solid cancers (ClinicalTrials.gov). Importantly, ECM proteins may be particularly promising candidates for the treatment of advanced and metastatic cancers because of their emerging role in regulating tumor cell dormancy. Ghajar et al. recently reported that, within the metastatic sites, quiescent endothelium rich in TSP-1 levels maintains cancer cell dormancy, while neovascular tips stimulate exit from dormancy through increased expression of periostin, tenascin, versican, and fibronectin [148]. In addition, osteoblast-derived osteopontin promotes tumor dormancy in acute lymphoblastic leukemia (ALL) [149]. However, it has been reported that ECM proteins, much like their receptors, act in a context-dependent manner. For instance, CCN4 promotes the progression of breast and colon cancer but inhibits the metastasis of melanoma [150–152]. Other members of the same family, such as CCN3, show differential effects in melanoma and glioma [153, 154]. Finally, osteopontin, shown to inhibit ALL relapse, promotes metastasis of many epithelial tumors [149, 155].

The most probable reason for such discrepancies is that ECM proteins and their receptors follow complex temporal and spatial patterns of expression. ECM is secreted by both cancer and stromal cells, and its contents are specific for different tissues and cancer types, as well as for different stages of progression found within the same tumor. Importantly, the development of technologies able to define specific ECM signatures or 'matrisomes' has started to shed some light on this complex issue. The comparison of ECM contents in lung and colon tissue reveal both common proteins, as well as organ-specific ECM signatures [156]. Furthermore, composition of the tumor ECM changes with the tumor's metastatic potential. Finally, it appears that both cancer and stromal cells evolve during cancer progression, contributing different combinations of ECM components along the way [157].

Taken together, although ECM-based therapeutics represent an exciting direction in the area of cancer treatment, additional studies will be necessary to further understand the functional complexity of the ECM and its components. Only by integrating these findings with our current treatment practice, will we be able to fully unlock the potential of these novel targets of cancer therapy.

References

1. Schaefer L, Schaefer RM. Proteoglycans: from structural compounds to signaling molecules. Cell Tissue Res. 2010;339:237–46. doi:10.1007/s00441-009-0821-y.
2. Wight TN. Cell biology of arterial proteoglycans. Arteriosclerosis. 1989;9:1–20.
3. Iozzo RV. Basement membrane proteoglycans: from cellar to ceiling. Nat Rev Mol Cell Biol. 2005;6:646–56. doi:nrm1702 [pii]10.1038/nrm1702.
4. Couchman JR, Pataki CA. An introduction to proteoglycans and their localization. J Histochem Cytochem. 2012;60:885–97. doi:10.1369/0022155412464638.
5. Rozario T, DeSimone DW. The extracellular matrix in development and morphogenesis: a dynamic view. Dev Biol. 2010;341:126–40. doi:S0012-1606(09)01285-8 [pii]10.1016/j.ydbio.2009.10.026.
6. Muiznieks LD, Keeley FW. Molecular assembly and mechanical properties of the extracellular matrix: a fibrous protein perspective. Biochim Biophys Acta. 2013;1832:866–75. doi:S0925-4439(12)00283-9 [pii]10.1016/j.bbadis.2012.11.022.

7. Wagenseil JE, Mecham RP. Vascular extracellular matrix and arterial mechanics. Physiol Rev. 2009;89:957–89. doi:89/3/957 [pii]10.1152/physrev.00041.2008.
8. Kielty CM, Sherratt MJ, Shuttleworth CA. Elastic fibres. J Cell Sci. 2002;115:2817–28.
9. Kostourou V, Papalazarou V. Non-collagenous ECM proteins in blood vessel morphogenesis and cancer. Biochim Biophys Acta. 2014;2014(02):018. doi:S0304-4165.1400079 8 pii-101016jbbagen.
10. Frantz C, Stewart KM, Weaver VM. The extracellular matrix at a glance. J Cell Sci. 2010;123:4195–4200. doi:123/24/4195 [pii]10.1242/jcs.023820.
11. Smith ML. et al. Force-induced unfolding of fibronectin in the extracellular matrix of living cells. PLoS Biol. 2007;5:e268. doi:07-PLBI-RA-0149 [pii]10.1371/journal.pbio.0050268.
12. Bornstein P, Sage EH. Matricellular proteins: extracellular modulators of cell function. Curr Opin Cell Biol. 2002;14:608–16. doi:S0955067402003617 [pii].
13. Nikolic I, Plate KH, Schmidt MH. EGFL7 meets miRNA-126: an angiogenesis alliance. J Angiogenesis Res. 2010;2:9. doi:10.1186/2040-2384-2-9.
14. Werneck CC, et al. Mice lacking the extracellular matrix protein MAGP1 display delayed thrombotic occlusion following vessel injury. Blood. 2008;111:4137–44. doi:10.1182/blood-2007-07-101733.
15. Bradshaw AD, et al. SPARC-null mice display abnormalities in the dermis characterized by decreased collagen fibril diameter and reduced tensile strength. J Invest Dermatol. 2003;120:949–55. doi:12241 [pii]101046j15231747200312241x.
16. Rios H, et al. periostin null mice exhibit dwarfism, incisor enamel defects, and an early-onset periodontal disease-like phenotype. Mol Cell Biol. 2005;25:11131–44. doi:25/24/11131 [pii]10.1128/MCB.25.24.11131–11144.2005.
17. Chiodoni C, Colombo MP, Sangaletti S. Matricellular proteins: from homeostasis to inflammation, cancer, and metastasis. Cancer Metastasis Rev. 2010;29:295–307. doi:10.1007/s10555-010-9221-8.
18. Wong GS, Rustgi AK. Matricellular proteins: priming the tumour microenvironment for cancer development and metastasis. Br J Cancer. 2013;108:755–61. doi:bjc2012592 [pii]10.1038/bjc.2012.592.
19. Bellon G, Martiny L, Robinet A. Matrix metalloproteinases and matrikines in angiogenesis. Crit Rev Oncol Hematol. 2004;49:203–20. doi:101016jcritrevonc200310004 S1040842803002762 [pii].
20. Hynes RO. The extracellular matrix: not just pretty fibrils. Science. 2009;326:1216–19. doi:326/5957/1216 [pii]10.1126/science.1176009.
21. Hood JD, Cheresh DA. Role of integrins in cell invasion and migration. Nat Rev Cancer. 2002;2:91–100. doi:10.1038/nrc727.
22. Stupack DG, Cheresh DA. ECM remodeling regulates angiogenesis: endothelial integrins look for new ligands. Sci STKE. 2002. doi:10.1126/stke.2002.119.pe72002/119/pe72002/119/pe7 [pii].
23. Tkachenko E, Rhodes JM, Simons M. Syndecans: new kids on the signaling block. Circ Res. 2005;96:488–500. doi:96/5/488 [pii]10.1161/01.RES.0000159708.71142.c8.
24. Hosokawa H, Ninomiya H, Kitamura Y, Fujiwara K, Masaki T. Vascular endothelial cells that express dystroglycan are involved in angiogenesis. J Cell Sci. 2002;115:1487–96.
25. Flamme I, Risau W. Induction of vasculogenesis and hematopoiesis in vitro. Development. 1992;116:435–9.
26. Carmeliet P. Mechanisms of angiogenesis and arteriogenesis. Nat Med. 2000;6:389–95. doi:10.1038/74651.
27. Herbert SP, Stainier DY. Molecular control of endothelial cell behaviour during blood vessel morphogenesis. Nat Rev. Mol Cell Biol. 2011;12:551–64. doi:nrm3176 [pii]10.1038/nrm3176.
28. Nitta T, et al. Size-selective loosening of the blood-brain barrier in claudin-5-deficient mice. J Cell Biol. 2003;161:653–60. doi:10.1083/jcb.200302070jcb.200302070 [pii].
29. Corada M, et al. Vascular endothelial-cadherin is an important determinant of microvascular integrity in vivo. Proc Natl Acad Sci U S A. 1999;96:9815–20.

30. Darland DC, D'Amore PA. Blood vessel maturation: vascular development comes of age. J Clin Invest. 1999;103:157–8. doi:10.1172/JCI6127.
31. Myers C, Charboneau A, Cheung I, Hanks D, Boudreau N. Sustained expression of homeobox D10 inhibits angiogenesis. Am J Pathol. 2002;161:2099–109. doi:S0002-9440(10)64488-4 [pii]10.1016/S0002-9440(10)64488-4.
32. Rhoads K, et al. A role for Hox A5 in regulating angiogenesis and vascular patterning. Lymphat Res Biol. 2005;3:240–52. doi:10.1089/lrb.2005.3.240.
33. Stoll SJ, Bartsch S, Augustin HG, Kroll J. The transcription factor HOXC9 regulates endothelial cell quiescence and vascular morphogenesis in zebrafish via inhibition of interleukin 8. Circ Res. 2011;108:1367–77. doi:CIRCRESAHA.111.244095 [pii]10.1161/CIRCRESAHA.111.244095.
34. Patel S, Leal AD, Gorski DH. The homeobox gene Gax inhibits angiogenesis through inhibition of nuclear factor-kappaB-dependent endothelial cell gene expression. Cancer Res. 2005;65:1414–1424. doi:65/4/1414 [pii]10.1158/0008-5472.CAN-04-3431.
35. Potter MD, Barbero S, Cheresh DA. Tyrosine phosphorylation of VE-cadherin prevents binding of p120- and beta-catenin and maintains the cellular mesenchymal state. J Biol Chem. 2005;280:31906–12. doi:M505568200 [pii]10.1074/jbc.M505568200.
36. Chen XL, et al. VEGF-induced vascular permeability is mediated by FAK. Dev Cell. 2012;22:146–57. doi:S1534-5807(11)00512-0 [pii]10.1016/j.devcel.2011.11.002.
37. Boudreau N, Andrews C, Srebrow A, Ravanpay A, Cheresh DA. Induction of the angiogenic phenotype by Hox D3. J Cell Biol. 1997;139:257–64.
38. Diehl F, Rossig L, Zeiher AM, Dimmeler S, Urbich C. The histone methyltransferase MLL is an upstream regulator of endothelial-cell sprout formation. Blood. 2007;109:1472–78. doi:blood-2006-08-039651 [pii]10.1182/blood-2006-08-039651.
39. Myers C, Charboneau A, Boudreau N. Homeobox B3 promotes capillary morphogenesis and angiogenesis. J Cell Biol. 2000;148:343–51.
40. Estrach S, et al. Laminin-binding integrins induce Dll4 expression and Notch signaling in endothelial cells. Circ Res. 2011;109:172–82. doi:CIRCRESAHA.111.240622 [pii]10.1161/CIRCRESAHA.111.240622.
41. Stenzel D, et al. Endothelial basement membrane limits tip cell formation by inducing Dll4/Notch signalling in vivo. EMBO Rep. 2011;12:1135–43. doi:embor2011194 [pii]10.1038/embor.2011.194.
42. del Toro R, et al. Identification and functional analysis of endothelial tip cell-enriched genes. Blood. 2010;116:4025–33. doi:blood-2010-02-270819 [pii]10.1182/blood-2010-02-270819.
43. Nichol D, et al. Impaired angiogenesis and altered Notch signaling in mice overexpressing endothelial Egfl7. Blood. 2010;116:6133–43. doi:10.1182/blood-2010-03-274860.
44. Schmidt MH, et al. Epidermal growth factor-like domain 7 (EGFL7) modulates Notch signalling and affects neural stem cell renewal. Nat Cell Biol. 2009;11:873–80. doi:10.1038/ncb1896.
45. Albig AR, Becenti DJ, Roy TG, Schiemann WP. Microfibril-associate glycoprotein-2 (MAGP-2) promotes angiogenic cell sprouting by blocking notch signaling in endothelial cells. Microvasc Res. 2008;76:7–14. doi:10.1016/j.mvr.2008.01.001.
46. Houck KA, Leung DW, Rowland AM, Winer J, Ferrara N. Dual regulation of vascular endothelial growth factor bioavailability by genetic and proteolytic mechanisms. J Biol Chem. 1992;267:26031–7.
47. Otrock ZK, Mahfouz RA, Makarem JA, Shamseddine AI. Understanding the biology of angiogenesis: review of the most important molecular mechanisms. Blood Cells Mol Dis. 2007;39:212–20. doi:S1079-9796(07)00077-0 [pii]10.1016/j.bcmd.2007.04.001.
48. Pepper MS. Role of the matrix metalloproteinase and plasminogen activator-plasmin systems in angiogenesis. Arterioscler Thromb Vasc Biol. 2001;21:1104–17.
49. Senger DR, et al. The alpha(1)beta(1) and alpha(2)beta(1) integrins provide critical support for vascular endothelial growth factor signaling, endothelial cell migration, and tumor angiogenesis. Am J Pathol. 2002;160:195–204. doi:S0002-9440(10)64363-5 [pii].

50. Klemke RL, et al. Regulation of cell motility by mitogen-activated protein kinase. J Cell Biol. 1997;137:481–92.
51. Lamalice L, Le Boeuf F, Huot J. Endothelial cell migration during angiogenesis. Circ Res. 2007;100:782–94. doi:100/6/782 [pii]10116101RES0000259593076611e.
52. De Smet F, Segura I, De Bock K, Hohensinner PJ, Carmeliet P. Mechanisms of vessel branching: filopodia on endothelial tip cells lead the way. Arterioscler Thromb Vasc Biol. 2009;29:639–49. doi:ATVBAHA.109.185165 [pii]10.1161/ATVBAHA.109.185165.
53. Dallas SL, Chen Q, Sivakumar P. Dynamics of assembly and reorganization of extracellular matrix proteins. Curr Topics Dev Biol. 2006;75:1–24. doi:10.1016/S0070-2153(06)75001-3.
54. Murphy-Ullrich JE, Hook M. Thrombospondin modulates focal adhesions in endothelial cells. J Cell Biol. 1989;109:1309–19.
55. Murphy-Ullrich JE, et al. Focal adhesion integrity is downregulated by the alternatively spliced domain of human tenascin. J Cell Biol. 1991;115:1127–36.
56. Greenwood JA, Murphy-Ullrich JE. Signaling of de-adhesion in cellular regulation and motility. Microsc Res Tech. 1998;43:420–32. doi:10.1002/(SICI)1097-0029(19981201)43:5<420::AID-JEMT8>3.0.CO;2-B [pii]10.1002/(SICI)1097-0029(19981201)43:5<420::AID-JEMT8>3.0.CO;2-B.
57. DiMilla PA, Barbee K, Lauffenburger DA Mathematical model for the effects of adhesion and mechanics on cell migration speed. Biophys J. 1991;60:15–37. doi:S0006-3495(91)82027-6 [pii]10.1016/S0006-3495(91)82027–6.
58. Palecek SP, Loftus JC, Ginsberg MH, Lauffenburger DA, Horwitz AF. Integrin-ligand binding properties govern cell migration speed through cell-substratum adhesiveness. Nature. 1997;385:537–40. doi:10.1038/385537a0.
59. Nikolic I, et al. EGFL7 ligates alphavbeta3 integrin to enhance vessel formation. Blood. 2013;121:3041–50. doi:blood-2011-11-394882 [pii]10.1182/blood-2011-11-394882.
60. Murphy-Ullrich JE. The de-adhesive activity of matricellular proteins: is intermediate cell adhesion an adaptive state? J Clin Invest. 2001;107:785–90. doi:10.1172/JCI12609.
61. Phng LK, et al. Nrarp coordinates endothelial Notch and Wnt signaling to control vessel density in angiogenesis. Dev Cell. 2009;16:70–82. doi:S1534-5807(08)00519-4 [pii]10.1016/j.devcel.2008.12.009.
62. Phng LK, Gerhardt H. Angiogenesis: a team effort coordinated by notch. Dev Cell. 2009;16:196–208. doi:S1534-5807(09)00043-4 [pii]10.1016/j.devcel.2009.01.015.
63. Wary KK, Mainiero F, Isakoff SJ, Marcantonio EE, Giancotti FG. The adaptor protein Shc couples a class of integrins to the control of cell cycle progression. Cell. 1996;87:733–43. doi:S0092-8674(00)81392-6 [pii].
64. Roovers K, Assoian RK. Integrating the MAP kinase signal into the G1 phase cell cycle machinery. Bioessays. 2000;22:818–26. doi:10.1002/1521-1878(200009)22:9<818::AID-BIES7>3.0.CO;2-6 [pii]10.1002/1521-1878(200009)22:9<818::AID-BIES7>3.0.CO;2-6.
65. Khwaja A, Rodriguez-Viciana P, Wennstrom S, Warne PH, Downward J. Matrix adhesion and Ras transformation both activate a phosphoinositide 3-OH kinase and protein kinase B/Akt cellular survival pathway. EMBO J. 1997;16:2783–93. doi:10.1093/emboj/16.10.2783.
66. Ilic D, et al. Extracellular matrix survival signals transduced by focal adhesion kinase suppress p53-mediated apoptosis. J Cell Biol. 1998;143:547–60.
67. Meredith JE Jr, Schwartz MA. Integrins, adhesion and apoptosis. Trends Cell Biol. 1997;7:146–50. doi:S0962-8924(97)01002-7 [pii]10.1016/S0962-8924(97)01002-7.
68. Bayless KJ, Davis GE. Sphingosine-1-phosphate markedly induces matrix metalloproteinase and integrin-dependent human endothelial cell invasion and lumen formation in three-dimensional collagen and fibrin matrices. Biochem Biophys Res Commun. 2003;312:903–13. doi:S0006291X03023659 [pii].
69. Gerhardt H, et al. VEGF guides angiogenic sprouting utilizing endothelial tip cell filopodia. J Cell Biol. 2003;161:1163–77. doi:10.1083/jcb.200302047jcb.200302047 [pii].
70. Davis GE, Koh W, Stratman AN. Mechanisms controlling human endothelial lumen formation and tube assembly in three-dimensional extracellular matrices. Birth Defects Res C Embryo Today. 2007;81:270–85. doi:10.1002/bdrc.20107.

71. Davis GE, Camarillo CW. An alpha 2 beta 1 integrin-dependent pinocytic mechanism involving intracellular vacuole formation and coalescence regulates capillary lumen and tube formation in three-dimensional collagen matrix. Exp Cell Res. 1996;224:39–51. doi:S0014-4827(96)90109-3 [pii]10.1006/excr.1996.0109.
72. Kamei M, et al. Endothelial tubes assemble from intracellular vacuoles in vivo. Nature. 2006;442: 453–456. doi:nature04923 [pii]10.1038/nature04923.
73. Blum Y, et al. Complex cell rearrangements during intersegmental vessel sprouting and vessel fusion in the zebrafish embryo. Dev Biol. 2008;316:312–22. doi:S0012-1606(08)00079-1 [pii]10.1016/j.ydbio.2008.01.038.
74. Strilic B, et al. The molecular basis of vascular lumen formation in the developing mouse aorta. Dev Cell. 2009;17:505–15. doi:S1534-5807(09)00349-9 [pii]10.1016/j.devcel.2009.08.011.
75. Whelan MC, Senger DR. Collagen I initiates endothelial cell morphogenesis by inducing actin polymerization through suppression of cyclic AMP and protein kinase A. J Biol Chem. 2003;278:327–34. doi:10.1074/jbc.M207554200M207554200 [pii].
76. Liu Y, Senger DR. Matrix-specific activation of Src and Rho initiates capillary morphogenesis of endothelial cells. FASEB J. 2004;18:457–68. doi:10.1096/fj.03-0948com18/3/457 [pii].
77. Drake CJ, Davis LA, Little CD. Antibodies to beta 1-integrins cause alterations of aortic vasculogenesis, in vivo. Dev Dyn. 1992;193:83–91. doi:10.1002/aja.1001930111.
78. Zovein AC, et al. Beta1 integrin establishes endothelial cell polarity and arteriolar lumen formation via a Par3-dependent mechanism. Dev Cell. 2010;18:39–51. doi:S1534-5807(09)00488-2 [pii]10.1016/j.devcel.2009.12.006.
79. Bayless KJ, Salazar R, Davis GE. RGD-dependent vacuolation and lumen formation observed during endothelial cell morphogenesis in three-dimensional fibrin matrices involves the alpha(v)beta(3) and alpha(5)beta(1) integrins. Am J Pathol. 2000;156:1673–83. doi:S0002-9440(10)65038-9 [pii].
80. Koh W, Mahan RD, Davis GE. Cdc42- and Rac1-mediated endothelial lumen formation requires Pak2, Pak4 and Par3, and PKC-dependent signaling. J Cell Sci. 2008;121:989–1001. doi:jcs.020693 [pii]10.1242/jcs.020693.
81. Parker LH, et al. The endothelial-cell-derived secreted factor Egfl7 regulates vascular tube formation. Nature. 2004;428:754–8. doi:10.1038/nature02416nature02416 [pii].
82. Charpentier MS, et al. CASZ1 promotes vascular assembly and morphogenesis through the direct regulation of an EGFL7/RhoA-mediated pathway. Dev Cell. 2013;25:132–43. doi:S1534-5807(13)00134-2 [pii]10.1016/j.devcel.2013.03.003.
83. Jakobsson L, Domogatskaya A, Tryggvason K, Edgar D, Claesson-Welsh L. Laminin deposition is dispensable for vasculogenesis but regulates blood vessel diameter independent of flow. FASEB J. 2008;22:1530–9. doi:fj.07-9617com [pii]101096fj07-9617com.
84. Jain RK. Molecular regulation of vessel maturation. Nat Med. 2003;9:685–93. doi:10.1038/nm0603-685 [pii]nm0603-685 [pii].
85. Stratman AN, Malotte KM, Mahan RD, Davis MJ, Davis GE. Pericyte recruitment during vasculogenic tube assembly stimulates endothelial basement membrane matrix formation. Blood. 2009;114:5091–101. doi:blood-2009-05-222364 [pii]10.1182/blood-2009-05-222364.
86. Zhou Z, et al. Deletion of laminin-8 results in increased tumor neovascularization and metastasis in mice. Cancer Res. 2004;64:4059–63. doi:10.1158/0008-5472.CAN-04-029164/12/4059 [pii].
87. Li Q, Olsen BR. Increased angiogenic response in aortic explants of collagen XVIII/endostatin-null mice. Am J Pathol. 2004;165:415–24. doi:S0002-9440(10)63307-X [pii]10.1016/S0002-9440(10)63307–X.
88. Mettouchi A, et al. Integrin-specific activation of Rac controls progression through the G(1) phase of the cell cycle. Mol Cell. 2001;8:115–27. doi:S1097-2765(01)00285-4 [pii].
89. Klein S, et al. Alpha 5 beta 1 integrin activates an NF-kappa B-dependent program of gene expression important for angiogenesis and inflammation. Mol Cell Biol. 2002;22:5912–22.
90. Hall-Glenn F, et al. CCN2/connective tissue growth factor is essential for pericyte adhesion and endothelial basement membrane formation during angiogenesis. PLoS One. 2012;7:e30562. doi:10.1371/journal.pone.0030562PONE-D-11-19485 [pii].

91. Yamaguchi N, et al. Endostatin inhibits VEGF-induced endothelial cell migration and tumor growth independently of zinc binding. EMBO J. 1999;18:4414–23. doi:10.1093/emboj/18.16.4414.
92. Taddei L, et al. Inhibitory effect of full-length human endostatin on in vitro angiogenesis. Biochem Biophys Res Commun. 1999;263:340–5. doi:10.1006/bbrc.1999.1342S0006-291X(99)91342-6 [pii].
93. Pasco S, et al. A specific sequence of the noncollagenous domain of the alpha3(IV) chain of type IV collagen inhibits expression and activation of matrix metalloproteinases by tumor cells. Cancer Res. 2000;60:467–73.
94. Adams JC. Thrombospondins: multifunctional regulators of cell interactions. Annu Rev Cell Dev Biol. 2001;17:25–51. doi:10.1146/annurev.cellbio.17.1.2517/1/25 [pii].
95. Bornstein P. Thrombospondins as matricellular modulators of cell function. J Clin Invest. 2001;107:929–34. doi:10.1172/JCI12749.
96. Hanahan D, Weinberg RA. The hallmarks of cancer. Cell. 2000;100:57–70.
97. Hanahan D, Folkman J. Patterns and emerging mechanisms of the angiogenic switch during tumorigenesis. Cell. 1996;86:353–64.
98. Morikawa S, et al. Abnormalities in pericytes on blood vessels and endothelial sprouts in tumors. Am J Pathol. 2002;160:985–1000. doi:10.1016/S0002-9440(10)64920-6.
99. Weis SM, Cheresh DA. Pathophysiological consequences of VEGF-induced vascular permeability. Nature. 2005;437:497–504. doi:10.1038/nature03987.
100. Bergers G, et al. Matrix metalloproteinase-9 triggers the angiogenic switch during carcinogenesis. Nat Cell Biol. 2000;2:737–44. doi:10.1038/35036374.
101. Foley CJ, et al. Matrix metalloprotease 1a deficiency suppresses tumor growth and angiogenesis. Oncogene. 2014;33:2264–72. doi:onc2013157 [pii]10.1038/onc.2013.157.
102. Nakamura T, et al. Stromal metalloproteinase-9 is essential to angiogenesis and progressive growth of orthotopic human pancreatic cancer in parabiont nude mice. Neoplasia. 2007;9:979–86.
103. Maity G, et al. Pancreatic tumor cell secreted CCN1/Cyr61 promotes endothelial cell migration and aberrant neovascularization. Sci Rep. 2014;4:4995. doi:srep04995 [pii]10.1038/srep04995.
104. Johnson L, et al. Anti-EGFL7 antibodies enhance stress-induced endothelial cell death and anti-VEGF efficacy. J Clin Investg. 2013;123:3997–4009. doi:67892 [pii]10.1172/JCI67892.
105. Kauppila S, Stenback F, Risteli J, Jukkola A, Risteli L. Aberrant type I and type III collagen gene expression in human breast cancer in vivo. J Pathol. 1998;186:262–8. doi:10.1002/(SICI)1096-9896(1998110)186:3<262::AID-PATH191>3.0.CO;2-3 [pii]10.1002/(SICI)1096-9896(1998110)186:3<262::AID-PATH191>3.0.CO;2-3.
106. Yeh YT, et al. Matrix stiffness regulates endothelial cell proliferation through septin 9. PLoS One. 2012;7:e46889. doi:10.1371/journal.pone.0046889PONE-D-12-16147 [pii].
107. Seaman S, et al. Genes that distinguish physiological and pathological angiogenesis. Cancer Cell. 2007;11:539–54. doi:S1535-6108(07)00144-4 [pii]10.1016/j.ccr.2007.04.017.
108. Tanaka K, Hiraiwa N, Hashimoto H, Yamazaki Y, Kusakabe M. Tenascin-C regulates angiogenesis in tumor through the regulation of vascular endothelial growth factor expression. Int J Cancer. 2004;108:31–40. doi:10.1002/ijc.11509.
109. Schenk S, Chiquet-Ehrismann R, Battegay EJ. The fibrinogen globe of tenascin-C promotes basic fibroblast growth factor-induced endothelial cell elongation. Mol Biol Cell. 1999;10:2933–43.
110. Brooks PC, et al. Integrin alpha v beta 3 antagonists promote tumor regression by inducing apoptosis of angiogenic blood vessels. Cell. 1994;79:1157–64. doi:0092-8674(94)90007-8 [pii].
111. Kim S, Harris M, Varner JA. Regulation of integrin alpha vbeta 3-mediated endothelial cell migration and angiogenesis by integrin alpha5beta1 and protein kinase A. J Biol Chem. 2000;275:33920–8. doi:10.1074/jbc.M003668200.

112. Senger DR, et al. Angiogenesis promoted by vascular endothelial growth factor: regulation through alpha1beta1 and alpha2beta1 integrins. Proc Natl Acad Sci U S A. 1997;94:13612–7.
113. Zhang Z, et al. alpha2beta1 integrin expression in the tumor microenvironment enhances tumor angiogenesis in a tumor cell-specific manner. Blood. 2008;111:1980–8. doi:blood-2007-06-094680 [pii]10.1182/blood-2007-06-094680.
114. Reynolds LE, et al. Enhanced pathological angiogenesis in mice lacking beta3 integrin or beta3 and beta5 integrins. Nat Med. 2002;8:27–34. doi:10.1038/nm0102-27nm0102-27 [pii].
115. Steri V, et al. Acute depletion of endothelial beta3-integrin transiently inhibits tumor growth and angiogenesis in mice. Circ Res. 2014;114:79–91. doi:CIRCRESAHA.114.301591 [pii]10.1161/CIRCRESAHA.114.301591.
116. Shi Q, et al. Evidence for circulating bone marrow-derived endothelial cells. Blood. 1998;92:362–7.
117. Asahara T, et al. Isolation of putative progenitor endothelial cells for angiogenesis. Science. 1997;275:964–7.
118. Patenaude A, Parker J, Karsan A. Involvement of endothelial progenitor cells in tumor vascularization. Microvasc Res. 2010;79:217–23. doi:10.1016/j.mvr.2010.01.007.
119. Gao D, et al. Endothelial progenitor cells control the angiogenic switch in mouse lung metastasis. Science. 2008;319:195–98. doi:319/5860/195 [pii]10.1126/science.1150224.
120. Nolan DJ, et al. Bone marrow-derived endothelial progenitor cells are a major determinant of nascent tumor neovascularization. Genes Dev. 2007;21:1546–58. doi:21/12/1546 [pii]10.1101/gad.436307.
121. Bouvard C, et al. alpha6-integrin subunit plays a major role in the proangiogenic properties of endothelial progenitor cells. Arterioscler Thromb Vasc Biol. 2010;30:1569–75. doi:ATVBAHA.110.209163 [pii]10.1161/ATVBAHA.110.209163.
122. Caiado F, Dias S. Endothelial progenitor cells and integrins: adhesive needs. Fibrogenesis Tissue Repair. 2012;5:4. doi:1755-1536-5-4 [pii]10.1186/1755-1536-5-4.
123. Walter DH, et al. Statin therapy accelerates reendothelialization: a novel effect involving mobilization and incorporation of bone marrow-derived endothelial progenitor cells. Circulation. 2002;105:3017–24.
124. Kokubo T, Uchida H, Choi ET. Integrin alpha(v)beta(3) as a target in the prevention of neointimal hyperplasia. J Vasc Surg. 2007;45 Suppl A:A33–8. doi:S0741-5214(07)00422-3 [pii]10.1016/j.jvs.2007.02.069.
125. Di Santo S, et al. Oxidized low density lipoprotein impairs endothelial progenitor cell function by downregulation of E-selectin and integrin alpha(v)beta5. Biochem Biophys Res Commun. 2008;373:528–32. doi:S0006-291X(08)01213-8 [pii]10.1016/j.bbrc.2008.06.066.
126. Igreja C, et al. Detailed molecular characterization of cord blood-derived endothelial progenitors. Exp Hematol. 2008;36:193–203. doi:S0301-472X(07)00558-9 [pii]10.1016/j.exphem.2007.09.001.
127. Wijelath ES, et al. Fibronectin promotes VEGF-induced CD34 cell differentiation into endothelial cells. J Vasc Surg. 2004;39:655–60. doi:10.1016/j.jvs.2003.10.042.
128. Yu Y, et al. CCN1 promotes the differentiation of endothelial progenitor cells and reendothelialization in the early phase after vascular injury. Basic Res Cardiol. 105:713–724. doi:10.1007/s00395-010-0117-0(2010).
129. Caiado F, et al. The role of fibrin E on the modulation of endothelial progenitors adhesion, differentiation and angiogenic growth factor production and the promotion of wound healing. Biomaterials. 2011;32:7096–105. doi:S0142-9612(11)00687-9 [pii]10.1016/j.biomaterials.2011.06.022.
130. Wang R, et al. Glioblastoma stem-like cells give rise to tumour endothelium. Nature. 2010;468:829–33. doi:nature09624 [pii]10.1038/nature09624.
131. Ricci-Vitiani L, et al. Tumour vascularization via endothelial differentiation of glioblastoma stem-like cells. Nature. 2010;468:824–8. doi:nature09557 [pii]10.1038/nature09557.

132. McGuire TF, Sajithlal GB, Lu J, Nicholls RD, Prochownik EV. In vivo evolution of tumor-derived endothelial cells. PLoS One. 2012;7:e37138. doi:10.1371/journal.pone.0037138PONE-D-12-03955 [pii].
133. Maniotis AJ, et al. Vascular channel formation by human melanoma cells in vivo and in vitro: vasculogenic mimicry. Am J Pathol. 1999;155:739–52. doi:10.1016/S0002-9440(10)65173-5.
134. Shirakawa K, et al. Absence of endothelial cells, central necrosis, and fibrosis are associated with aggressive inflammatory breast cancer. Cancer Res. 2001;61:445–51.
135. Sood AK, et al. The clinical significance of tumor cell-lined vasculature in ovarian carcinoma: implications for anti-vasculogenic therapy. Cancer Biol Ther. 2002;1:661–4. doi:316 [pii].
136. van der Schaft DW, et al. Tumor cell plasticity in Ewing sarcoma, an alternative circulatory system stimulated by hypoxia. Cancer Res. 2005;65:11520–8. doi:65/24/11520 [pii]10.1158/0008-5472.CAN-05-2468.
137. Watt FM, Huck WT. Role of the extracellular matrix in regulating stem cell fate. Nat Rev Mol Cell Biol. 2013;14:467–73. doi:10.1038/nrm3620.
138. Pezzolo A, et al. Oct-4+/Tenascin C+ neuroblastoma cells serve as progenitors of tumor-derived endothelial cells. Cell Res. 2011;21:1470–86. doi:10.1038/cr.2011.38.
139. Mourad-Zeidan AA, Melnikova VO, Wang H, Raz A, Bar-Eli M. Expression profiling of Galectin-3-depleted melanoma cells reveals its major role in melanoma cell plasticity and vasculogenic mimicry. Am J Pathol. 2008;173:1839–52. doi:S0002-9440(10)61567-2 [pii]10.2353/ajpath.2008.080380.
140. Seftor EA, et al. Epigenetic transdifferentiation of normal melanocytes by a metastatic melanoma microenvironment. Cancer Res. 2005;65:10164–9. doi:65/22/10164 [pii]10.1158/0008-5472.CAN-05-2497.
141. Seftor RE, et al. Cooperative interactions of laminin 5 gamma2 chain, matrix metalloproteinase-2, and membrane type-1-matrix/metalloproteinase are required for mimicry of embryonic vasculogenesis by aggressive melanoma. Cancer Res. 2001;61:6322–7.
142. Goodman SL, Picard M. Integrins as therapeutic targets. Trends Pharmacol Sci. 2012;33:405–12. doi:S0165-6147(12)00057-0 [pii]10.1016/j.tips.2012.04.002.
143. Bach-Gansmo T, et al. Integrin receptor imaging of breast cancer: a proof-of-concept study to evaluate 99mTc-NC100692. J Nucl Med. 2006;47:1434–9. doi:47/9/1434 [pii].
144. Cheresh DA, Stupack DG. Tumor angiogenesis: putting a value on plastic GEMMs. Circ Res. 2014;114:9–11. doi:CIRCRESAHA.113.302812 [pii]10.1161/CIRCRESAHA.113.302812.
145. Reynolds AR, et al. Stimulation of tumor growth and angiogenesis by low concentrations of RGD-mimetic integrin inhibitors. Nat Med. 2009;15:392–400. doi:nm.1941 [pii]10.1038/nm.1941.
146. Dornhofer N, et al. Connective tissue growth factor-specific monoclonal antibody therapy inhibits pancreatic tumor growth and metastasis. Cancer Res. 2006;66:5816–27. doi:66/11/5816 [pii]10.1158/0008-5472.CAN-06-0081.
147. Neri D, Bicknell R. Tumour vascular targeting. Nat Rev Cancer. 2005;5:436–46. doi:nrc1627 [pii]10.1038/nrc1627.
148. Ghajar CM, et al. The perivascular niche regulates breast tumour dormancy. Nat Cell Biol. 2013;15:807–17. doi:ncb2767 [pii]10.1038/ncb2767.
149. Boyerinas B, et al. Adhesion to osteopontin in the bone marrow niche regulates lymphoblastic leukemia cell dormancy. Blood. 2013;121:4821–31. doi:blood-2012-12-475483 [pii]10.1182/blood-2012-12-475483.
150. Xie D, Nakachi K, Wang H, Elashoff R, Koeffler HP. Elevated levels of connective tissue growth factor, WISP-1, and CYR61 in primary breast cancers associated with more advanced features. Cancer Res. 2001;61:8917–23.
151. Tian C, et al. Overexpression of connective tissue growth factor WISP-1 in Chinese primary rectal cancer patients. World J Gastroenterol. 2007;13:3878–82.

152. Hashimoto Y, et al. Expression of the Elm1 gene, a novel gene of the CCN (connective tissue growth factor, Cyr61/Cef10, and neuroblastoma overexpressed gene) family, suppresses In vivo tumor growth and metastasis of K-1735 murine melanoma cells. J Exp Med. 1998;187:289–96.
153. Gupta N, et al. Inhibition of glioma cell growth and tumorigenic potential by CCN3 (NOV). Mol Pathol. 2001;54:293–9.
154. Vallacchi V, et al. CCN3/nephroblastoma overexpressed matricellular protein regulates integrin expression, adhesion, and dissemination in melanoma. Cancer Res. 2008;68:715–23. doi:68/3/715 [pii]10.1158/0008-5472.CAN-07-2103.
155. Shevde LA, Das S, Clark DW, Samant RS. Osteopontin: an effector and an effect of tumor metastasis. Curr Mol Med. 2010;10:71–81. doi:CMM#07 [pii].
156. Naba A, et al. The matrisome: in silico definition and in vivo characterization by proteomics of normal and tumor extracellular matrices. Mol Cell Proteomics. 2012;11:M111.014647. doi:M111.014647 [pii]10.1074/mcp.M111.014647.
157. Naba A, Clauser KR, Lamar JM, Carr SA, Hynes RO. Extracellular matrix signatures of human mammary carcinoma identify novel metastasis promoters. Elife. 2014;3:e01308.
158. Heino J. The collagen receptor integrins have distinct ligand recognition and signaling functions. Matrix Biol. 2000;19:319–23. doi:S0945-053X(00)00076-7 [pii].
159. Kern A, Eble J, Golbik R, Kuhn K. Interaction of type IV collagen with the isolated integrins alpha 1 beta 1 and alpha 2 beta 1. Eur J Biochem. 1993;215:151–9.
160. Fukuda K, Koshihara Y, Oda H, Ohyama M, Ooyama T. Type V collagen selectively inhibits human endothelial cell proliferation. Biochem Biophys Res Commun. 1988;151:1060–8. doi:S0006-291X(88)80473-X [pii].
161. Kielty CM, Whittaker SP, Grant ME, Shuttleworth CA. Attachment of human vascular smooth muscles cells to intact microfibrillar assemblies of collagen VI and fibrillin. J Cell Sci. 1992;103(Pt 2):445–51.
162. Midwood KS, Schwarzbauer JE. Elastic fibers: building bridges between cells and their matrix. Curr Biol. 2002;12:R279–R81. doi:S096098220200800X..
163. Jovanovic J, et al. alphaVbeta6 is a novel receptor for human fibrillin-1. Comparative studies of molecular determinants underlying integrin-rgd affinity and specificity. J Biol Chem. 2007;282:6743–51. doi:M607008200 [pii]10.1074/jbc.M607008200.
164. Timpl R, Sasaki T, Kostka G, Chu ML. Fibulins: a versatile family of extracellular matrix proteins. Nat Rev Mol Cell Biol. 2003;4:479–89. doi:10.1038/nrm1130nrm1130 [pii].
165. Zou L, Cao S, Kang N, Huebert RC, Shah VH. Fibronectin induces endothelial cell migration through beta1 integrin and Src-dependent phosphorylation of fibroblast growth factor receptor-1 at tyrosines 653/654 and 766. J Biol Chem. 2012;287:7190–202. doi:M111.304972 [pii]10.1074/jbc.M111.304972.
166. Wang J, Milner R. Fibronectin promotes brain capillary endothelial cell survival and proliferation through alpha5beta1 and alphavbeta3 integrins via MAP kinase signalling. J Neurochem. 2006;96:148–59. doi:JNC3521 [pii]10.1111/j.1471-4159.2005.03521.x.
167. Lawler PR, Lawler J. Molecular basis for the regulation of angiogenesis by thrombospondin-1 and—2. Cold Spring Harb Perspect Med. 2012;2:a006627. doi:10.1101/cshperspect.a006627a006627 [pii].
168. Chaqour B. Molecular control of vascular development by the matricellular proteins () and (). Trends Dev Biol. 2013;7:59–72.
169. Van Obberghen-Schilling E, et al. Fibronectin and tenascin-C: accomplices in vascular morphogenesis during development and tumor growth. Int J Dev Biol. 2011;55:511–25. doi:103243eo [pii]10.1387/ijdb.103243eo.
170. Williams KJ. Arterial wall chondroitin sulfate proteoglycans: diverse molecules with distinct roles in lipoprotein retention and atherogenesis. Curr Opin Lipidol. 2001;12:477–87.

Chapter 8
Vascular Endothelial Growth Factor-A-Induced Vascular Permeability and Leukocyte Extravasation

Mia Phillipson, Gustaf Christoffersson, Lena Claesson-Welsh and Michael Welsh

Abbreviations

VEGFA	Vascular endothelial growth factor A
VEGFR	Vascular endothelial growth factor receptor
FAK	Focal adhesion kinase
PI3K	Phosphatidyl-inositol 3′kinase
eNOS	Endothelial nitric oxide synthase
TSAd	T cell specific adapter protein
Shb	Src homology-2 domain protein B
GAB1	Grb2-associated binding protein 1
IQGAP1	IQ motif-containing GTPase activating protein 1
VVOs	Vesiculo-vacuolar organelles
VE-cadherin	Vascular endothelial-cadherin
NO	Nitric oxide
PAK	p21-activated kinase
HGF	Hepatocyte growth factor

M. Welsh (✉) · M. Phillipson · G. Christoffersson
Department of Medical Cell Biology, Uppsala University, Husargatan 3, P.O. Box 571, 75123 Uppsala, Sweden
e-mail: Michael.welsh@mcb.uu.se

M. Phillipson
Department of Medical Cell Biology, Uppsala University, Husargatan 3, P.O. Box 571, 75123 Uppsala, Sweden
e-mail: mia.phillipson@mcb.uu.se

G. Christoffersson
Department of Medical Cell Biology, Uppsala University, Husargatan 3, P.O. Box 571, 75123 Uppsala, Sweden
e-mail: gustaf.christoffersson@mcb.uu.se

L. Claesson-Welsh
Department of Immunology, Genetics and Pathology, Uppsala University, Dag Hammarskjöldsv. 20, 75185 Uppsala, Sweden
e-mail: lena.welsh@igp.uu.se

M. HH. Schmidt, S. Liebner (eds.), *Endothelial Signaling in Development and Disease,* DOI 10.1007/978-1-4939-2907-8_8

N-WASP	Neural Wiskott–Aldrich syndrome protein
Tiam	T-cell lymphoma invasion and metastasis
VE-PTP	Vascular endothelial-protein tyrosine phosphatase
ERK	Extracellular signal-regulated kinase
CXCR	Chemokine C-X-C motif ligand receptor
MMP	Matrix metalloproteinase
ICAM-1	Intercellular adhesion molecule-1
PSGL1	P-selectin glycoprotein ligand 1
SDF-1	Stromal cell-derived factor-1
ROS	Reactive oxygen species
BRB	Blood-retinal barrier
RCE	Retinal capillary cell
RPE	Retinal pigment epithelial cell

8.1 Introduction

The two hallmark responses to the main angiogenic factor vascular endothelial growth factor-A (VEGFA) are increased angiogenesis and vascular permeability [1]. Indeed, the vascular permeability response to VEGFA is so prominent that this molecule was originally described as vascular permeability factor (VPF) [2, 3]. Vascular permeability can be categorized according to its functional implications and cues responsible for promoting this response. Under basal conditions, the vasculature is primarily responsible for supplying oxygen and nutrients, and disposing of waste products. These compounds have a high capacity for diffusion over the membrane due to their small sizes, and thus no specific need for increased vascular permeability is required to perform these functions. However, effusion of larger compounds to the extravascular space requires an acute increase in vascular permeability, a process that can be elicited by growth factors, cytokines, or chemokines, which actively modify the function of the vasculature [4, 5]. Examples of factors responsible for acutely increased vascular permeability during hypoxia and inflammation, respectively, are VEGFA and bradykinin and/or histamine [1, 4, 6–9]. Whether the increased vascular permeability serves the same purpose when initiated by different mediators during diverse settings remains unknown. The acute permeability response can easily be recorded by studying leakage of high molecular weight tracers to the extravascular space [10]. If the acute response persists, which, for example, happens in tumors, wound healing, and chronic inflammatory diseases, the permeability increase will transition to a chronic phase that sometimes becomes pathological. Increased permeability implies that larger molecules become diffusible over the endothelial barrier, and thus an exudate of plasma will leave the vessels and immerse the extravascular space. During this process, fibrin clots containing deposits of plasma factors can be formed and they then function as gels forming a provisional additional matrix for angiogenic sprouts to grow on [11, 12]. In addition to the extravasation of plasma proteins, blood leukocytes will leave the vasculature and infiltrate the surrounding tissue. Although a considerable amount of knowledge has accumulated on the mechanisms of induced vascular permeability and its regulation, many aspects of the roles of this response for physiological and pathological processes remain unresolved. Presumably, extravascular deposits of plasma proteins and infiltrating leukocytes will promote the repair of damaged tissue by supporting neoangiogenesis, and, in addition, infiltrating leukocytes will help to clear out damaged cells and dissolve the fibrin clots. These assumptions seem plausible and suggest that vascular permeability is indeed important in physiologically relevant repair processes; nevertheless, precise data defining a role of vascular permeability response in such scenarios are sparse.

In the following text, mechanisms of VEGFA-induced vascular permeability and relevant signal transduction schemes downstream of VEGF receptor-2 (VEGFR2) will be summarized. This will be followed by a description of VEGFA-induced vascular permeability *in vivo*. Finally, mechanisms of VEGFA-stimulated leukocyte extravasation and vascular permeability in disease will be reviewed, followed by concluding remarks.

8.2 Mechanisms of Increased Vascular Permeability

The major part of the vascular permeability response is due to leakage that takes place in postcapillary venules [13, 14], but VEGFA-induced leakage has also been found to occur in capillaries and muscular venules [15]. Numerous mechanisms responsible for VEGFA-induced leakage have been proposed. Initially, caveolae were thought to be responsible for vascular permeability [16] but this suggestion was contradicted by the finding that the caveolin-1 knockout mouse is capable of mounting an increased vascular permeability response despite the absence of caveolae [17]. In another model, permeability is achieved by passage through a stacked system of intracellular vesicular organelles that traverse the endothelial cytoplasm [10]. These have been named vesiculo-vacuolar organelles (VVOs), and by juxtaposition or fusion they create a continuous channel that allows passage of high molecular weight compounds. Their formation is not dependent on the presence of caveolae and they allow passage of ferritin [10]. How these organelles fuse or allow opening of the diaphragms between adjacent vesicles in a regulated manner during the permeability response is at present largely unknown but could reflect a mechanical contraction of the endothelial cell.

Retraction of endothelial cells represents another potential mechanism, giving rise to increased vascular permeability [13]. The action of intracellular motor proteins will thus cause cell contraction, which creates gaps at the cell boundary and allows passage of high molecular weight compounds through these. However, the cell retraction hypothesis has been challenged and the cell shape changes observed have been attributed to a natural recoil process occurring when cell–cell junctions are disassembled [18, 19].

Plasticity of endothelial cell junctions is an additional mechanism of control of vascular permeability [4] (see Chap. 6 by Bravi and Lampugnani). The main structures on endothelial cells for cell–cell contacts are adherens junctions, and these are primarily organized through homophilic complexes of vascular endothelial (VE)-cadherin [20] between two adjacent endothelial cells [21, 22]. The intracellular sites of such complexes bind α-catenin, β-catenin, and p120-catenin [23], allowing regulation of gene expression via β-catenin and the generation of an interface that connects junctions with the cytoskeleton. Expression of a VE-cadherin/α-catenin fusion protein in endothelial cells effected a tighter association of the cytoskeleton to this fusion protein, and a reduced permeability response, indicating that an association of the cytoskeleton to adherens junctions causes junction stabilization [24]. VEGFA-induced vascular permeability causes disassembly of VE-cadherin from adherens junctions *in vivo* [25]. This effect may be partial, generating loose junctions that are partly unfolded, or complete, forming gaps between the cells [4, 26]. VE-cadherin may interact directly with VEGFR2 [20, 27–29] or vascular endothelial-protein tyrosine phosphatase (VE-PTP) [30–32], and these associations are thought to regulate junction integrity. VEGFR2 stimulation will dissociate VE-cadherin [23, 25], partly or completely, from the complexes that build up junctions and initiate VE-cadherin endocytosis. The ensuing fate of the endocytosed VE-cadherin is not clear; it may recycle to the membrane or be degraded. Disruption of adherens junctions may further affect intracellular processes such as microtubule growth via Src and phospholipase C-γ-dependent dephosphorylation of end-binding protein 3 [33]. In addition to adherens junctions, some endothelial cells have specialized tight junctions, as yet another barrier preventing permeability.

Vascular permeability is additionally regulated by vasodilation, which is a process primarily regulated by precapillary arterioles [34]. Proximal vessel dilation or dilation at the site of vascular permeability will increase the local blood pressure and flow, thus allowing facilitated efflux of plasma components. Nitric oxide (NO) has been implicated in vascular permeability, possibly by causing vasodilation and thereby increased blood flow [35].

8.3 Vascular Endothelial Growth Factor Receptor-2 Signal Transduction in the Regulation of Vascular Permeability

Signal transduction downstream of VEGFR2 has been investigated extensively, and pathways that are likely to participate in the regulation of the vascular permeability response will be described in this section. The subsequent part will describe current knowledge on *in vivo* regulation of VEGFA-induced vascular permeability.

The Src family of tyrosine kinases, Src and Yes, have been shown to exert an important role in VEGFA-induced vascular permeability [27, 28]. VEGFR2 activates these kinases by various means. *In vitro* data suggest direct binding of Src to tyrosine Y1059 in the kinase domain of active VEGFR2 as a possible mode of Src activation [36]. In addition, activation occurs via binding to the T cell specific adapter protein (TSAd) downstream of VEGFR2 *in vitro* and *in vivo* [37, 38]. A third possible mechanism of interaction involves binding to the adapter Src -homology 2 domain protein B (Shb) [39–41]. Finally, Src phosphorylation of the IQ motif-containing GTPase-activating protein (IQGAP1) has been reported to exert VEGFA-dependent effects [36, 42].

The effects of Src activation on the vascular permeability response are pleiotropic. Src kinases may directly phosphorylate VE-cadherin [28, 43] and β−catenin [44, 45], thus modifying the properties of adherens junctions and the intracellular localization of VE-cadherin. Src kinases may also phosphorylate and activate the tyrosine kinase Axl [46] or the Rac guanine nucleotide exchange factor Vav2 [47], which may induce vascular permeability via downstream signaling through Akt and Rac (see paragraphs on Rac and Akt below in this section). In addition, a complex between Shb and Src [41] may activate focal adhesion kinase (FAK), which participates in the regulation of vascular permeability.

FAK plays a significant role for VEGFA-induced vascular permeability, and exerts this effect via numerous mechanisms [48, 49]. One is direct phosphorylation of β-catenin on residue Y142, leading to disruption of adherens junctions [49]. Another is FAK-dependent activation of Rac [50], which by various means promotes the vascular permeability response (see paragraph on Rac below in this section). FAK activation may also stimulate PI3K activity [51] or exert direct effects on the cytoskeleton. Contrary to these effects, FAK may also tighten the endothelial barrier by phosphorylating N-WASP (neural Wiskott–Aldrich syndrome protein) [52], which then forms a complex with cortical actin and p120-catenin. The molecular mechanism(s) responsible for VEGFR2-induced FAK activation are not fully understood but may involve a direct association between VEGFR2 and clustered integrins [53–55] or RhoA-dependent stimulation [56]. Alternatively, the Shb adapter protein may mediate FAK activation [41] via its association with Src [40, 56]. VEGFA may also control FAK activity via regulation of protein tyrosine phosphatases [57].

In addition to FAK, the Axl tyrosine kinase has recently been found to operate downstream of Src [46]. Phosphatidyl-inositol 3′kinase (PI3K) is a downstream effector of Axl that promotes vascular permeability via diverse mechanisms (see paragraph on PI3K below in this section).

The Rac family G-protein members are established regulators of the vascular permeability response to VEGFA and, as mentioned, function downstream of FAK. Rac may cause serine phosphorylation at S665 of VE-cadherin via its downstream kinase p21-activated kinase (PAK), making the latter dissociate from adherens junctions and internalize [47]. Alternatively, Rac may alter the cytoskeleton effecting a cellular retraction [58]. A third possible mode of action of Rac with respect to induction of vascular permeability is through the generation of reactive oxygen species [59]. On the other hand, active Rac may reduce vessel leakiness under certain conditions [60, 61], possibly due to a promotion of the attachment of the adherens junction complex to the cytoskeleton, suggesting a duality in the effects of Rac on vascular permeability, depending on the specific condition at which Rac is stimulated. The lung endothelial barrier exhibits a complex mode of regulation, depending on VEGFA, hepatocyte growth factor (HGF), and stretch [62]. HGF and physiological stretch tightens the barrier via Rac activation, whereas VEGFA and excessive stretch loosens the barrier involving Rho activation [62]. The pleiotropy in the effect of Rac on endothelial leakiness may reflect the multifaceted regulation of effector mechanisms exerted by Rac—one related to the cytoskeleton, another to VE-cadherin phosphorylation and internalization, and a third dependent on the generation of reactive oxygen species. There are several means by which Rac activity may be regulated. Src-dependent phosphorylation of Vav2 will increase Rac activity [47, 63]. Tiam (T-cell lymphoma invasion and metastasis) is another guanine exchange factor for Rac, and can be stimulated by FAK [64] or PI3K [65].

PI3K is an important signaling intermediate downstream of VEGFR2 [66]. VEGFA may activate PI3K via Axl [46], FAK [51], Shb [41], IQGAP1 [36], or by direct binding of the PI3K p85 subunit to pY-1175 on VEGFR2 [67]. Activation of PI3K generates increased synthesis of phosphatidyl-inositol 3′-phosphates, and these phospholipids activate a number of downstream effectors such as Rho family G-protein (Rho, Rac, Cdc42) guanine nucleotide exchange factors (see Tiam above) affecting the cytoskeleton or Akt. The latter exerts regulation of multiple responses via phosphorylation of substrate proteins. The Akt target most relevant for the permeability response appears to be endothelial nitric oxide synthase (eNOS) [68].

The gas nitric oxide (NO) plays an important role for maintaining vascular homeostasis in the retina (see Chap. 12 by Hammes). Increased activity of eNOS participates in the VEGFA-induced vascular permeability response [68] by increasing the production of NO. This effect results from Akt-dependent phosphorylation of this enzyme [69, 70]. Other kinases that activate eNOS in the context of VEGFA-induced vascular permeability are protein kinase A [71] and Ca^{2+}/calmodulin-dependent kinase [72]. Activity of eNOS may also be regulated by reactive oxygen species [73] or Hsp90 [74]. The reason why active eNOS causes vascular permeability is not fully understood but this action probably involves multiple mechanisms. One established effect of NO is vasodilation that will increase the local blood flow. Another target effect of NO is S-nitrosylation of β-catenin which will cause its dissociation from VE-cadherin and consequently the disassembly of adherens junctions [75].

VE-PTP interacts with both VEGFR2 and VE-cadherin, and is thus considered a regulator of adherens junctions [30–32]. The precise mode of action of VE-PTP in regulating vascular permeability is poorly understood but it was recently shown that VE-PTP participates in the dephosphorylation of VEGFR2 and VE-cadherin, and that a trimeric complex forms between VEGFR2, VE-PTP, and the angiopoietin-1 receptor Tie2 [76]. Dissociation of VE-PTP from VE-cadherin is necessary for VEGFA-induced vascular permeability [77], and loss of VE-PTP in zebrafish will induce breakdown of adherens junctions and vascular leakage [78].

The Shb adapter protein binds to tyrosine 1175 in activated human VEGFR-2, and thus transduces, in a Src-dependent manner, certain signals in response to VEGFA stimulation [41], including activation of FAK and PI3K [41]. Since Shb binds Src in endothelial cells [40], activation of FAK may involve the generation of a trimeric signaling complex consisting of Shb-Src-FAK. *Shb* knockout endothelial cells display reduced VEGFA-dependent activation of FAK, myosin light-chain kinase, extracellular signal-regulated kinase (ERK), Akt, and Rac1 [29, 79], and thus the phenotype presents a signaling signature that may influence vascular permeability in several ways. The association between VEGFR2 and VE-cadherin is normally reduced upon the addition of VEGFA [23], but when Shb is absent, such a dissociation does not occur [29].

TSAd is another SH2 domain-containing adapter molecule that binds tyrosine 951 in VEGFR-2 after receptor activation [38], and thus confers VEGFA stimulation of Src activity since TSAd also binds Src via its proline-rich motif [37]. Consequently, certain Src-dependent responses, including disruption of adherens junctions, show a dependence on TSAd-mediated signaling via TSAd complex formation with VEGFR2, VE-cadherin, and c-Src [37].

IQGAP1 becomes phosphorylated and stimulated by Src subsequent to VEGFA activation [36]. This will reduce the localization of IQGAP1 and VE-cadherin to adherens junctions and reactive oxygen species-dependent tyrosine phosphorylation of VE-cadherin [42]. IQGAP1 may also exist in a complex with Rac1, Src, p47phox, and cortactin, thus suggesting the possibility that IQGAP1 may modify junctions via Rac1 signaling and generation of reactive oxygen species [80].

GAB1 is an adapter protein downstream of VEGFR2 that, upon tyrosine phosphorylation, binds and activates Grb2, SHP2, the p85 subunit of PI3K, and phospholipase C-γ. The increase of PI3K activity is likely to play a role for VEGFA-stimulated vascular permeability [81], although such an effect was not observed in the GAB1 knockout *in vivo*.

8.4 VEGFA-Induced Vascular Permeability *In Vivo*

The description of VEGFR2 signal transduction outlined above is mostly based on *in vitro* studies. These have been complemented with *in vivo* work using models of genetic alterations or chemical inhibitors that have specified the involvement of cer-

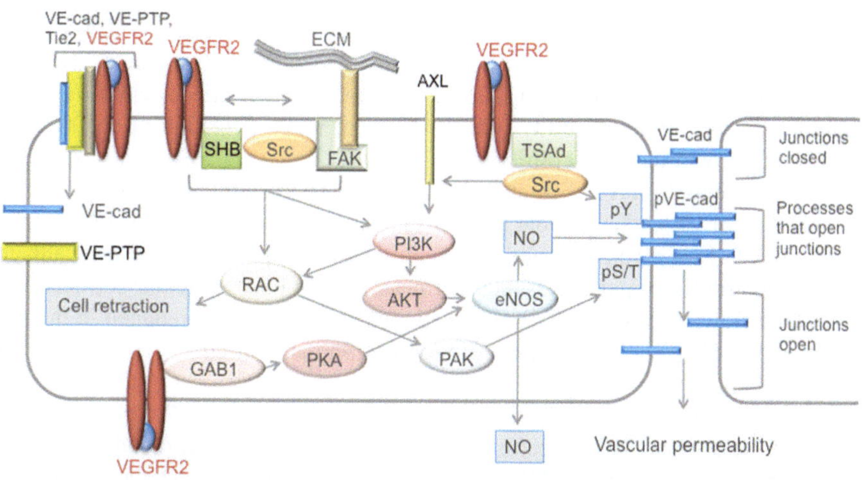

Fig. 8.1 Summary of signal transduction pathways operating downstream of vascular endothelial growth factor-A (VEGFA) in endothelial cells of relevance for vascular permeability *in vivo*. See text for details. Figure is based on Claesson-Welsh and Welsh [7]

tain signaling mechanisms in VEGFA vascular permeability. A summary of signal transduction pathways described *in vivo*, of relevance to the vascular permeability response to VEGFA, is given in Fig. 8.1.

In vivo studies of vascular permeability have implicated the participation of Src family kinases in this process. Experiments using Src kinase-deficient mice revealed reduced VEGFA-induced vascular permeability [27] and subsequently this was found to depend on phosphorylation of VE-cadherin [28] that destabilized adherens junctions. Consequently, tissue edema was reduced and recovery after myocardial infarction improved in the absence of Src activity.

Axl is another tyrosine kinase that contributes to VEGFA-stimulated vascular permeability *in vivo* [46]. Axl is downstream of TSAd/Src and is thought to regulate vascular permeability by activation of PI3K.

In vivo studies using endothelial-specific FAK gene inactivation have demonstrated the requirement of FAK for VEGFA-induced vascular permeability [48, 49].

Studies directly addressing the role of Rac1 in vascular permeability *in vivo* are sparse since mice deficient in endothelial Rac1 die *in utero* due to vascular abnormalities [82]. However, the absence of Rac1 activity in endothelial cells will diminish endothelial cell-driven vasodilation [83]. This effect may, in part, be a consequence of reduced eNOS activation, leading to less blood flow after induced hind-limb ischemia. However, increased endothelial cell Rac1 activity may under certain conditions give less vascular leakage, suggesting multiple modes of action of this signaling component in the vascular permeability response [60].

Absence of the p110γ isoform of PI3K causes diminished vascular permeability in response to VEGFA [84]. On the other hand, inactivation of class 1A PI3K activity leads to increased dye leakage in the neovasculature, suggesting a complex mode of action of PI3K in regulation of this process [85]. Increased activity of the PI3K downstream signaling intermediate Akt in endothelial cells causes chronically increased vascular permeability, with edema as a consequence [68]. This effect is at least partly due to vessel dilation and stimulation of eNOS. Conflicting results have been obtained on the effects of *Akt* deficiency. Whereas one report described reduced VEGFA-induced vascular permeability [86] as a consequence of *Akt*1 deficiency, another study noted an increased response [87]. The deficient VEGFA-stimulated vascular permeability was attributed to weak activation of eNOS, with decreased production of NO [86].

Shb is essential for VEGFA-stimulated vascular permeability *in vivo* [79]. The dissociation of VE-cadherin from adherens junctions that occurs *in vivo* in response to VEGFA fails as a consequence of *Shb* deficiency [26], providing a mechanistic explanation for the impaired vascular permeability response. Vascular permeability and blood flow in experimentally-induced muscle ischemia are reduced in the absence of *Shb* [26], suggesting a contribution of Shb and vascular permeability for recovery in this situation.

TSAd is required for VEGFA-induced vascular permeability [37, 38], both in the skin and the trachea. This effect is explained by impaired VEGFA-induced disassembly of adherens junctions [37]. TSAd signaling appears not to have a major effect on eNOS activity, and histamine-induced vascular permeability was unaffected by the absence of TSAd [37].

No data directly implicating IQGAP1 in the VEGFA-induced vascular permeability response are at present available, although IQGAP1 was shown to be required for macrophage infiltration into damaged muscle tissue during post-ischemic recovery [88].

GAB1 is necessary for post-ischemic recovery, VEGFA-induced angiogenesis and vascular permeability *in vivo* [71]. Although GAB1 is normally thought to mediate PI3K activation, the *in vivo* response of the knockout primarily depends on diminished activation of protein kinase A, leading to reduced activation of eNOS [71].

Several of the signal transduction responses downstream of VEGFR-2 ultimately cause activation of eNOS, as described above. This is an important effector since absence of eNOS will abolish the vascular permeability response [89]. As mentioned above, the mechanisms behind the permeability response of NO are poorly understood but may involve increased vasodilation and direct disruption of adherens junctions.

8.5 Modulation of VEGFA-Induced Vascular Permeability by Other Factors

Angiopoietin-1 is an angiogenic factor, operating via its receptor Tie2, which is known to reduce VEGFA-stimulated vascular permeability [90], although the underlying mechanism remains unclear. One proposed model involves RhoA-dependent sequestration of Src onto mDia (diaphanous homolog), thus reducing the phosphorylation of S665 on VE-cadherin in an Rac-dependent manner [91]. However, the effects of angiopoietin-1 on Rac1 activity are contradictory since another study described increased Rac1 activity in response to this ligand, causing reduced VEGFA-stimulated vascular leakage [60]. Apparently, Rac1 may stimulate or inhibit vascular permeability, depending on the local conditions. Angiopoietin-1 may also inhibit VEGFA-induced vascular permeability via protein kinase C zeta-dependent phosphorylation of an inhibitory site on eNOS [92]. Tie2 and VE-PTP interact on the cell surface but downregulation of such complexes does not affect the VE-cadherin staining pattern [93]. However, VE-PTP-dependent dephosphorylation of VEGFR2 requires angiopoietin-1-activated Tie2 [76], suggesting a VE-PTP-dependent antagonism between VEGFA and angiopoietin-1 regulating junction permeability. In addition, angiopoietin-1 decreases the phosphorylation of VE-cadherin at Y658, presumably via the action of VE-PTP, since VE-cadherin phosphorylation was elevated in VE-PTP-deficient cells.

The phospholipid sphingosine-1-phosphate is an extracellular ligand operating via G-protein-coupled receptors that influences the vasculature [94]. Sphingosine-1-phosphate may stimulate VEGFR2, resulting in Akt and eNOS activation [95]. *In vivo,* sphingosine-1-phosphate reduces vascular leakage by increasing Rac1 activity in a manner similar to that of angiopoietin-1 [60]. Alternatively, sphingosine-1-phosphate may stabilize vessels by inhibition of VEGFR2 signaling and retention of VE-cadherin at junctions after VEGFA stimulation [96].

8.6 VEGFA and Leukocyte Extravasation

VEGFA has been demonstrated to recruit leukocytes from the circulation to tissues, using different *in vivo* models. When VEGFA overexpression was induced in selected organs in a transgenic system, bone marrow-derived mononuclear myeloid cells were recruited to these specific sites [97]. Furthermore, neutrophils to accumulated at the site of intramuscularly transplanted pancreatic islets in a VEGFA-dependent manner [98, 99]. The revascularization process of these transplanted islets was dependent on recruitment of a distinct circulating neutrophil ($CXCR4^{hi}$, $MMP\text{-}9^{hi}$) population [98, 99]. The relationship between VEGFA, neoangiogenesis, and neutrophil extravasation is illustrated in Fig. 8.2. VEGFA production in transplanted islets promotes the extravasation of a subset of neutrophils that participate in the revascularization process.

How tissue-derived VEGFA exerts its effect on leukocytes in circulation is still not clear, but diverse mechanisms have been suggested and may act in concert, potentiating each other. Studies performed *in vitro* on human umbilical vein endothelial cells (HUVECs) or colonic microvasculature demonstrated upregulation of endothelial adhesion molecules (ICAM-1 and P-selectin) in response to VEGFA [100–102], while VEGFA-induced leukocyte adhesion was shown to be dependent on CD18 [103, 104]. The expression of VEGFR1 on monocytes [97, 105] implies direct effects of VEGFA on cells in circulation. Indeed, VEGF-induced adhesion and migration of isolated monocytes, neutrophils, and T cells have been demonstrated [100, 101, 103, 106–108]. VEGFA is also known to induce expression of other chemotactic agents such as stromal cell-derived factor-1 (SDF-1). Transgenic induction of VEGFA expression induced expression of SDF-1 by fibroblasts, which, in the tissue, positioned recruited leukocytes perivascularly relative to the angiogenic vessels [97]. A role for SDF-1 in mobilization of circulating leukocytes at hypoxic sites has been reported [109]. In addition, multiple tumor-derived chemoattractants, including VEGFA, recruit $CD11b^{+}Gr1^{+}$leukocytes by activating the PI3-kinase isoform p110γ, which results in activation of the integrin $\alpha_4\beta_1$ on the leukocyte, and concomitant tumor invasion [110].

Leukocyte extravasation can occur transcellularly through the thin endothelial cells, or paracellularly through endothelial junctions [111–113]. The extravasation process is mechanistically not yet well-defined and the route of choice might depend on stimulus, type of leukocyte, and location of the vascular bed. The importance of the paracellular route was demonstrated in two genetic models where the endothelial junctions were kept sealed by maintaining VE-cadherin in a constitutively active state [24, 77]. These models revealed greatly impaired leukocyte extravasation in response to inflammatory stimuli at different sites, demonstrating that opening of endothelial junctions is essential for leukocyte recruitment out of the vasculature.

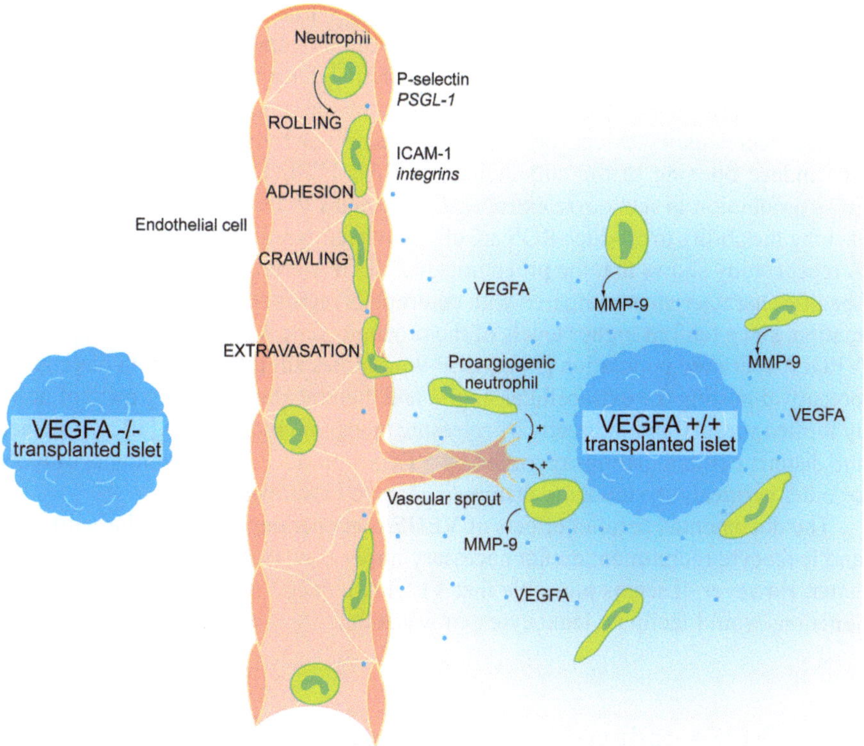

Fig. 8.2 Schematic view illustrating neutrophil extravasation in relation to local vascular endothelial growth factor-A (*VEGFA*) production after islet transplantation to the cremaster muscle. Different steps of the leukocyte extravasation process (leukocyte rolling following capture, adhesion, crawling, and extravasation) are depicted that sequentially result in leukocyte migration towards the site of VEGFA production, which, in this scenario, is the transplanted islet. Leukocyte capture/rolling requires P-selectin on the endothelial cells and the P-selectin glycoprotein ligand-1 (PSGL1) expressed on the leukocyte. Expression of the endothelial cell adhesion molecule ICAM-1 and leukocyte integrins probably participates in this process. Consequently, an enrichment of a subpopulation of neutrophils (CXCR4^{hi}, MMP-9^{hi}) that assist neoangiogenesis occurs in the tissue surrounding the islet. The mechanisms by which they exert their proangiogenic effect are not completely understood, although they are likely to involve release of matrix metalloproteinase (MMP)-9. Islets devoid of VEGFA production fail to induce this process

Whether opening of endothelial junctions during leukocyte extravasation inevitably results in a permeability increase remains to be shown.

Edema formation and accumulation of tissue leukocytes are cardinal signs of inflammation and these occur concomitantly with increased vascular permeability. Intuitively, an association between these processes is commonly made but several studies prove them to be both spatially and temporally uncoupled [114, 115], and leukocyte extravasation can occur at sites distant from formed endothelial gaps [116–118]. The formation of endothelial cell domes during leukocyte diapedesis has been implicated as a mechanism for controlling vascular permeability and endothelial barrier integrity [119]. Whether this is true for VEGFA-induced leukocyte recruitment has not yet been experimentally established. Thus, increased permeability identified by the hitherto applied methods appears not to be a prerequisite for leukocyte extravasation during inflammatory conditions.

Adherent leukocytes can produce and release reactive oxygen species (ROS) that induce opening of the endothelial junctions [120, 121]. The contribution of ROS production in leukocyte extravasation remains to be defined, and leukocytes lacking the ability to produce ROS are efficient in emigrating out of the vasculature. A recent study shows that the proangiogenic VEGFA-recruited neutrophils possess specific characteristics compared with neutrophils recruited to an infectious inflammation, since tenfold higher levels of the proangiogenic MMP-9 were found in the VEGFA-recruited population [99]. MMP-9 promotes angiogenesis both directly by potently digesting the extracellular matrix and thereby allowing for growth of newly formed vessels, and indirectly by releasing matrix-bound VEGFA [122]. Whether the distinct cell populations recruited by VEGFA produce similar amounts of ROS or other permeability-inducing factors remains to be shown.

The documented parallel effects of VEGFA on opening of endothelial junctions and leukocyte recruitment do not necessary imply contribution of the former to the latter. However, it seems plausible that VEGFA-induced weakening of endothelial junctions would facilitate leukocyte extravasation.

8.7 VEGFA-Induced Vascular Permeability and Disease

8.7.1 Vascular Permeability in Tumors

The tumor vasculature shows numerous abnormalities, such as poor perfusion, high vessel turnover, increased vessel tortuosity, and vascular leakage [123]. Anti-angiogenic treatment by VEGFA-blocking regimens inhibits or reverses, in many instances, the abnormal tumor vessel phenotype and causes vascular 'normalization' [123]. Such strategies commonly inhibit tumor growth, and VEGFA inhibition is at present an accepted clinical practice for treatment of tumors such as glioblastomas unresponsive to other treatments, metastatic colorectal cancers, metastatic renal cell cancer, some nonsmall cell lung cancers, hepatocellular tumors, and neuroendocrine tumors (www.cancer.gov/cancertopics/factsheet/Therapy/angiogenesis-inhibitors). The reason why this therapy has proven relatively successful has nevertheless not been fully resolved since it has been shown that the VEGFA-inhibited tumor vasculature displays improved vascular function [123]. A possible explanation for this dichotomy is that the VEGFA-blocking treatments, in addition to reducing tumor angiogenesis, simultaneously inhibit VEGFA-induced vascular permeability, and that the latter contributes to tumor growth and dissemination. The angiogenic factor angiopoietin-1 decreases vascular permeability *in vivo* [124]. When tumors were treated with an angiopoietin-2 inhibitor (L1-7(N)), tumor vessel 'normalization' was observed but this effect was reversed by inhibition of angiopoietin-1, suggesting that angiopoietin-1 can reduce tumor vascular permeability by antagonizing angiopoietin-2 [125]. A consequence of vascular permeability could be leukocyte extravasation, and it has been shown that many tumors expand in a manner dependent on infiltration of class II macrophages (M2) [126, 127], which confer a pro-angiogenic phenotype. Indeed, VEGFA has been shown to cause selective extravasation of proangiogenic leukocytes [99], and such an effect could promote tumor expansion.

8.7.2 Vascular Permeability in Retinal Disease

The vasculature in the eye is protected by the blood–retinal barrier (BRB), which is maintained by tight junctions between retinal capillary endothelial (RCE) cells on the one hand, and retinal pigment epithelial (RPE) cells on the other, which form the inner and outer BRB, respectively [128]. The tight junctions of RCE cells are formed by intercellular communications between RCE and glial cells [129]. Thereby, organization of the BRB resembles the blood–brain barrier. Loss of normal BRB function is a common feature of many retinal degenerative disorders that are leading causes of visual dysfunction. Such diseases include age-related macular degeneration, diabetic retinopathy, and retinal vein occlusions [130] (see Chap. 12 by Hammes). Patients with age-related macular degeneration present focal ischemia of the outer retina, which induces VEGFA production and angiogenesis, resulting in hyperpermeable vessels and inflammation. Prolonged elevation of blood sugar levels in diabetic patients causes endothelial apoptosis, basement membrane thickening, and pericyte loss, accompanied by increased VEGFA synthesis and vascular permeability. Retinal vein occlusions can be attributed to hemodynamic disturbances, such as increased coagulation and impaired flow properties, resulting in ischemia and increased VEGFA synthesis (see Stewart [130] for details). Therefore, common aspects of many vascular eye diseases are ischemia, increased VEGFA production, and excess vascular permeability [131]. The excess permeability has been attributed to both the overstimulated, abnormal vasculature and to changes in the phoshorylation of tight junction proteins such as occludin and zona occludens protein-1 (ZO1) [132]. Bevacizumab, a VEGFA neutralizing antibody or Lucentis, a VEGFA neutralizing Fab2 fragment, administered by intravitreal injection, have been used successfully for the treatment of ocular diseases, resulting in preservation and even gain of visual acuity.

Figure 8.3 summarizes VEGFA-induced vascular permeability in the context of pathophysiological processes, and suggests possible modes of pharmacological intervention.

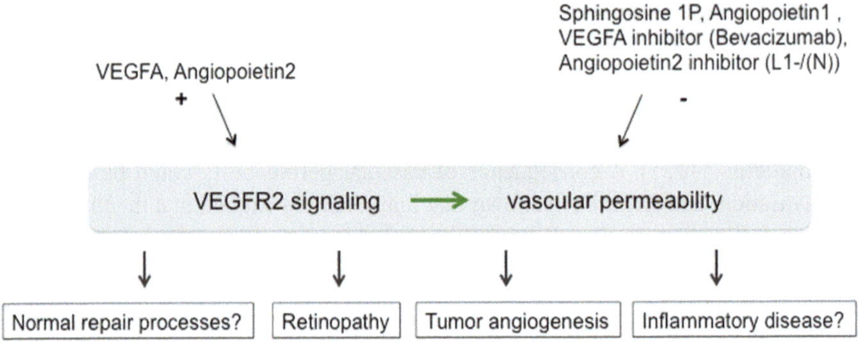

Fig. 8.3 Vascular endothelial growth factor-A (*VEGFA*)-induced vascular permeability in the context of disease. Potential sites of pharmacological intervention include VEGFA/VEGF receptor-2 (*VEGFR2*) antagonists/inhibitors, angiopoietin-1, angiopoietin-2 inhibitor, sphingosine-1 phosphate, and intracellular VEGFR2 signaling. See texts for details

8.8 Conclusions

Numerous signaling mechanisms participate in the response, conferring vascular permeability in response to VEGFA and significant redundancy exists between the various pathways. Presumably, precise control of the characteristics and degree of stimulation of vascular permeability can be achieved by differentially orchestrating the individual signaling pathways. In addition, this pleiotropy ascertains that the permeability response will not be suppressed in case one signaling scheme fails. Taken together, this indicates the importance of vascular permeability for the tissue repair response to ischemia. It is still uncertain what the precise beneficial effects of vascular permeability for the repair processes are, but they could involve extravasation of both plasma proteins and leukocytes. A factor that complicates specific elucidation of the beneficial contribution of vascular permeability to tissue restoration is that most signal transduction pathways of importance in this context are also essential for tissue angiogenesis [7], making delineation of the relative importance of vascular permeability versus angiogenesis challenging. In fact, one may speculate that these are two inseparable aspects of the angiogenic repair process. Inhibition of tumor angiogenesis by VEGFA-blocking regimens has been assumed to be the main targets of such treatments, but it is conceivable that inhibition of VEGFA-induced vascular permeability may be an equally important component.

Acknowledgments The study was supported by grants from Diabetes Wellness Sweden (MP), the Novo Nordisk foundation (MP), the Swedish Cancer Fund (MW, LCW), the Swedish Research Council (MP, MW, LCW), the Swedish Diabetes Fund (MP, MW), the Ragnar Söderberg foundation (MP), the Knut and Alice Wallenberg Foundation (MP, LCW), and the Family Ernfors Fund (MP, MW).

References

1. Ferrara N. Vascular endothelial growth factor: basic science and clinical progress. Endocr Rev. 2004;25(4):581–611.
2. Keck PJ, Hauser SD, Krivi G, Sanzo K, Warren T, Feder J, et al. Vascular permeability factor, an endothelial cell mitogen related to PDGF. Science. 1989;246(4935):1309–12.
3. Senger DR, Perruzzi CA, Feder J, Dvorak HF. A highly conserved vascular permeability factor secreted by a variety of human and rodent tumor cell lines. Cancer Res. 1986;46(11):5629–32.
4. McDonald DM, Thurston G, Baluk P. Endothelial gaps as sites for plasma leakage in inflammation. Microcirculation. 1999;6(1):7–22.
5. Dvorak HF. Angiogenesis: update 2005. J Thromb Haemost. 2005;3(8):1835–42.
6. Sprague AH, Khalil RA. Inflammatory cytokines in vascular dysfunction and vascular disease. Biochem Pharmacol. 2009;78(6):539–52.
7. Claesson-Welsh L, Welsh M. VEGFA and tumour angiogenesis. J Intern Med. 2013;273: 114–27.
8. Maurer M, Bader M, Bas M, Bossi F, Cicardi M, Cugno M, et al. New topics in bradykinin research. Allergy. 2011;66(11):1397–406.
9. Kumar P, Shen Q, Pivetti CD, Lee ES, Wu MH, Yuan SY. Molecular mechanisms of endothelial hyperpermeability: implications in inflammation. Expert Rev Mol Med. 2009;11:e19..
10. Nagy JA, Benjamin L, Zeng H, Dvorak AM, Dvorak HF. Vascular permeability, vascular hyperpermeability and angiogenesis. Angiogenesis. 2008;11(2):109–19.
11. Dvorak HF, Quay SC, Orenstein NS, Dvorak AM, Hahn P, Bitzer AM, et al. Tumor shedding and coagulation. Science. 1981;212(4497):923–4.
12. Nagy JA, Dvorak AM, Dvorak HF. Vascular hyperpermeability, angiogenesis, and stroma generation. Cold Spring Harb Perspect Med. 2012;2(2):a006544.
13. Majno G, Shea SM, Leventhal M. Endothelial contraction induced by histamine-type mediators: an electron microscopic study. J Cell Biol. 1969;42(3):647–72.
14. Kohn S, Nagy JA, Dvorak HF, Dvorak AM. Pathways of macromolecular tracer transport across venules and small veins. Structural basis for the hyperpermeability of tumor blood vessels. Lab Invest. 1992;67(5):596–607.
15. Roberts WG, Palade GE. Increased microvascular permeability and endothelial fenestration induced by vascular endothelial growth factor. J Cell Sci. 1995; 108(Pt 6):2369–79.
16. Simionescu M, Simionescu N, Palade GE. Differentiated microdomains on the luminal surface of capillary endothelium: distribution of lectin receptors. J Cell Biol. 1982;94(2):406–13.
17. Schubert W, Frank PG, Woodman SE, Hyogo H, Cohen DE, Chow CW, et al. Microvascular hyperpermeability in caveolin-1 (-/-) knock-out mice. Treatment with a specific nitric-oxide synthase inhibitor, L-NAME, restores normal microvascular permeability in Cav-1 null mice. J Biol Chem. 2002;277(42):40091–8.
18. Adamson RH, Zeng M, Adamson GN, Lenz JF, Curry FE. PAF- and bradykinin-induced hyperpermeability of rat venules is independent of actin-myosin contraction. Am J Physiol Heart Circ Physiol. 2003;285(1):H406–17.
19. Waschke J, Drenckhahn D, Adamson RH, Curry FE. Role of adhesion and contraction in Rac 1-regulated endothelial barrier function in vivo and in vitro. Am J Physiol Heart Circ Physiol. 2004;287(2):H704–11.
20. Carmeliet P, Lampugnani MG, Moons L, Breviario F, Compernolle V, Bono F, et al. Targeted deficiency or cytosolic truncation of the VE-cadherin gene in mice impairs VEGF-mediated endothelial survival and angiogenesis. Cell. 1999;98(2):147–57.
21. Dejana E. Endothelial cell-cell junctions: happy together. Nat Rev Mol Cell Biol. 2004;5(4):261–70.
22. Dejana E, Tournier-Lasserve E, Weinstein BM. The control of vascular integrity by endothelial cell junctions: molecular basis and pathological implications. Dev Cell. 2009;16(2):209–21.
23. Rudini N, Dejana E. Adherens junctions. Curr Biol. 2008;18(23):R1080–2.

24. Schulte D, Kuppers V, Dartsch N, Broermann A, Li H, Zarbock A, et al. Stabilizing the VE-cadherin-catenin complex blocks leukocyte extravasation and vascular permeability. Embo J. 2011;30(20):4157–70.
25. Cai J, Wu L, Qi X, Shaw L, Li Calzi S, Caballero S, et al. Placenta growth factor-1 exerts time-dependent stabilization of adherens junctions following VEGF-induced vascular permeability. PLoS One. 2011;6(3):e18076.
26. Christoffersson G, Zang G, Zhuang ZW, Vagesjo E, Simons M, Phillipson M, et al. Vascular adaptation to a dysfunctional endothelium as a consequence of Shb deficiency. Angiogenesis. 2012;15(3):469–80.
27. Eliceiri BP, Paul R, Schwartzberg PL, Hood JD, Leng J, Cheresh DA. Selective requirement for Src kinases during VEGF-induced angiogenesis and vascular permeability. Mol Cell. 1999;4(6):915–24.
28. Weis S, Shintani S, Weber A, Kirchmair R, Wood M, Cravens A, et al. Src blockade stabilizes a Flk/cadherin complex, reducing edema and tissue injury following myocardial infarction. J Clin Invest. 2004;113(6):885–94.
29. Zang G, Christoffersson G, Tian G, Harun-Or-Rashid M, Vagesjo E, Phillipson M, et al. Aberrant association between vascular endothelial growth factor receptor-2 and VE cadherin in response to vascular endothelial growth factor-a in Shb-deficient lung endothelial cells. Cell Signal. 2013;25:85–92.
30. Nawroth R, Poell G, Ranft A, Kloep S, Samulowitz U, Fachinger G, et al. VE-PTP and VE-cadherin ectodomains interact to facilitate regulation of phosphorylation and cell contacts. Embo J. 2002;21(18):4885–95.
31. Nottebaum AF, Cagna G, Winderlich M, Gamp AC, Linnepe R, Polaschegg C, et al. VE-PTP maintains the endothelial barrier via plakoglobin and becomes dissociated from VE-cadherin by leukocytes and by VEGF. J Exp Med. 2008;205(12):2929–45.
32. Mellberg S, Dimberg A, Bahram F, Hayashi M, Rennel E, Ameur A, et al. Transcriptional profiling reveals a critical role for tyrosine phosphatase VE-PTP in regulation of VEGFR2 activity and endothelial cell morphogenesis. Faseb J. 2009;23(5):1490–502.
33. Komarova YA, Huang F, Geyer M, Daneshjou N, Garcia A, Idalino L, et al. VE-cadherin signaling induces EB3 phosphorylation to suppress microtubule growth and assemble adherens junctions. Mol Cell 2012;48(6):914–25.
34. Davis MJ. Perspective: physiological role(s) of the vascular myogenic response. Microcirculation. 2012;19(2):99–114. [Review].
35. Casey DP, Joyner MJ. Local control of skeletal muscle blood flow during exercise: Influence of available oxygen. J Appl Physiol 2011;111(6):1527–38.
36. Meyer RD, Sacks DB, Rahimi N. IQGAP1-dependent signaling pathway regulates endothelial cell proliferation and angiogenesis. PLoS One. 2008;3(12):e3848.
37. Sun Z, Li X, Massena S, Kutschera S, Padhan N, Gualandi L, et al. VEGFR2 induces c-Src signaling and vascular permeability in vivo via the adaptor protein TSAd. J Exp Med. 2012;209(7):1363–77.
38. Matsumoto T, Bohman S, Dixelius J, Berge T, Dimberg A, Magnusson P, et al. VEGF receptor-2 Y951 signaling and a role for the adapter molecule TSAd in tumor angiogenesis Embo J. 2005;24(13):2342–53.
39. Anneren C, Lindholm CK, Kriz V, Welsh M. The FRK/RAK-SHB signaling cascade: a versatile signal-transduction pathway that regulates cell survival, differentiation and proliferation. Curr Mol Med. 2003;3(4):313–24.
40. Holmqvist K, Cross M, Riley D, Welsh M. The Shb adaptor protein causes Src-dependent cell spreading and activation of focal adhesion kinase in murine brain endothelial cells. Cell Signal. 2003;15(2):171–9.
41. Holmqvist K, Cross MJ, Rolny C, Hagerkvist R, Rahimi N, Matsumoto T, et al. The adaptor protein shb binds to tyrosine 1175 in vascular endothelial growth factor (VEGF) receptor-2 and regulates VEGF-dependent cellular migration. J Biol Chem. 2004;279(21):22267–75.
42. Yamaoka-Tojo M, Tojo T, Kim HW, Hilenski L, Patrushev NA, Zhang L, et al. IQGAP1 mediates VE-cadherin-based cell-cell contacts and VEGF signaling at adherence junctions linked to angiogenesis. Arterioscler Thromb Vasc Biol. 2006;26(9):1991–7.

43. Wallez Y, Cand F, Cruzalegui F, Wernstedt C, Souchelnytskyi S, Vilgrain I, et al. Src kinase phosphorylates vascular endothelial-cadherin in response to vascular endothelial growth factor: identification of tyrosine 685 as the unique target site. Oncogene. 2007;26(7):1067–77.
44. Huber AH, Weis WI. The structure of the beta-catenin/E-cadherin complex and the molecular basis of diverse ligand recognition by beta-catenin. Cell. 2001;105(3):391–402.
45. Roura S, Miravet S, Piedra J, Garcia de Herreros A, Dunach M. Regulation of E-cadherin/Catenin association by tyrosine phosphorylation. J Biol Chem. 1999;274(51):36734–40.
46. Ruan GX, Kazlauskas A. Axl is essential for VEGF-A-dependent activation of PI3K/Akt. Embo J. 2012;31(7):1692–703.
47. Gavard J, Gutkind JS. VEGF controls endothelial-cell permeability by promoting the beta-arrestin-dependent endocytosis of VE-cadherin. Nat Cell Biol. 2006;8(11):1223–34.
48. Lee J, Borboa AK, Chun HB, Baird A, Eliceiri BP. Conditional deletion of the focal adhesion kinase FAK alters remodeling of the blood-brain barrier in glioma. Cancer Res. 2010;70(24):10131–40.
49. Chen XL, Nam JO, Jean C, Lawson C, Walsh CT, Goka E, et al. VEGF-induced vascular permeability is mediated by FAK. Dev Cell. 2012;22(1):146–57.
50. Parsons JT. Focal adhesion kinase: the first ten years. J Cell Sci. 2003; 116(Pt 8):1409–16.
51. Qi JH, Claesson-Welsh L. VEGF-induced activation of phosphoinositide 3-kinase is dependent on focal adhesion kinase. Exp Cell Res. 2001;263(1):173–82.
52. Rajput C, Kini V, Smith M, Yazbeck P, Chavez A, Schmidt T, et al. Neural Wiskott-Aldrich syndrome protein (N-WASP)-mediated p120-catenin interaction with Arp2 actin complex stabilizes endothelial adherens junctions. J Biol Chem. 2013;288(6):4241–50.
53. Masson-Gadais B, Houle F, Laferriere J, Huot J. Integrin alphavbeta3, requirement for VEGFR2-mediated activation of SAPK2/p38 and for Hsp90-dependent phosphorylation of focal adhesion kinase in endothelial cells activated by VEGF. Cell Stress Chaperones. 2003;8(1):37–52.
54. Lakshmikanthan S, Sobczak M, Chun C, Henschel A, Dargatz J, Ramchandran R, et al. Rap1 promotes VEGFR2 activation and angiogenesis by a mechanism involving integrin alphavbeta. Blood. 2011;118(7):2015–26.
55. Mahabeleshwar GH, Feng W, Reddy K, Plow EF, Byzova TV. Mechanisms of integrin-vascular endothelial growth factor receptor cross-activation in angiogenesis. Circ Res. 2007;101(6):570–80.
56. Le Boeuf F, Houle F, Huot J. Regulation of vascular endothelial growth factor receptor 2-mediated phosphorylation of focal adhesion kinase by heat shock protein 90 and Src kinase activities. J Biol Chem. 2004;279(37):39175–85.
57. Abdelsaid MA, El-Remessy AB. S-glutathionylation of LMW-PTP regulates VEGF-mediated FAK activation and endothelial cell migration. J Cell Sci. 2012; 125(Pt 20):4751–60.
58. Ridley AJ. Life at the leading edge. Cell. 2011;145(7):1012–22.
59. Monaghan-Benson E, Burridge K. The regulation of vascular endothelial growth factor-induced microvascular permeability requires Rac and reactive oxygen species. J Biol Chem. 2009;284(38):25602–11.
60. Hoang MV, Nagy JA, Senger DR. Active Rac1 improves pathologic VEGF neovessel architecture and reduces vascular leak: mechanistic similarities with angiopoietin-1. Blood. 2011;117(5):1751–60.
61. Waschke J, Baumgartner W, Adamson RH, Zeng M, Aktories K, Barth H, et al. Requirement of Rac activity for maintenance of capillary endothelial barrier properties. Am J Physiol Heart Circ Physiol. 2004;286(1):H394–401.
62. Birukova AA, Moldobaeva N, Xing J, Birukov KG. Magnitude-dependent effects of cyclic stretch on HGF- and VEGF-induced pulmonary endothelial remodeling and barrier regulation. Am J Physiol Lung Cell Mol Physiol. 2008;295(4):L612–23.
63. Garrett TA, Van Buul JD, Burridge K. VEGF-induced Rac1 activation in endothelial cells is regulated by the guanine nucleotide exchange factor Vav2. Exp Cell Res. 2007;313(15):3285–97.

64. Elias BC, Bhattacharya S, Ray RM, Johnson LR. Polyamine-dependent activation of Rac1 is stimulated by focal adhesion-mediated Tiam1 activation. Cell Adh Migr. 2010;4(3):419–30.
65. Rivard N. Phosphatidylinositol 3-kinase: a key regulator in adherens junction formation and function. Front Biosci. 2009;14:510–22.
66. Jiang BH, Liu LZ. PI3K/PTEN signaling in angiogenesis and tumorigenesis. Adv Cancer Res. 2009;102:19–65.
67. Dayanir V, Meyer RD, Lashkari K, Rahimi N. Identification of tyrosine residues in vascular endothelial growth factor receptor-2/FLK-1 involved in activation o phosphatidylinositol 3-kinase and cell proliferation. J Biol Chem. 2001;276(21):17686–92.
68. Phung TL, Ziv K, Dabydeen D, Eyiah-Mensah G, Riveros M, Perruzzi C, et al. Pathological angiogenesis is induced by sustained Akt signaling and inhibited by rapamycin. Cancer Cell. 2006;10(2):159–70.
69. Dimmeler S, Fleming I, Fisslthaler B, Hermann C, Busse R, Zeiher AM. Activation of nitric oxide synthase in endothelial cells by Akt-dependent phosphorylation. Nature. 1999;399(6736):601–5.
70. Fulton D, Gratton JP, McCabe TJ, Fontana J, Fujio Y, Walsh K, et al. Regulation of endothelium-derived nitric oxide production by the protein kinase Akt. Nature. 1999;399(6736):597–601.
71. Lu Y, Xiong Y, Huo Y, Han J, Yang X, Zhang R, et al. Grb-2-associated binder 1 (Gab1) regulates postnatal ischemic and VEGF-induced angiogenesis through the protein kinase A-endothelial NOS pathway. Proc Natl Acad Sci U S A. 2011;108(7):2957–62.
72. Schneider JC, El Kebir D, Chereau C, Lanone S, Huang XL, De Buys Roessingh AS, et al. Involvement of Ca2+/calmodulin-dependent protein kinase II in endothelial NO production and endothelium-dependent relaxation. Am J Physiol Heart Circ Physiol. 2003;284(6): H2311–9.
73. Maron BA, Michel T. Subcellular localization of oxidants and redox modulation of endothelial nitric oxide synthase. Circ J. 2012;76(11):2497–512.
74. Duval M, Le Boeuf F, Huot J, Gratton JP. Src-mediated phosphorylation of Hsp90 in response to vascular endothelial growth factor (VEGF) is required for VEGF receptor-2 signaling to endothelial NO synthase. Mol Biol Cell. 2007;18(11):4659–68.
75. Thibeault S, Rautureau Y, Oubaha M, Faubert D, Wilkes BC, Delisle C, et al. S-nitrosylation of beta-catenin by eNOS-derived NO promotes VEGF-induced endothelial cell permeability. Mol Cell. 2010;39(3):468–76.
76. Hayashi M, Majumdar A, Li X, Adler J, Sun Z, Vertuani S, et al. VE-PTP regulates VEGFR2 activity in stalk cells to establish endothelial cell polarity and lumen formation. Nat Commun. 2013;4:1672.
77. Broermann A, Winderlich M, Block H, Frye M, Rossaint J, Zarbock A, et al. Dissociation of VE-PTP from VE-cadherin is required for leukocyte extravasation and for VEGF-induced vascular permeability in vivo. J Exp Med. 2011;208(12):2393–401.
78. Carra S, Foglia E, Cermenati S, Bresciani E, Giampietro C, Lora Lamia C, et al. Ve-ptp modulates vascular integrity by promoting adherens junction maturation. PLoS One. 2012;7(12):e51245.
79. Funa NS, Kriz V, Zang G, Calounova G, Akerblom B, Mares J, et al. Dysfunctional microvasculature as a consequence of shb gene inactivation causes impaired tumor growth. Cancer Res. 2009;69(5):2141–8.
80. Usatyuk PV, Gorshkova IA, He D, Zhao Y, Kalari SK, Garcia JG, et al. Phospholipase D-mediated activation of IQGAP1 through Rac1 regulates hyperoxia-induced p47phox translocation and reactive oxygen species generation in lung endothelial cells. J Biol Chem. 2009;284(22):15339–52.
81. Dance M, Montagner A, Yart A, Masri B, Audigier Y, Perret B, et al. The adaptor protein Gab1 couples the stimulation of vascular endothelial growth factor receptor-2 to the activation of phosphoinositide 3-kinase. J Biol Chem. 2006;281(32):23285–95.
82. Tan W, Palmby TR, Gavard J, Amornphimoltham P, Zheng Y, Gutkind JS. An essential role for Rac1 in endothelial cell function and vascular development. Faseb J. 2008;22(6): 1829–38.

83. Sawada N, Salomone S, Kim HH, Kwiatkowski DJ, Liao JK. Regulation of endothelial nitric oxide synthase and postnatal angiogenesis by Rac1. Circ Res. 2008;103(4):360–8.
84. Serban D, Leng J, Cheresh D. H-ras regulates angiogenesis and vascular permeability by activation of distinct downstream effectors. Circ Res. 2008;102(11):1350–8.
85. Yuan TL, Choi HS, Matsui A, Benes C, Lifshits E, Luo J, et al. Class 1A PI3K regulates vessel integrity during development and tumorigenesis. Proc Natl Acad Sci U S A. 2008;105(28):9739–44.
86. Ackah E, Yu J, Zoellner S, Iwakiri Y, Skurk C, Shibata R, et al. Akt1/protein kinase Balpha is critical for ischemic and VEGF-mediated angiogenesis. J Clin Invest. 2005;115(8): 2119–27.
87. Chen J, Somanath PR, Razorenova O, Chen WS, Hay N, Bornstein P, et al. Akt1 regulates pathological angiogenesis, vascular maturation and permeability in vivo. Nat Med. 2005;11(11):1188–96.
88. Urao N, Razvi M, Oshikawa J, McKinney RD, Chavda R, Bahou WF, et al. IQGAP1 is involved in post-ischemic neovascularization by regulating angiogenesis and macrophage infiltration. PLoS One. 2010;5(10):e13440.
89. Fukumura D, Gohongi T, Kadambi A, Izumi Y, Ang J, Yun CO, et al. Predominant role of endothelial nitric oxide synthase in vascular endothelial growth factor-induced angiogenesis and vascular permeability. Proc Natl Acad Sci U S A. 2001;98(5):2604–9.
90. Thurston G, Suri C, Smith K, McClain J, Sato TN, Yancopoulos GD, et al. Leakage-resistant blood vessels in mice transgenically overexpressing angiopoietin-1. Science. 1999;286(5449):2511–4.
91. Gavard J, Patel V, Gutkind JS. Angiopoietin-1 prevents VEGF-induced endothelial permeability by sequestering Src through mDia. Dev Cell. 2008;14(1):25–36.
92. Oubaha M, Gratton JP. Phosphorylation of endothelial nitric oxide synthase by atypical PKC zeta contributes to angiopoietin-1-dependent inhibition of VEGF-induced endothelial permeability in vitro. Blood. 2009;114(15):3343–51.
93. Winderlich M, Keller L, Cagna G, Broermann A, Kamenyeva O, Kiefer F, et al. VE-PTP controls blood vessel development by balancing Tie-2 activity. J Cell Biol. 2009;185(4): 657–71.
94. Takuwa Y, Okamoto Y, Yoshioka K, Takuwa N. Sphingosine-1-phosphate signaling and biological activities in the cardiovascular system. Biochim Biophys Acta. 2008;1781(9): 483–8.
95. Tanimoto T, Jin ZG, Berk BC. Transactivation of vascular endothelial growth factor (VEGF) receptor Flk-1/KDR is involved in sphingosine 1-phosphate-stimulated phosphorylation of Akt and endothelial nitric-oxide synthase (eNOS). J Biol Chem. 2002;277(45):42997–3001.
96. Gaengel K, Niaudet C, Hagikura K, Lavina B, Muhl L, Hofmann JJ, et al. The sphingosine-1-phosphate receptor S1PR1 restricts sprouting angiogenesis by regulating the interplay between VE-cadherin and VEGFR2. Dev Cell. 2012;23(3):587–99.
97. Grunewald M, Avraham I, Dor Y, Bachar-Lustig E, Itin A, Jung S, et al. VEGF-induced adult neovascularization: recruitment, retention, and role of accessory cells. Cell. 2006;124(1): 175–89.
98. Christoffersson G, Henriksnas J, Johansson L, Rolny C, Ahlstrom H, Caballero-Corbalan J, et al. Clinical and experimental pancreatic islet transplantation to striated muscle: establishment of a vascular system similar to that in native islets. Diabetes. 2010;59(10):2569–78.
99. Christoffersson G, Vagesjo E, Vandooren J, Liden M, Massena S, Reinert RB, et al. VEGF-A recruits a proangiogenic MMP-9-delivering neutrophil subset that induces angiogenesis in transplanted hypoxic tissue. Blood. 2012;120(23):4653–62.
100. Zhang H, Issekutz AC. Growth factor regulation of neutrophil-endothelial cell interactions. J Leukoc Biol. 2001;70(2):225–32.
101. Zittermann SI, Issekutz AC. Endothelial growth factors VEGF and bFGF differentially enhance monocyte and neutrophil recruitment to inflammation. J Leukoc Biol. 2006;80(2):247–57.

102. Kim I, Moon SO, Kim SH, Kim HJ, Koh YS, Koh GY. Vascular endothelial growth factor expression of intercellular adhesion molecule 1 (ICAM-1), vascular cell adhesion molecule 1 (VCAM-1), and E-selectin through nuclear factor-kappa B activation in endothelial cells. J Biol Chem. 2001;276(10):7614–20.
103. Goebel S, Huang M, Davis WC, Jennings M, Siahaan TJ, Alexander JS, et al. VEGF-A stimulation of leukocyte adhesion to colonic microvascular endothelium: implications for inflammatory bowel disease. Am J Physiol Gastrointest Liver Physiol. 2006;290(4): G648–54.
104. Chidlow JH, Jr., Glawe JD, Alexander JS, Kevil CG. VEGF(1)(6)(4) differentially regulates neutrophil and T cell adhesion through ItgaL- and ItgaM-dependent mechanisms. Am J Physiol Gastrointest Liver Physiol. 2010;299(6):G1361–7.
105. Lohela M, Bry M, Tammela T, Alitalo K. VEGFs and receptors involved in angiogenesis versus lymphangiogenesis. Curr Opin Cell Biol. 2009;21(2):154–65.
106. Ancelin M, Chollet-Martin S, Herve MA, Legrand C, El Benna J, Perrot-Applanat M. Vascular endothelial growth factor VEGF189 induces human neutrophil chemotaxis in extravascular tissue via an autocrine amplification mechanism. Lab Invest. 2004;84(4): 502–12.
107. Barleon B, Sozzani S, Zhou D, Weich HA, Mantovani A, Marme D. Migration of human monocytes in response to vascular endothelial growth factor (VEGF) is mediated via the VEGF receptor flt-1. Blood. 1996;87(8):3336–43.
108. Czepluch FS, Olieslagers S, van Hulten R, Voo SA, Waltenberger J. VEGF-A-induced chemotaxis of CD16+ monocytes is decreased secondary to lower VEGFR-1 expression. Atherosclerosis. 2011;215(2):331–8.
109. Ceradini DJ, Kulkarni AR, Callaghan MJ, Tepper OM, Bastidas N, Kleinman ME, et al. Progenitor cell trafficking is regulated by hypoxic gradients through HIF-1 induction of SDF-1. Nat Med. 2004;10(8):858–64.
110. Schmid MC, Avraamides CJ, Dippold HC, Franco I, Foubert P, Ellies LG, et al. Receptor tyrosine kinases and TLR/IL1Rs unexpectedly activate myeloid cell PI3kgamma, a single convergent point promoting tumor inflammation and progression. Cancer Cell. 2011;19(6):715–27.
111. Vestweber D. Relevance of endothelial junctions in leukocyte extravasation and vascular permeability. Ann N Y Acad Sci. 2012;1257:184–92.
112. Phillipson M, Kubes P. The neutrophil in vascular inflammation. Nat Med. 2011;17(11): 1381–90.
113. Nourshargh S, Hordijk PL, Sixt M. Breaching multiple barriers: leukocyte motility through venular walls and the interstitium. Nat Rev Mol Cell Biol. 2010;11(5):366–78.
114. Kim MH, Curry FR, Simon SI. Dynamics of neutrophil extravasation and vascular permeability are uncoupled during aseptic cutaneous wounding. Am J Physiol Cell Physiol. 2009;296(4):C848–56.
115. He P. Leucocyte/endothelium interactions and microvessel permeability: coupled or uncoupled? Cardiovasc Res. 2010;87(2):281–90.
116. Hurley JV. Acute Inflammation: The Effect of Concurrent Leucocytic Emigration and Increased Permeability on Particle Retention by the Vascular Wall. Br J Exp Pathol. 1964;45:627–33.
117. Lewis RE, Granger HJ. Diapedesis and the permeability of venous microvessels to protein macromolecules: the impact of leukotriene B4 (LTB4). Microvasc Res. 1988;35(1):27–47.
118. Baluk P, Bolton P, Hirata A, Thurston G, McDonald DM. Endothelial gaps and adherent leukocytes in allergen-induced early- and late-phase plasma leakage in rat airways. Am J Pathol. 1998;152(6):1463–76.
119. Phillipson M, Kaur J, Colarusso P, Ballantyne CM, Kubes P. Endothelial domes encapsulate adherent neutrophils and minimize increases in vascular permeability in paracellular and transcellular emigration. PLoS One. 2008;3(2):e1649.
120. van Buul JD, Mul FP, van der Schoot CE, Hordijk PL. ICAM-3 activation modulates cell-cell contacts of human bone marrow endothelial cells. J Vasc Res. 2004;41(1):28–37.

121. Wang Q, Doerschuk CM. Neutrophil-induced changes in the biomechanical properties of endothelial cells: roles of ICAM-1 and reactive oxygen species. J Immunol. 2000;164(12):6487–94.
122. Vandooren J, Van den Steen PE, Opdenakker G. Biochemistry and molecular biology of gelatinase B or matrix metalloproteinase-9 (MMP-9): The next decade. Crit Rev Biochem Mol Biol. 2013;48(3):222-72
123. Jain RK. Normalization of tumor vasculature: an emerging concept in antiangiogenic therapy. Science. 2005;307(5706):58–62.
124. Baffert F, Le T, Thurston G, McDonald DM. Angiopoietin-1 decreases plasma leakage by reducing number and size of endothelial gaps in venules. Am J Physiol Heart Circ Physiol. 2006;290(1):H107–18.
125. Falcon BL, Hashizume H, Koumoutsakos P, Chou J, Bready JV, Coxon A, et al. Contrasting actions of selective inhibitors of angiopoietin-1 and angiopoietin-2 on the normalization of tumor blood vessels. Am J Pathol. 2009;175(5):2159–70.
126. Allavena P, Sica A, Garlanda C, Mantovani A. The Yin-Yang of tumor-associated macrophages in neoplastic progression and immune surveillance. Immunol Rev. 2008;222:155–61.
127. Cassetta L, Cassol E, Poli G. Macrophage polarization in health and disease. ScientificWorldJournal. 2011;11:2391–402.
128. Barar J, Asadi M, Mortazavi-Tabatabaei SA, Omidi Y. Ocular drug delivery: impact of in vitro cell culture models. J Ophthalmic Vis Res. 2009;4(4):238–52.
129. Gardner TW, Antonetti DA, Barber AJ, Lieth E, Tarbell JA. The molecular structure and function of the inner blood-retinal barrier. Penn State Retina Research Group. Doc Ophthalmol. 1999;97(3–4):229–37.
130. Stewart MW. The expanding role of vascular endothelial growth factor inhibitors in ophthalmology. Mayo Clin Proc. 2012;87(1):77–88.
131. Miller JW, Le Couter J, Strauss EC, Ferrara N. Vascular endothelial growth factor a in intraocular vascular disease. Ophthalmology. 2013;120(1):106–14.
132. Antonetti DA, Barber AJ, Hollinger LA, Wolpert EB, Gardner TW. Vascular endothelial growth factor induces rapid phosphorylation of tight junction proteins occludin and zonula occluden 1. A potential mechanism for vascular permeability in diabetic retinopathy and tumors. J Biol Chem. 1999;274(33):23463–7.

Chapter 9
Cytochrome P450-Derived Lipid Mediators and Vascular Responses

Ingrid Fleming

Abbreviations

$[Ca^{2+}]_i$	Intracellular Ca^{2+} concentration
BK_{Ca}	Large conductance Ca^{2+}-activated K^+ channels
COX	Cyclooxygenase
CREB	cAMP-response element-binding protein
CYP	Cytochrome P450
DHA	Docosahexenoic acid
DHET	Dihydroxyeicosatrienoic acid
EDHFs	Endothelium-derived hyperpolarizing factors
EET	Epoxyeicosatrienoic acid
EGF	Epidermal growth factor
EPA	Eicosapentaenoic acid
FABPs	Fatty acid-binding proteins
HETE	Hydroxyeicosatetraenoic acid
K_{Ca}	Ca^{2+}-dependent K^+ channels
MKP-1	MAP kinase phosphatase-1
MMP	Matrix metalloproteinase
NFκB	Nuclear factor κB
NO	Nitric oxide
PI3-K	Phosphatidylinositol 3-kinase
PKA	Protein kinase A
PPAR	Peroxisome proliferator-activated receptor
PUFA	Polyunsaturated fatty acid
sEH	Soluble epoxide hydrolase;
TRP channels	Transient receptor potential channels

I. Fleming (✉)
Institute for Vascular Signalling, Goethe University, Theodor-Stern-Kai 7, 60590 Frankfurt/am Main, Germany
e-mail: fleming@em.uni-frankfurt.de

M. HH. Schmidt, S. Liebner (eds.), *Endothelial Signaling in Development and Disease*, DOI 10.1007/978-1-4939-2907-8_9

9.1 The Cytochrome P450 (CYP)/Soluble Epoxide Hydrolase Axis

Cytochrome P450 (CYP) enzymes are membrane-bound, heme-containing terminal oxidases in a multi-enzyme complex that also includes an flavin adenine dinucleotide/flavin mononucleotide (FAD/FMN)-containing nicotinamide adenine dinucleotide phosphate (NADPH)-CYP reductase and cytochrome b_5. CYP enzymes oxidize, peroxidize and/or reduce cholesterol, vitamins, steroids, xenobiotics, and numerous pharmacological substances in an oxygen- and NADPH-dependent manner. Some isoforms are fairly specific in their choice of substrates but many catalyze a large number of chemical reactions and can use an almost unlimited number of biologically occurring and synthetic compounds. Hepatic CYP enzymes are responsible for the metabolism of xenobiotica and many pharmaceuticals, but they also utilize endogenous compounds as substrates, such as cholesterol and fatty acids. Even though many CYP isozymes can oxidize a spectrum of ω-6 and ω-3 polyunsaturated fatty acids (PUFAs) such as retinoic acid, linoleic acid, eicosapentaenoic acid (EPA), and docosahexenoic acid (DHA; Fig. 9.1), they are often referred to as the third pathway of arachidonic acid metabolism, mainly because more is known about the biological actions of these products [1].

Interest in the vascular actions of CYP enzymes followed reports that the epoxides of arachidonic acid (the epoxyeicosatrienoic acids [EETs]) were endothelium-derived hyperpolarizing factors (EDHFs) [2, 3], while the 20-hydroxyeicosatetraenoic acid (HETE) generated by ω-hydroxylases belonging to the CYP4A family were potent vasoconstrictors (for review see Harder et al. and Imig et al. [4, 5]). Initial reports focused on the effects of arachidonic acid metabolites on membrane potential, but it is now generally appreciated that these compounds mediate a number of membrane potential-independent effects and regulate angiogenesis [6, 7]. The arachidonic acid-metabolizing CYP enzymes with prominent roles in vascular regulation are the epoxygenases of the CYP2 gene family (e.g. CYP2B, 2C8, 2C9, 2C10, and 2J2 in humans; 2C34 in pigs; 2C11, 2C23, and 2J4 in rats) and the arachidonic acid ω-hydroxylases belonging to the CYP4A family which form subterminal and ω-terminal HETEs [8, 9].

Epoxide generation is thought to be determined by both the level of epoxygenase expression and the availability of the PUFA substrate, which in the case of arachidonic acid is determined by the activity of phospholipases such as phospholipase A_2. Intracellular levels of the epoxides are tightly regulated and metabolism occurs relatively rapidly by hydrolysis, β-oxidation, and chain elongation [10]. The soluble epoxide hydrolase (sEH) is the most important epoxide-metabolizing enzyme that generates dihydroxy fatty acids (or diols). For a long time, the latter were considered to be less active that the parent epoxides but recent evidence has challenged this assumption (see section 9.6). There are, of course, exceptions to every rule and some epoxides are not great sEH substrates—the best-studied exception is probably 5,6-EET which is more rapidly metabolized by cyclooxygenases (COXs) [11, 12].

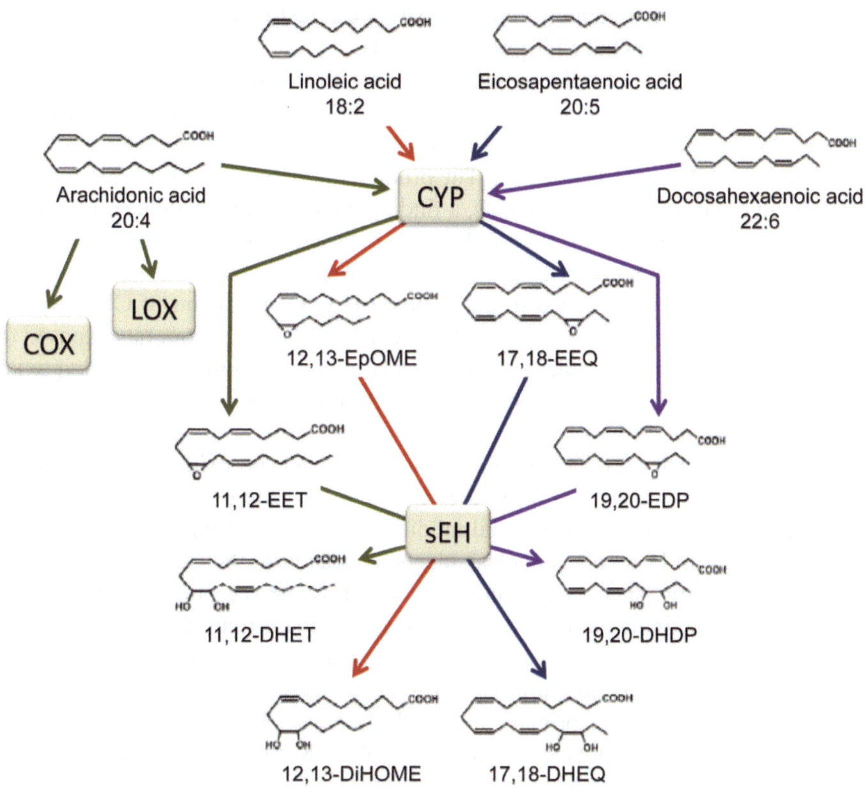

Fig. 9.1 Cytochrome P450 (*CYP*)-dependent metabolism of n-6 (arachidonic and linoleic acid) and n-3 (eicosapentaenoic and docosahexaenoic acid), and metabolism of the epoxides generated to the corresponding diols by the soluble epoxide hydrolase (*sEH*). *DHDP* dihydroxydocosapentaenoic acid, *DHEQ* dihydroxyeicosatetraenoic acid, *DHET* dihydroxyeicosatrienoic acid, *DiHOME* dihydroxyoctadecenoic acid, *EDP* epoxydocosapentaenoic acid, *EEQ* epoxyeicosatetraenoic acid, *EET* epoxyeicosatrienoic acid, *EpOME* epoxyoctadecenoic acid. Reproduced from Fleming [9], with permission

9.2 Regulation of CYP Expression and Activity

EET production may change as a consequence of altered CYP expression (by induction or repression) or altered activity. Little is known about the regulation of vascular CYP expression, and although CYP2C protein has been convincingly demonstrated in native endothelial cells, messenger RNA (mRNA) and protein levels rapidly decrease after cell isolation, so that in passaged cultured endothelial cells, mRNA can only be detected using reverse transcriptase-polymerase chain reaction (RT-PCR) [3, 13]. Such findings indicate that CYP2C proteins are relatively unstable and that transcriptional processes play an important role in determining CYP expression levels, and at the same time highlight the importance of physiological stimuli in the control of CYP levels. Indeed, the exposure of cultured endothelial cells to either cyclic stretch or fluid shear stress can restore CYP2C protein expression as well as endothelial EET production [14].

The expression of several CYP enzymes is modulated by changes in oxygen tension; for example, hypoxia downregulates CYP2J2 [15] but upregulates CYP2C8/9 expression in cultured human endothelial cells [16], while transient cerebral ischemia induces CYP2C11 in rats [17]. The promoter regions of several CYP2C genes contain hypoxia-responsive elements, a finding which may explain the observation that the myogenic contraction, as well as the constrictor response to phenylephrine, is attenuated in mesenteric arteries from rats exposed to hypoxia for 48 h [18]. This phenomenon can be attributed to the hypoxia-induced induction of CYP expression as CYP2C protein was elevated above control in arteries from rats exposed to hypoxia, and both the vascular smooth muscle hyperpolarization and the hypoxia-induced decrease in the myogenic response were normalized by the CYP2C inhibitor sulfaphenazole [19].

Given the size of the CYP family of proteins, it is not surprising that there is considerable inter-isoform variation in the regulation of gene expression and mRNA stability, as well as post-translational modification of the CYP protein. Regulation of the CYP2 family involves nuclear receptors related to the steroid hormone receptor superfamily, such as the constitutive androstane receptor and the retinoic acid receptor. Retinoic acid is a CYP2C (CYP2C8) substrate, thus the regulation of CYP2C8 expression by a receptor that is activated by an endogenous substrate such as retinoic acid is not surprising. CYP2C9, which is highly homologous to CYP2C8, is inducible in primary human hepatocytes by xenobiotics, including dexamethasone and phenobarbital [20]. The CYP2C8 and 2C9 promoters contain a glucocorticoid-responsive element that is recognized and transactivated by human glucocorticoid receptor [21–23]. Identification of this functional element provides a rational mechanistic basis for the induction of CYP2C protein, and an increase in EDHF-mediated responses in porcine coronary arteries by cortisol [24]. Retinoic acid is not the only CYP substrate that affects CYP protein levels, and the expression of many of the CYP enzymes can be induced by a substrate excess. Indeed, a number of cardiovascular drugs currently in clinical use are metabolized, at least in part, by CYP2C family members. While this process mainly occurs in the liver and is associated with the induction of the metabolizing enzyme, the expression of CYP2C enzymes in endothelial cells can also be affected. For example, the HMG-CoA reductase inhibitor fluvastatin, which is metabolized by CYP2C9 in the liver [25, 26], also increases CYP2C expression in cultured and native porcine coronary artery endothelial cells [13]. The L-type Ca^{2+} channel blocker nifedipine also elicits a very marked increase in the expression of CYP2C in porcine coronary arteries, and enhances EDHF-mediated responses [13, 27].

Once the protein is expressed, CYP activity is thought to be determined mainly by the availability of its substrates [28]. Since phospholipase A_2 inhibitors attenuate CYP-dependent EDHF responses, the activation cascade is thought to involve a stimulus-induced increase in intracellular Ca^{2+}, followed by the activation of phospholipase A_2, which then liberates the PUFA substrate (i.e. arachidonic acid) from membrane phospholipids. The increase in substrate immediately results in the activation of CYP enzymes (when expressed) and the generation of vasoactive products. While this sequence of events is certainly plausible, it is highly likely that additional mechanisms, such as phosphorylation, play a role in regulating CYP activity. Indeed, some CYP enzymes (CYP2B1, 2B2, and 2E1) are reported to be phosphorylated by protein kinase A (PKA), and the consequences of CYP phosphorylation range from the regulation of activity [29] and subcellular localization [30, 31] to proteasome degradation [30, 32].

An additional mechanism thought to modulate CYP activity is nitrosation by nitric oxide (NO), which can interact with CYP enzymes in two ways. NO reversibly binds to the heme moiety of CYP enzymes, forming iron-nitrosyl complexes, and it can irreversibly react with cysteine residues [33]. Both NO–CYP adducts are enzymatically inactive *in vitro*. As endothelial CYP enzymes of the 2C family were found to be inhibited by NO, the role of EETs in the regulation of vascular tone in the healthy vasculature which constantly generates NO was suggested to be of minor importance compared with that in circumstances of an endothelial dysfunction in which the bioavailability of NO is impaired [34]. However, there are clear physiological consequences of EET activation in endothelial cells (e.g. on Akt, PKA, and transient receptor potential [TRP] channels) that can be demonstrated, even in the presence of a fully functional endothelial NO synthase. Thus, whether or not physiologically relevant (low nmol/L) levels of NO really affect CYP epoxygenase activity *in vivo*, remains to be determined.

9.3 The Soluble Epoxide Hydrolase

The sEH protein is a homodimer composed of two 60 kDa monomers joined by a proline-rich bridge [35], with each monomer consisting of an N-terminal domain that displays lipid phosphatase activity and a larger C-terminal that processes classical α/β-hydrolase activity [36, 37]. Surprisingly little is known about the mechanisms that regulate sEH activity. There have been a number of studies linking changes in sEH expression with inflammatory or hormonal stimuli [38, 39]. Two tyrosine residues (Tyr383 and Tyr466) in the active site of the hydrolase are reportedly essential for enzyme activity [40], and these were recently shown to be nitrated by peroxynitrite *in vitro* and *in vivo* in mouse models of type 1 and type 2 diabetes, leading to a decrease in sEH activity [41]. It is currently only possible to speculate about the involvement of sEH tyrosine nitration in the amplification of inflammation associated with diabetes, but at least one sEH polymorphism, which results in decreased enzymatic activity, has previously been associated with human insulin resistance [42]. The sEH was also recently reported to be nitrosated in leptin-stimulated wild-type but not endothelial NO synthase knockout mice, suggesting that the effects of NO on PUFA metabolism may be partly related to the modulation of sEH activity [43].

Inhibition or deletion of the sEH increases tissue and circulating levels of the PUFA epoxides at the same time as decreasing diol production, and has pronounced effects on blood pressure [44, 45], inflammation [46], progenitor cell proliferation, angiogenesis and vascular repair [47]. The particular effectiveness of sEH inhibitors against hypertension associated with activation of the renin-angiotensin system is most likely related to the fact that angiotensin II markedly increases sEH expression *in vivo* [39]. Interestingly, hypoxia does the opposite and markedly downregulates sEH promoter activity and thus protein expression in the lung [48]. There are other examples of hypertension being associated with elevated sEH expression and/or activity, such as the spontaneously hypertensive rat. In these animals, elevated sEH expression is linked to an increase in the renal metabolism of EETs to dihydroxyeicosatrienoic acids (DHETs), and sEH inhibitors blunt the development of hypertension [44]. Initial reports also documented that $sEH^{-/-}$ mice have lower blood pressure and elevated EET levels than their wild-type littermates [49]. However, the blood pressure phenotype now seems to be controversial as the loss of the hydrolase can be compensated by elevated concentrations of the pressor and vasoconstrictor eicosanoid, 20-HETE, as well as increased lipoxygenase-derived hydroxylation and prostanoid production [50]. Despite the lack of alteration in blood pressure, hearts from these $sEH^{-/-}$ animals show improved recovery of left ventricular contractility and less infarction than hearts from wild-type mice after ischemia [51], and have a survival advantage following acute systemic inflammation [50]. Several of the metabolites generated by the sEH, such as the DHETs generated from the EETs, are also biologically active but generally less so than the parent epoxides. However, the DHETs are not as readily incorporated into membrane lipids as the EETs, and the latter are thought to be the form in which the majority of endothelium-derived EETs leave the cell [52].

The exact physiological role of the lipid phosphatase activity associated with the N-terminal domain of the sEH is currently unclear as there are currently no selective inhibitors of this domain (sEH inhibitors act on the hydrolase domain and do not affect the phosphatase activity [36]). However, the lipid phosphatase has been associated with cholesterol-related disorders, peroxisome proliferator-activated receptor (PPAR) activity, and the isoprenoid/cholesterol biosynthesis pathway [53]. Indeed, in addition to demonstrating enhanced circulating EET levels [49], male $sEH^{-/-}$ mice exhibit decreased plasma cholesterol and testosterone levels [54]. Moreover, it seems that isoprenoid pyro- and monophosphates are substrates for the N-terminal domain of the enzyme [55, 56], and these lipid phosphates are metabolic precursors of cholesterol biosynthesis and are also utilized for isoprenylation of small G-proteins involved in multiple cell signaling pathways [57]. Lysophosphatidic acids are involved in regulating cell survival, apoptosis, motility, shape, differentiation, gene transcription, and malignant transformation, and are reportedly excellent substrates for the lipid phosphatase [58, 59]. However, to what extent this can affect physiology/pathophysiology needs to be determined.

It is interesting to note that even though most current sEH literature attributes the hydrolase domain to the cardiovascular effects seen in humans, the human sEH single nucleotide polymorphism most often associated with cardiovascular disease (R287Q) encodes a protein with significantly lower rather than elevated hydrolase activity [60]. Thus, solely incriminating the hydrolase domain for adverse cardio- and pulmonary-vascular effects seems premature and highlights the importance of further investigating the independent roles of the hydrolase and phosphatase domains. Indeed, some aspects of the phenotype of $sEH^{-/-}$ mice (e.g. pulmonary vascular muscularization) cannot be reproduced by chronic sEH inhibitor treatment, which may be an indirect indication of a physiological role for the phosphatase domain [61].

9.4 How do Lipid Epoxides Initiate Cellular Signaling?

Most is known about the actions of the epoxides or arachidonic acid or EETs for which several modes of signal initiation have been proposed. One of them involves the transactivation of the epidermal growth factor (EGF) receptor in endothelial cells, and the activation of this particular signaling pathway has been linked to cell proliferation and angiogenesis [62–64]. For actions other than angiogenesis, a separate mechanism has been proposed as a high-affinity EET binding site was reported to exist on monocytes and U937 cells [65–67]. Competition studies showed a specific high-affinity binding of 14,15- and 11,12-EET to a receptor that seems to be protein in nature [66, 68]. In addition, in isolated membranes, [^{3}H]-14,15-EET binding was found to be specific, reversible, and saturable, and the ligand was not displaced by antagonists of the thromboxane, platelet-activating factor, or leukotriene receptors. However, binding was inhibited by 14,15- and 11,12-EETs, but not by inactive analogs of 14,15-EET or 15-HETE. Importantly, ligand binding was inhibited by GTPγS, indicating that the binding site or receptor is coupled to a G protein. Such findings are in agreement with other reports indicating the involvement of a G protein in the actions of the EETs [69, 70]. One characteristic of many EET-induced cellular responses such as gap junctional communication [71] and TRP channel translocation [72] is their ability to increase intracellular cyclic adenosine monophosphate (cAMP) levels and activate PKA [66, 73]. Moreover, an EET analog that is able to induce the complete relaxation of bovine coronary arteries also does so by increasing cAMP levels [74, 75]. Putting the evidence of a protein receptor on cell membranes together with that indicating a reliance on cAMP/PKA for EET-induced signaling, the existence of a Gαs-coupled EET receptor has been postulated [76]. However, to-date no specific EET receptor has been identified.

Many lipids also interact with intracellular fatty acid receptors such as the PPARs, and the EETs are no different. For example, ω-hydroxylated 14,15-EET and 14,15-DHET [77] are reported to bind with a high affinity to PPAR-α, while the EETs generated in endothelial cells in response to fluid shear stress increase PPAR-γ transcriptional activity [78]. Additional intracellular receptors for CYP products have not yet been identified, but one possibility is that these oxidized fatty acids bind to fatty acid binding proteins (FABPs) such as heart type (H) FABP [79], which in turn mediate some of the physiologically relevant actions of these intermediates, possibly including the activation of PPARs [80].

A further proposed mechanism involves the incorporation of EETs into the plasma membrane, where they associate with effector molecules such as small G proteins [69] or change the lipid bilayer order, fluidity, and volume, and thereby regulate the flux of ions (e.g. Ca^{2+}) across the membrane [81]. Certainly, the EETs can be esterified to phosphatidylcholine, phosphatidylethanolamine, and phosphatidylinositols [28]. Moreover, at least in the pancreas, it has been suggested that the long-chain acyl-coenzyme A (acyl-CoA) synthetase 4 activates EETs to form EET-CoAs that are incorporated into glycerophospholipids [82].

Given that enhanced EET production has frequently been correlated with an increase in intracellular cAMP levels, it is not entirely surprising that increased CYP expression and EET production are associated with activation of the cAMP-response element-binding protein (CREB) which underlies the EET-induced expression of COX-2 [83]. However, the first transcription factor reported to be regulated by CYP-derived EETs was nuclear factor κB (NF-κB) [84]. Indeed, the EET-dependent inhibition of the I-κB kinase led to classification of the EETs as anti-inflammatory mediators. However, this classification is complicated by the fact that some CYP epoxygenases can generate physiologically relevant levels of superoxide anions which tend to activate NF-κB, and thus functionally antagonize the inhibitory effects of EETs and promote the expression of adhesion molecules on endothelial cells [85]. The reason why CYP epoxygenases of the 2C family generate superoxide anions [85, 86] while the 2J enzymes do not [87, 88] is currently unclear but is probably related to substrate binding and metabolism. However, the differential ability to generate free radicals accounts for the disparate effects of these isozymes on vascular protection.

Other transcription factors that are reported to be modulated by EETs/DHETs are PPARα [77, 89, 90] and FOXO3a [91]. While the nuclear localization of FOXO3a is regulated by the EET-dependent activation of Akt [91], much less is known about the mechanisms involved in the EET-dependent activation of PPARα, or indeed the consequences of this effect.

9.5 CYP and Cardiovascular Function

9.5.1 Vascular Reactivity

The realization that EETs, especially 11,12- and 14,15-EET, can activate large conductance Ca^{2+}-activated K^+ channels (BK_{Ca}) on vascular smooth muscle cells to elicit hyperpolarization and relaxation led to their identification as a class of EDHF [2, 3]. The latter term is now recognized as an oversimplification as there are three principal mechanisms linked to the EDHF phenomenon: (i) an increase in endothelial $[Ca^{2+}]_i$ following cell stimulation triggers the synthesis of a metabolite which is essential for the subsequent EDHF-mediated responses; (ii) K^+, released from endothelial cells via Ca^{2+}-dependent K^+ (K_{Ca}) channels, induces smooth muscle hyperpolarization by activating inwardly rectifying K^+ channels and/or the Na^+/K^+-

ATPase on vascular smooth muscle cells; and (iii) endothelial cell hyperpolarization is transmitted to the vascular smooth muscle via gap junctions. The strengths and weaknesses of the arguments for each of these specific types of EDHF has been discussed at length [92] but each of them appears to be valid in certain vascular beds. Interestingly, all of these mechanisms can be modulated by EETs.

In endothelial cells, the activation of K_{Ca} channels by EETs is preceded by an increase in intracellular Ca^{2+} levels that can be accounted for by an increased open probability of nonselective cation channels of the TRP family. How this happens was initially attributed to the presence of an arachidonic acid-binding site in some of the TRP channels that can be activated by the parent lipid [93, 94] as well as the EETs [93, 95]. However, while relatively high concentrations of the EETs may affect TRP channels directly, more physiological concentrations activate TRP channels in a PKA-dependent manner that involves their translocation to caveolin-rich areas in the plasma membrane [72, 96]. There appear to be regioisomer-specific differences in EET-induced TRP channel translocation and activation as 5,6-EET, but not 11,12-EET, can activate TRPV4 in endothelial cells [93, 95], a phenomenon that underlies the EDHF-dependent, flow-induced vasodilatation [96]. On the other hand, 11,12-EET, but not 14,15-EET or 5,6-EET, enhance the bradykinin-induced capacitive Ca^{2+} influx in endothelial cells by stimulating the translocation of TRPC6 and TRPC3 to caveolin-rich areas in the plasma membrane [72].

9.5.2 Pulmonary Circulation

While increasing intracellular Ca^{2+} in endothelial cells elicits vasodilatation, the same process in vascular smooth muscle cells does exactly the opposite. This means that when EETs activate TRPC6 channels in pulmonary smooth muscle cells, an increase in pulmonary vascular tone would be expected. The fact that activation of TRPC6 channels plays a role in regulating hypoxic pulmonary vasoconstriction (a physiological mechanism by which pulmonary arteries constrict in hypoxic lung areas in order to redirect blood flow to areas with greater oxygen supply) was demonstrated using mice lacking the channel. Indeed, in pulmonary vascular smooth muscle cells from these animals, hypoxia completely failed to cause Ca^{2+} entry. It should be noted here that the TRPC6 is reported to primarily conduct Na^{+}, and Ca^{2+} follows secondarily through voltage-gated Ca^{2+} channels or by the Na^{+}/Ca^{2+} exchanger. In line with the disturbed Ca^{2+} entry, these animals completely lacked the initial, acute phase of hypoxia-induced pulmonary constriction [97]. Moreover, $TRPC6^{-/-}$ mice did not respond to 11,12-EET, although the eicosanoid induced a pronounced increase in pulmonary pressure in $TRPC6^{+/-}$ littermates. Furthermore, inhibition of the sEH potentiated the hypoxic pulmonary vasoconstriction in the heterozygous mice, but had no effect in the $TRPC6^{-/-}$ mice [48]. In line with the functional data, hypoxia and 11,12-EET caused the translocation of TRPC6 to caveolae in isolated pulmonary vascular smooth muscle cells. In addition, hypoxia-induced translocation of the channel could be prevented by pretreating the cells with an EET antagonist [48]. More recently [98], the site for pulmonary oxygen sensing

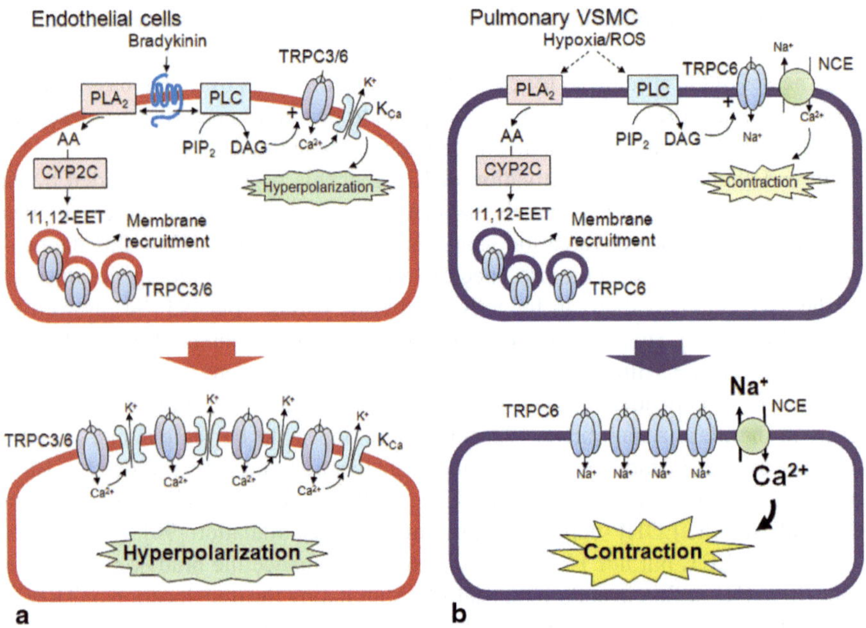

Fig. 9.2 Proposed mechanism for the differential consequences of epoxyeicosatrienoic acid (*EET*)-induced transient receptor potential (*TRP*) channel activation in the systemic and pulmonary circulations. **a** In the systemic circulation, EETs are generated in endothelial cells in response to stimulation (e.g. by bradykinin) following the activation of phospholipase A_2 and cytochrome P450 (*CYP*)2C epoxygenases. *EET*-induced activation of protein kinase A (*PKA*) results in the translocation of *TRP* channels to the plasma membrane to potentiate the activation on K_{Ca} channels initiated by the phospholipase C (*PLC*)-induced conversion of phosphatidylinositol 4,5-bisphosphate (*PIP_2*) to diacylglycerol (*DAG*). The overall consequence is hyperpolarization and vasodilatation. **b** While activating *TRPC6* in endothelial cells elicits vasodilatation, the same process in vascular smooth muscle cells does exactly the opposite as the TRPC6 channels in vascular smooth muscle cells (*VSMC*) primarily conduct Na^+, Ca^{2+} follows secondarily through voltage-gated Ca^{2+} channels or by the Na^+, Ca^{2+} exchanger (*NCE*). Reproduced from Loot and Fleming [99], with permission

was identified at alveolocapillary level, from which the hypoxic signal is propagated as endothelial membrane depolarization to upstream arterioles in an EET- and Cx40-dependent manner (Fig. 9.2; [99]).

The evidence for a vasodilator role of the arachidonic acid epoxides in humans is, needless to say, indirect and relies on the use of CYP inhibitors that cannot be guaranteed to be completely selective. That said, sulfaphenazole is one of the most selective inhibitors available for CYP2C9 [100], and while several studies failed to demonstrate any effects of sulfaphenazole on forearm vasodilatation in healthy subjects [86, 101, 102], a component of the flow-induced vasodilatation of skeletal muscle arterioles [103] and the radial artery [104–106], both of which have been shown to express CYP2C protein, is attenuated by the CYP inhibitor. Clearly, however, disease can affect responses as forearm vasodilator responses to acetylcholine could be blunted by CYP inhibitors in patients with hypercholesterolemia and reduced NO-dependent vasodilatation [107].

9.5.3 Hypertension and Atherosclerosis

Inhibition of the sEH increases intracellular levels of EETs, and thus prolongs their vasodilator and anti-inflammatory actions. Indeed, pharmacological inhibition of the sEH prevents angiotensin II-induced hypertension in rats and mice, and protects the kidney from hypertension-induced damage [44, 45, 52]. Furthermore, in humans increased sEH activity was associated with more advanced endothelial dysfunction and vascular inflammation [108].

Although the link between the sEH and cholesterol metabolism would make it logical to look at atherosclerosis, the situation is somehow less clear. Certainly, polymorphisms of the sEH have been linked with the risk of atherosclerosis and coronary heart disease [109–111]. Why this is the case is not known, but the initial report that sEH inhibitors can attenuate smooth muscle cell proliferation [112] most probably represented an off-target effect of the substance used [113]. In addition, some of the animal studies failed to deliver consistent results, and although inhibition of the sEH was reported to attenuate atherosclerosis, abdominal aortic aneurysm formation, and dyslipidemia by some researchers [114, 115], our group has been unable to detect clear effects. Furthermore, the effects on vascular remodeling are inconsistent with inhibition of the sEH preventing vascular remodeling in an inflammatory model but not in a blood flow-dependent model of neointima formation [116].

Most of the studies performed to date have focused on vascular smooth muscle cells, and the fact that monocytes express the sEH and a number of CYP enzymes has been largely overlooked. However, this area deserves much more attention as human and murine macrophages within atherosclerotic plaques express CYP2S1, a largely extrahepatic epoxygenase [117]. Interestingly, enzyme expression increased during monocyte differentiation to macrophages, and could be detected in classically activated or M1 macrophages and macrophages present in atherosclerotic plaques and inflamed tonsils, but not in macrophages polarized towards the M2 or alternatively-activated phenotype. Although the enzyme was able to accept several substrates and to generate bioactive epoxides from arachidonic acid, linoleic acid and EPA in an NADPH-dependent manner, perhaps from the macrophage polarization point of view the most relevant substrates seem to be prostaglandins G_2 and H_2 [117]. The resulting decrease in the immunomodulator prostaglandin E2 (PGE_2) would certainly be expected to result in a macrophage subtype with attenuated angiogenic potential, but whether or not the CYP2S1 product 12(S)-hydroxyheptadeca-5Z,8E,10E-trienoic acid actively contributes to inflammation remains to be determined.

9.6 Angiogenesis and Cancer

Given the fact that the activation of K_{Ca} channels has been linked to endothelial cell proliferation [118–120], and EETs activate K_{Ca} channels, it would seem logical to assume that K_{Ca} activation would play a role in EET-induced proliferation. However, although the activation of K_{Ca} channels has been linked to endothelial cell proliferation induced by basic fibroblast growth factor [118], this mechanism appears to not be involved in the EET-induced proliferation of endothelial cells. The first hint that EETs may affect cell signaling and proliferation was obtained in renal epithelial cells [62, 121] and, soon afterwards, 'authentic EDHF' recovered from the luminal incubate of rhythmically stretched coronary arteries was found to activate a number of kinases, whose function was closely linked with endothelial cell proliferation [122]. Activation of these MAP kinases could be inhibited by treatment with CYP inhibitors, as well as by antisense oligonucleotides directed against CYP2C, and could be mimicked by the treatment of endothelial cells with 11,12-EET or by overexpression of CYP2C8 [122]. More detailed analysis of the mechanisms involved revealed that CYP epoxygenase-derived metabolites of arachidonic acid are able to transactivate the EGF receptor [123, 124]. 14,15-EET was initially suggested to act as a second messenger following activation of the EGF receptor; however, it appears that 14,15-EET can also elicit the release of heparin-binding EGF-like growth factor from a renal epithelial cell line via a process involving the activation of matrix metalloproteinases (MMP) [123]. Although the MMP involved has not yet been identified, a very similar mechanism seems to be responsible for the transactivation of the EGF receptor in endothelial cells [124]. The EET-mediated activation of the EGF receptor leads, in turn, to the activation of the kinase Akt and an enhanced expression of cyclin D1. All four EET regioisomers have been reported to elicit an increase in Akt phosphorylation and cell proliferation in murine endothelial cells, but only the proliferative effects of 5,6- and 14,15-EET are reportedly sensitive to a phosphatidylinositol 3-kinase (PI3-K) inhibitor, whereas the 8,9- and 11,12-EET-induced increase in [^{3}H] thymidine incorporation seems to be dependent on the activation of the p38 MAP kinase [125]. In contrast, in bovine aortic endothelial cells 8,9-, 11,12-, and 14,15-EET-induced cell proliferation can be attenuated by MEK, ERK, and PI3-K inhibition [126]. Other signaling pathways also contribute to the increase in cyclin D1 expression, including the MAP kinase phosphatase-1 (MKP-1), which decreases c-Jun N-terminal kinase (JNK) activity [127]. Activation of Akt by EETs also induces phosphorylation and therefore inhibition of the forkhead factors FOXO1 and FOXO 3a, and subsequently a decrease in the expression of the cyclin-dependent kinase inhibitor p27^{kip1} [91]. The involvement of this mechanism in the CYP2C9-induced endothelial cell proliferation could be demonstrated by the transfection of CYP2C9-overexpressing cells with either a dominant negative Akt or a constitutively active FOXO3a, both of which inhibit CYP2C9-induced endothelial cell proliferation [91]. Although there is a precedent for the negative regulation of JNK after activation of Akt, inasmuch as Akt has been reported to phosphorylate and inactivate the kinase SEK1 and thus inactivate its substrate JNK [128], it remains unclear whether these pathways are linked to each other or are simply activated in parallel.

The first link between EETs and angiogenesis was obtained in co-cultures of astrocytes and endothelial cells. EETs released from astrocytes increased thymidine incorporation into endothelial cells and elicited the formation of capillary-like structures [129, 130]. Moreover, overexpression of CYP2C9 in, and/or the application of, 11,12- or 14,15-EET to monocultures of endothelial cells was associated with angiogenesis [124, 131]. *In vivo* data rapidly followed to support these *in vitro* findings, and EETs induced angiogenesis in the chick chorioallantoic membrane [124], as well as in EET-impregnated Matrigel plugs in adult rats [131] and an ischemic rat hindlimb model in which the overexpression of different CYP isozymes, including CYP2C11 and 2J2, was found to increase muscle capillary density [126]. Furthermore, tumor growth and metastasis can be increased by sEH inhibition in transgenic mice with high vascular EET levels, i.e. animals that overexpress either the human CYP2C8 or human CYP2J2 specifically in Tie-2-expressing cells, or that were treated with high concentrations of 14,15-EET [132]. All in all, such evidence indicated that activation of the CYP/sEH axis is linked with the promotion of angiogenesis; however, the latter models were somewhat artificial and focused on the products of arachidonic acid metabolism, largely ignoring the biological actions of other lipids that feed into the same CYP/sEH axis.

It was partly to assess the role of the sEH in angiogenesis under more physiological conditions that we determined the effects of the global and induced deletion of the sEH, as well as its pharmacological inhibition in vascular repair after ischemia and in the postnatal murine retina. We found that sEH deletion and inhibition resulted in markedly decreased angiogenesis [133] and vascular repair [47], and provided some of the first experimental data that linked the defect not to the accumulation of a PUFA epoxide but to the lack of diol production. To identify such lipids, liquid chromatography–tandem mass spectrometry (LC–MS/MS)-based lipid profiling approaches are used to screen for the PUFA epoxides or diols most affected by the deletion of the sEH, and, to date, biological activities have been attached to the DHA-derived diol 19,20-dihydroxydocosapentaenoic acid [133] and the linoleic acid-derived diol 12,13-dihydroxyoctadecenoic acid [47]. Interestingly, the signaling pathways targeted by the diols are distinct, as, while the defective vascular repair in $sEH^{-/-}$ mice could be attributed to altered Wnt signaling followed by attenuated progenitor cell proliferation and mobilization [47], defects in the retina could be linked to the translocation of presenilin 1 out of lipid rafts and the subsequent inhibition of the γ-secretase [133]. This means that the take-home message with respect to angiogenesis is that the ω-3/ω-6 profile of a particular tissue is likely to determine the overall effects on angiogenesis. Certainly, while EETs have been well-defined as angiogenic mediators [132, 134], a DHA-derived epoxide was recently reported to inhibit angiogenesis by preventing phosphorylation of the vascular endothelial growth factor receptor-2 (VEGFR2) [135]. This is of relevance since the lipids that feed into the CYP/sEH axis are largely provided by the diet, and regulating dietary intake of specific lipids, e.g. the fish oils EPA and DPA, has been linked with altered epoxide and diol profiles, as well as protection against vascular inflammation and cancer. On the other hand, increased dietary intake of linoleic acid is generally associated with inflammation and increased risk. It will therefore be interesting to determine to what extent diet can alter the influence of the CYP/sEH axis on angiogenesis and tumor growth [136, 137], as well as the development of the cardiovascular complications associated with the metabolic syndrome [136, 138].

To date, the CYP enzymes linked to angiogenesis have included the human 2C8/2C9 and 2J to enzymes, as well as the rat 2C11 and mouse 2c44 isoforms, all of which are epoxygenases. CYP1B1 is worth mentioning at this point, even though the enzyme is an estrogen-metabolizing CYP hydroxylase. CYP1B1 induction is an important factor in determining risk associated with hormone-mediated cancers, in particular as CYP1B1 is induced by hypoxia [139], probably because its expression is regulated by the AMP-activated protein kinase (AMPK) [140], and is involved in the metabolism of some clinically relevant anticancer agents [141]. In addition, CYP1B1 is tightly regulated by the angiogenic microRNA miR-27b [142–144]. The link to this particular microRNA is interesting as it has previously been described as a "regulator hub in lipid metabolism" [145]. Indeed, miR-27b levels are significantly upregulated by a high-fat diet and hepatic miR-27b and its target genes are inversely altered in a mouse model of dyslipidemia and atherosclerosis [145]. Whether or not CYP1B1 is involved in the latter is unclear but the enzyme has recently been linked with protection against angiotensin II-induced hypertension in female mice [146].

What makes CYP1B1 of interest in angiogenesis is that its deletion impaired revascularization in a model of oxygen-induced retinopathy in mice [147]. This effect was linked with a decrease in the expression of the endothelial NO synthase [148], as well as a corresponding increase in intracellular oxidative stress and increased production of thrombospondin-2, an endogenous inhibitor of angiogenesis [147, 149]. Interestingly, estrogen-induced angiogenesis has also been attributed to changes in endothelial NO synthase, thrombospondin, and free radical generation, making it tempting to speculate that CYP1B1 may actually mediate the effects of the hormone. Certainly, the CYP1B1-derived metabolites of β-estradiol promote angiogenesis in uterine artery endothelial cells [150]. Rather intriguingly, residues 41–48 of human CYP1B1 are part of a mitochondrial import signal, and the cleavage of CYP1B1 by serine proteases results in its targeting to mitochondria, which is associated with oxidative stress and mitochondrial dysfunction [151]. Given that angiogenic endothelial cells undergo changes in metabolism, the so-called Warburg effect [152], it will be interesting to determine whether or not CYP1B1 can also alter endothelial cell metabolism and mitochondrial function. Effects on CYP1B1 may also explain the antiangiogenic actions of the antidiabetic drug metformin, which prevents the tumor cell supernatant–induced upregulation of CYP1B1 in endothelial cells [140].

Acknowledgments The author acknowledges the work of the many groups whose work it has not been possible to cite here because of space limitations. Work performed in the author's own laboratory was supported by the Deutsche Forschungsgemeinschaft (SFB-TR 23, A6 and Exzellenzcluster 147 'Cardio-Pulmonary System').

References

1. Konkel A, Schunck WH. Role of cytochrome P450 enzymes in the bioactivation of polyunsaturated fatty acids. Biochim Biophys Acta. 2011;1814:210–22.
2. Campbell WB, Gebremedhin D, Pratt PF, Harder DR. Identification of epoxyeicosatrienoic acids as endothelium-derived hyperpolarizing factors. Circ Res. 1996;78:415–23.
3. Fisslthaler B, Popp R, Kiss L, Potente M, Harder DR, Fleming I, Busse R. Cytochrome P450 2C is an EDHF synthase in coronary arteries. Nature. 1999;401:493–7.
4. Harder DR, Narayanan J, Gebremedhin D. Pressure-induced myogenic tone and role of 20-HETE in mediating autoregulation of cerebral blood flow. Am J Physiol Heart Circ Physiol. 2011;300:H1557–65.
5. Imig JD, Simpkins AN, Renic M, Harder DR. Cytochrome P450 eicosanoids and cerebral vascular function. Expert Rev Mol Med. 2011;13:e7.
6. Michaelis UR, Fleming I. From endothelium-derived hyperpolarizing factor (EDHF) to angiogenesis: epoxyeicosatrienoic acids (EETs) and cell signaling. Pharmacol Ther. 2006;111:584–95.
7. Chen L, Ackerman R, Guo AM. 20-HETE in neovascularization. Prostaglandins Other Lipid Mediat. 2012;98:63–8.
8. Guengerich FP. Human cytochrome P450 enzymes. In: Ortiz de Montellano PR, editor. Cytochrome P450: structure, mechanism, and biochemistry. 2 ed. New York: Plenum Press; 1995. pp. 473–535.
9. Fleming I. The cytochrome P450 pathway in angiogenesis and endothelial cell biology. Cancer Metastasis Rev. 2011;30:541–55.
10. Imig JD. Epoxides and soluble epoxide hydrolase in cardiovascular physiology. Physiol Rev. 2012;92:101–30.
11. Oliw EH, Benthin G. On the metabolism of epoxyeicosatrienoic acids by ram seminal vesicles: isolation of 5(6)epoxy-prostaglandin F1a. Biochem Biophys Res Commun. 1985;126:1090–6.
12. Nüsing RM, Schweer H, Fleming I, Zeldin DC, Wegmann M. Epoxyeicosatrienoic acids affect electrolyte transport in renal tubular epithelial cells: dependence on cyclooxygenase and cell polarity. Am J Physiol Renal Physiol. 2007;293:F288–98.
13. Fisslthaler B, Hinsch N, Chataigneau T, Popp R, Kiss L, Busse R, Fleming I. Nifedipine increases cytochrome P4502C expression and EDHF-mediated responses in coronary arteries. Hypertension. 2000;36:270–5.
14. Fisslthaler B, Popp R, Michaelis UR, Kiss L, Fleming I, Busse R. Cyclic stretch enhances the expression and activity of coronary endothelium-derived hyperpolarizing factor synthase. Hypertension. 2001;38:1427–32.
15. Marden NY, Fiala-Beer E, Xiang SH, Murray M. Role of activator protein-1 in the down-regulation of the human CYP2J2 gene in hypoxia. Biochem J. 2003;373:669–80.
16. Michaelis UR, Fisslthaler B, Barbosa-Sicard E, Falck JR, Fleming I, Busse R. Cytochrome P450 epoxygenases 2C8 and 2C9 are implicated in hypoxia-induced endothelial cell migration and angiogenesis. J Cell Sci. 2005;118:5489–98.
17. Alkayed NJ, Goyagi T, Joh HD, Klaus J, Harder DR, Traystman RJ, Hurn PD. Neuroprotection and P450 2C11 upregulation after experimental transient ischemic attack. Stroke. 2002;33:1677–84.
18. Earley S, Walker BR. Endothelium-dependent blunting of myogenic responsiveness after chronic hypoxia. Am J Physiol Heart Circ Physiol. 2002;283:H2202–9.
19. Earley S, Pastuszyn A, Walker BR. Cytochrome P-450 epoxygenase products contribute to attenuated vasoconstriction after chronic hypoxia. Am J Physiol Heart Circ Physiol. 2003;285:H127–36.
20. Gerbal-Chaloin S, Pascussi JM, Pichard-Garcia L, Daujat M, Waechter F, Fabre JM, Carrere N, Maurel P. Induction of CYP2C genes in human hepatocytes in primary culture. Drug Metab Dispos. 2001;29:242–51.

21. Gerbal-Chaloin S, Daujat M, Pascussi JM, Pichard-Garcia L, Vilarem MJ, Maurel P. Transcriptional regulation of CYP2C9 gene. Role of glucocorticoid receptor and constitutive androstane receptor. J Biol Chem. 2002;277:209–17.
22. de Morais SMF, Schweikl H, Blaisdell J, Goldstein JA. Gene structure and upstream regulatory regions of human CYP2C9 and CYP2C18. Biochem Biophys Res Commun. 1993;194:194–201.
23. Ged C, Beaune P. Isolation of the human cytochrome P-450 IIC8 gene: multiple glucocorticoid responsive elements in the 5′ region. Biochim Biophys Acta. 1991;1088:433–5.
24. Bauersachs J, Christ M, Ertl G, Michaelis UR, Fisslthaler B, Busse R, Fleming I. Cytochrome P450 2C expression and EDHF-mediated relaxation in porcine coronary arteries is increased by cortisol. Cardiovasc Res. 2002;54:669–75.
25. Transon C, Leemann T, Dayer P. In vitro comparative inhibition profiles of major human drug metabolising cytochrome P450 isozymes (CYP2C9, CYP2D6 and CYP3A4) by HMG-CoA reductase inhibitors. Eur J Clin Pharmacol. 1996;50:209–15.
26. Fischer V, Johanson L, Heitz F, Tullman R, Graham E, Baldeck JP, Robinson WT. The 3-hydroxy-3-methylglutaryl coenzyme A reductase inhibitor fluvastatin: effect on human cytochrome P-450 and implications for metabolic drug interactions. Drug Metab Dispos. 1999;27:410–6.
27. Qiu WP, Hu Q, Paolocci N, Ziegelstein RC, Kass DA. Differential effects of pulsatile versus steady flow on coronary endothelial membrane potential. Am J Physiol Heart Circ Physiol. 2003;285:H341–6.
28. Capdevila JH, Falck JR, Harris RC. Cytochrome P450 and arachidonic acid bioactivation. Molecular and functional properties of the arachidonate monooxygenase. J Lipid Res. 2000;41:163–81.
29. Oesch-Bartlomowicz B, Richter B, Becker R, Vogel S, Padma PR, Hengstler JG, Oesch F. cAMP-dependent phosphorylation of CYP2B1 as a functional switch for cyclophosphamide activation and its hormonal control in vitro and in vivo. Int J Cancer. 2001;94:733–42.
30. Korsmeyer KK, Davoll S, Figueiredo-Pereira ME, Correia MA. Proteolytic degradation of heme-modified hepatic cytochromes P450: a role for phosphorylation, ubiquitination, and the 26S proteasome? Arch Biochem Biophys. 1999;365:31–44.
31. Anandatheerthavarada HK, Biswas G, Mullick J, Sepuri NB, Otvos L, Pain D, Avadhani NG. Dual targeting of cytochrome P4502B1 to endoplasmic reticulum and mitochondria involves a novel signal activation by cyclic AMP-dependent phosphorylation at Ser128. EMBO J. 1999;18:5494–504.
32. Eliasson E, Mkrtchian S, Ingelman-Sundberg M. Hormone- and substrate-regulated intracellular degradation of cytochrome P450 (2E1) involving MgATP-activated rapid proteolysis in the endoplasmic reticulum membranes. J Biol Chem. 1992;267:15765–9.
33. Minamiyama Y, Takemura S, Imaoka S, Funae Y, Tanimoto Y, Inoue M. Irreversible inhibition of cytochrome P450 by nitric oxide. J Pharmacol Exp Ther. 1997;283:1479–85.
34. Bauersachs J, Popp R, Hecker M, Sauer E, Fleming I, Busse R. Nitric oxide attenuates the release of endothelium-derived hyperpolarizing factor. Circulation. 1996;94:3341–7.
35. Beetham JK, Tian T, Hammock BD. cDNA cloning and expression of a soluble epoxide hydrolase from human liver. Arch Biochem Biophys. 1993;305:197–201.
36. Newman JW, Morisseau C, Harris TR, Hammock BD. The soluble epoxide hydrolase encoded by EPXH2 is a bifunctional enzyme with novel lipid phosphate phosphatase activity. Proc Natl Acad Sci U S A. 2003;100:1558–63.
37. Cronin A, Mowbray S, Durk H, Homburg S, Fleming I, Fisslthaler B, Oesch F, Arand M. The N-terminal domain of mammalian soluble epoxide hydrolase is a phosphatase. Proc Natl Acad Sci U S A. 2003;100:1552–7.
38. Newman JW, Morisseau C, Hammock BD. Epoxide hydrolases: their roles and interactions with lipid metabolism. Prog Lipid Res. 2005;44:1–51.
39. Ai D, Fu Y, Guo D, Tanaka H, Wang N, Tang C, Hammock BD, Shyy JYJ, Zhu Y. Angiotensin II up-regulates soluble epoxide hydrolase in vascular endothelium in vitro and in vivo. Proc Natl Acad Sci U S A. 2007;104:9018–23.

40. Yamada T, Morisseau C, Maxwell JE, Argiriadi MA, Christianson DW, Hammock BD. Biochemical evidence for the involvement of tyrosine in epoxide activation during the catalytic cycle of epoxide hydrolase. J Biol Chem. 2000;275:23082–8.
41. Barbosa-Sicard E, Frömel T, Keserü B, Brandes RP, Morisseau C, Hammock BD, Braun T, Krüger M, Fleming I. Inhibition of the soluble epoxide hydrolase by tyrosine nitration. J Biol Chem. 2009;284:28156–63.
42. Ohtoshi K, Kaneto H, Node K, Nakamura Y, Shiraiwa T, Matsuhisa M, Yamasaki Y. Association of soluble epoxide hydrolase gene polymorphism with insulin resistance in type 2 diabetic patients. Biochem Biophys Res Commun. 2005;331:347–50.
43. Doulias PT, Greene JL, Greco TM, Tenopoulou M, Seeholzer SH, Dunbrack RL, Ischiropoulos H. Structural profiling of endogenous S-nitrosocysteine residues reveals unique features that accommodate diverse mechanisms for protein S-nitrosylation. Proc Natl Acad Sci U S A. 2010;107:16958–63.
44. Imig JD. Epoxide hydrolase and epoxygenase metabolites as therapeutic targets for renal diseases. Am J Physiol Renal Physiol. 2005;289:F496–503.
45. Jung O, Brandes RP, Kim IH, Schweda F, Schmidt R, Hammock BD, Busse R, Fleming I. Soluble epoxide hydrolase is a main effector of angiotensin II-induced hypertension. Hypertension. 2005;45:759–65.
46. Liu JY, Li N, Yang J, Li N, Qiu H, Ai D, Chiamvimonvat N, Zhu Y, Hammock BD. Metabolic profiling of murine plasma reveals an unexpected biomarker in rofecoxib-mediated cardiovascular events. Proc Natl Acad Sci U S A. 2010;107:17017–22.
47. Frömel T, Jungblut B, Hu J, Trouvain C, Barbosa-Sicard E, Popp R, Liebner S, Dimmeler S, Hammock BD, Fleming I. Soluble epoxide hydrolase regulates hematopoietic progenitor cell function via generation of fatty acid diols. Proc Natl Acad Sci U S A. 2012;109:9995–10000.
48. Keserü B, Barbosa-Sicard E, Popp R, Fisslthaler B, Dietrich A, Gudermann T, Hammock BD, Falck JR, Weissmann N, Busse R, Fleming I. Epoxyeicosatrienoic acids and the soluble epoxide hydrolase are determinants of pulmonary artery pressure and the acute hypoxic pulmonary vasoconstrictor response. FASEB J. 2008;22:4306–15.
49. Sinal CJ, Miyata M, Tohkin M, Nagata K, Bend JR, Gonzalez FJ. Targeted disruption of soluble epoxide hydrolase reveals a role in blood pressure regulation. J Biol Chem. 2000;275:40504–10.
50. Luria A, Weldon SM, Kabcenell AK, Ingraham RH, Matera D, Jiang H, Gill R, Morisseau C, Newman JW, Hammock BD. Compensatory mechanism for homeostatic blood pressure regulation in Ephx2 gene-disrupted mice. J Biol Chem. 2007;282:2891–8.
51. Seubert JM, Sinal CJ, Graves J, Degraff LM, Bradbury JA, Lee CR, Goralski K, Carey MA, Luria A, Newman JW, Hammock BD, Falck JR, Roberts H, Rockman HA, Murphy E, Zeldin DC. Role of soluble epoxide hydrolase in postischemic recovery of heart contractile function. Circ Res. 2006;99:442–50.
52. Morisseau C, Hammock BD. Impact of soluble epoxide hydrolase and epoxyeicosanoids on human health. Annu Rev Pharmacol Toxicol. 2013;53:37–58.
53. Enayetallah AE, Luria A, Luo B, Tsai HJ, Sura P, Hammock BD, Grant DF. Opposite regulation of cholesterol levels by the phosphatase and hydrolase domains of soluble epoxide hydrolase. J Biol Chem. 2008;283:36592–8.
54. Luria A, Morisseau C, Tsai HJ, Yang J, Inceoglu B, De Taeye B, Watkins SM, Wiest MM, German JB, Hammock BD. Alteration in plasma testosterone levels in male mice lacking soluble epoxide hydrolase. Am J Physiol Endocrinol Metab. 2009;297:E375–83.
55. Enayetallah AE, Grant DF. Effects of human soluble epoxide hydrolase polymorphisms on isoprenoid phosphate hydrolysis. Biochem Biophys Res Commun. 2006;341:254–60.
56. Tran KL, Aronov PA, Tanaka H, Newman JW, Hammock BD, Morisseau C. Lipid sulfates and sulfonates are allosteric competitive inhibitors of the N-terminal phosphatase activity of the Mammalian soluble epoxide hydrolase. Biochemistry. 2005;44:12179–87.
57. Kovacs WJ, Olivier LM, Krisans SK. Central role of peroxisomes in isoprenoid biosynthesis. Prog Lipid Res. 2002;41:369–91.

58. Oguro A, Imaoka S. Lysophosphatidic acids are new substrates for the phosphatase domain of soluble epoxide hydrolase. J Lipid Res. 2012;53:505–12.
59. Morisseau C, Schebb NH, Dong H, Ulu A, Aronov PA, Hammock BD. Role of soluble epoxide hydrolase phosphatase activity in the metabolism of lysophosphatidic acids. Biochem Biophys Res Commun. 2012;419:796–800.
60. Przybyla-Zawislak BD, Srivastava PK, Vazquez-Matias J, Mohrenweiser HW, Maxwell JE, Hammock BD, Bradbury JA, Enayetallah AE, Zeldin DC, Grant DF. Polymorphisms in human soluble epoxide hydrolase. Mol Pharmacol. 2003;64:482–90.
61. Keserü B, Barbosa-Sicard E, Schermuly RT, Tanaka H, Hammock BD, Weissmann N, Fisslthaler B, Fleming I. Hypoxia-induced pulmonary hypertension: comparison of soluble epoxide hydrolase deletion vs. inhibition. Cardiovasc Res. 2010;85:232–40.
62. Chen JK, Wang D-W, Falck JR, Capdevila J, Harris RC. Transfection of an active cytochrome P450 arachidonic acid epoxygenase indicates that 14,15-epoxyeicosatrienoic acid functions as an intracellular messenger in response to epidermal growth factor. J Biol Chem. 1999;274:4764–9.
63. Chen J-K, Capdevila J, Harris RC. Heparin-binding EGF-like growth factor mediates the biological effects of P450 arachidonate epoxygenase metabolites in epithelial cells. Proc Natl Acad Sci U S A. 2002;99:6029–34.
64. Michaelis UR, Fisslthaler B, Medhora M, Harder D, Fleming I, Busse R. Cytochrome P450 2C9-derived epoxyeicosatrienoic acids induce angiogenesis via cross-talk with the epidermal growth factor receptor (EGFR). FASEB J. 2003;17:770–2.
65. Wong PY, Lin KT, Yan YT, Ahern D, Iles J, Shen SY, Bhatt RK, Falck JR. 14(R),15(S)-epoxyeicosatrienoic acid (14(R),15(S)-EET) receptor in guinea pig mononuclear cell membranes. J Lipid Mediat. 1993;6:199–208.
66. Wong PY, Lai PS, Falck JR. Mechanism and signal transduction of 14 (R), 15 (S)-epoxyeicosatrienoic acid (14,15-EET) binding in guinea pig monocytes. Prostaglandins Other Lipid Mediat. 2000;62:321–33.
67. Wong PY, Lai PS, Shen SY, Belosludtsev YY, Falck JR. Post-receptor signal transduction and regulation of 14(R),15(S)- epoxyeicosatrienoic acid (14,15-EET) binding in U-937 cells. J Lipid Mediat Cell Signal. 1997;16:155–69.
68. Snyder GD, Krishna UM, Falck JR, Spector AA. Evidence for a membrane site of action for 14,15-EET on expression of aromatase in vascular smooth muscle. Am J Physiol Heart Circ Physiol. 2002;283:H1936–42.
69. Li PL, Campbell WB. Epoxyeicosatrienoic acids activate K^+ channels in coronary smooth muscle through a guanine nucleotide binding protein. Circ Res. 1997;80:877–84.
70. Node K, Ruan XL, Dai J, Yang SX, Graham L, Zeldin DC, Liao JK. Activation of Gas mediates induction of tissue-type plasminogen activator gene transcription by epoxyeicosatrienoic acids. J Biol Chem. 2001;276:15983–9.
71. Popp R, Brandes RP, Ott G, Busse R, Fleming I. Dynamic modulation of interendothelial gap junctional communication by 11,12-epoxyeicosatrienoic acid. Circ Res. 2002;90:800–6.
72. Fleming I, Rueben A, Popp R, Fisslthaler B, Schrodt S, Sander A, Haendeler J, Falck JR, Morisseau C, Hammock BD, Busse R. Epoxyeicosatrienoic acids regulate Trp channel dependent Ca^{2+} signaling and hyperpolarization in endothelial cells. Arterioscler Thromb Vasc Biol. 2007;27:2612–8.
73. Fukao M, Mason HS, Kenyon JL, Horowitz B, Keef KD. Regulation of BKca channels expressed in human embryonic kidney 293 cells by epoxyeicosatrienoic acid. Mol Pharmacol. 2001;59:16–23.
74. Falck JR, Krishna UM, Reddy YK, Kumar PS, Reddy KM, Hittner SB, Deeter C, Sharma KK, Gauthier KM, Campbell WB. Comparison of vasodilatory properties of 14,15-EET analogs: structural requirements for dilation. Am J Physiol Heart Circ Physiol. 2003;284:H337–49.
75. Yang W, Tuniki VR, Anjaiah S, Falck JR, Hillard CJ, Campbell WB. Characterization of epoxyeicosatrienoic acid binding site in U937 membranes using a novel radiolabeled agonist, 20-^{125}I-14,15-epoxyeicosa-8(Z)-enoic acid. J Pharmacol Exp Ther. 2008;324:1019–27.

76. Inceoglu B, Schmelzer KR, Morisseau C, Jinks SL, Hammock BD. Soluble epoxide hydrolase inhibition reveals novel biological functions of epoxyeicosatrienoic acids (EETs). Prostaglandins Other Lipid Mediat. 2007;82:42–9.
77. Cowart LA, Wei S, Hsu MH, Johnson EF, Krishna MU, Falck JR, Capdevila JH. The CYP4A isoforms hydroxylate epoxyeicosatrienoic acids to form high affinity peroxisome proliferator-activated receptor ligands. J Biol Chem. 2002;277:35105–12.
78. Liu Y, Zhang Y, Schmelzer K, Lee TS, Fang X, Zhu Y, Spector AA, Gill S, Morisseau C, Hammock BD, Shyy JY. The antiinflammatory effect of laminar flow: the role of PPARg, epoxyeicosatrienoic acids, and soluble epoxide hydrolase. Proc Natl Acad Sci U S A. 2005;102:16747–52.
79. Widstrom RL, Norris AW, Spector AA. Binding of cytochrome P450 monooxygenase and lipoxygenase pathway products by heart fatty acid-binding protein. Biochemistry. 2001;40:1070–6.
80. Wolfrum C, Borrmann CM, Borchers T, Spener F. Fatty acids and hypolipidemic drugs regulate peroxisome proliferator-activated receptors alpha- and gamma-mediated gene expression via liver fatty acid binding protein: a signaling path to the nucleus. Proc Natl Acad Sci U S A. 2001;98:2323–8.
81. Capdevila JH, Falck JR, Harris RC. Cytochrome P450 and arachidonic acid bioactivation. Molecular and functional properties of the arachidonate monooxygenase. J Lipid Res. 2000;41:163–81.
82. Klett EL, Chen S, Edin ML, Li LO, Ilkayeva O, Zeldin DC, Newgard CB, Coleman RA. Diminished acyl-CoA synthetase isoform 4 activity in INS 832/13 cells reduces cellular epoxyeicosatrienoic acid levels and results in impaired glucose-stimulated insulin secretion. J Biol Chem. 2013;288:21618–29.
83. Michaelis UR, Falck JR, Schmidt R, Busse R, Fleming I. Cytochrome P4502C9-derived epoxyeicosatrienoic acids induce the expression of cyclooxygenase-2 in endothelial cells. Arterioscler Thromb Vasc Biol. 2005;25:321–6.
84. Node K, Huo Y, Ruan X, Yang B, Spiecker M, Ley K, Zeldin DC, Liao JK. Anti-inflammatory properties of cytochrome P450 epoxygenase-derived eicosanoids. Science. 1999;285:1276–9.
85. Fleming I, Michaelis UR, Bredenkötter D, Fisslthaler B, Dehghani F, Brandes RP, Busse R. Endothelium-derived hyperpolarizing factor synthase (cytochrome P450 2C9) is a functionally significant source of reactive oxygen species in coronary arteries. Circ Res. 2001;88:44–51.
86. Fichtlscherer S, Dimmeler S, Breuer S, Busse R, Zeiher AM, Fleming I. Inhibition of cytochrome P450 2C9 improves endothelium-dependent, nitric oxide-mediated vasodilatation in patients with coronary artery disease. Circulation. 2004;109:178–83.
87. Wu S, Moomaw CR, Tomer KB, Falck JR, Zeldin DC. Molecular cloning and expression of CYP2J2, a human cytochrome P450 arachidonic acid epoxygenase highly expressed in heart. J Biol Chem. 1996;271:3460–8.
88. Seubert J, Yang B, Bradbury JA, Graves J, Degraff LM, Gabel S, Gooch R, Foley J, Newman J, Mao L, Rockman HA, Hammock BD, Murphy E, Zeldin DC. Enhanced postischemic functional recovery in CYP2J2 transgenic hearts involves mitochondrial ATP-sensitive K^+ channels and p42/p44 MAPK pathway. Circ Res. 2004;95:506–14.
89. Fang X, Hu S, Watanabe T, Weintraub NL, Snyder GD, Yao J, Liu Y, Shyy JYJ, Hammock BD, Spector AA. Activation of peroxisome proliferator-activated receptor alpha by substituted urea-derived soluble epoxide hydrolase inhibitors. J Pharmacol Exp Ther. 2005;314:260–70.
90. Fang X, Hu S, Xu B, Snyder G, Harmon S, Yao J, Liu Y, Sangras B, Falck J, Weintraub NL, Spector AA. 14,15-Dihydroxyeicosatrienoic acid activates peroxisome proliferator activated receptor alpha. Am J Physiol Heart Circ Physiol. 2006;290:H55–63.
91. Potente M, Fisslthaler B, Busse R, Fleming I. 11,12-Epoxyeicosatrienoic acid-induced inhibition of FOXO factors promotes endothelial proliferation by down-regulating $p27^{Kip1}$. J Biol Chem. 2003;278:29619–25.
92. Busse R, Edwards G, Feletou M, Fleming I, Vanhoutte PM, Weston AH. EDHF: bringing the concepts together. Trends Pharmacol Sci. 2002;23:374–80.

93. Watanabe H, Vriens J, Prenen J, Droogmans G, Voets T, Nilius B. Anandamide and arachidonic acid use epoxyeicosatrienoic acids to activate TRPV4 channels. Nature. 2003;424:434–8.
94. Zheng X, Zinkevich NS, Gebremedhin D, Gauthier KM, Nishijima Y, Fang J, Wilcox DA, Campbell WB, Gutterman DD, Zhang DX. Arachidonic acid-induced dilation in human coronary arterioles: convergence of signaling mechanisms on endothelial TRPV4-mediated Ca^{2+} entry. J Am Heart Assoc. 2013;2:e000080.
95. Vriens J, Owsianik G, Fisslthaler B, Suzuki M, Janssens A, Voets T, Morisseau C, Hammock BD, Fleming I, Busse R, Nilius B. Modulation of the Ca^{2+} permeable cation channel TRPV4 by cytochrome P450 epoxygenases in vascular endothelium. Circ Res. 2005;97:908–15.
96. Loot AE, Popp R, Fisslthaler B, Vriens J, Nilius B, Fleming I. Role of cytochrome P450-dependent transient receptor potential V4 activation in flow-induced vasodilatation. Cardiovasc Res. 2008;80:445–52.
97. Weissmann N, Dietrich A, Fuchs B, Kalwa H, Ay M, Dumitrascu R, Olschewski A, Storch U, Schnitzler M, Ghofrani HA, Schermuly RT, Pinkenburg O, Seeger W, Grimminger F, Gudermann T. Classical transient receptor potential channel 6 (TRPC6) is essential for hypoxic pulmonary vasoconstriction and alveolar gas exchange. Proc Natl Acad Sci U S A. 2006;103:19093–8.
98. Wang L, Yin J, Nickles HT, Ranke H, Tabuchi A, Hoffmann J, Tabeling C, Barbosa-Sicard E, Chanson M, Kwak BR, Shin HS, Wu S, Isakson BE, Witzenrath M, de Wit C, Fleming I, Kuppe H, Kuebler WM. Hypoxic pulmonary vasoconstriction requires connexin 40-mediated endothelial signal conduction. J Clin Invest. 2012;122:4218–30.
99. Loot AE, Fleming I. Cytochrome P450-derived epoxyeicosatrienoic acids and pulmonary hypertension: central role of transient receptor potential (TRP) C6 channels. J Cardiovasc Pharmacol. 2011;57:140–7.
100. Khojasteh S, Prabhu S, Kenny J, Halladay J, Lu A. Chemical inhibitors of cytochrome P450 isoforms in human liver microsomes: a re-evaluation of P450 isoform selectivity. Eur J Drug Metab Pharmacokinet. 2011;36:1–16.
101. Passauer J, Büssemaker E, Lassig G, Pistrosch F, Fauler J, Gross P, Fleming I. Baseline blood flow and bradykinin-induced vasodilator responses in the human forearm are insensitive to the CYP 2C9 inhibitor sulfaphenazole. Clin Sci (Lond). 2003;105:513–8.
102. Passauer J, Pistrosch F, Lässig G, Herbrig K, Büssemaker E, Gross P, Fleming I. Nitric oxide- and EDHF-mediated arteriolar tone in uremia is unaffected by selective inhibition of vascular cytochrome P450 2C9. Kidney Int. 2005;67:1907–12.
103. Hillig T, Krustrup P, Fleming I, Osada T, Saltin B, Hellsten Y. Cytochrome P450 2C9 plays an important role in the regulation of exercise-induced skeletal muscle blood flow and oxygen uptake in humans. J Physiol (Lond). 2003;546:307–14.
104. Bellien J, Iacob M, Gutierrez L, Isabelle M, Lahary A, Thuillez C, Joannides R. Crucial role of NO and endothelium-derived hyperpolarizing factor in human sustained conduit artery flow-mediated dilatation. Hypertension. 2006;48:1088–94.
105. Bellien J, Joannides R, Iacob M, Arnaud P, Thuillez C. Evidence for a basal release of a cytochrome-related endothelium-derived hyperpolarizing factor in the radial artery in humans. Am J Physiol Heart Circ Physiol. 2006;290:H1347–52.
106. Fischer D, Landmesser U, Spiekermann S, Hilfiker-Kleiner D, Hospely M, Muller M, Busse R, Fleming I, Drexler H. Cytochrome P450 2C9 is involved in flow-dependent vasodilation of peripheral conduit arteries in healthy subjects and in patients with chronic heart failure. Eur J Heart Fail. 2007;9:770–5.
107. Ozkor MA, Murrow JR, Rahman AM, Kavtaradze N, Lin J, Manatunga A, Quyyumi AA. Endothelium-derived hyperpolarizing factor determines resting and stimulated forearm vasodilator tone in health and in disease. Circulation. 2011;123:2244–53.
108. Schuck RN, Theken KN, Edin ML, Caughey M, Bass A, Ellis K, Tran B, Steele S, Simmons BP, Lih FB, Tomer KB, Wu MC, Hinderliter AL, Stouffer GA, Zeldin DC, Lee CR. Cytochrome P450-derived eicosanoids and vascular dysfunction in coronary artery disease patients. Atherosclerosis. 2013;227:442–8.

109. Fornage M, Hinojos CA, Nurowska BW, Boerwinkle E, Hammock BD, Morisseau CH, Doris PA. Polymorphism in soluble epoxide hydrolase and blood pressure in spontaneously hypertensive rats. Hypertension. 2002;40:485–90.
110. Wei Q, Doris PA, Pollizotto MV, Boerwinkle E, Jacobs J, Siscovick DS, Fornage M. Sequence variation in the soluble epoxide hydrolase gene and subclinical coronary atherosclerosis: interaction with cigarette smoking. Atherosclerosis. 2007;190:26–34.
111. Lee CR, North KE, Bray MS, Fornage M, Seubert JM, Newman JW, Hammock BD, Couper DJ, Heiss G, Zeldin DC. Genetic variation in soluble epoxide hydrolase (EPHX2) and risk of coronary heart disease: The Atherosclerosis Risk in Communities (ARIC) study. Hum Mol Genet. 2006;15:1640–9.
112. Davis BB, Thompson DA, Howard LL, Morisseau C, Hammock BD, Weiss RH. Inhibitors of soluble epoxide hydrolase attenuate vascular smooth muscle cell proliferation. Proc Natl Acad Sci U S A. 2002;99:2222–7.
113. Davis BB, Morisseau C, Newman JW, Pedersen TL, Hammock BD, Weiss RH. Attenuation of vascular smooth muscle cell proliferation by 1-cyclohexyl-3-dodecyl urea is independent of soluble epoxide hydrolase inhibition. J Pharmacol Exp Ther. 2006;316:815–21.
114. Ulu A, Davis BB, Tsai HJ, Kim IH, Morisseau C, Inceoglu B, Fiehn O, Hammock BD, Weiss RH. Soluble epoxide hydrolase inhibitors reduce the development of atherosclerosis in apolipoprotein E-knockout mouse model. J Cardiovasc Pharmacol. 2008;52:314–23.
115. Zhang LN, Vincelette J, Cheng Y, Mehra U, Chen D, Anandan SK, Gless R, Webb HK, Wang YX. Inhibition of soluble epoxide hydrolase attenuated atherosclerosis, abdominal aortic aneurysm formation, and dyslipidemia. Arterioscler Thromb Vasc Biol. 2009;29:1265–70.
116. Revermann M, Schloss M, Barbosa-Sicard E, Mieth A, Liebner S, Morisseau C, Geisslinger G, Schermuly RT, Fleming I, Hammock BD, Brandes RP. Soluble epoxide hydrolase deficiency attenuates neointima formation in the femoral cuff model of hyperlipidemic mice. Arterioscler Thromb Vasc Biol. 2010;30:909–14.
117. Frömel T, Kohlstedt K, Popp R, Yin X, Barbosa-Sicard E, Thomas AC, Liebertz R, Mayr M, Fleming I. Cytochrome P4502S1: a novel monocyte/macrophage fatty acid epoxygenase in human atherosclerotic plaques. Basic Res Cardiol. 2013;108:1–12.
118. Wiecha J, Munz B, Wu Y, Noll T, Tillmanns H, Waldecker B. Blockade of Ca^{2+}-activated K^+ channels inhibits proliferation of human endothelial cells induced by basic fibroblast growth factor. J Vasc Res. 1998;35:363–71.
119. Faehling M, Koch ED, Raithel J, Trischler G, Waltenberger J. Vascular endothelial growth factor-A activates Ca^{2+}-activated K^+ channels in human endothelial cells in culture. Int J Biochem Cell Biol. 2001;33:337–46.
120. Wolfram Kuhlmann CR, Wiebke LD, Schaefer CA, Kerstin MA, Backenkohler U, Neumann T, Tillmanns H, Erdogan A. Lysophosphatidylcholine-induced modulation of Ca^{2+}-activated K^+ channels contributes to ROS-dependent proliferation of cultured human endothelial cells. J Mol Cell Cardiol. 2004;36:675–82.
121. Chen JK, Falck JR, Reddy KM, Capdevila J, Harris RC. Epoxyeicosatrienoic acids and their sulfonimide derivatives stimulate tyrosine phosphorylation and induce mitogenesis in renal epithelial cells. J Biol Chem. 1998;273:29254–61.
122. Fleming I, Fisslthaler B, Michaelis UR, Kiss L, Popp R, Busse R. The coronary endothelium-derived hyperpolarizing factor (EDHF) stimulates multiple signalling pathways and proliferation in vascular cells. Pflugers Arch Eur J Physiol. 2001;442:511–8.
123. Chen J-K, Capdevila J, Harris RC. Heparin-binding EGF-like growth factor mediates the biological effects of P450 arachidonate epoxygenase metabolites in epithelial cells. Proc Natl Acad Sci U S A. 2002;99:6029–34.
124. Michaelis UR, Fisslthaler B, Medhora M, Harder D, Fleming I, Busse R. Cytochrome P450 2C9-derived epoxyeicosatrienoic acids induce angiogenesis via cross-talk with the epidermal growth factor receptor (EGFR). FASEB J. 2003;17:770–2.
125. Pozzi A, Macias-Perez I, Abair T, Wey S, Su Y, Zent R, Falk JR, Capdevila JH. Charaterization of 5,6-and 8,9-epoxyeicosatrienoic acids (5,6- and 8,9-EET) as potent in vivo angiogenic lipids. J Biol Chem. 2005;280:27138–46.

126. Wang Y, Wei X, Xiao X, Hui R, Card JW, Carey MA, Wang DW, Zeldin DC. Arachidonic acid epoxygenase metabolites stimulate endothelial cell growth and angiogenesis via mitogen-activated protein kinase and phosphatidylinositol 3-kinase/Akt signaling pathways. J Pharmacol Exp Ther. 2005;314:522–32.
127. Potente M, Michaelis UR, Fisslthaler B, Busse R, Fleming I. Cytochrome P450 2C9-induced endothelial cell proliferation involves induction of mitogen-activated protein (MAP) kinase phosphatase-1, inhibition of the c-Jun N-terminal kinase, and up-regulation of cyclin D1. J Biol Chem. 2002;277:15671–6.
128. Park HS, Kim MS, Huh SH, Park J, Chung J, Kang SS, Choi EJ. Akt (protein kinase B) negatively regulates SEK1 by means of protein phosphorylation. J Biol Chem. 2002;277:2573–8.
129. Munzenmaier DH, Harder DR. Cerebral microvascular endothelial cell tube formation: role of astrocytic epoxyeicosatrienoic acid release. Am J Physiol Heart Circ Physiol. 2000;278:H1163–7.
130. Zhang C, Harder DR. Cerebral capillary endothelial cell mitogenesis and morphogenesis induced by astrocytic epoxyeicosatrienoic acid. Stroke. 2002;33:2957–64.
131. Medhora M, Daniels J, Mundey K, Fisslthaler B, Busse R, Jacobs ER, Harder DR. Epoxygenase-driven angiogenesis in human lung microvascular endothelial cells. Am J Physiol Heart Circ Physiol. 2003;284:H215–24.
132. Panigrahy D, Edin ML, Lee CR, Huang S, Bielenberg DR, Butterfield CE, Barnés CM, Mammoto A, Mammoto T, Luria A, Benny O, Chaponis DM, Dudley AC, Greene ER, Vergilio JA, Pietramaggiori G, Scherer-Pietramaggiori SS, Short SM, Seth M, Lih FB, Tomer KB, Yang J, Schwendener RA, Hammock BD, Falck JR, Manthati VL, Ingber DE, Kaipainen A, D'Amore PA, Kieran MW, Zeldin DC. Epoxyeicosanoids stimulate multiorgan metastasis and tumor dormancy escape in mice. J Clin Invest. 2012;122:178–91.
133. Hu J, Popp R, Frömel T, Ehling M, Awwad K, Adams RH, Hammes HP, Fleming I. Müller glia cells regulate Notch signaling and retinal angiogenesis via the generation of 19,20-dihydroxydocosapentaenoic acid. J Exp Med. 2014;211:281–95.
134. Grocott-Mason R, Fort S, Lewis MJ, Shah AM. Myocardial relaxant effect of endogenous nitric oxide in isolated ejecting hearts. Am J Physiol. 1994;266:H1699–705.
135. Zhang G, Panigrahy D, Mahakian LM, Yang J, Liu JY, Stephen Lee KS, Wettersten HI, Ulu A, Hu X, Tam S, Hwang SH, Ingham ES, Kieran MW, Weiss RH, Ferrara KW, Hammock BD. Epoxy metabolites of docosahexaenoic acid (DHA) inhibit angiogenesis, tumor growth, and metastasis. Proc Natl Acad Sci U S A. 2013;110:6530–5.
136. Pauwels EK. The protective effect of the Mediterranean diet: focus on cancer and cardiovascular risk. Med Princ Pract. 2011;20:103–11.
137. Gerber M. Omega-3 fatty acids and cancers: a systematic update review of epidemiological studies. Br J Nutr. 2012;107(Suppl 2):S228–39.
138. Robinson L, Mazurak V. N-3 polyunsaturated fatty acids: relationship to inflammation in healthy adults and adults exhibiting features of metabolic syndrome. Lipids. 2013;48:319–32.
139. Kirwan RP, Felice L, Clark AF, O'Brien CJ, Leonard MO. Hypoxia regulated gene transcription in human optic nerve lamina cribrosa cells in culture. Invest Ophthalmol Vis Sci. 2012;53:2243–55.
140. Dallaglio K, Bruno A, Cantelmo AR, Esposito AI, Ruggiero L, Orecchioni S, Calleri A, Bertolini F, Pfeffer U, Noonan DM, Albini A. Paradoxic effects of metformin on endothelial cells and angiogenesis. Carcinogenesis. 2014;doi:10.1093/carcin/bgu 001.
141. Sissung TM, Price DK, Sparreboom A, Figg WD. Pharmacogenetics and regulation of human cytochrome P450 1B1: implications in hormone-mediated tumor metabolism and a novel target for therapeutic intervention. Mol Cancer Res. 2006;4:135–50.
142. Tsuchiya Y, Nakajima M, Takagi S, Taniya T, Yokoi T. MicroRNA regulates the expression of human cytochrome P450 1B1. Cancer Res. 2006;66:9090–8.
143. Devlin AH, Thompson P, Robson T, McKeown SR. Cytochrome P450 1B1 mRNA untranslated regions interact to inhibit protein translation. Mol Carcinog. 2010;49:190–9.
144. Chuturgoon AA, Phulukdaree A, Moodley D. Fumonisin B1 modulates expression of human cytochrome P450 1b1 in human hepatoma (Hepg2) cells by repressing Mir-27b. Toxicol Lett. 2014;227:50–5.

145. Vickers KC, Shoucri BM, Levin MG, Wu H, Pearson DS, Osei-Hwedieh D, Collins FS, Remaley AT, Sethupathy P. MicroRNA-27b is a regulatory hub in lipid metabolism and is altered in dyslipidemia. Hepatology. 2013;57:533–42.
146. Jennings B, George L, Pingili A, Khan N, Estes A, Fang X, Gonzalez F, Malik K. Estrogen metabolism by cytochrome P450 1B1 modulates the hypertensive effect of angiotensin II in female mice. Hypertension. 2014;64:134–40.
147. Tang Y, Scheef EA, Wang S, Sorenson CM, Marcus CB, Jefcoate CR, Sheibani N. CYP1B1 expression promotes the proangiogenic phenotype of endothelium through decreased intracellular oxidative stress and thrombospondin-2 expression. Blood. 2009;113:744–54.
148. Tang Y, Scheef EA, Gurel Z, Sorenson CM, Jefcoate CR, Sheibani N. CYP1B1 and endothelial nitric oxide synthase combine to sustain proangiogenic functions of endothelial cells under hyperoxic stress. Am J Physiol Cell Physiol. 2010;298:C665–78.
149. Palenski TL, Gurel Z, Sorenson CM, Hankenson KD, Sheibani N. Cyp1B1 expression promotes angiogenesis by suppressing NF-kB activity. Am J Physiol Cell Physiol. 2013;305:C1170–84.
150. Jobe SO, Ramadoss J, Koch JM, Jiang Y, Zheng J, Magness RR. Estradiol-17b and its cytochrome P450- and catechol-O-methyltransferase-derived metabolites stimulate proliferation in uterine artery endothelial cells: role of estrogen receptor-alpha versus estrogen receptor-beta. Hypertension. 2010;55:1005–11.
151. Bansal S, Leu A, Gonzalez FJ, Guengerich FP, Roy Chowdhury A, Anandathheerthavarada HK, Avadhani NG. Mitochondrial targeting of cytochrome P450 (CYP) 1B1 and its role in polycyclic aromatic hydrocarbon-induced mitochondrial dysfunction. J Biol Chem. 2014;289:9936–51.
152. Polet F, Feron O. Endothelial cell metabolism and tumour angiogenesis: glucose and glutamine as essential fuels and lactate as the driving force. J Intern Med. 2013;273:156–65.

Chapter 10
The Role of Coagulation Factor Signaling in Angiogenesis and Vascular Remodeling

Christoph Reinhardt, Davit Manukyan and Wolfram Ruf

10.1 The Coagulation System

Hemostasis is the physiological process that arrests bleeding and prevents hemorrhage following injury of the vessel wall. Hemostasis involves the major functions of vascular constriction, platelet adhesion/aggregation, and blood coagulation. The coagulation system is a tightly regulated protease network that is essential for the integrity of a high-pressure vascular circuit [1]. To keep this system in a quiescent state, anticoagulant mechanisms have evolved that are localized to the microvascular endothelium [2, 3]. If the balance of procoagulant and anticoagulant mechanisms is perturbed, the result is either a hemorrhagic diathesis with bleeding episodes or a thrombophilic state that favors the occurrence of arterial or venous thrombosis. In addition to their hemostatic function, clotting factors prevent the dissemination of invading microbes in a process termed immunothrombosis [4]. Infection leading to sepsis syndrome and severe trauma can be associated with excessive activation of the coagulation system and increased vascular permeability [5]. As a consequence, disseminated intravascular coagulation (DIC) can occur and plasmatic coagulation factors are consumed. This imbalance favors platelet activation and depletion, resulting in microbleeds and petechiae.

C. Reinhardt (✉) · D. Manukyan · W. Ruf
Center for Thrombosis and Hemostasis, University Medical Center Mainz,
Langenbeckstrasse 1, 55131 Mainz, Germany
e-mail: Christoph.Reinhardt@unimedizin-mainz.de

D. Manukyan
Institute of Clinical Chemistry and Laboratory Medicine, University Medical Center Mainz,
Langenbeckstrasse 1, 55131 Mainz, Germany
e-mail: Davit.Manukyan@unimedizin-mainz.de

W. Ruf
Department of Immunology and Microbial Science, The Scripps Research Institute,
La Jolla, CA, USA
e-mail: ruf@uni-mainz.de

M. HH. Schmidt, S. Liebner (eds.), *Endothelial Signaling in Development and Disease*, DOI 10.1007/978-1-4939-2907-8_10

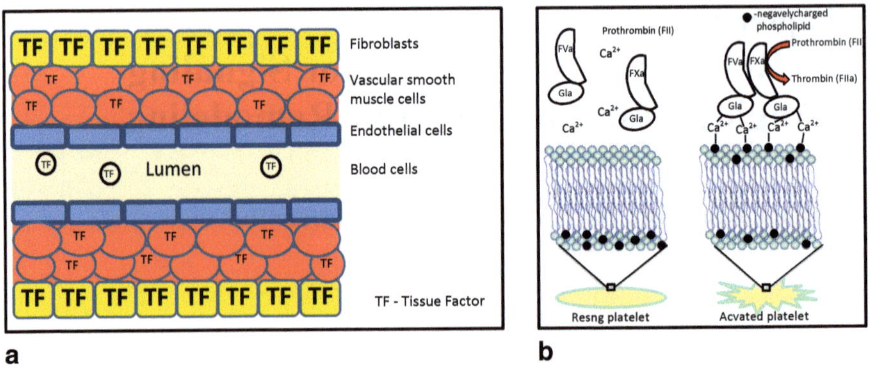

Fig. 10.1 **a** Distribution of tissue factor (*TF*) in the vessel wall. **b** Disturbance of the membrane asymmetry in the membrane of activated platelets enables Ca^{2+}-dependent binding of coagulation factors to negatively charged phospholipids (phosphatidylserine) via γ-carboxyglutamic acid (*Gla*) residues situated in Gla domains

Upon vascular injury, activation of the coagulation cascade by the membrane receptor tissue factor (TF) that is expressed on fibroblasts of the adventitial layer (Fig. 10.1a) leads to fibrin deposition and thrombus stabilization [6]. After vascular injury, components of the subendothelial matrix, e.g. collagen, laminin, and vitronectin, come into contact with circulating platelets. Primary platelet adhesion to the subendothelial matrix is mediated via von Willebrand factor (vWF) through the glycoprotein-Ib-V-IX receptor complex [7]. Following adhesion, platelets become activated, their granules are released, and platelet aggregation occurs [8]. In addition, the membrane asymmetry of resting platelets is disturbed following platelet activation, and thus a procoagulant membrane surface is provided by exposure of phosphatidylserine (PS) which is a prerequisite for formation of Ca^{2+}-dependent formation of membrane complexes via the γ-carboxyglutamic acid (Gla) domains of clotting factors (Fig. 10.1b) [9].

Coagulation can be started by two pathways (Fig. 10.2):

1. The *intrinsic pathway*, which is triggered by contact activation of coagulation factor XII.
2. The *extrinsic pathway*, which is initiated by the transmembrane glycoprotein TF.

A substantial body of evidence implicates the contact pathway that has no essential role in hemostasis in the development of thrombosis (Fig. 10.2). The intrinsic pathway is initiated by factor XII in a reaction involving high molecular weight kininogen (HMWK) and plasma kallikrein (PK), collectively referred to as plasma contact system [10]. Contact with negatively charged surfaces induces a conformational change in FXII zymogen, resulting in small amounts of active FXII (FXIIa). FXIIa cleaves PK to generate active kallikrein, which in turn reciprocally activates additional FXII [11]. Physiologically relevant surfaces such as extracellular RNA [12] and platelet-derived polyphosphates [13] were shown to trigger FXII activation. FXIIa cleaves factor XI (FXI), leading to activation of FIX by FXIa in the intrinsic pathway (Fig. 10.2). Importantly, FXI activation is further augmented by thrombin via a feedback mechanism [14].

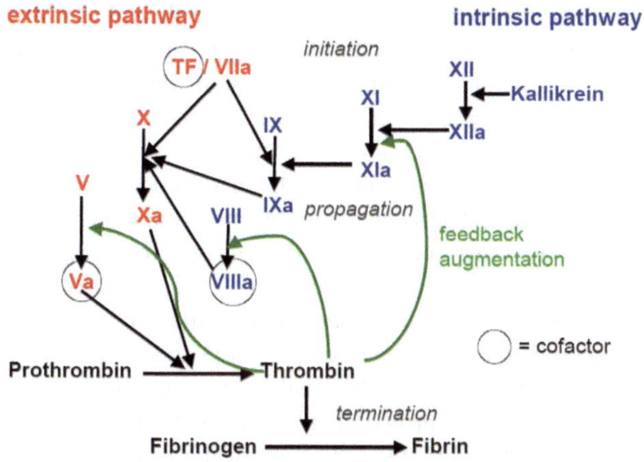

Fig. 10.2 Schematic view of the coagulation network subdivided in an *initiation* phase, a *propagation* phase, and a *termination* phase. Tissue factor, FVa, and FVIIIa act as cofactors to augment the zymogen activation of coagulation factors

The coagulation initiator TF is constitutively expressed in highly vascularized organs such as the placenta, brain, heart, kidney, and lung [15]. TF is especially expressed in the adventitial and medial layers of the vessel wall by fibroblasts and vascular smooth muscle cells [16] (Fig. 10.1a). Endothelial cells under physiological conditions are probably devoid of TF [6], and TF was originally thought to be exclusively located in the extravascular compartment without any contact with the plasma coagulation factors. Perivascular TF thus forms a hemostatic envelope and is ready to activate coagulation when vascular integrity is perturbed [16]. In contrast, in thrombosis the initiation of blood coagulation can be triggered by TF that is present in human blood under physiological conditions [17]. This pool of TF has been termed circulating or blood-borne TF [18]. Microparticles that originate from monocytes were identified as major TF carriers in blood [19].

Class 2 of the cytokine receptor superfamily comprises the interferon-α receptor, interferon-γ receptor, interleukin (IL)-10 receptor, and the coagulation initiator TF [20]. The crystal structure of the extracellular region of the coagulation initiator TF revealed that it consists of two fibronectin type III-like domains with a binding site for FVII that lies at the interface region and involves residues from domain 1 and an extended loop of immunoglobulin-like domain 2 [21–23]. TF is the cofactor for the plasma serine protease factor VIIa. The procoagulant activity of factor VIIa dramatically increases after binding of its membrane receptor TF [24]. The TF/VIIa binary complex then activates the serine protease zymogens FIX and FX by limited proteolysis [25, 26]. Subsequently, FIXa binds its cofactor VIIIa and activates more FX in the so-called Xase complex (Fig. 10.2). Together with its cofactor FVa, FXa forms a complex on the surface of activated platelets, the so-called prothrombinase complex which activates prothrombin [27] (Fig. 10.2). Ca^{2+}, together with negatively charged phospholipids (PS), mediates the anchoring of coagulation proteases on the surface of activated platelets, which is crucial for the formation of the functional coagulation complexes. The protease thrombin formed initially in coagulation is a pivotal enzyme [28] that activates FV, FVIII, and FXI, and thereby amplifies its own formation (Fig. 10.2). Cleavage of fibrinogen by thrombin leads to the formation of fibrin polymers (Fig. 10.2), and thrombin also activates FXIII, a transglutaminase that crosslinks fibrin and thus stabilizes the fibrin meshwork.

Thrombin is not only involved in the regulation of procoagulant mechanisms but also initiates fibrinolytic and anticoagulant systems that counteract fibrin formation when it binds to the vascular endothelium through specific receptors [29]. Thrombin bound to thrombomodulin on the endothelial cell surface has altered macromolecular substrate specificity and is no longer procoagulant, but now activates protein C [30]. Activation of protein C is augmented by the endothelial protein C receptor (EPCR) [31]. Activated protein C (APC) binds to protein S on the surface of activated cells and this complex then proteolytically inactivates factors Va and VIIIa, thus halting coagulation by disabling the formation of the prothrombinase (Va/Xa/prothormbin) and the intrinsic Xase complex (VIIIa/IXa/X) [32]. Furthermore, APC can exert additional anticoagulant action by inactivation of the tissue plasminogen activator inhibitor, thus enhancing fibrinolysis [33]. However, the protein C pathway not only regulates coagulation but the protein C/EPCR complex also activates protease-activated receptor (PAR)-1 [34], and thereby inhibits inflammatory signaling and induces cell survival of endothelial cells [35–36].

10.2 Activation of Protease-Activated Receptors (PARs) by Coagulation Proteases

Thrombin, the central trypsin-like serine protease that converts fibrinogen to fibrin also promotes activation of PAR 1 and PAR4 on human platelets, and thus connects coagulation and platelet activation [37]. The first thrombin receptor (PAR1) was cloned in 1991 [38] and three other members of this receptor family have been identified [39–41] (Fig. 10.3). Members of the PAR family of heptahelical G-protein-coupled receptors (GPCRs) are expressed on various cell types, including endothelial cells, smooth muscle cells, fibroblasts, epithelial cells, mast cells, neutrophils, monocytes, and macrophages [42–48]. These receptors not only mediate the cellular actions of the central coagulation protease thrombin on platelets but they also fulfill important nonhemostatic functions in development, play a role in tumor biology, regulation of inflammatory responses, and mediate remodeling and tissue repair processes. In contrast to human platelets, mouse platelet thrombin signaling is mediated via PAR3 and PAR4 [49], and therefore phenotypes of PAR1-deficient animals reflect only on PAR1 signaling in extravascular and vascular cells other than platelets. In addition to their function in thrombosis and hemostasis, the coagulation proteases and their cellular receptors are involved in a myriad of vascular signaling processes that ensure the maintenance of vascular development, endothelial function, and vascular tone [50].

The initiation complex of coagulation that is formed by the membrane receptor TF and its ligands, the coagulation proteases Factor VIIa and Factor Xa, elicits the proteolytic activation of PAR2, whereas Factor Xa also activates PAR1 [51]. The seven-transmembrane GPCRs PAR1 and PAR2 serve partially redundant functions as vascular detectors of coagulation activation [52]. In the vasculature, expression of the thrombin receptors PAR1 and PAR2 was detected on human endothelial cells, smooth muscle cells, and macrophages [43, 53–56]. PAR1 is required for developmental angiogenesis [57–59]. In contrast, PAR2 signaling regulates postnatal and pathological angiogenesis [60, 61]. Furthermore, PAR1 and PAR2 are involved in microvascular remodeling [62, 63]. Dysregulation of PAR-mediated coagulation factor signaling under pathologic conditions results in increased angiogenesis (e.g. in malignancy of solid tumors) [64] or an augmented innate immune response (e.g. during the pathogenesis of sepsis or viral infections) [65, 66]. Furthermore, activation of PAR1 and PAR2 signaling lowers vascular tone via activation of the nitric oxide (NO)-synthase pathway [67]. The regulation of vascular tone [43] and permeability [68, 69] is predominantly caused by PAR signaling in endothelial cells.

The PARs are activated via limited proteolysis by serine proteases close to the N-terminus of the receptor [48] (Fig. 10.3). The newly formed N-terminus functions as a tethered peptide agonist that binds intramolecularly to the seven-transmembrane helix bundle of the receptor, and thus affects G-protein activation. Transactivation

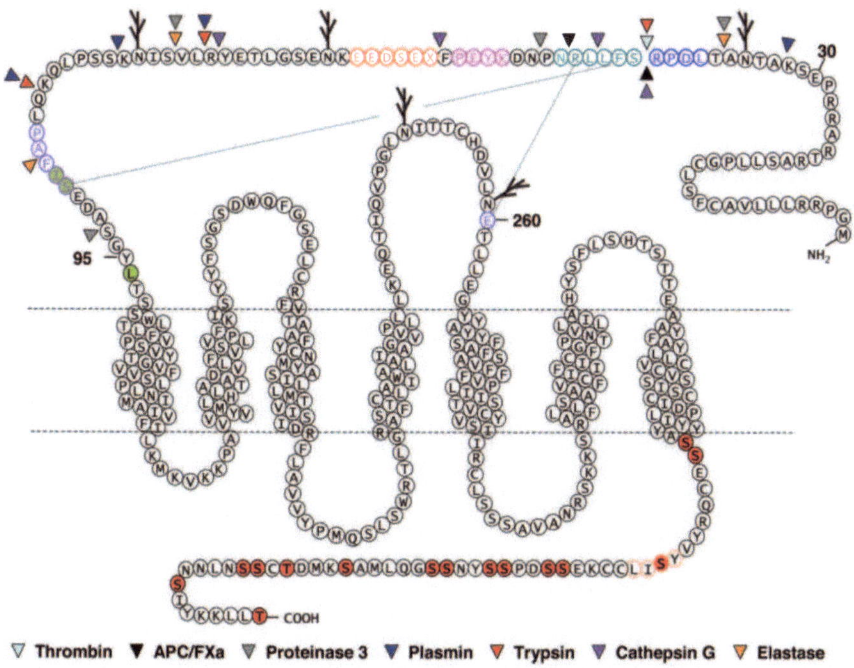

Fig. 10.3 Cleavage sites of serine proteases at the N-terminus of protease-activated receptor (*PAR*)-1

of another PAR has also been documented for PAR1 or PAR3 which have tethered ligands capable of activating PAR2 in heterodimeric complexes [70–72].

The G-protein-coupled PAR downstream signaling can be exemplified with PAR1, which couples to members of the $G_{12/13}$, G_q, and G_i families [73–75]. The α-subunits of G_{12} and G_{13} bind Rho guanine-nucleotide exchange factors that activate small G-proteins and induce Rho-dependent cytoskeletal responses involved in the platelet shape change [76], the regulation of permeability [77], and migration of endothelial cells [78]. $G\alpha_q$ activates phospholipase Cβ and triggers the hydrolysis of phosphoinositides, resulting in calcium mobilization and activation of protein kinase C. This activates Ca^{2+}-regulated kinases, phosphatases, guanine-nucleotide exchange factors, and mitogen-activated protein kinases (MAPKs), resulting in cellular responses that range from granule secretion to integrin activation, aggregation and transcriptional responses in endothelial and mesenchymal cells [79]. $G\alpha_i$ inhibits adenylate cyclase and thus promotes platelet responses. $G\beta_\gamma$ subunits can activate phosphoinositide 3-kinase (PI3K) and other lipid-modifying enzymes, protein kinases, and ion channels [80, 81]. Thus, PAR1 activation results in pleiotropic effects and the diverse actions which thrombin exerts on different cell types [49].

Several serine proteases can activate PAR1 at the same cleavage site (i.e. the Arg^{41}–Ser^{42} bond) (Table 10.1). For instance, the primary protease that activates PAR1 is thrombin, but the receptor is also activated by FXa and the anticoagulant serine protease APC [51]. However, APC has effects opposite to thrombin in barrier protection, regulation of endothelial inflammation, and apoptosis [82, 83]. These distinct signaling properties have recently been shown to be caused by biased agonism, since APC, but not thrombin, can also cleave PAR1 at Arg46, generating an alternative tethered ligand sequence [82, 84]. Thrombin cleavage of PAR1 results in proinflammatory effects, whereas the N-terminus of PAR1 that is generated by APC cleavage at Arg46 acts as a biased agonist for cytoprotective effects [84]. Matrix metalloproteinases (MMPs) also activate PAR1 at a distinct cleavage site [85, 86], but it is unclear whether this produces a biased agonist response, as shown for APC cleavage.

Table 10.1 The activating proteases of protease-activated receptors (PARs)

Activating proteases	PAR1	PAR2	PAR3	PAR4
Coagulation and fibrinolysis system	Thrombin Plasmin FVIIa, FXa Activated protein C	FVIIa, FXa Activated protein C	Thrombin	Thrombin plasmin
Others	Trypsin Tryptase Cathepsin G Granzyme A Matrix metalloproteinase-1 Gingipain-R	Trypsin I, IV Mast cell Tryptase Matriptase Epitheliasin Proteinase 3 Acrosin Kallikrein 5, 6, 14		Trypsin Cathepsin G Gingipain-R

Table 10.2 The inactivating proteases of protease-activated receptors (PARs)

Inactivating proteases	PAR1	PAR2	PAR3	PAR4
Coagulation and fibri-nolysis system	Plasmin	Plasmin		
Others	Trypsin Cathepsin G Elastase Chymotrypsin	Plasmin Cathepsin G Proteinase 3 Elastase	Cathepsin G Elastase	

Several proteases activate a number of PARs; for example, APC can activate both PAR1, PAR2 [87], and PAR3 [71]. PAR cleavage does not necessarily lead to receptor activation if cleavage occurs at a site that amputates the tethered ligand sequence (e.g. Cathepsin G cleavage of PAR1). The proteases that either activate or inactivate PARs are listed in Table 10.1 and Table 10.2, respectively [50, 88].

10.3 Coagulation Factor and PAR1 Deficiency in Genetic Mouse Models Results in Arrested Vascular Development

The central role of the initiation of coagulation by TF and of thrombin generation for embryonic development of a functional vasculature has been demonstrated with mice that are deficient in coagulation factors and with mouse models that lack PARs. TF plays an indispensable role in establishing and maintaining vascular integrity in the embryo at days 9.5–10.5 when embryonic and extraembryonic vasculatures initially form and fuse (Fig. 10.4a). Targeted disruption of the murine TF gene results in defective yolk sac vessels and embryos, with severe growth and development retardation at day E10.5 [89, 90]. Approximately 85 % of the $TF^{-/-}$ embryos die before E11.5, and embryos that manage to escape this developmental bottleneck do not survive gestation [90]. Mouse embryogenesis critically depends on the TF extracellular domain but is independent of its cytoplasmic domain [91]. Day E9.5 $TF^{-/-}$ embryos showed extremely pale yolk sacs, highly enlarged pericardial sacs, and poorly developed forebrain structures. Embryonic red blood cells are not retained in the yolk sac vessels in $TF^{-/-}$ embryos, suggesting that TF is required to maintain vascular integrity [90, 92]. At E10.5, $TF^{-/-}$ embryos were significantly smaller than their $TF^{+/-}$ and $TF^{+/+}$ littermates [92]. Interestingly, plasma clotting time and bleeding time of $TF^{+/-}$ mice was similar to $TF^{+/+}$ mice, suggesting that half-normal amounts of TF are sufficient for both hemostatic function and vascular development [91]. The TF cytoplasmic domain contains a conserved protein kinase C phosphorylation site [93] that is phosphorylated by p38 (X-S*/T*-P-X) [94]. However, deficiency of the TF cytoplasmic domain does not result in abnormal embryonic development [95]. This further supports the notion that the cofactor function of the extracellular domain is critical for vascular development.

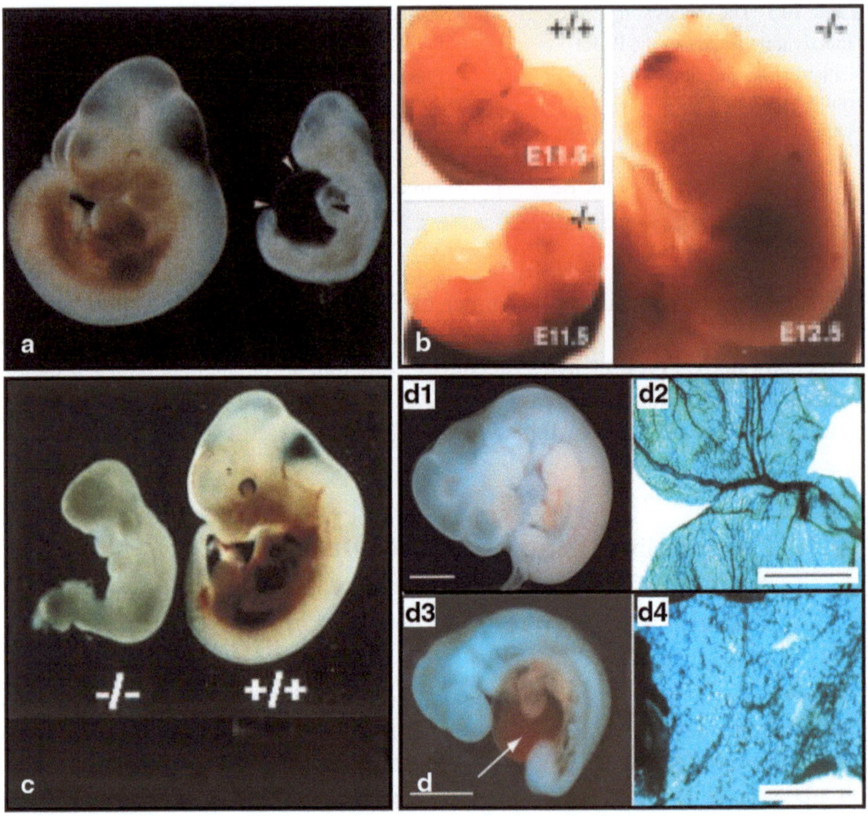

Fig. 10.4 Similar vascular phenotypes of tissue factor (*TF*), FX, thrombin, and protease-activated receptor (PAR)-1-deficient mouse embryos. Vascular development in mice with defective thrombin signaling is perturbed between E9.5 and E12.5, leading to vascular leakage, growth retardation, and embryonic lethality. **a** Day E10.5 $TF^{+/+}$ (*left*) and $TF^{-/-}$ (*right*) embryos dissected from their yolk sacs. An obvious developmental arrest was noted with pale appearance, massively enlarged pericardial sac (*white arrow heads*) and distended heart (*black arrow head*) of the mutant embryo ([92]). **b** Embryos deficient in FX show normal gross morphology compared with wild-type controls at E11.5 and E12.5 but indicate the occurrence of bleeding ([96]). **c** E12.5 embryos deficient in prothrombin (FII) showed pale appearance and developmental arrest. **d** $PAR1^{-/-}$ ($F2r^{-/-}$) embryos at E10.5 (*d3*) show developmental delay and dilated, blood-filled pericardial cavity compared with $PAR1^{+/-}$ ($F2r^{+/-}$) littermate embryos (*d1*). β-Galactosidase-stained yolk sacs of $PAR1^{+/-}$ embryos (*d2*) and $PAR1^{-/-}$ embryos (*d4*) carrying a TIE2p/e-*LacZ* transgene that allows visualization of the vascular endothelium. Scale bars, 1 mm

Similar to $TF^{-/-}$ mice, mice that are deficient in coagulation factor X suffer partial embryonic lethality between E11.5 and E12.5 and fatal neonatal bleeding [96] (Fig. 10.4b). At day E12.5, the percentage of non-resorbed $FX^{-/-}$ embryos was only 17%. The majority of factor $X^{-/-}$ neonates die before postnatal day 5, and show intraabdominal, subcutaneous, or massive intracranial bleeding. In contrast to $TF^{-/-}$ mice, embryos deficient in FVII develop normally but succumb to early intraabdominal or intracranial bleeding later in life [97]. However, survival of FVII-deficient embryos has been shown to be due to materno-fetal transfer of FVII [98].

Fifty percent of factor V-deficient embryos lacking the cofactor that is required to form the prothrombinase complex die at mid-gestation with bleeding and vascular abnormalities in the yolk sac [99]. Those $FV^{-/-}$ embryos that survive the developmental arrest suffer fatal postnatal bleeding. Prothrombin-deficient mice ($FII^{-/-}$) also show partial embryonic lethality, with more than one-half of the $FII^{-/-}$ embryos dying between E9.5 and E11.5 [100] (Fig. 10.4c). Again, bleeding into the yolk sac cavity was observed. One-quarter of the $FII^{-/-}$ mice survived to term but died within a few days after birth due to hemorrhage. Since fibrinogen-deficient embryos [101], as well as NF-E2-deficient embryos with very low platelet counts [102], do not show signs of abnormal development, it is unlikely that defective clot formation is responsible for the impaired vascular development observed in TF, FX, FV, and prothrombin-deficient mice. The similarity of embryonic phenotypes of FV, FX, and prothrombin-deficient mice suggests a pivotal role of thrombin generation and formation of the prothrombinase complex (FVa/FXa/prothrombin) in embryonic vascular development.

Only $PAR1^{-/-}$ mice ($F2r^{-/-}$ thrombin-receptor-deficient mice) show vascular defects resulting in embryonic death, whereas no abnormalities in vascular development were observed in $PAR2^{-/-}$, $PAR3^{-/-}$, or $PAR4^{-/-}$ mice [50]. Until gestational day E8, $PAR1^{-/-}$ mice develop normally but their growth is retarded at E9 compared with $PAR1^{+/-}$ control mice [103]. At E9.5, gross bleeding (22 %) and microscopic bleeding (66 %) becomes apparent, and half of the $PAR1^{-/-}$ embryos die due to vascular defects (Fig. 10.4d) and half survive with no apparent defects in the vasculature [56]. Similar to the coagulation factor knockouts, day E9.5 $PAR1^{-/-}$ embryos with development abnormalities have yolk sacs that lack blood-filled vessels (Fig. 10.4d). A dilated pericardial sac is observed in embryos with bleeding, a sign of cardiovascular failure. Pericardial bleeding becomes prominent at E10.5, and 52 % of $PAR1^{-/-}$ embryos die by day E12.5 [57]. Endothelium-specific expression of PAR1 under control of the TIE2 promoter prevented embryonic death of PAR1-deficient mice. This suggests that PAR1 signaling in endothelial cells is crucial for normal vascular development [57].

The similar vascular phenotypes of TF, FX, FV, prothrombin, and PAR1-deficient mice indicate that thrombin-mediated signaling via endothelial PAR1 is indispensable for embryonic vascular development at mid-gestation [57]. The central role of thrombin-mediated PAR1 signaling in embryonic blood vessel development is further supported by the finding that combined deficiency of PAR4, required for platelet activation, and fibrinogen recapitulates the hemostatic defect of prothrombin-deficient mice but not the early embryonic lethality that also characterizes prothrombin-deficient mice [104]. In addition to vascular development, partially redundant signaling of PAR1 and PAR2 is required for neural tube closure in the mouse embryo which is mediated by a local network of membrane-tethered proteases [105].

10.4 Regulatory Mechanisms of Coagulation Factor Signaling

Binding of FVII to TF, the allosteric catalytic activation of FVIIa, and negatively charged phosphatidylserine supporting binding of γ-carboxyglutamyl residues in the Gla domains of substrates provide effective mechanisms for localizing coagulation initiation to membrane surfaces. The procoagulant function of TF is tightly regulated by the TF pathway inhibitor (TFPI) [106], the physiologic inhibitor of the initiator complex (TF/VIIa/Xa). The Kunitz domain 2 of TFPI forms a complex with FXa, and Kunitz domain 1 with FVIIa, and TFPI thus inhibits coagulation initiation as well as TF-induced signaling via PARs [107]. However, the Kunitz-type inhibitor TFPI is prone to degradation by leukocyte serine proteases [108]. It can be degraded by neutrophil elastase and cathepsin G in neutrophil extracellular traps (NETs) of platelet-neutrophil conjugates [4].

TF function is also regulated by post-translational modifications of the receptor [109]. TF has two disulfide bridges at positions Cys49–Cys57 and Cys186–Cys209. The Cys186–Cys209 disulfide is required for function [110]. It lies at the end of the fibronectin type III-like domain 2 and links adjacent strands of an antiparallel β sheet [21]. This disulfide bond is exposed to solvent and is prone to reduction [21]. TF with a reduced Cys186–Cys209 disulfide has very low procoagulant activity but can be converted to the functional receptor by formation of the C186–C209 disulfide bond by oxidoreductase-mediated dithiol-disulfide exchange reactions [111–113]. The redox state of the TF Cys186–Cys209 disulfide determines the signaling specificity of the binary complex (TF/VIIa) [111]. Whereas signaling of the binary complex (TF/VIIa) can occur with a broken TF Cys186-Cys209 disulfide bond, ternary complex signaling (TF/VIIa/Xa) requires this disulfide [111].

TF interacts with α3β1 and α6β1 integrins [114], and integrin ligation of TF has been shown to support TF/VIIa proteolytic signaling through PAR2 [115]. In contrast to full-length TF, the alternatively spliced variant of TF regulates angiogenesis independent of PAR2 and FVIIa by direct integrin ligation of αvβ3 or α6β1 [116]. Integrin function is also regulated by the TF cytoplasmic domain. PAR2 activation by a PAR2 agonist and TF signaling complexes leads to phosphorylation of the cytoplasmic domain of human TF at position Ser 253 and Ser 258 [116]. The phosphorylation of Ser 253 requires PKCα [117], and p38 phosphorylates Ser 258 [94], in a process that is regulated by thioester modification of TF at the intracellular Cys 245 residue [118]. TF cytoplasmic domain signaling regulates integrin function in cell migration. It can exert negative effects as well as TF/FVIIa and phosphorylation-mediated positive effects [114] in postnatal angiogenesis [62, 63].

10.5 PAR Signaling Regulates Expression of Angiogenic Factors and Vascular Function

The (patho)physiologic responses of coagulation factor-activated PAR signaling in the vasculature are mediated via changes in the expression profile of a myriad of autocrine and paracrine factors in a cell-type-specific manner.

10.5.1 Cellular Effects of PAR1 Signaling on Angiogenesis

Endothelial Cells Thrombin induces PAR1-dependent angiogenesis in various angiogenesis models (e.g. chick chorioallantoic membrane assay, matrigel plug assay) [119, 120]. In endothelial cells [121, 122], thrombin exerts mitogenic and antiapoptotic effects via PAR1-induced release of heparin-binding epidermal growth factor (HB-EGF) [123]. Thrombin treatment is accompanied by the induction of vascular endothelial growth factor (VEGF) and angiopoietin-2 (Ang2) transcripts [119], and by potentiation of VEGF activity on endothelial cells by upregulation of its receptor VEGFR-2 [124]. Thrombin stimulation upregulates growth-regulated oncogene-α in endothelial cells, and this chemokine can induce the expression of VEGF, Ang2, and VEGFR-2 [125].

In human late endothelial progenitors, PAR1 activation affects angiogenesis by upregulating the SDF-1/CXCR4 system [126]. In addition to its effects on cells of the vascular wall, thrombin also regulates the expression of proangiogenic cytokines via activation of PAR1 in mononuclear cells [127]. Moreover, PAR1-independent effects of thrombin on cytoprotection are mediated by $\alpha_5\beta_3$ and $\alpha_5\beta_1$ integrins [123]. Thrombin stimulation upregulates $\alpha_v\beta_3$ expression in human umbilical vein endothelial cells (HUVECs), and thus contributes to endothelial cell adhesion, migration, and survival [128].

Vascular Smooth Muscle Cells PAR1 activation by thrombin increases the release of VEGF in human vascular smooth muscle cells. This effect is dependent on the intracellular Ca^{2+} concentration and on extracellular signal-regulated kinase (ERK1/2) signaling [129]. Hypoxia-inducible factor (HIF)-1α, a regulator of VEGF induction [130], can be upregulated under nonhypoxic conditions by a thrombin-dependent mechanism involving reactive oxygen species production in endothelial cells and smooth muscle cells [131, 132].

10.5.2 Cellular Effects of PAR2 Signaling on Angiogenesis

Endothelial Cells Angiogenesis is supported by PAR2 signaling in several angiogenesis models (hindlimb ischemia, mouse retina model, matrigel assay) [61, 133, 134]. Incubation of HUVECs with a PAR2 agonist peptide elicits a dose- and time-dependent mitogenic response [135]. Small interfering RNA (siRNA) silencing of TF in human microvascular endothelial cells inhibits the formation of stable, tube-like structures, whereas overexpression of TF, or stimulation with a PAR2 agonist, rescue the loss of TF in a matrigel assay [134]. In endothelial cells, TF can regulate PAR-mediated microvessel formation by induction of chemokine ligand 2 (CCL2 or monocyte chemotactic protein-1 [MCP-1]) [136]. Of note, CCL2 is typically induced by PAR1 signaling [34]. In this *in vitro* assay, CCL2 mediates the angiogenic effect of TF by recruiting smooth muscle cells toward endothelial cells, and facilitates the maturation of newly formed microvessels. In the retinal vasculature, the proangiogenic properties of PAR2 depend on TNF-α and subsequent induction of the receptor tyrosine kinase TIE2 via the MEK/ERK pathway [137]. Furthermore, in FVIIa-stimulated HUVECs overexpressing TF and PAR2, the transcription factor cyclic AMP responsive binding protein (CREB) is phosphorylated and is thus activated [138].

10.5.3 Cellular Effects of PAR Signaling on Vascular Function

PAR1 Signaling In addition to angiogenesis and vascular remodeling, PAR signaling influences vascular function through effects on both endothelium and vascular smooth muscle cells. Treatment of confluent HUVEC monolayers with thrombin leads to a change in the shape of the endothelial cells, resulting in gaps in the monolayer and exposure of the subendothelial matrix [139]. TFLLRNPNDK, a PAR1-selective activating peptide, caused hypotension and heart rate decreases in wild-type mice. These effects were absent in $PAR1^{-/-}$ mice [59]. The stimulation of PAR1 induces endothelium-dependent relaxation in human coronary artery rings, but the endothelium-dependent relaxation was attenuated with the severity of atherosclerotic lesions [140]. PAR1 activation changes endothelial expression of endothelin-converting enzyme (ECE) [141]. Furthermore, thrombin may induce NO production in endothelial cells [142]; however, long-term treatment of endothelial cells with thrombin downregulates endothelial NO synthase (eNOS) expression [141, 143].

PAR2 Signaling PAR2 activation by the selective activating peptides SLIGKV-NH_2 or SLIGRL cause arterial and venous dilation, leading to hypotension in humans and mice [59, 67, 144]. The SLIGRL-induced response was absent in PAR2-deficient mice, and these effects are reduced by inhibition of NO or prostaglandin synthesis.

10.6 Roles of PARs in Physiologic and Pathologic Processes

10.6.1 PAR Signaling in Wound Healing and Remodeling

The vascular phenotypes of PAR1-deficient mice clearly demonstrate an important role of coagulation factor-mediated PAR1 signaling in developmental angiogenesis [145]. In addition to developmental angiogenesis, PAR1-dependent signaling is also relevant in postnatal vascular remodeling of the small intestinal mucosa where epithelial coagulation factor signaling and PAR1 expression is triggered by the gut microbiota and upregulates expression of angiopoietin-1 (Ang1) [63]. Furthermore, activation of PAR signaling improves wound healing in the skin [146–149], and a proangiogenic function of APC that promotes cutaneous wound healing has been demonstrated [150]. Upon myocardial infarction, stimulation of PAR2 signaling alters the endothelial cell phenotype and induces expression of angiogenic chemokines [151]. Interestingly, PAR2-deficient mice were protected from postinfarction remodeling and showed less impairment of heart function [152]. PARs also contribute to fibrotic processes in a number of organs [153–155]. In cardiac fibroblasts, PAR1 signaling leads to transactivation of the EGF receptor (EGFR) pathway, which may contribute to cardiac remodeling [156].

10.6.2 PAR Signaling in Malignancy

A prothrombotic state and cancer-associated thrombosis is frequently observed in cancer patients [157], and TF procoagulant activity is crucial for metastatic tumor dissemination [158, 159]. Coagulation factor signaling via PARs promotes neovascularization and tumor cell migration, and thus contributes to growth and metastasis of many solid tumors [160–162].

Expression and activation of PAR1 has been associated with tumor progression in prostate cancer, breast cancer, lung cancer, gastric cancer, colon cancer, melanoma, and ovarian cancer [163–170]. Breast carcinoma cell invasion is promoted by PAR1 activation, leading to persistent transactivation of EGFR and ErbB2/HER2 [171]. Inhibition of PAR1 signaling results in inhibition of Akt survival pathways in breast cancer cells, and thus suppresses tumor survival and metastasis [172]. Microarray studies have been performed to assess the transcriptional program induced via PAR1- and PAR2-dependent coagulation factor signaling [173, 174]. In this context, PAR1 mediates the induction of IL-8 and VEGF in prostate cancer cells [175]. PAR2 activation regulates VEGF and IL-8 expression in the breast cancer cell line MDA-MB-231 cells via MAPK pathways [176, 177]. Both the TF/VIIa/Xa complex and PAR2 are essential for FVIIa- and FXa-induced signaling, migration, and invasion of breast cancer cells [178, 179]. Formation of the TF/VIIa/Xa complex induces activation of the mammalian target of rapamycin (mTOR) pathway that regulates migration of human breast cancer cells [180]. In breast cancer development, cooperation of TF cytoplasmic domain and PAR2 signaling occurs [181], and TF phosphorylation correlates with PAR2 expression and is associated with relapse, indicating prognostic significance [182].

10.6.3 PAR Signaling in Innate Immunity

Recent studies have demonstrated that PAR signaling plays a role in the regulation of the innate immune response. In a mouse model of LPS-induced sterile infection, PAR1-mediated coagulation factor signaling amplifies inflammation and lethality through sphingosine-1-phosphate downstream signaling in dendritic cells [65]. It has recently been shown that PAR1 contributes to the innate immune response during viral infection [183]. Furthermore, it has been demonstrated that TF-dependent PAR2 signaling on dendritic cells suppresses antigen-specific $CD4^+$ T-cell priming [184]. In gingival epithelial cells, proteases of *Porphyromonas gingivalis* induce human β-defensin-2 expression via epithelial PAR2 activation [185]. MyD88-dependent, PAR2 agonist peptide-induced nuclear factor-kappa B (NFκB) signaling is augmented by agonist peptide-dependent physical interaction of toll-like receptor 4 (TLR4) with PAR2 [186]. Interestingly, in high-fat diet-induced obesity the genetic ablation of TF-dependent PAR2 signaling reduced adipose tissue macrophage inflammation, and inhibition of macrophage TF signaling ameliorated insulin resistance [187]. This indicates a role for coagulation factor signaling in metabolic syndrome development.

10.7 Perspective

While the essential role of TF as a coagulation initiator is firmly established, the coagulation cascade plays important additional roles in embryonic and postnatal angiogenesis through the activation of PAR. More recent data expand this view and document much broader roles of coagulation protease signaling in physiological and pathological processes in the context of vascular diseases, wound healing, tissue remodeling, and cancer. The role of coagulation factor-mediated PAR signaling in the regulation of innate and adaptive immunity under various conditions is yet to be defined. In future, these areas of investigation will be aided by an increasing number of excellent molecular tools and genetic animal models.

References

1. Furie B, Furie BC. Mechanisms of thrombus formation. N Engl J Med. 2008;359:938–49.
2. Broze GJ Jr. Tissue factor pathway inhibitor. Thromb Haemost. 1995;74:90–3.

3. Esmon CT, Esmon NL. The link between vascular features and thrombosis. Annu Rev Physiol. 2011;73:503–14.
4. Massberg S, Grahl L, von Bruehl ML, Manukyan D, et al. Reciprocal coupling of coagulation and innate immunity via neutrophil serine proteases. Nat Med. 2010;16:887–96.
5. Opal SM, Esmon C. Bench-to-bedside review: functional relationships between coagulation and the innate immune response and their respective roles in the pathogenesis of sepsis. Crit Care. 2003;7:22–8.
6. Østerud B, Bjørklid E. Sources of tissue factor. Semin Thromb Hemost. 2006;32:11–23.
7. Ruggeri ZM. The role of von Willebrand factor in thrombus formation. Thromb Res. 2007;120:S5–9.
8. Stegner D, Nieswandt B. Platelet receptor signaling in thrombus formation. J Mol Med. 2011;89:109–21.
9. Zwaal RF, Comfurius P, Bevers EM. Mechanisms and function of changes in membrane-phospholipid asymmetry in platelets and erythrocytes. Biochem Soc Trans. 1993;21:248–53.
10. Renné T. The procoagulant and proinflammatory plasma contact system. Semin Immunopathol. 2012;34:31–41.
11. Colman RW, Marder VJ, Clowes AW, George JN, Goldhaber SZ. Hemostasis and thrombosis: basic principles and clinical practice. 5th ed. Philadelphia: Lippincott Williams & Wilkins; 2006.
12. Kannemeier C, Shibamiya A, Nakazawa F, Trusheim H, Ruppert C, Markart P, Song Y, Tzima E, Kennerknecht E, Niepmann M, von Bruehl ML, Sedding D, Massberg S, Günther A, Engelmann B, Preissner KT. Extracellular RNA constitutes a natural procoagulant cofactor in blood coagulation. Proc Natl Acad Sci U S A. 2007;104:6388–93.
13. Müller F, Mutch NJ, Schenk WA, Smith SA, Esterl L, Spronk HM, Schmidbauer S, Gahl WA, Morrissey JH, Renné T. Platelet polyphosphates are proinflammatory and procoagulant mediators in vivo. Cell. 2009;139:1143–56.
14. Matafonov A, Sarilla S, Sun MF, Sheehan JP, Serebrov V, Verhamme IM, Gailani D. Activation of factor XI by products of prothrombin activation. Blood. 2011;118:437–45.
15. Fleck RA, Rao LV, Rapaport SI, Varki N. Lokalization of human tissue factor antigen by immunostaining with monospecific, polyclonal anti-human tissue factor antibody. Thromb Res. 1990;59:421–37.
16. Drake TA, Morrissey JH, Edgington TS. Selective cellular expression of tissue factor in human tissues. Implications for disorders of hemostasis and thrombosis. Am J Pathol. 1989;134:1087–97.
17. Giesen PL, Rauch U, Bohrmann B, Kling D, Roqué M, Fallon JT, Badimon JJ, Himber J, Riederer MA, Nemerson Y. Blood-borne tissue factor: another view of thrombosis. Proc Natl Acad Sci U S A. 1999;96:2311–5.
18. Reinhardt C. New locations of intravascular tissue factor: indications. Hämostaseologie. 2007;27:55–8.
19. Müller I, Klocke A, Alex M, Kotzsch M, Luther T, Morgenstern E, Zieseniss S, Zahler S, Preissner K, Engelmann B. Intravascular tissue factor initiates coagulation via circulating microvesicles and platelets. FASEB J. 2003;17:476–8.
20. Bazan JF. Structural design and molecular evolution of a cytokine receptor superfamily. Proc Natl Acad Sci U S A. 1990;87:6934–8.
21. Harlos K, Martin DM, O'Brien DP, Jones EY, Stuart DI, Polikarpov I, Miller A, Tuddenham EG, Boys CW. Crystal structure of the extracellular region of human tissue factor. Nature. 1994;370:662–6.
22. Ruf W, Edgington TS. Two sites in the tissue factor extracellular domain mediate the recognition of the ligand factor VIIa. Proc Natl Acad Sci U S A. 1991;88:8430–4.
23. Ruf W, Schullek JR, Stone MJ, Edgington TS. Mutational mapping of functional residues in tissue factor: identification of factor VII recognition determinants in both structural modules of the predicted cytokine receptor homology domain. Biochemistry. 1994;33:1565–72.
24. Nemerson Y, Esnouf MP. Activation of a proteolytic system by a membrane lipoprotein: mechanism of action of tissue factor. Proc Natl Acad Sci U S A. 1973;70:310–4.

25. Østerud B, Rapaport SI. Activation of factor IX by the reaction product of tissue factor and factor VII: additional pathway for initiating blood coagulation. Proc Natl Acad Sci U S A. 1977;74:5260–4.
26. Jesty J, Spencer AK, Nakashima Y, Nemerson Y, Konigsberg W. The activation of coagulation factor X. Identity of cleavage sites in the alternative activation pathways and characterization of the COOH-terminal peptide. J Biol Chem. 1975;250:4497–504.
27. Mann KG, Jenny RJ, Krishnaswamy S. Cofactor proteins in the assembly and expression of blood clotting enzyme complexes. Annu Rev Biochem. 1988;57:915–56.
28. Bode W. The structure of thrombin, a chameleon-like proteinase. J Thromb Haemost. 2005;3:2379–88.
29. Griffin JH. Blood coagulation. The thrombin paradox. Nature. 1995;378:337–8.
30. Esmon CT. The regulation of natural anticoagulant pathways. Science. 1987;235:1348–52.
31. Stearns-Kurosawa DJ, Kurosawa S, Mollica JS, Ferrell GL, Esmon CT. The endothelial cell protein C receptor augments protein C activation by the thrombin-thrombomodulin complex. Proc Natl Acad Sci U S A. 1996;93:10212–6.
32. Esmon CT. The protein C pathway. Crit Care Med. 2000;28:S44–8.
33. Van Hinsbergh VW, Bertina RM, van Wijngaarden A, van Tilburg NH, Emeis JJ, Haverkate F. Activated protein C decreases plasminogen activator-inhibitor activity in endothelial cell-conditioned medium. Blood. 1985;65:444–51.
34. Riewald M, Petrovan RJ, Donner A, Mueller BM, Ruf W. Activation of endothelial cell protease activated receptor 1 by the protein C pathway. Science. 2002;296:1880–2.
35. Grey ST, Tsuchida A, Hau H, Orthner CL, Salem HH, Hancock WW. Selective inhibitory effects of the anticoagulant activated protein C on the responses of human mononuclear phagocytes to LPS, IFN-gamma, or phorbol ester. J Immunol. 1994;153:3664–72.
36. Mosnier LO, Zlokovic BV, Griffin JH. The cytoprotective protein C pathway. Blood. 2007;109:3161–72.
37. Sambrano GR, Weiss EJ, Zheng YW, Huang W, Coughlin SR. Role of thrombin signalling in platelets in haemostasis and thrombosis. Nature. 2001;413:74–8.
38. Vu TK, Hung DT, Wheaton VI, Coughlin SR. Molecular cloning of a functional thrombin receptor reveals a novel proteolytic mechanism of receptor activation. Cell. 1991;64:1057–68.
39. Nystedt S, Emilsson K, Wahlestedt C, Sundelin J. Molecular cloning of a potential proteinase activated receptor. Proc Natl Acad Sci U S A. 1994;91:9208–12.
40. Ishihara H, Connolly AJ, Zeng D, Kahn ML, Zheng YW, Timmons C, Tram T, Coughlin SR. Protease-activated receptor 3 is a second thrombin receptor in humans. Nature. 1997;386:502-6.
41. Kahn ML, Zheng YW, Huang W, Bigornia V, Zeng D, Moff S, Farese RV Jr, Tam C, Coughlin SR. A dual thrombin receptor system for platelet activation. Nature. 1998;394:690–4.
42. Kong W, McConaloque K, Khitin LM, Hollenberg MD, Payan DG, Böhm SK, Bunnett NW. Luminal trypsin may regulate enterocytes through proteinase-activated receptor 2. Proc Natl Acad Sci U S A. 1997;94:8884–9.
43. Nelken NA, Soifer SJ, O'Keefe J, Vu TK, Charo IF, Coughlin SR. Thrombin receptor expression in normal and atherosclerotic human arteries. J Clin Invest. 1992;90:1614–21.
44. Schechter NM, Brass LF, Lavker RM, Jensen PJ. Reaction of mast cell proteases tryptase and chymase with protease activated receptors (PARs) on keratinocytes and fibroblasts. J Cell Physiol. 1998;176:365–73.
45. D'Andrea MR, Rogahan CJ, Andrade-Gordon P. Localization of protease-activated receptors-1 and -2 in human mast cells: indications for an amplified mast cell degranulation cascade. Biotech Histochem. 2000;75:85–90.
46. Bizios R, Lai L, Fenton JW 2nd, Malik AB. Thrombin-induced chemotaxis and aggregation of neutrophils. J Cell Physiol. 1986;128:485–90.
47. Colognato R, Slupsky JR, Jendrach M, Burysek L, Syrovets T, Simmet T. Differential expression and regulation of protease-activated receptors in human peripheral monocytes and monocyte-derived antigen-presenting cells. Blood. 2003;102:2645–52.

48. D'Andrea MR, Derian CK, Santulli RJ, Andrade-Gordon P. Differential expression of protease-activated receptors-1 and -2 in stromal fibroblasts of normal, benign, and malignant human tissues. Am J Pathol. 2001;158:2031–41.
49. Coughlin SR. Thrombin signaling and protease-activated receptors. Nature. 2000;407:258–64.
50. Hirano K. The roles of proteinase-activated receptors in the vascular physiology and pathophysiology. Arterioscler Thromb Vasc Biol. 2007;27:27–36.
51. Steinberg SF. The cardiovascular actions of protease-activated receptors. Mol Pharmacol. 2005;67:2–11.
52. Pawlinski R, Mackman N. Tissue factor, coagulation proteases, and protease-activated receptors in endotoxemia and sepsis. Crit Care Med. 2004;32:S293–7.
53. Molino M, Raghunath PN, Kuo A, Ahuya M, Hoxie JA, Brass LF, Barnathan ES. Differential expression of functional protease-activated receptor-2 (PAR-2) in human vascular smooth muscle cells. Arterioscler Thromb Vasc Biol. 1998;18:825–32.
54. Sabri A, Short J, Guo J, Steinberg SF. Protease-activated receptor-1 mediated DNA synthesis in cardiac fibroblasts is via epidermal growth factor receptor transactivation: distinct PAR-1 signaling pathways in cardiac fibroblasts and cardiomyocytes. Circ Res. 2002;91:532–9.
55. Mirza H, Yatsula V, Bahou WF. The proteinase activated receptor-2 (PAR-2) mediates mitogenic responses in human vascular endothelial cells. J Clin Invest. 1996;97:1705–14.
56. Colognato R, Slupsky JR, Jendrach M, Burysek L, Syrovets T, Simmet T. Differential expression and regulation of the protease-activated receptors in human peripheral monocytes and monocyte-derived antigen-presenting cells. Blood. 2003;102:2645–52.
57. Griffin CT, Srinivasan Y, Zheng YW, Huang W, Coughlin SR. A role for thrombin receptor signaling in endothelial cells during embryonic development. Science. 2001;293:1666–70.
58. Darrow AL, Fung-Leung WP, Ye RD, Santulli RJ, Cheung WM, Derian CK, Burns CL, Damiano BP, Zhou L, Keenan CM, Peterson PA, Andrade-Gordon P. Biological consequences of thrombin receptor deficiency in mice. Thromb Haemost. 1996;76:860–6.
59. Damiano BP, Cheung WM, Santulli RJ, Fung-Leung WP, Ngo K, Ye RD, Darrow AL, Derian CK, de Garavilla L, Andrade-Gordon P. Cardiovascular responses mediated by protease-activated receptor-2 (PAR-2) and thrombin receptor (PAR-1) are distinguished in mice deficient in PAR-2 and PAR-1. J Pharmacol Exp Ther. 1999;288:671–8.
60. Ruf W, Disse J, Carneiro-Lobo TC, Yokota N, Schaffner F. Tissue factor and cell signalling in cancer progression and thrombosis. J Thromb Haemost. 2011;9 Suppl. 1:306–15.
61. Uusitalo-Jarvinen H, Kurokawa T, Mueller BM, Andrads-Gordon P, Friedlander M, Ruf W. Role of protease activated receptor 1 and 2 signaling in hypoxia-induced angiogenesis. Arterioscler Thromb Vasc Biol. 2007;27:1456–62.
62. Belting M, Dorrell MI, Sandgren S, Aguilar E, Ahamed J, Dorfleutner A, Carmeliet P, Mueller BM, Friedlander M, Ruf W. Regulation of angiogenesis by tissue factor cytoplasmic domain signaling. Nat Med. 2004;10:502–9.
63. Reinhardt C, Bergentall M, Greiner TU, Schaffner F, Östergren-Lundén G, Petersen LC, Ruf W, Bäckhed F. Tissue factor and PAR1 promote microbiota-induced intestinal vascular remodelling. Nature. 2012;483:627–31.
64. Ruf W, Yokata N, Schaffner F. Tissue factor in cancer progression and angiogenesis. Thromb Res. 2010;125:S36–8.
65. Niessen F, Schaffner F, Furlan-Freguia C, Pawlinski R, Bhattacharjee G, Chun J, Derian CK, Andrade-Gordon P, Rosen H, Ruf W. Dendritic cell PAR-1-S1P3 signalling couples coagulation and inflammation. Nature. 2008;452:654–8.
66. Antoniak S, Owens AP 3rd, Baunacke M, Williams JC, Lee RD, Weithäuser A, Sheridan PA, Malz R, Luyendyk JP, Esserman DA, Trejo J, Kirchhofer D, Blaxall BC, Pawlinski R, Beck MA, Rauch U, Mackman N. PAR-1 contributes to the innate immune response during viral infection. J Clin Invest. 2013;123:1310–22.
67. Emilsson K, Wahlestedt C, Sun MK, Nystedt S, Owman C, Sundelin J. Vascular effects of proteinase-activated receptor 2 agonist peptide. J Vasc Res. 1997;34:267–72.

68. Camerer E, Regard JB, Cornelissen I, Srinivasan Y, Duong DN, Palmer D, Pham TH, Wong JS, Pappu R, Coughlin SR. Sphingosine-1-phosphate in the plasma compartment regulates basal and inflammation-induced vascular leak in mice. J Clin Invest. 2009;119:1871–9.
69. Kaneider NC, Leger AJ, Agarwal A, Nguyen N, Perides G, Derian C, Covic L, Kuliopulos A. 'Role reversal' for the receptor PAR1 in sepsis-induced vascular damage. Nat Immunol. 2007;8:1303–12.
70. O'Brien PJ, Prevost N, Molino M, Hollinger MK, Woolkalis MJ, Woulfe DS, Brass LF. Thrombin responses in human endothelial cells. Contributions from receptors other than PAR1 include the transactivation of PAR2 by thrombin-cleaved PAR1. J Biol Chem. 2000;275:13502–9.
71. Madhusudhan T, Wang H, Straub BK, Gröne E, Zhou Q, Shazad K, Müller-Krebs S, Schwenger V, Gerlitz B, Grinnell BW, Griffin JH, Reiser J, Gröne HJ, Esmon CT, Nawroth PP, Isermann B. Cytoprotective signaling by activated protein C requires protease-activated receptor-3 in podocytes. Blood. 2012;119:874–83.
72. Lin H, Trejo J. Transactivation of the PAR1-PAR2 heterodimer by thrombin elicits β-arrestin-mediated endosomal signaling. J Biol Chem. 2013;288:11203–15.
73. Hung DT, Wong YH, Vu TK, Coughlin SR. The cloned platelet thrombin receptor couples to at least two distinct effectors to stimulate phosphoinositide hydrolysis and inhibit adenylyl cyclase. J Biol Chem. 1992;267:20831–4.
74. Offermanns S, Laugwitz KL, Spicher K, Schultz G. G proteins of the G12 family are activated via thromboxane A2 and thrombin receptors in human platelets. Proc Natl Acad Sci U S A. 1994;91:504–8.
75. Barr AJ, Brass LF, Manning DR. Reconstitution of receptors and GTP-binding regulatory proteins (G proteins) in Sf9 cells. A direct evaluation of selectivity in receptor.G protein coupling. J Biol Chem. 1997;272:2223–9.
76. Klages B, Brandt U, Simon MI, Schultz G, Offermanns S. Activation of G12/G13 results in shape change and Rho/Rho-kinase-mediated myosin light chain phosphorylation in mouse platelets. J Cell Biol. 1999;144:745–54.
77. Vouret-Craviari V, Boquet P, Pouysségur J, Van Obberghen-Schilling E. Regulation of the actin cytoskeleton by thrombin in human endothelial cells: role of Rho proteins in endothelial barrier function. Mol Biol Cell. 1998;9:2639–53.
78. Offermanns S, Mancino V, Revel JP, Simon MI. Vascular system defects and impaired cell chemokinesis as a result of Galpha13 deficiency. Science. 1997;275:533–6.
79. Taylor SJ, Chae HZ, Rhee SG, Exton JH. Activation of the beta 1 isozyme of phospholipase C by alpha subunits of the Gq class of G proteins. Nature. 1991;350:516–8.
80. Stoyanov B, Volinia S, Hanck T, Rubio L, Loubtchenkow M, Malek D, Stoyanova S, Vanhaesebroeck B, Dhand R, Nürnberg B, et al. Cloning and characterization of a G protein-activated human phosphoinositide-3 kinase. Science. 1995;269:690–3.
81. Clapham DE, Neer EJ. G protein beta gamma subunits. Annu Rev Pharmacol Toxicol. 1997;37:167–203.
82. Schuepbach RA, Feistritzer C, Fernández JA, Griffin JH, Riewald M. Protection of vascular barrier integrity by activated protein C in murine models depends on protease-activated receptor-1. Thromb Haemost. 2009;101:724–33.
83. Riewald M, Ruf W. Protease-activated receptor-1 signaling by activated protein C in cytokine-perturbed endothelial cells is distinct from thrombin signaling. J Biol Chem. 2005;280:19808–14.
84. Mosnier LO, Sinha RK, Burnier L, Bouwens EA, Griffin JH. Biased agonism of protease-activated receptor 1 by activated protein C caused by noncanonical cleavage at Arg46. Blood. 2012;120:5237–46.
85. Boire A, Covic L, Agarwal A, Jacques S, Sherifi S, Kuliopulos A. PAR1 is a matrix metalloprotease-1 receptor that promotes invasion and tumorigenesis of breast cancer cells. Cell. 2005;120:303–13.
86. Austin KM, Covic L, Kuliopulos A. Matrix metalloproteases and PAR1 activation. Blood. 2013;121:431–9.

87. Riewald M, Petrovan RJ, Donner A, Mueller BM, Ruf W. Activation of endothelial cell protease activated receptor 1 by the protein C pathway. Science. 2002;296:1880–2.
88. Rothmeier AS, Ruf W. Protease-activated receptor 2 signaling in inflammation. Semin Immunopathol. 2012;34:133–49.
89. Carmeliet P, Mackman N, Moons L, Luther T, Gressens P, Van Vlaenderen I, Demunck H, Kasper M, Breier G, Evrard P, Müller M, Risau W, Edgington T, Collen D. Role of tissue factor in embryonic blood vessel development. Nature. 1996;383:73–5.
90. Toomey JR, Kratzer KE, Lasky NM, Stanton JJ, Broze GJ Jr. Targeted disruption of the murine tissue factor gene results in embryonic lethality. Blood. 1996;88:1583–7.
91. Parry GC, Mackman N. Mouse embryogenesis requires the tissue factor extracellular domain but not the cytoplasmic domain. J Clin Invest. 2000;105:1547–54.
92. Bugge TH, Xiao Q, Kombrinck KW, Flick Holmbäck K, Danton MJ, Colbert MC, Witte DP, Fujikawa K, Davie EW, Degen JL. Fatal embryonic bleeding events in mice lacking tissue factor, the cell-associated initiator of blood coagulation. Proc Natl Acad Sci U S A. 1996;93:6258–63.
93. Zioncheck TF, Roy S, Vehar GA. The cytoplasmic domain of tissue factor is phosphorylated by a protein kinase C-dependent mechanism. J Biol Chem. 1992;267:3561–4.
94. Ettelaie C, Elkeeb AM, Maraveyas A, Collier ME. P38α phosphorylates serine 258 within the cytoplasmic domain of tissue factor and prevents its incorporation into cell-derived microparticles. Biochim Biophys Acta. 2013;1833:613–21.
95. Melis E, Moons L, De Mol M, Herbert JM, Mackman N, Collen D, Carmeliet P, Dewerchin M. Targeted deletion of the cytosolic domain of tissue factor in mice does not affect development. Biochem Biophys Res Commun. 2001;286:580–6.
96. Dewerchin M, Liang Z, Moons L, Carmeliet P, Castellino FJ, Collen D, Rosen ED. Blood coagulation factor X deficiency causes partial embryonic lethality and fatal neonatal bleeding in mice. Thromb Haemost. 2000;83:185–90.
97. Rosen ED, Chan JC, Idusogie E, Clotman F, Vlasuk G, Luther T, Jalbert LR, Albrecht S, Zhong L, Lissens A, Schoonjans L, Moons L, Collen D, Castellino FJ, Carmeliet P. Mice lacking factor VII develop normally but suffer fatal perinatal bleeding. Nature. 1997;390:290–4.
98. Chan JCY, Carmeliet P, Moons L, Roosen ED, Hunang Z-F, Broze GJ Jr, Collen D, Castellino FJ. Factor VII deficiency rescues the intrauterine lethality in mice associated with a tissue factor pathway inhibitor deficit. J Clin Invest. 1999;103:475–82.
99. Cui J, O'Shea KS, Purkayastha A, Saunders TL, Ginsburg D. Fatal hemorrhage and incomplete block to embryogenesis in mice lacking coagulation factor V. Nature. 1996;384:66–8.
100. Sun WY, Witte DP, Degen JL, Colbert MC, Burkart MC, Holmbäck K, Xiao Q, Bugge TH, Degen SJ. Prothrombin deficiency results in embryonic and neonatal lethality in mice. Proc Natl Acad Sci U S A. 1998;95:7597–602.
101. Suh TT, Holmbäck K, Jensen NJ, Daugherty CC, Small K, Simon DI, Potter S, Degen JL. Resolution of spontaneous bleeding events but failure of pregnancy in fibrinogen-deficient mice. Genes Dev. 1995;9:2020–33.
102. Shivdasani RA, Orkin SH. Erythropoiesis and globin gene expression in mice lacking the transcription factor NF-E2. Proc Natl Acad Sci U S A. 1995;92:8690–4.
103. Connolly AJ, Ishihara H, Kahn ML, Farese RV Jr, Coughlin SR. Role of the thrombin receptor in development and evidence for a second receptor. Nature. 1996;381:516–9.
104. Camerer E, Duong DN, Hamilton JR, Coughlin SR. Combined deficiency of protease-activated receptor-4 and fibrinogen recapitulates the hemostatic defect but not the embryonic lethality of prothrombin deficiency. Blood. 2004;103:152–4.
105. Camerer E, Barker A, Duong DN, Ganesan R, Kataoka H, Cornelissen I, Darragh MR, Hussain A, Zheng Y-W, Srinivasan Y, Brown C, Xu S-M, Regard JB, Lin C-Y, Craik CS, Kirchhofer D, Coughlin SR. Local protease signaling contributes to neural tube closure in the mouse embryo. Dev Cell. 2010;18:25–38.
106. Broze GJ Jr. The rediscovery and isolation of TFPI. J Thromb Haemost. 2003;1:1671–5.

107. Ahamed J, Belting M, Ruf W. Regulation of tissue factor-induced signaling by endogenous and recombinant tissue factor pathway inhibitor 1. Blood. 2005;105:2384–91.
108. Higuchi DA, Wun TC, Likert KM, Broze GJ Jr. The effect of leukocyte elastase on tissue factor pathway inhibitor. Blood. 1992;79:1712–9.
109. Butenas S, Amblo-Krudysz J, Mann KG. Posttranslational modifications of tissue factor. Front Biosci. 2012;4:381–91.
110. Rehemtulla A, Ruf W, Edgington TS. The integrity of the cysteine 186-cysteine 209 bond of the second disulfide loop of tissue factor is required for binding of factor VII. J Biol Chem. 1991;266:10294–9.
111. Ahamed J, Versteeg HH, Kerver M, Chen VM, Mueller BM, Hogg PJ, Ruf W. Disulfide isomerization switches tissue factor from coagulation to cell signaling. Proc Natl Acad Sci U S A. 2006;103:13932–7.
112. Reinhardt C, von Brühl ML, Manukyan D, Grahl L, Lorenz M, Altmann B, Dlugai S, Hess S, Konrad I, Orschiedt L, Mackman N, Ruddock L, Massberg S, Engelmann B. Protein disulfide isomerase acts as an injury response signal that enhances fibrin generation via tissue factor activation. J Clin Invest. 2008;118:1110–22.
113. Liang HP, Brophy TM, Hogg PJ. Redox properties of the tissue factor Cys186-Cys209 disulfide bond. Biochem J. 2011;437:455–60.
114. Dorfleutner A, Hintermann E, Tarui T, Takada Y, Ruf W. Cross-talk of integrin alpha3beta1 and tissue factor in cell migration. Mol Biol Cell. 2004;15:4416–25.
115. Versteeg HH, Schaffner F, Kerver M, Petersen HH, Ahamed J, Felding-Habermann B, Takada Y, Mueller BM, Ruf W. Inhibition of tissue factor signaling suppresses tumor growth. Blood. 2008;111:190–9.
116. van den Berg YW, van den Hengel LG, Myers HR, Ayachi O, Jordanova E, Ruf W, Spek CA, Reitsma PH, Bogdanov VY, Versteeg HH. Alternatively spliced tissue factor induces angiogenesis through integrin ligation. Proc Natl Acad Sci U S A. 2009;106:19497–502.
117. Ahamed J, Ruf W. Protease-activated receptor 2-dependent phosphorylation of the tissue factor cytoplasmic domain. J Biol Chem. 2004;279:23038–44.
118. Dorfleutner A, Ruf W. Regulation of tissue factor cytoplasmic domain phosphorylation by palmitoylation. Blood. 2003;102:3998–4005.
119. Caunt M, Huang YQ, Brooks PC, Karpatkin S. Thrombin induces neoangiogenesis in the chick chorioallantoic membrane. J Thromb Haemost. 2003;1:2097–102.
120. Haralabopoulos GC, Grant DS, Kleinman HK, Maragoudakis ME. Thrombin promotes endothelial cell alignment in Matrigel in vitro and angiogenesis in vivo. Am J Physiol. 1997;273:C239–45.
121. Minami T, Sugiyama A, Wu SQ, Abid R, Kodama T, Arid WC. Thrombin and phenotypic modulation of the endothelium. Arterioscler Thromb Vasc Biol. 2004;24:41–53.
122. McLaughlin JN, Shen L, Holinstat M, Brooks JD, Dibenedetto E, Hamm HE. Functional selectivity of G protein signaling by agonist peptides and thrombin for the protease-activated receptor-1. J Biol Chem. 2005;280:25048–59.
123. Zania P, Papaconstantinou M, Flordellis CS, Maragoudakis ME, Tsopanoglou NE. Thrombin mediates mitogenesis and survival of human endothelial cells through distinct mechanisms. Am J Physiol Cell Physiol. 2008;294:C1215–26.
124. Tsopanoglou NE, Maragoudakis ME. On the mechanism of thrombin-induced angiogenesis. Potentiation of vascular endothelial growth factor activity on endothelial cells by up-regulation of its receptors. J Biol Chem. 1999;274:23969–76.
125. Caunt M, Hu L, Tang T, Brooks PC, Ibrahim S, Karpatkin S. Growth-regulated oncogene is pivotal in thrombin-induced angiogenesis. Cancer Res. 2006;66:4125–32.
126. Smedja DM, Bièche I, Uzan G, Bompais H, Muller L, Boisson-Vidal C, Vidaud M, Aiach M, Gaussem P. PAR-1 activation on human late endothelial-progenitor cells enhances angiogenesis in vitro with upregulation of the SDF-1/CXCR4 system. Arterioscler Thromb Vasc Biol. 2005;25:2321–7.
127. Naldini A, Carney DH, Pucci A, Pasquali A, Carraro F. Thrombin regulates the expression of proangiogenic cytokines via proteolytic activation of protease-activated receptor-1. Gen Pharmacol. 2000;35:255–9.

128. Tsopanoglou NE, Andriopoulou P, Maragoudakis ME. On the mechanism of thrombin-induced angiogenesis: involvement of alphavbeta3-integrin. Am J Physiol Cell Physiol. 2002;283:C1501–10.
129. Arisato T, Sarker KP, Kawahara K, Nakata M, Hashiguchi T, Osame M, Kitajima I, Maruyama I. The agonist of the protease-activated receptor-1 (PAR) but not PAR3 mimics thrombin-induced vascular endothelial growth factor release in human smooth muscle cells. Cell Mol Life Sci. 2003;60:1716–24.
130. Forsythe JA, Jiang BH, Iyer NV, Agani F, Leung SW, Koos RD, Semenza GL. Activation of vascular endothelial growth factor gene transcription by hypoxia-inducible factor. Mol Cell Biol. 1996;16:4604–13.
131. Görlach A, Diebold I, Schini-Kerth VB, Berchner-Pfannschmidt U, Roth U, Brandes RP, Kietzmann T, Busse R. Thrombin activates the hypoxia-inducible factor-1 signaling pathway in vascular smooth muscle cells: role of the p22(phox)-containing NADPH oxidase. Circ Res. 2001;89:47–54.
132. BelAiba RS, Djordjevic T, Bonello S, Flügel D, Hess J, Kietzmann T, Görlach A. Redox-sensitive regulation of the HIF pathway under non-hypoxic conditions in pulmonary artery smooth muscle cells. Biol Chem. 2004;385:249–57.
133. Milia AF, Salis MB, Stacca T, Pinna A, Madeddu P, Trevisani M, Geppetti P, Emanueli C. Protease-activated receptor-2 stimulates angiogenesis and accelerates hemodynamic recovery in a mouse model of hindlimb ischemia. Circ Res. 2002;91:346–52.
134. Arderiu G, Peña E, Aledo R, Badimon L. Tissue factor-Akt signaling triggers microvessel formation. J Thromb Haemost. 2012;10:1895–905.
135. Mirza H, Yatsula V, Bahou WF. The proteinase activated receptor-2 (PAR-2) mediates mitogenic responses in human vascular endothelial cells. J Clin Invest. 1996;97:1705–14.
136. Arderiu G, Peña E, Aledo R, Juan-Babot O, Badimon L. Tissue factor regulates microvessel formation and stabilization by induction of chemokine (C-C motif) ligand 2 expression. Arterioscler Thromb Vasc Biol. 2011;31:2607–15.
137. Zhu T, Sennlaub F, Beauchamp MH, Fan L, Joyal JS, Checchin D, Nim S, Lachapelle P, Sirinyan M, Hou X, Bossolasco M, Rivard GE, Heveker N, Chemtomb S. Proangiogenic effects of protease-activated receptor 2 are tumor necrosis factor-alpha and consecutively Tie2 dependent. Arterioscler Thromb Vasc Biol. 2006;26:744–50.
138. Versteeg HH, Borensztain KS, Kerver ME, Ruf W, Reitsma PH, Spek CA, Peppelenbosch MP. TF:VIIa-specific activation on CREB upregulates proapoptotic proteins via protease-activated receptor-2. J Thromb Haemost. 2008;6:1550–7.
139. Loposata M, Dovnarsky DK, Shin HS. Thrombin-induced gap formation in confluent endothelial cell monolayers in vitro. Blood. 1983;62:549–56.
140. Ku DD, Dai J. Expression of thrombin receptors in human atherosclerotic coronary arteries leads to an exaggerated vasoconstrictory response in vitro. J Cardiovasc Pharmacol. 1997;30:649–57.
141. Eto M, Barandiér C, Rathgeb L, Kozai T, Joch H, Yang Z, Lüscher TF. Thrombin supresses endothelial nitric oxide synthase and upregulates endothelin-converting enzyme-1 expression by distinct pathways: role of Rho/ROCK and mitogen-activated protein kinase. Circ Res. 2001;89:583–90.
142. Mizuno O, Kobayashi S, Hirano K, Nishimura J, Kubo C, Kanaide H. Stimulus-specific alteration of the relationship between cytosolic Ca(2+) transients and nitric oxide production in endothelial cells in vivo. Br J Pharmacol. 2000;130:1140–6.
143. Ming XF, Viswambharan H, Barandier C, Ruffieux J, Kaibuchi K, Rusconi S, Yang Z. Rho GTPase/Rho kinase negatively regulates endothelial nitric oxide synthase phosphorylation through the inhibition of protein kinase B/Akt in human endothelial cells. Mol Cell Biol. 2002;22:8467–77.
144. Robin J, Kharbanda R, McLean P, Campbell R, Vallance P. Protease-activated receptor 2-mediated vasodilatation in humans in vivo: role of nitric oxide and prostanoids. Circulation. 2003;107:954–9.
145. Carmeliet P. Biomedicine. Clotting factors build blood vessels. Science. 2001;293:1602–4.

146. Julovi SM, Xue M, Dervish S, Sambrook PN, March L, Jackson CJ. Protease activated receptor-2 mediates activated protein C-induced cutaneous wound healing via inhibition of p38. Am J Pathol. 2011;179:2233–42.
147. Xue M, Campbell D, Sambrook PN, Fukudome K, Jackson CJ. EndotheliaL protein C receptor and protease-activated receptor-1 mediate induction of wound-healing phenotype in human keratinocytes by activated protein C. J Invest Dermatol. 2005;125:1279–85.
148. Strukova SM, Dugina TN, Christov IV, Lange M, Markvicheva EA, Kuptsova S, Zubov VP, Glusa E. Immobilized thrombin receptor agonist peptide accelerates wound healing in mice. Clin Appl Thromb Hemost. 2001;7:325–9.
149. Borensztajn K, Stiekema J, Nijmeijer S, Reitsma PH, Peppelenbosch MP, Spek CA. Factor Xa stimulates proinflammatory and profibrotic responses in fibroblasts via protease-activated receptor-2 activation. Am J Pathol. 2008;172:309–20.
150. Jackson CJ, Xue M, Thompson P, Davey RA, Whitmont K, Smith S, Buisson-Legendre N, Sztynda T, Furphy LJ, Cooper A, Sambrook P, March L. Activated protein C prevents inflammation yet stimulates angiogenesis to promote cutaneous wound healing. Wound Repair Regen. 2005;13:284–94.
151. Somasundaram P, Ren G, Nagar H, Kraemer D, Mendoza L, Michael LH, Caughey GH, Entman ML, Frangogiannis NG. Mast cell tryptase may modulate endothelial cell phenotype in healing myocardial infarcts. J Pathol. 2005;205:102–11.
152. Antoniak S, Rojas M, Spring D, Bullard TA, Verrier ED, Blaxall BC, Mackman N, Pawlinski R. Protease-activated receptor 2 deficiency reduces cardiac ischemia/reperfusion injury. Arterioscler Thromb Vasc Biol. 2010;30:2136–42.
153. Howell DC, Johns RH, Lasky JA, Shan B, Scotton CJ, Laurent GJ, Chambers RC. Absence of proteinase-activated receptor-1 signaling affords protection from bleomycin-induced lung inflammation and fibrosis. Am J Pathol. 2005;166:1353–65.
154. Gaca MD, Zhou X, Benyon RC. Regulation of hepatic stellate cell proliferation and collagen synthesis by proteinase-activated receptors. J Hepatol. 2002;36:362–9.
155. Masamune A, Kikuta K, Satoh M, Suzuki N, Shimosegawa T. Protease-activated receptor-2-mediated proliferation and collagen production of rat pancreatic stellate cells. J Pharmacol Exp Ther. 2005;312:651–8.
156. Sabri A, Short J, Guo J, Steinberg SF. Protease-activated receptor-1-mediated DNA synthesis in cardiac fibroblasts is via epidermal growth factor receptor transactivation: distinct PAR-1 signaling pathways in cardiac fibroblasts and cardiomyocytes. Circ Res. 2002;91:532–9.
157. ten Cate H, Falanga A. Overview of the postulated mechanisms linking cancer and thrombosis. Pathophysiol Haemost Thromb. 2008;36:122–30.
158. Mueller BM, Ruf W. Requirement for binding of catalytically active factor VIIa in tissue factor-dependent experimental metastasis. J Clin Invest. 1998;101:1372–8.
159. Bromberg ME, Bailly MA, Konigsberg WH. Role of protease-activated receptor 1 in tumor metastasis promoted by tissue factor. Thromb Haemost. 2001;86:1210–4.
160. Yin YJ, Salah Z, Maoz M, Even Ram SC, Ochayon S, Neufeld G, Katzav S, Bar-Shavit R. Oncogenic transformation induces tumor angiogenesis: a role for PAR1 activation. FASEB J. 2003;17:163–74.
161. Even-Ram SC, Maoz M, Pokroy E, Reich R, Katz BZ, Gutwein P, Altevogt P, Bar-Shavit R. Tumor cell invasion is promoted by activation of protease-activated receptor-1 in cooperation with alpha vbeta 5 integrin. J Biol Chem. 2001;276:10952–62.
162. Jiang X, Bailly MA, Panetti TS, Cappello M, Konigsberg WH, Bromberg ME. Formation of tissue factor-factor VIIa-factor Xa complex promotes cellular signaling and migration of human breast cancer cells. J Thromb Haemost. 2004;2:93–101.
163. Kaushal V, Kohli M, Dennis RA, Siegel ER, Chiles WW, Mukunyadzi P. Thrombin receptor expression is upregulated in prostate cancer. Prostate. 2006;66:273–82.
164. Yin YJ, Salah Z, Grisaru-Granovsky S, Cohen I, Even Ram SC, Maoz M, Uziely B, Peretz T, Bar-Shavit R. Human protease-activated receptor 1 expression in malignant epithelia: a role in invasiveness. Arterioscler Thromb Vasc Biol. 2003;23:940–4.

165. Hernández NA, Correa E, Avila EP, Vela TA, Pérez VM. PAR1 is selectively over expressed in high grade breast cancer patients: a cohort study. J Transl Med. 2009;7:47.
166. Zhu L, Wang X, Wu J, Mao D, Xu Z, He Z, Yu A. Cooperation of protease-activated receptor 1 and integrin αvβ5 in thrombin-mediated lung cancer cell invasion. Oncol Rep. 2012;28:553–60.
167. Fujimoto D, Hirono Y, Goi T, Katayama K, Matsukawa S, Yamaguchi A. The activation of proteinase-activated receptor-1 (PAR1) mediates gastric cancer cell proliferation and invasion. BMC Cancer. 2010;10:433.
168. Darmoul D, Gratio V, Devaud H, Peiretti F, Laburthe M. Activation of proteinase-activated receptor 1 promotes human colon cancer cell proliferation through epidermal growth factor receptor transactivation. Mol Cancer Res. 2004;2:514–22.
169. Depasquale I, Thompson WD. Prognosis in human melanoma: PAR-1 expression is superior to other coagulation components and VEGF. Histopathology. 2008;52:500–9.
170. Grisaru-Granovsky S, Salah Z, Maoz M, Pruss D, Beller U, Bar-Shavit R. Differential expression of protease-activated receptor 1 (Par1) and pY397FAK in benign and malignant human ovarian tissue samples. Int J Cancer. 2005;113:372–8.
171. Arora P, Cuevas BD, Russo A, Johnson GL, Trejo J. Persistent transactivation of EGFR and ErbB2/HER2 by protease-activated receptor-1 promotes breast carcinoma cell invasion. Oncogene. 2008;27:4434–45.
172. Yang E, Boire A, Agerwal A, Nguyen N, O'Callaghan K, Tu P, Kuliopulos A, Covic L. Blockade of PAR1 signaling with cell-penetrating pepducins inhibits Akt survival pathways in breast cancer cells and suppresses tumor survival and metastasis. Cancer Res. 2009;69:6223–31.
173. Albrektsen T, Sørensen BB, Hjortø GM, Fleckner J, Rao LV, Petersen LC. Transcriptional program induced by factor VIIa-tissue factor, PAR1 and PAR2 in MDA-MB-231 cells. J Thromb Haemost. 2007;5:1588–97.
174. Petersen LC. Microarray studies of factor VIIa-activated cancer cells. Thromb Res. 2008;122:S11–3.
175. Liu J, Schuff-Werner P, Steiner M. Thrombin/thrombin receptor (PAR-1)-mediated induction of IL-8 and VEGF expression in prostate cancer cells. Biochem Biophys Res Commun. 2006;343:183–9.
176. Liu Y, Mueller BM. Protease-activated receptor-2 regulates vascular endothelial growth factor expression in MDA-MB-231 cells via MAPK pathways. Biochem Biophys Res Commun. 2006;344:1263–70.
177. Hjorte GM, Petersen LC, Albrektsen T, Sorensen BB, Norby PL, Mandal SK, Pendurthi UR, Rao LV. Tissue factor-factor VIIa-specific up-regulation of IL-8 expression in MDA-MB-231 cells is mediated by PAR-2 and results in increased cell migration. Blood. 2004;103:3029–37.
178. Jiang X, Bailly MA, Panetti TS, Cappello M, Konigsberg WH, Bromberg ME. Formation of tissue factor-factor VIIa-factor Xa complex promotes cellular signaling and migration of human breast cancer cells. J Thromb Haemost. 2004;2:93–101.
179. Morris DR, Ding Y, Ricks TK, Gullapalli A, Wolfe BL, Trejo J. Protease-activated receptor-2 is essential for factor VIIa and Xa-induced signaling, migration, and invasion of breast cancer cells. Cancer Res. 2006;66:307–14.
180. Jiang X, Zhu S, Panetti TS, Bromberg ME. Formation of tissue factor-factor VIIa-factor Xa complex induces activation of the mTOR pathway which regulates migration of human breast cancer cells. Thromb Haemost. 2008;100:127 –33.
181. Schaffner F, Versteeg HH, Schillert A, Yokota N, Petersen LC, Mueller BM, Ruf W. Cooperation of tissue factor cytoplasmic domain and PAR2 signaling in breast cancer development. Blood. 2010;116:6106–13.
182. Rydén L, Grabau D, Schaffner F, Jönsson PE, Ruf W, Belting M. Evidence for tissue factor phosphorylation and its correlation with protease-activated receptor expression and the prognosis of primary breast cancer. Int J Cancer. 2010;126:2330–40.

183. Antoniak S, Owens AP 3rd, Baunacke M, Williams JC, Lee RD, Weithäuser A, Sheridan PA, Malz R, Luyendyk JP, Esserman DA, Trejo J, Kirchhofer D, Blaxall BC, Pawlinski R, Beck MA, Rauch U, Mackman N. PAR-1 contributes to the innate immune response during viral infection. J Clin Invest. 2013;123:1310–22.
184. Shrivastava S, Ma L, Tham el-L, McVey JH, Chen D, Dorling A. Protease-activated receptor-2 signalling by tissue factor on dendritic cells suppresses antigen-specific $CD4^+$ T-cell priming. Immunology. 2013;139:219–26.
185. Chung WO, Hansen SR, Rao D, Dale BA. Protease-activated receptor signaling increases epithelial antimicrobial peptide expression. J Immunol. 2004;173:5165–70.
186. Rallabhandi P, Nhu QM, Toshchakov VY, Piao W, Medvedev AE, Hollenberg MD, Fasano A, Vogel SN. Analysis of proteinase-activated receptor 2 and TLR4 signal transduction: a novel paradigm for receptor cooperativity. J Biol Chem. 2008;283:24314–25.
187. Badeanlou L, Furlan-Freguia C, Yang G, Ruf W, Samad F. Tissue factor-protease-activated receptor 2 signaling promotes diet-induced obesity and adipose inflammation. Nat Med. 2011;17:1490–7.

Part III
Pathophysiology of the Vascular System

Chapter 11
Axon Guidance Factors in Developmental and Pathological Angiogenesis

Robert Luck and Carmen Ruiz de Almodovar

11.1 Introduction

Development of a vascular network, similar to the development of the nervous system, requires precise, localized, and controlled signals from the environment. Investigations of the last decade have shown that the vascular and nervous systems are not only similar in their anatomical pattern but they also grow in a similar way and use the same set of signaling molecules to develop and find their final target. Indeed, at the tip of a growing axon there is a specialized structure termed the *growth cone* that extents filopodia to sense the environment and lead the axon path. In a similar way, at the tip of a growing blood vessel there is a specialized endothelial cell (EC) known as the *endothelial tip cell* that also extends filopodia, senses the different guidance signals, and guides the vessel to its final destination. Thus, in the last decade a new research concept termed '*the neurovascular link*', focuses on studying both systems together, their cellular and molecular similarities, and their crosstalk communication in health and disease. Here, we summarize the so far known role of the canonical axon guidance molecules in the vascular system, during development and in disease conditions.

11.2 Netrins

11.2.1 Introduction

Netrins were first described as guidance molecules for axons exerting both attractive and repulsive functions, depending on the receptors to which they bind. Netrin1, 3, 4, G1, and G2 have been described in mammals, and Netrin2 has been found in

C. Ruiz de Almodovar (✉) · R. Luck
Biochemistry Center, Heidelberg University, Im Neuenheimer Feld 328, 69120 Heidelberg, Germany
e-mail: carmen.ruizdealmodovar@bzh.uni-heidelberg.de

M. HH. Schmidt, S. Liebner (eds.), *Endothelial Signaling in Development and Disease*, DOI 10.1007/978-1-4939-2907-8_11

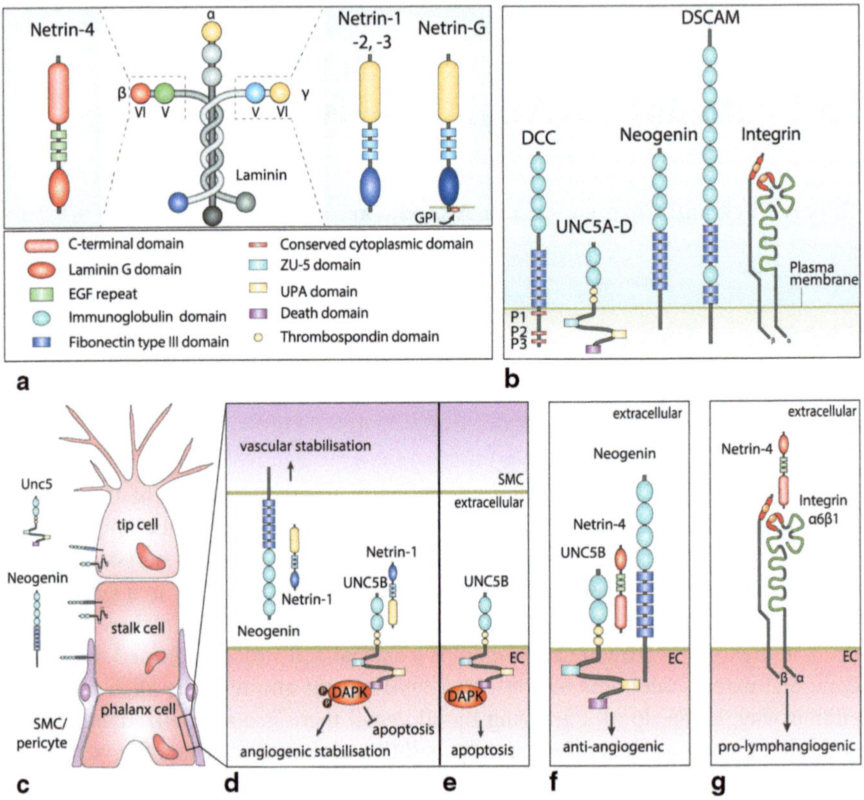

Fig. 11.1 Netrins and their receptors in the vascular system. **a** Netrin family members show a structural homology to the β-laminin (Netrin4) or γ-laminin (Netrin1, 2, 3, and G) subunits. **b** DCC and Unc5A–D are the main netrin receptors. Neogenin, DSCAM, or integrins also bind to netrins and can function in a co-receptor complex. **c** An angiogenic sprout with an endothelial tip cell followed by stalk and phalanx endothelial cells (*ECs*). Smooth muscle cells (*SMCs*) and pericytes cover stalk and phalanx cells. Unc5 and neogenin are expressed on sprouting ECs as well as mural cells. **d** Unbound Unc5B in ECs activates death-associated protein (*DAP*) kinase signaling that induces apoptosis. **e** Netrin1 binding to Unc5B in ECs inhibits apoptosis signaling and leads to the stabilization of the angiogenic sprout. Netrin1/neogenin interactions on SMCs induce their proliferation and migration, which further enhances vascular stabilization. **f** Netrin4 blocks angiogenesis by binding to Unc5B/neogenin heterodimers in ECs. **g** Netrin4/integrin signaling in lymphatic endothelial cells induces lymphangiogenesis. (Adapted with permission from Macmillan Publishers Ltd: Mehlen et al. [48], copyright 2011; Cirulli and Yebra [174], copyright 2007; Kitajewski [175], copyright 2011)

chicken. The amino acid sequences of Netrin1, Netrin2 (for chicken), and Netrin3 are similar to the γ-chain of the laminin molecule, whereas Netrin4 and NetrinGs show a higher overlap with the β-laminin structure [1] (Fig. 11.1a). Netrin1, 2, 3, and 4 are described as secreted molecules, whereas NetrinGs are membrane-bound and anchored by glycosylphosphatydylinositol (GPI) [2, 3] (Fig. 11.1a). As a result of their separate structural properties, NetrinGs have different functional mechanisms [4]. The receptors for netrins (with the exception of NetrinGs) are members of the DCC (deleted in colorectal cancer) and Unc5 family [5]; however, interactions with other receptors, including the DCC paralog neogenin, DSCAM, and integrins, are also described [5–9] (Fig. 11.1b). NetrinGs bind NetrinG ligands (NGLs), which are single-pass transmembrane proteins unrelated to the DCC or Unc5 family [5, 10]. In the nervous system, DCC alone has been shown to mediate axonal chemoattraction upon Netrin1 binding, whereas the binding of Netrin1 to a complex of DCC with Unc5 or to Unc5 homodimers mainly results in a repulsive effect [11]. In *Drosophila m.*, there is scientific evidence indicating that homodimeric Unc5 receptors serve as the functional unit for short-range repulsion. In contrast, with greater distance to the Netrin1 source according to long-range signals, Unc5 members seem to form heterodimeric complexes with DCC receptors for mediating the repelling effect [12]. Moreover, members of the DCC and Unc5 family belong to a group of so-called 'dependence receptors', which induce apoptosis in the absence of their ligand [13, 14]. Therefore, netrins have currently acquired importance as general survival factors, in addition to their known role in neuronal guidance.

Netrin1 is the best-studied member of the netrin family. It guides different neuronal populations to distinct locations, and controls central nervous system (CNS) development [15–21]. One of the best-known roles for Netrin1 is in the developing spinal cord, where floor-plate-derived Netrin1 attracts ventral-directed commissural axons and regulates their midline crossing [4]. Netrin4 and NetrinGs are also found in the CNS, and also in non-neuronal tissue [22, 23]. Functionally, Netrin4 was found to promote neurite outgrowth in cortical explants, whereas NetrinG1 induces the outgrowth of thalamocortical axons via NGL1 binding [1, 5, 24].

11.2.2 Netrins and Netrin Receptors in the Vascular System

Expression of Netrin Receptors in Endothelial Cells In addition to the nervous system, there is accumulating literature showing a potent effect of netrins during vascular development and angiogenesis. Netrin1 and Netrin4 are the best-studied members, whereas a functional role of Netrin3 and NetrinGs in the vascular system lacks robust investigations. Interestingly, similar to what occurs in neurons, the effect of Netrin1 and Netrin4 stimulation in the vascular system depends on the EC origin and is linked to differential netrin receptor expression. Generally, ECs do not express DCC but express low levels of Unc5B, C, and D, as well as neogenin [25–29] (Fig. 11.1c). However, human arterial ECs (HUAECs) express higher levels of Unc5B than human umbilical vein ECs (HUVECs) [one of the most commonly used cells for studying angiogenesis] [27]. This higher expression of Unc5B in arterial ECs was also shown *in vivo* in the developing mouse retina where Unc5B expression was detected in ECs of arteries but not veins [27]. Apart from the basal expression levels, Unc5 receptor expression in ECs seems to be regulated by angiogenic factors such as vascular endothelial growth factor-A (VEGFA, hereon termed VEGF) [30].

Function of Unc5B and Netrin1 in Physiological Angiogenesis Two different vascular phenotypes have been described in Unc5B mutant mice. On the one hand, it has been shown that Netrin1-induced Unc5B activation in ECs led to an antiangiogenic effect similar to the repulsive effect observed in the nervous system. Genetic Unc5B depletion led to increased, but collapsed, arterial branches accompanied by abnormal endothelial tip-cell navigation and extensive filopodia extension in different vascular beds of the mouse embryo (in the CNS vasculature and the retina), as well as in intersegmental vessels of Unc5B morphant zebrafish [27]. These results suggest that, in the absence of Unc5B, there is reduced repulsion of angiogenic blood vessels by Netrin1. Similarly, studies in chicken embryos show that Netrin1 inhibits angiogenic sprouting of Unc5B-positive vessels [31]. Netrin1 stimulation of HUAECs *in vitro* also results in a Unc5B-dependent reduction of migration and increased filopodia retraction [27].

On the other hand, *in vitro* and *in vivo* experiments by Li's group showed that Netrin1 positively stimulated angiogenesis and that the knockdown of Netrin1 in zebrafish resulted in inhibition of the parachordal vessel, rather than in increased vascular branching [28, 29]. Moreover, analysis of another Unc5B mutant mouse line revealed normal blood vessel branching in the hindbrain, cranial, and intersomitic vessels [27]. The generation of an EC-specific Unc5B mutant mouse line showed that endothelial specific ablation of Unc5B causes embryonic lethality and that Unc5B is required for proper development of the vascular bed of the placenta but not the hindbrain or yolk sac, which appeared normal [32].

Consistent with an angiogenic role of Netrin1, different EC lines such as mouse brain ECs, human microvascular ECs (HMVECs), human aortic endothelial cells (HAECs), or HUVECs showed a positive angiogenic response to Netrin1 stimulation with increased sprouting, proliferation, migration, and tube-formation; however, no known netrin receptors could be identified in these cell lines [25, 28, 29, 33]. Interestingly, vascular smooth muscle cells (SMCs) express neogenin (Fig. 11.1c), and Netrin1 stimulation *in vitro* leads to a positive effect on SMC adhesion, proliferation, and migration (Fig. 11.1d) [28]. Moreover, *in vivo* experiments such as xenografts, chicken chorioallantoic membrane (CAM) assays, and corneal pocket assays result in increased angiogenesis in response to Netrin1, as well as in a synergistic effect together with VEGF [25, 28].

As mentioned previously, Unc5 members and DCC are known to function as dependence receptors, which can be activated both in the presence and absence of their ligands. In the absence of Netrin1, DCC and Unc5 unbound receptors activate a proapoptotic cascade, leading to cell death via activation of caspases [34]. In this regard, during zebrafish development, Netrin1 prevents ECs from apoptosis in a Unc5B-dependent manner, whereby genetic knockdown of Netrin1 leads to increased cell death of ECs and vascular defects that can be rescued by Unc5B inhibition [35]. Netrin1 inhibits *in vitro* Unc5B downstream caspase signaling and decreases the level of apoptosis in HUAECs and HUVECs that could be mimicked by Unc5B silencing [35] (Fig. 11.1d, e). Netrin1/Unc5B signaling leads to inactivation of the death-associated protein kinase (DAPK) and thus blockage of apoptosis [35] (Fig. 11.1d). Regarding DCC, to date only one study has found a role for DCC in ECs where Netrin1 induces EC proliferation and migration *in vitro* via activation of DCC in aortic ECs and an increase in nitric oxide production [36]. No link between DCC and its dependence function has yet been described.

Apart for acting directly on ECs, Netrin1 can also control lymphangiogenesis (the formation of lymphatic vessels) via an interplay between muscle pioneers, motoneuron axons, and ECs [37].

Altogether, it is clear that Netrin1 and Unc5B are important for physiological angiogenesis and for proper formation of the developing vasculature. Future studies will be required to better understand their functions (i.e. survival, proangiogenic, or antiangiogenic) within specific vascular beds and whether a pro- or antisurvival function can result in a pro- or antiangiogenic effect.

Unc5B and Netrin1 in Pathological Angiogenesis A role for Netrin1 and Unc5B has also been identified in pathological angiogenesis using different disease mouse models. After postnatal angiogenesis, Unc5B becomes downregulated and remains absent in quiescent ECs [38]. However, Unc5B becomes re-expressed in adult ECs of sprouting vessels upon subjecting adult mice to oxygen-induced ischemic retinopathy (OIR). Netrin1 also becomes upregulated in the retina upon OIR, and intravitreal lentivirus delivery of Netrin1 small hairpin RNA (shRNA) inhibited retinal neovascularization, indicating Netrin1 as a potential target for different retinopathies or other ocular neovascular diseases [39].

Unc5B is also re-expressed in sprouting vessels after matrigel or tumor implants [38]. In these two experimental models, Netrin-1, via Unc5B, leads to repulsion of ECs and, as a consequence, to inhibition of matrigel neovascularization or inhibition of tumor angiogenesis (using Netrin1-containing matrigel plugs or tumor cells expressing Netrin1) [38]. However, another study showed that angiogenesis in matrigel plugs containing Netrin1 was enhanced compared with control plugs [25] (Table 11.1).

In a mouse model of hindlimb ischemia, where the iliac artery was ligated, intramuscular delivery of Netrin1 or Netrin4 into the ischemic gastrocnemius resulted in improved hindlimb perfusion and greater vascular density in Netrin1-injected muscles when compared with control injected muscles [29]. In a rat model of limb

Table 11.1 Role of axon guidance factors in tumor angiogenesis

Axon guidance factors	Inhibition of tumor angiogenesis and/or lymphangiogenesis	Promotion of tumor angiogenesis and/or lymphangiogenesis
Netrins	Netrin1/Unc5B [38] Netrin4/neogenin [47]	Netrin1/Unc5B [25] Netrin4 [44]
Slits	–	Slit2/Robo1 [68, 70, 71]
Ephrins	–	EphrinB2/EphB4 [95, 97, 100, 104–106] EphA2 [108–110]
Semaphorins	Sema3A/Nrp1 [137, 155–157] Sema3F/Nrp2 [158–160] Sema3E [128, 162, 163] Sema3G [163]	Sema3C [164] Sema 4D; Sema4D (cleaved)/PlexinB1 [166–170]

Nrp neuropilin

ischemia, it was consistently shown that transplantation of mesenchymal stem cells (MSCs), together with Netrin1, directly into the muscle of ischemic limbs, improves the effectiveness of MSCs and increases the neocapillary network [40]. Notably, in another study, hindlimb ischemia in adult mice (in this case ligating the femoral artery) did not lead to re-expression of Unc5B in sprouting ECs, thus suggesting that the observed improvement in the study described above might be due to Netrin1 signaling in a Unc5B-independent manner [38]. Altogether, these studies suggest Netrin1 as a possible candidate for therapeutic neovascularization. In addition, Netrin1 was found to be upregulated in patients with proliferating diabetic retinopathy (PDR), which is characterized by vascular malformations [41].

Role of Netrin4 in Physiological and Pathological Angiogenesis Although little is known about Netrin4 signaling in the nervous system, its contribution in the vascular context has been considerably investigated; however, its signaling mechanisms still remain intricate.

Angiogenic ECs in zebrafish express Netrin4, and genetic knockdown in zebrafish using morpholinos results in vascular defects, suggesting a proangiogenic effect for Netrin4 in developmental angiogenesis [42]. Similar to Netrin1, Netrin4 stimulation increased proliferation, migration, survival, and tube formation in several EC lines, including HUVECs, HMVECs, and HUAECs *in vitro* [29, 42, 43]. However, the receptor mediating these proangiogenic effects remains unclear. In contrast, another study proposed an antiangiogenic role for Netrin4. In this study, it was described that Netrin4 binds neogenin, and that a neogenin/Unc5B heteroreceptor complex mediates the *in vitro* antiangiogenic effects of Netrin4 [26] (Fig. 11.1f). ECs of lymph vessels also express Netrin4. *In vitro* experiments using human lymphatic ECs revealed the same promoting effects on proliferation, migration, survival, and tube formation as with Netrin1 [44]. Although lymphatic ECs express Unc5B and neogenin, their single- or dual-receptor knockdown was insufficient to block the cellular response to Netrin4 [44]. A later study identified integrin $\alpha6\beta1$ as a possible Netrin4 receptor in lymphatic ECs, showing the ability to bind Netrin4 (Fig. 11.1g). Consistently, the blockage of either subunit was also able to abolish the lymphatic EC migration, adhesion, and signaling induced by Netrin4 *in vitro* [45].

Several *in vivo* applications support the proangiogenic capacity of Netrin4 in blood and lymphatic vessels under pathological conditions [44]. Netrin4 administration in different stroke and ischemia models promotes neovascularization and positively affects behavior recovery without affecting integrity of the blood–brain barrier [29, 46]. It is important to mention that stroke itself upregulated Netrin4 within the ischemic core [46]. Netrin4 overexpression in mouse skin induces lymphangiogenesis *in vivo*. Moreover, using Netrin4 overexpressing breast cancer models, it was shown that Netrin4 promoted tumor lymphangiogenesis and tumor metastasis by stimulating lymphatic permeability [44]. However, whether Netrin4 may act differently on ECs or lympathic ECs from different tumor entities remains to be determined. In support of this, in *in vivo* matrigel angiogenesis assays, Netrin4 overexpression in colon cancer cells negatively affects angiogenesis in a neogenin-dependent manner, leading to reduced tumor growth and increased tumor apoptosis [47] (Table. 11.1). Although further research is still needed to understand the effects of Netrin4 in different angiogenic pathological conditions, the studies mentioned above highlight Netrin4 as a possible therapeutic target for targeting angiogenesis and lymphangiogenesis.

In summary, netrin signaling contributes to vascular development and angiogenesis, not only in a direct fashion via acting on ECs but also in a cellular interplay via acting on SMCs. Netrins exert proangiogenic and also inhibitory effects, perhaps regulated by the receptor composition. In addition, this might differ, dependent on the tissue and the specific endothelial subset. We are only just beginning to fully unravel the signaling properties of netrins in ECs. Moreover, the roles of other netrin family members (i.e. Netrin3 and NetrinGs) still need to be evaluated.

11.3 Slits and Robo Receptors

11.3.1 Introduction

Slits are secreted molecules found throughout invertebrate and vertebrate species. There are three described genes in mammals (Slit1, 2, and3), as well as several splice variants of all three members [48]. They contain leucin-rich repeats (LRRs), endothelial growth factor (EGF) repeats, and a laminin G-like domain (Fig. 11.2a). The known receptors for slits belong to the Robo family. In mammals, there are four Robo receptors, Robo1, 2, 3, and 4 (Fig. 11.2b), together with their described isoforms Dutt1 (for Robo1), Robo2_tv2 (for Robo2, described in zebrafish and rat), as well as Robo3.1 and Robo3.2 (for Robo3) [48]. While Robo1, 2, and 3 are similar and are mainly expressed in the nervous system, Robo4 appears to be special, showing a different molecular structure accompanied with a specific expression pattern within the vascular compartment in mammals [49–51] (Fig. 11.2a). In addition, controversy exists regarding the binding of slits to Robo4 (see below).

Slit/Robo signaling is involved in different neuronal processes, such as axon guidance, axon fasciculation, and dendritic branching [52, 53]. In the developing nervous system of invertebrates and vertebrates, slits were described as exerting a repulsive effect on axons [54, 55]. It was thereby found that Slit/Robo signaling regulates the midline crossing of commissural axon projections by a mechanism that involves differential expression of Robo receptors in pre- and postcrossing axons [53, 56]. Interestingly, interactions between Robo1 and the netrin receptor DCC have been identified in commissural axons [57].

11.3.2 Slits and Robo Receptors in the Vascular System

There is a growing body of literature indicating that Slit/Robo functions are not simply restricted to axonal guidance, and do indeed also play a role in the vascular system. Since Robo4 is specifically expressed in the vascular system, the majority

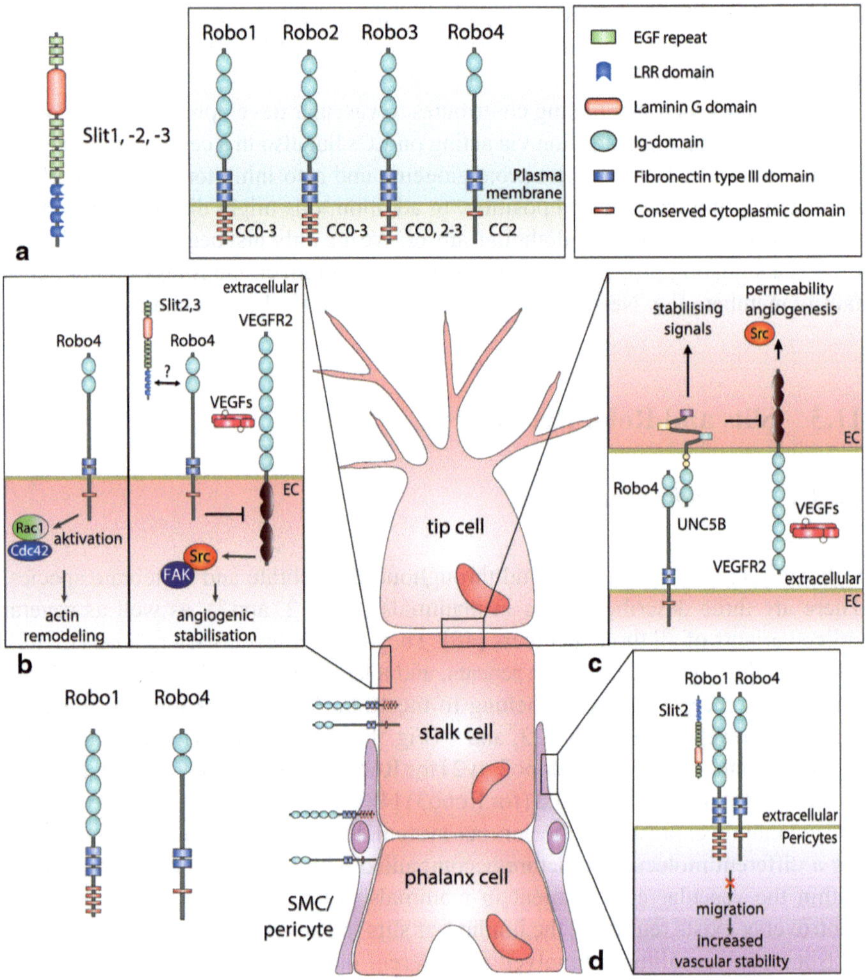

Fig. 11.2 Slits and Robos in the vascular system. **a** Different domains of slits and Robo receptors. Robo4 (and also, in some cases, Robo1) is expressed on the angiogenic endothelium and on smooth muscle cells (*SMCs*), but not in endothelial tip cells (see scheme of sprouting blood vessel). **b** *Left*: Robo4 in zebrafish ECs regulates actin remodeling in an Rac1 Cdc42-dependent manner. *Right*: Slit2 and 3 binding to Robo4 blocks vascular endothelial growth factor receptor-2 (*VEGFR2*)-mediated Src and focal adhesion kinase (*FAK*) signaling, leading to vascular stabilization. Controversies exist regarding the binding of Slit2 and 3 to Robo4. **c** Robo4 trans-interacts with Unc5B and suppresses vascular endothelial growth factor (*VEGF*) signaling via VEGFR2 by blocking VEGFR2-dependent Src kinase activation. **d** Robo4 interacts in *cis* with Robo1 to transduce Slit2 signals, leading to reduced pericyte migration and therefore increased vascular stability. (Adapted with permission from Macmillan Publishers Ltd: Mehlen et al. [48], copyright 2011; Kitajewski [175], copyright 2011)

of studies have focused on its endothelial-specific function. We summarise the main findings below.

In situ hybridization of whole-mount mouse embryos shows that Robo4 expression starts in large axial vessels and shifts from the dorsal aorta to the intersomitic vessels and capillaries as development proceeds [50]. Robo4 is also expressed in the vasculature of adult tissues [50]. Robo4 knockout mice are fertile and viable, and show normal patterning of blood vessels in different embryonic structures, indicating that Robo4 is either dispensable for developmental angiogenesis or its absence is compensated by yet an unknown mechanism [58]. In contrast, studies in zebrafish show that Robo4 depletion via morpholinos injection, and also overexpression, leads to reduced and defective formation of intersomitic vessels via a mechanism that involves Cdc42 and Rac1 Rho GTPases [59] (Fig. 11.2b).

Slit1, 2 and 3 have been proposed as ligands for Robo4 in ECs [50, 58, 60, 61]; however, other studies challenge whether slits really bind to Robo4, raising the question of potential, still unknown, ligands or a ligand-independent activation of Robo4. Here we describe studies showing activation of Robo4 by slits, as well as studies where slits did not bind to Robo4.

Function in Blood Vessel Growth and Stabilization Analysis of Slit3 knockout mice revealed that these mice present defects in the vascular system during embryonic development. While the vasculature of diaphragms from E15.5 wild-type mouse embryos have a uniform vessel architecture, including regular vascular patterning and branching, diaphragms from Slit3 null mice show a dramatic reduction in blood vessel density and branches [61]. Consistent with this *in vivo* result, *in vitro* experiments with HUVECs, and *ex vivo* experiments using the rat aortic ring assay, show that Slit3 promotes EC proliferation, migration, and motility via Robo4 signaling, and also induces remodeling of the actin cytoskeleton [61].

During the growth of a new blood vessel sprout, stalk cells follow behind the leading angiogenic endothelial tip cell and participate in maturation and lumen formation of the new growing vessel [58, 62]. Phalanx cells are the most quiescent ECs in the vessels and are covered by pericytes and form tight junctions with each other (Fig. 11.2). A detailed analysis of Robo4 in the developing retina vasculature revealed that stalk cells express Robo4, while tips cells are Robo4-negative [58] (Fig. 11.2), suggesting that Robo4 might not function as a typical guidance receptor in the endothelium but rather as factor-regulating vascular stability. In support of this concept, even though Robo4 knockout mice do not show obvious differences in vascular patterning, Robo4-deficient animals possess a higher basal level of vascular permeability [58]. Moreover, when subjecting these mice to oxygen-induced retinopathy, they respond with increased vascular leakage compared with wild-type mice [58]. Robo4 stabilizing function is strengthened by results showing that Robo4 (in this case activated by Slit2 or Slit3) inhibited VEGF-induced vascular destabilization and blocked VEGF-induced EC migration, tube formation, and permeability *in vitro*. These effects were due to a Robo4-dependent blockage of VEGF-induced activation of the Src-family kinases [58, 63] (Fig. 11.2b). Another study in the mouse mammary gland also found a role for Slit/Robo4 signaling in restricting VEGF function by downregulating VEGFR2 activation of Src and focal adhesion kinase (FAK) [62] (Fig. 11.2b).

Interestingly, direct binding of slits to Robo4 is questioned by structural analysis as well as functional investigations of the binding properties of slits [64–66]. In a protein–protein interaction screen to identify Robo4 binding partners, no association of slits to Robo4 was detected. In contrast, an interaction with Unc5B, a Netrin1 receptor (see above), was identified [64] (Fig. 11.2c). Robo4 and Unc5B are co-expressed in ECs, and stimulation of Unc5B-transfected porcine aorta ECs (PAECs) with the extracellular domain of Robo4 (sRobo4) led to Unc5B internalization and cell retraction. Robo4/Unc5B signaling leads to the association of Src kinase to Unc5B and its activation. Moreover, Robo4/Unc5B signaling counteracts VEGF-induced Src activation by sequestering Src away from VEGFR2 [64] (Fig. 11.2c). Consistently, the intraperitoneal injection of an antibody that specifically blocks Robo4/Unc5B interaction led to increased angiogenesis and vascular hyperpermeability *in vivo*. Since Unc5B is expressed on all cells of the angiogenic sprout (see above), and the expression of Robo4 is excluded from the tip cell, their trans-interaction might represent an additional mechanism regulating tip and stalk cells and, consequently, the angiogenic sensitivity of the vascular sprout (Fig. 11.2c).

Altogether, the studies described above clearly demonstrate the function of Robo4 signaling for vascular stabilization, either by acting as a ligand for Unc5B or by acting as a receptor for slits. These two possible scenarios are not exclusive when considering that other studies have shown the formation of heterodimers between Robo1 and Robo4 in HUVECs [67], thus opening the possibility that slits binding to Robo1–Robo4 complexes also leads to the described effects via a Robo4-dependent, Unc5B-independent mechanism. Moreover, *in vitro* studies showed that Slit2 via activation of Robo1 and Robo4 inhibits pericyte migration, which might be interpreted as an additional mechanism for stabilizing endothelial integrity [51] (Fig. 11.2d). These results suggest that signal transduction of Slit/Robo4 interaction depends on Robo1 regulating vascular stabilization and thereafter influencing endothelial sprouting.

Role of Slits and Robos in Tumor Angiogenesis A function for slits and Robos in tumor angiogenesis has also been described. In this regard, it was found that cells of several solid tumors, including malignant melanoma, rectal mucinous adenocarcinoma, breast invasive carcinoma, stomach squamous carcinoma, and hepatocellular carcinoma, express and secrete Slit2 [68]. Interestingly, Slit2 expression in human carcinoma tumor sections seemed to increase in the center of the tumor compared with the periphery [68]. This, as well as hypoxia studies in choriocarcinoma cells, leads to the assumption that Slit2 expression is dependent on oxygen levels [69]. In *in vitro* settings, and in contrast to other published studies, Slit2 was able to attract HUVECs and promote tube formation in a Robo1-PI3K-dependent manner [68]. Consistently, in a xenograft tumor model with human malignant melanoma cells that express Slit2, treatment with a functional blocking antibody against the first immunoglobulin (Ig) motif of Robo1 led to decreased microvessel density and tumor weight when compared with IgG control-treated mice [68]. Another cancer model of chemical-induced squamous cell carcinogenesis also linked high Slit2 expression to increased angiogenesis, in a Robo1-dependent manner [70] (Table 11.1). Similar effects have been described in the lymphatic endothelium, where Robo1 is also expressed [71]. Slit2 stimulation thereby sufficiently enhanced the migration and tube formation of lymphatic ECs *in vitro*. Consistently, *in vivo* tumor models with mice overexpressing Slit2 showed increased tumor lymphangiogenesis and lymph node metastasis [71].

Robo4, which is strongly expressed in highly active endothelium, is also regulated by hypoxia [49]. However, to date no studies have been conducted proposing a role for Robo4 in tumor angiogenesis. Thus, whether Robo4 also plays a role in tumor angiogenesis, or whether Robo4 is only required in healthy conditions and Robo1 in pathological circumstances, remains to be determined. In support of the second hypothesis, a quantitative analysis of Robo1, Robo4, and Slit2 messenger RNA (mRNA) in colorectal cancer tissue versus normal tissue revealed that while Robo1 was significantly upregulated in tumor tissue, no changes where observed for Robo4 or Slit2 [72].

Role in Pre-Eclampsia It has also been suggested that the Slit/Robo pathway contributes to pre-eclampsia, a disease occurring during pregnancy that is characterized by high blood pressure, endothelial dysfunction, and impaired angiogenesis [69]. Analysis of slits and Robo receptors showed that Slit3, Robo1, and Robo4 are expressed in the endothelium of placental villi, and Slit2 and Robo1 are expressed in the syncytiotrophoblast [69]. In pre-eclamptic placentas, Robo1 and Robo4 are significantly increased compared with healthy controls, thus suggesting a potential pathological role for the Slit/Robo pathway in this disease [69]. It is not yet known whether there is a functional effect of the increased expression of Robo1 and Robo4.

Taken together, the available evidence suggests that Slit/Robo signaling in the vasculature possesses a stabilizing and maturating role. Further studies would need to clarify the situations where they might have an antiangiogenic or proangiogenic effect.

11.4 Ephrins and Eph Receptors

11.4.1 Introduction

Eph receptors are named after the *E*rythropoietin-*p*roducing human *h*epatocellular carcinoma cell line in which they were discovered. Eph receptors are divided into two subgroups—EphAs (EphA1–10 in vertebrates) and EphBs (EphB1–6 in vertebrates)—depending on the sequence homology of their extracellular domain [73, 74] (Fig. 11.3a). Eph receptors interact with the membrane-bound proteins termed ephrins (greek 'ephros' for controller), which are also divided into two subclasses—EphrinAs and EphrinBs (Fig. 11.3a). The five homologous members EphrinA1–5 are anchored in the plasma membrane by a GPI linker, and bind almost exclusively to EphA receptors [73]. In contrast, EphrinBs are transmembrane proteins possessing a C-terminal. The name PDZ is derived from the first three proteins in which these domains were found: PSD-95 (a 95 kDa protein involved in signaling in the post-synaptic density), Dlg (the *Drosophila* discs large protein), and ZO1 (the zonula occludens 1 protein involved in maintaining epithelial cell polarity). PDZ motif, and mainly signal via EphBs [73] (Fig. 11.3a). However, reports show that members of the A and B classes interact with each other, as exemplified by EphB2 and EphrinA5 binding [75].

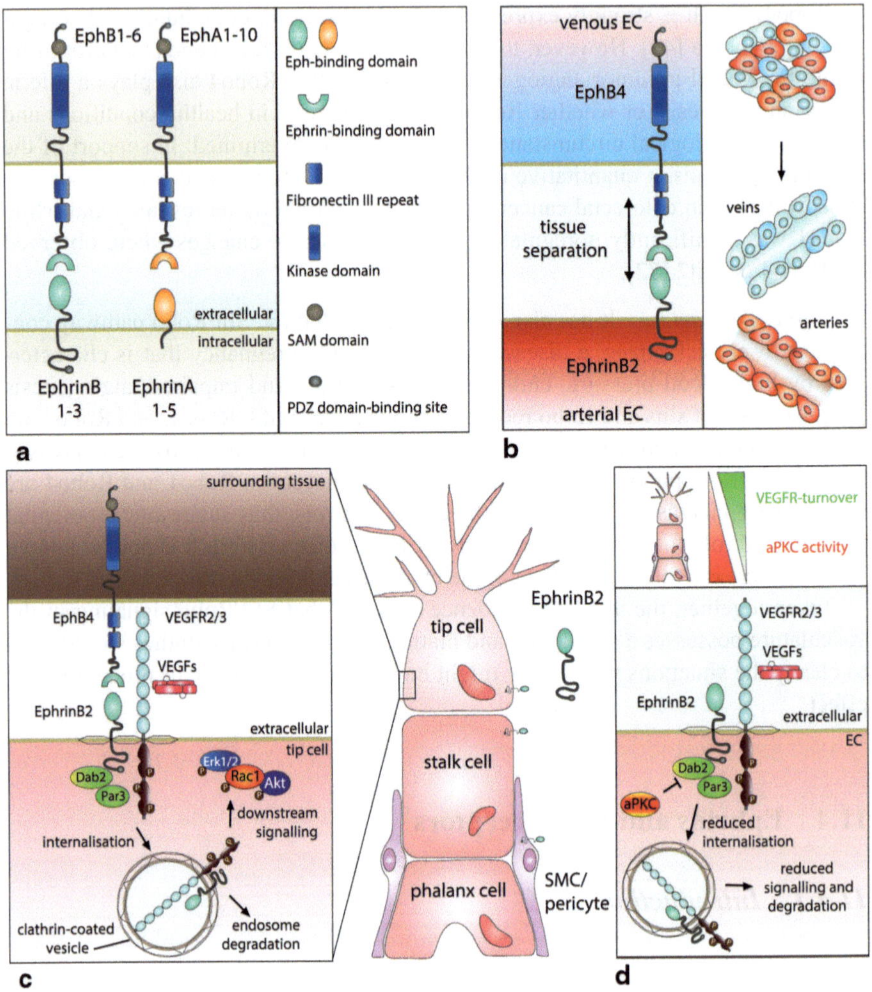

Fig. 11.3 Ephrins and Ephs in the vascular system. **a** Ephrins and their Eph receptors are membrane-bound molecules. EphrinAs interact preferentially with EphAs, and EphrinBs with EphBs. **b** EphB4 is primarily found on venous endothelial cells (*ECs*), whereas EphrinB2 expressed on arterial ECs; their signaling controls arterial-venous separation. **c** EphrinB2 is expressed in the angiogenic endothelium, as well as in the mural compartment, and also interacts with vascular endothelial growth factor receptor-2 (*VEGFR2*) and -3 in ECs. Upon EphB4 binding, EphrinB2 recruits Dab2 and Par3 effector proteins, leading to clathrin-dependent endocytosis of the EphrinB2–VEGFR complex. Thereby, VEGFR internalization serves as a crucial step for downstream VEGFR signaling, and directly affects angiogenesis. **d** VEGFR2 internalization is controlled by activated atypical protein kinase C (aPKC). Activation of aPKC is higher in stalk cells than in tip cells, thus leading to reduced internalization of the complex and reduced VEGF signaling in stalk cells. (Adapted with permission from Macmillan Publishers Ltd: Kitajewski [175], copyright 2011)

Eph/Ephrin signaling in neurons is extensively investigated and is involved in many contexts throughout the development of the CNS. Based on their structural characteristics, Eph/Ephrin signaling is considered a short-range neuronal wiring signal. Eph/Ephrin signaling controls the formation of topographic maps [76, 77] and, in particular, is essential in pre- and postsynaptic development, as well as synaptic plasticity [78, 79]. The molecular mechanisms through which Eph/Ephrin signaling takes place are very sophisticated and include special features such as receptor clustering and the ability to signal bidirectionally and autonomously [80]. The high degree of possible receptor/ligand combinations, as well as interactions with co-receptors, increases the signal diversity of Eph receptors and ephrins.

11.4.2 Eph Receptors and Ephrins in Vascular Development

In the field of angiogenesis, EphA2 and EphrinA1 are among the best-studied members of the Eph/EphrinA classes, even though almost all EphAs are expressed in ECs (for example, in human brain microvascular ECs) [81]. EphrinA1 is expressed at sites of developmental angiogenesis, and soluble EphA2-Fc inhibits EC migration, sprouting, survival, and corneal angiogenesis induced by VEGF [82, 83]. EphA2 knockout mice initially showed a normal vascular development [84]; however, more detailed studies have described that EphA2-deficient mice possess fewer and enlarged lung capillaries and increased numbers of endothelial sprouts, together with less pericyte coverage [85]. In addition, EphA2-deficient mice hyperreact to bacterial infection of the lung and ovalbumin sensitization by enhanced neoangiogenesis, immune cellular infiltration, and cytokine release [85]. Taken together, these investigations point to a functional role of EphA2 and EphrinA1 signaling in maintaining vascular integrity by stabilizing intracellular connection not only between ECs but also endothelial and mural cells.

EphB4 and EphrinB2 are the best-studied B-class members in angiogenesis, particularly because of their distinctive expression pattern, with higher expression of EphrinB2 in arterial ECs and EphB4 in the venous endothelium [86–88] (Fig. 11.3b). This specific expression pattern, together with the fact that EphB4/EphrinB2 function as a receptor-ligand pair able to signal bidirectionally, outlines the possibility of controlling cellular segregation and boundary formation as functional features of B-class signaling in ECs [89, 90]. An example of this is found in the zebrafish model organism, where the dorsal aorta and caudal vein derive from the same primordial structure that subsequently segregates into arterial and vein ECs. Angioblast cells migrate ventrally from this primary vessel to form the caudal vein in an EphB4/EphrinB2-dependent manner, whereas the remaining cells later form the dorsal aorta [91]. In this regard, a loss-of-function study in mice specifically targeting endothelial EphB4 pointed out that the dorsal aorta of the embryos was enlarged, whereas the formation of cardinal veins was negatively disturbed [90]. More importantly, it was observed that venous ECs were misallocated in arteries [90]. These findings highlight the prominent role of EphB4/EphrinB2 signaling for intercellular communication and tissue boundary formation (Fig. 11.3b).

The generation and evaluation of different EphrinB2 transgenic mouse lines demonstrated the essential role of EphrinB2 for proper vascular development. EphrinB2 knockout mice, or mice with ubiquitous overexpression of EphrinB2, are embryonic or neonatal lethal, respectively, due to severe vascular defects [92, 93]. The lethal phenotype of the ubiquitous EphrinB2 deletion could be rescued by crossing this line with an endothelial EphrinB2 overexpressing mouse line [92, 93], indicating that EphrinB2 expression in the endothelium is essential for survival. Analysis of a mouse line where EphrinB2 overexpression in the endothelium is temporally controlled revealed that overexpression of EphrinB2 during early development also leads to severe vascular defects and embryonic lethality. The severity of the phenotype decreases when EphrinB2 overexpression is induced in the more advanced developmental stages [88]. Consistent with these findings, the endothelial-specific knockout of EphrinB2 under the Cdh5 (vascular endothelial [VE]-cadherin) promoter showed a less-developed vascular system [88]. Therefore, these loss-of-function and gain-of-function studies are evidence that EphrinB2 is essential for vascular development in a specific temporal frame and proangiogenic manner. Interestingly, transgenic mice lacking the cytoplasmic domain of EphrinB2 also die during embryonic development due to defects in vasculogenesis and angiogenesis (similar to the EphrinB2 null mutants) [94], thus showing the EphrinB2 intracellular domain and reverse signaling is important for vascular development. The requirement of the EphrinB2 intracellular domain for developmental angiogenesis was also confirmed with a transgenic mouse line containing deletion of a single valine residue in the cytoplasmic domain (EphrinB2ΔV), which disrupts PDZ-dependent reverse signaling. EphrinB2ΔV showed impaired retina and brain vascularization with reduced vessel branching and vascular sprouts [95]. Similarly, *in vitro* experiments showed that EphrinB2 is a potent regulator of EC behavior by regulating the actin cytoskeleton through Rho-family small GTPases [96]. In addition, *in vitro* analyses show that the membrane-bound extracellular EphB4 domain is sufficient to induce EphrinB2-dependent EC invasion, proliferation, and overall survival [97] (Fig. 11.3c). While these studies clearly highlight the importance of EphrinB2 reverse signaling in developmental angiogenesis, the function of EphB4 forward signaling (upon EphrinB2 binding) in this process still remains unknown. Nevertheless, ablation of the EphB4 gene leads to embryonic lethality before E11.5 [98].

EphrinB2 is upregulated in ECs in response to angiogenic factors such as VEGF, fibroblast growth factor (FGF), hepatocyte growth factor (HGF), or activin receptor-like kinase-1 (Alk1) [a member of the transforming growth factor (TGF)-β family], whereas angiopoietin-1 (Ang1) attenuates its expression levels [87, 99]. *In vivo* and *in vitro* experiments show that EphrinB2 controls developmental angiogenesis and lymphangiogenesis by controlling EC sprouting activity rather than EC proliferation [88, 100]. EphrinB2 co-localizes and co-immunoprecipitates with VEGFR2 and VEGFR3. EphrinB2 reverse signaling (activated by EphB4) plays a crucial role in VEGFR2 and VEGFR3 receptor internalization, phosphorylation, and downstream signaling (Rac1, Akt, and Erk1/2) after VEGF stimulation [88, 95] (Fig. 11.3c). A recent study showed that a protein complex containing EphrinB2, the clathrin-associated sorting protein Dab2, and the cell polarity regulator Par3 associates with VEGFR2 in the clathrin-coated vesicle and mediates VEGFR2 trafficking towards the early endosome [101] (Fig. 11.3c, d). Atypical PKC (aPKC) acts as a negative regulator of this process by phosphorylating Dab2 and reducing VEGFR2 endocytosis [101] (Fig. 11.3d). Interestingly, sprouting ECs at the leading front of the mouse retina have lower levels of activated aPKC and higher rates of VEGFR endocytosis and turnover compared with nonsprouting ECs (Fig. 11.3d). Consistent with this model, EC-specific knockout mice for Dab2, Par3, and aPKC show vascular phenotypes that resemble those of EphrinB2 mutants [101].

Expression of EphrinB2 does not only localize to ECs but also to pericytes or SMCs [102, 103] (Fig. 11.3). EphrinB2 deficiency in the mural cell compartment led to embryonic vascular malformation and perinatal lethality [89, 103]. Analysis of the vasculature of these mice showed disrupted vessels and defects in the spatial organization of pericytes and SMCs, as revealed by microvessels showing scattered coverage with mural cells [102, 103]. *In vitro*, EphrinB2 expression in SMCs participated in SMC adhesion, migration, polarization, and formation of focal adhesions via a molecular mechanism involving Crk and p130(CAS) [89, 103].

In summary, EphrinB2 and EphB4 signals are essential for cellular segregation and vascular development, affecting ECs as well as mural cells in a proangiogenic fashion. Moreover, their interaction with VEGFRs, as well as their essential contribution to receptor turnover and signal transduction, makes them an integrative unit for different pathways in the concept of a neurovascular link.

11.4.3 Ephrins and Eph Receptors in Tumor Angiogenesis

Several studies report a role for EphrinB2 and EphB4 in tumor angiogenesis. In malignant brain tumors, EphrinB2 and EphB4 are expressed in both ECs and tumor cells [104]. EphB4 expressed in tumor cells leads to enhanced tumor angiogenesis via EphrinB2 reverse signaling in the endothelium [97] (Table. 11.1). Consistently, EphrinB2 functional blocking antibodies cause a reduction in angiogenesis and lymphangiogenesis in xenografted mice and a concomitant reduction in tumor growth [105]. Confirmation of the role of EphrinB2 reverse signaling in tumor angiogenesis was demonstrated using an orthotopic glioma tumor model and a skin heterotopic tumor model in the already-mentioned EprhinB2ΔV mice [95]. In these tumor models, angiogenesis and tumor growth was severely reduced compared with tumors in control mice. Moreover, similar to the studies using an EphrinB2 functional blocking antibody, tumor blood vessels in EprhinB2ΔV mice were devoid of sprouts and filopodia [95, 105]. EphrinB2 reverse signaling in tumor vessels was also shown to promote enlargement of blood vessels, the interaction between ECs and mural cells, and reduction of vessel permeability, leading to more stable tumor vessels [97, 104]. A soluble monomeric extracellular domain of EphB4 was shown to act as an antagonist of EphB4/EphrinB2 signaling and inhibited angiogenesis *in vitro* and *in vivo* as well as tumor growth in xenograft tumor models [100, 106].

With respect to other Ephrins and Eph members, few studies have involved EphA2 and its ligand EphrinA1 in tumor angiogenesis. EphrinA1 and EphA2 are expressed in tumor cells and tumor ECs [107]. Tumors grown in EphA2-deficient mice are smaller and less vascularized compared with tumors in wild-type mice [108, 109]. In addition, injection of soluble EphA2 and EphA3-Fc into tumor-bearing mice results in reduced tumor growth and reduced tumor angiogenesis [110, 111].

11.5 Semaphorins and Their Receptors

11.5.1 Introduction

Semaphorins (Sema) are membrane-bound, soluble signaling molecules that are able to signal long range via diffusion gradients, as well as short range via cell-to-cell contact. Dependent on their molecular structure, semaphorins are grouped into eight subclasses, with members of classes 3–7 expressed in vertebrates [112] (Fig. 11.4a). Semaphorins were first described in the nervous system where they function mainly as chemorepellents during axon guidance and neuronal wiring. However, similar to netrins and slits, there are reports highlighting a bifunctional role, whereby semaphorins can either repel or attract axonal growth cones, and also harbor both capacities [113]. Semaphorin signaling often induces cytoskeletal remodeling, affecting cellular properties such as shape, migration, and cellular connectivity [114]. The main receptors of semaphorins are plexins and neuropilins (Nrps). The nine members of the plexin family are divided into four subclasses (A, B, C, and D), whereas the group of neuropilins has two representative receptors, Nrp1 and Nrp2 [115] (Fig. 11.4a). Plexins are able to directly interact with membrane-bound semaphorins, whereas soluble semaphorin members (class 3 semaphorins only) often bind to neuropilins and signal through a heterodimer complex of neuropilin and plexin (with the exception of Sema3E, which binds directly to PlexinD1 and can induce signaling independent of Nrp1) [115]. The fact that neuropilins are known to be a co-receptor for other tyrosine kinase receptors (such as VEGFRs) makes semaphorin signaling an ideal pathway for studying their neurovascular properties [116].

Class 3 semaphorins are the best-studied semaphorins in the vascular system. Therefore, in this section we will focus on describing the fundamental aspects and functions of this group of molecules and their receptors in blood vessels. The emerging role of class 4 and 6 semaphorins in the vascular system will be briefly mentioned.

11.5.2 Class 3 Semaphorins in the Developing Vascular System

Class 3 Semaphorins and Plexins Despite their main function as axon repellents in neurons, in the vascular system only Sema3s are mainly described as antiangiogenic, whereas the literature refers to members of the other subclasses primarily in

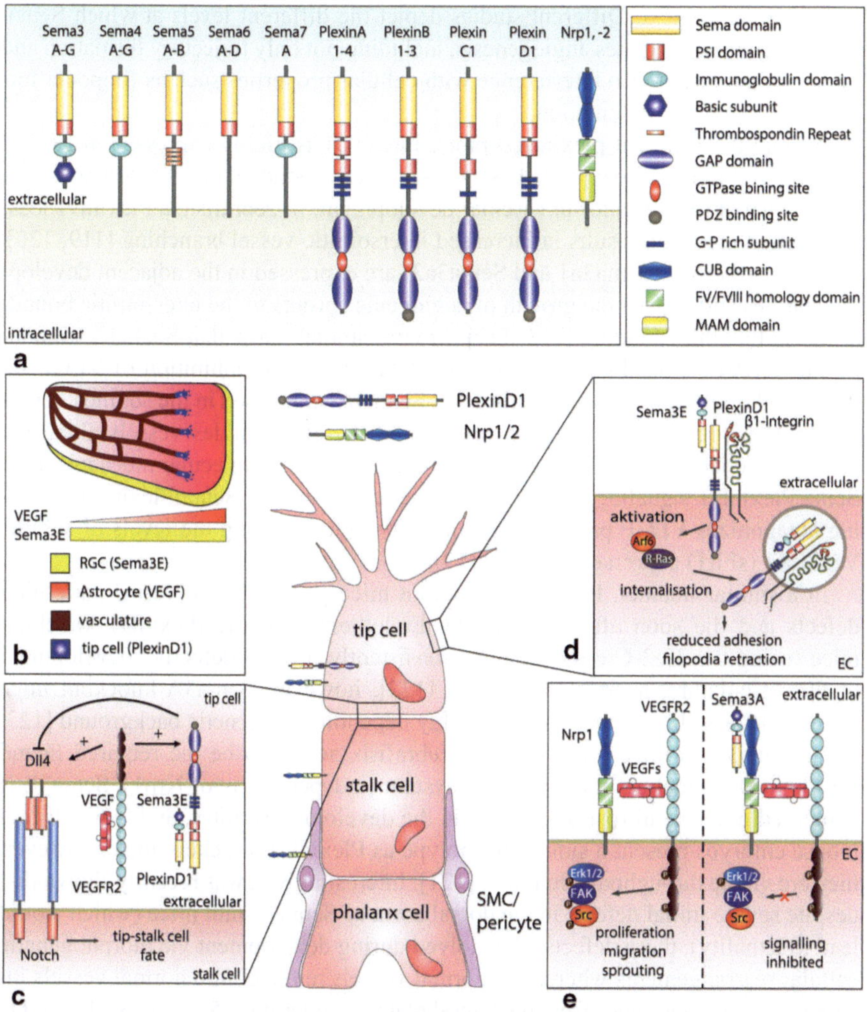

Fig. 11.4 Semaphorins and their receptors. **a** Class 3 semaphorins are secreted molecules, whereas members of the other semaphorin classes are membrane-bound. Semaphorin and plexin receptors possess a specific Sema domain. Neuropilins also interact with semaphorins. Nrp1 and 2 are expressed in the vasculature and in smooth muscle cells (SMCs)/pericytes, whereas the expression of PlexinD1 is restricted to the endothelial tip cell (see scheme of blood vessel sprout). **b** In the developing retina, Sema3E is evenly expressed by retinal ganglion cells (RGC) in deeper layers. The angiogenic vasculature grows in the direction of a VEGF gradient that is generated by astrocytes, showing a higher concentration in the avascular border regions. PlexinD1 is expressed in endothelial tip cells, which restricts the Sema3E effect locally to the angiogenic front. **c** VEGF/VEGFR2 signaling positively controls PlexinD1 expression in the endothelial tip cell. In turn, Sema3E/PlexinD1 signaling negatively regulates Dll4 in tip cells and thus controls tip-stalk cell fate. **d** PlexinD1 interacts with β1-integrins and, on Sema3E binding, activates Arf6 and R-Ras, leading to internalization of the receptor complex, reduced adhesion, and filopodia retraction. **e** Neuropilins not only serve as a receptor for semaphorins but also bind to VEGF ligands. Nrp1 plays a crucial role in mediating VEGF signaling by interacting with VEGFR2. The presence of Nrp1 modulates VEGFR2 downstream signaling pathways (including Erk1/2, FAK, and Src). Sema3A binding to Nrp1 does not block VEGFR2 phosphorylation but inhibits its downstream signaling. (Adapted with permission from Macmillan Publishers Ltd: Kitajewski [175], copyright 2011)

an activating context. Different studies depict the different levels at which Sema/Plexin signaling regulates angiogenesis, including not only trajectory formation and vessel guidance but also interference with cellular properties such as filopodia formation or receptor presentation.

Among the different plexin receptors, PlexinD1 is highly expressed in ECs of developing blood vessels [117, 118]. Gene targeting studies in zebrafish show the importance of this receptor in vascular development. In zebrafish, a PlexinD1 loss-of-function mutation results in increased intersomitic vessel branching [119, 120]. Sema3A orthologs (Sema3a1 and Sema3a2) are expressed in the adjacent developing somites and restrict the growth of angiogenic sprouts to the intersomitic boundaries via PlexinD1 binding [118, 119]. *In vitro* studies show that Sema3A stimulation of HUVECs results in loss of actin stress fibers and inhibition of VEGF-induced cell migration, thus suggesting that Sema3A, expressed in the somites, repels PlexinD1-expressing intersomitic vessels and therefore guides vessels along the somite boundaries [118]. Further studies reporting on the molecular mechanisms of Sema/PlexinD1 signaling in intersomitic vessels show that Sema/PlexinD1 signaling antagonizes VEGF proangiogenic activity by controlling the levels of soluble VEGFR1 (sFlt1) expressed by ECs [119].

In a similar manner, PlexinD1 knockout mice present blood vessel patterning defects and die soon after birth [121]. Endothelial-specific PlexinD1 knockout mice (using the Tie2-Cre mouse line) consistently present defective development of the vasculature, heart, and skeleton [122]; however, Sema3A knockout mice show none or mild vascular defects, highly depending on genetic background [123, 124]. Sema3E in mice, in contrast to zebrafish, seems to be the required ligand for PlexinD1 signaling in the vascular system (independently of Nrp1). Sema3E is expressed in a caudal-rostral gradient in the developing somites, and Sema3E null mouse embryos present a similar phenotype as PlexinD1 knockout mice; however, they are viable throughout adulthood [121]. Interestingly, it was recently shown that despite severe initial defects in the dorsal aorta of Sema3E null mice (which would lead to lethality), these defects are resolved during development via morphogenetic cellular rearrangements whereby abnormal vessels remodel into normal vessels instructed by guidance cues from the lateral plate mesoderm [125]. Sema3E/PlexinD1 signaling in the vasculature has also been investigated in the developing mouse retina, where astrocyte-derived VEGF induces PlexinD1 expression in ECs of the angiogenic front (Fig. 11.4b). Even though Sema3E expression is evenly distributed in the retinal ganglion cell layer, local PlexinD1 expression at the angiogenic front restricts Sema3E signaling to the sprouting front of the retina vasculature [126, 127] (Fig. 11.4bb). Further investigation into the molecular mechanisms revealed that Sema3E/PlexinD1 signaling negatively regulates Dll4 expression in the retinal vasculature and that, in turn, Sema3E/PlexinD1 signaling is required for a balanced ratio of tip and stalk cells [127] (Fig. 11.4c).

Apart from the *in vivo* studies described above, *in vitro* experiments show that, similar to their repulsive role in the nervous system, Sema3E/PlexinD1 signals induce the collapse of the actin skeleton [126]. Mechanistically, Sema3E activation of PlexinD1 in ECs *in vitro* activates Arf6, leading to integrin-β1 internalization and, consequently, to cell retraction by negatively regulating cell–extracellular matrix adhesive interactions [128] (Fig. 11.4d).

Finally, several plexins have been shown to functionally affect angiogenesis and to be tightly interconnected with VEGF signaling. EC-specific knockdown of PlexinA1 and PlexinA4, for example, inhibits VEGF- and FGF-induced angiogenesis *in vitro* by remodeling the actin cytoskeleton, whereby PlexinA4, in particular, directly interacts with VEGFR2 and FGF receptor-2 [129].

Class 3 Semaphorins and Neuropilins: Link to Vascular Endothelial Growth Factor Signaling The second main receptors of semaphorins are neuropilins (Fig. 11.4a). Neuropilins do not only bind to semaphorins but also to VEGF ligands. Thereby, Nrp1 binding of VEGFA165, A121 (*in vitro*), B, and C, and placental growth factor (PlGF), as well as Nrp2 binding to VEGFA165, A145, and C, and platelet-derived growth factor, are described [130].

Nrp1 knockdown in zebrafish, similar to VEGFR2 knockdown, negatively affects the establishment of an intersegmental circulation, whereby the Nrp1 and VEGFR2 double knockdown synergistically aggravates this phenotype [131]. Similarly, Nrp1 knockout mice are embryonically lethal due to impaired heart development and defective EC migration [132, 133]. Endothelial-specific knockout of Nrp1 also leads to underdeveloped and enlarged vessels [134, 135]. Transgenic mice expressing an Nrp1 mutant form unable to bind Sema3A ($Nrp1^{Sema-}$), yet still able to bind VEGF, show no vascular phenotype, indicating that Sema3A/Nrp signaling is not required during vascular development but VEGF/Nrp1 signaling is essential [134]. Consistently, as mentioned above, Sema3A knockout mice present no, or very mild (depending on the genetic background), vascular phenotype, but do present strong axonal patterning defects [123].

Structural studies showed that Sema3A and VEGF do not compete for binding to Nrp1, but have different binding sites in the extracellular domain of Nrp1 [136] (Fig. 11.4e). To further understand the function of Nrp1, functional blocking antibodies specific to Sema- or VEGF-binding domains of Nrp1 were generated (anti-$Nrp1^{A}$ anti-$Nrp1^{B}$, respectively) [137]. Both antibodies prevent Nrp1 complex formation with VEGFR2 but have little effect on VEGFR2-mediated events. Application of each of the antibodies reduced neovascularization in the rat corneal pocket assay and vascular remodeling in neonatal mouse retinas *in vivo* (anti-$Nrp1^{B}$ showed a more potent effect) [137]. These antibodies also reduced VEGF-induced EC migration and sprouting *in vitro*, thus indicating that, despite the fact that VEGFR2 signaling still occurs, an Nrp1/VEGFR2 complex is required for proper angiogenesis.

In vivo and *vitro* experiments show that Sema3A has antiangiogenic properties and inhibits VEGF-induced EC proliferation, migration, and survival [138] (Fig. 11.4e). Sema3A stimulation blocks VEGF-induced FAK and Src (but not Erk) [139] without affecting VEGFA165-induced phosphorylation of VEGFR2 [140], thus indicating that Sema3A blocks VEGF-mediated effects downstream of the receptor activation (Fig. 11.4e). Sema3A stimulation of HUVECs induces rapid collapsing of the actin cytoskeleton, which might be one of the key explanations for the reduced effect of VEGF-induced angiogenesis and proliferation *in vitro* [140]. Interestingly, Sema3A induces Nrp1-dependent vascular permeability by itself and even potentiates VEGF-induced vascular leakage *in vivo* via phosphorylation of VE-cadherin [139].

Neuropilins are single-pass transmembrane proteins possessing a short intracellular C-terminal domain that contains a PDZ domain. The generation of a transgenic mouse line lacking the Nrp1 intracellular domain demonstrated that this domain is not essential for developmental angiogenesis but is important for spatial separation of retinal arteries and veins [141] and for VEGF-dependent developmental and adult arteriogenesis [142]. Arteriogenesis defects were due to the absence of a PDZ-dependent interaction between Nrp1, VEGFR2, and synectin, which delayed the endocytic trafficking of VEGFR2 upon VEGF binding and led to reduced ERK activation, which is required for proper arteriogenesis [142]. Interestingly, Nrp1 endocytosis in response to VEGFA165 and Sema3C stimulation appears to be different, whereby VEGFA165 induced clathrin-mediated uptake and Sema3C lipid raft-dependent endocytosis [143], indicating that Nrp1 trafficking is determined by its ligand. A role for Sema3D/Nrp1 signaling in pulmonary venous patterning was also described. During development, Sema3D-expressing cells form a boundary that restricts the pulmonary endothelium expressing Nrp1 [144]. Sema3D consistently acts as a migratory repulsive cue via Nrp1 signaling in ECs *in vitro* [144].

Nrp2 forms complexes with PlexinA1–4, PlexinD1, VEGFR2, and VEGFR3. Different semaphorins, VEGFA, PlGF, and VEGFC bind Nrp2 [115]. In the vascular system, Nrp2 is mainly expressed in vein and lymphatic ECs. Consequently, Nrp2 null mice show fewer and smaller lymphatic vessels [145]. Application of a functional blocking antibody for Nrp2 to mouse embryos recapitulates the lymphatic defects shown in Nrp2 null mice [146]. Apart from acting via VEGF signaling, Nrp2 also seems to affect the vascular system upon binding to class 3 semaphorins. Sema3G is expressed in the angiogenic vasculature and regulates angiogenic sprouting in an autocrine manner. Sema3G signaling via Nrp2 was found not only on ECs but also on paracrine SMCs. Interestingly, full-length Sema3G binds to Nrp2 and acts on SMCs to induce migration, whereas processed cleaved Sema3G can bind Nrp1 and Nrp2 but are not able to induce SMC migration [147]. Sema3F, which also binds Nrp2, inhibits VEGFA165-induced HUVEC proliferation and tube formation *in vitro*, as well as VEGFA165-induced vascularization of alginate cell encapsulated plugs *in vivo* [148].

11.5.3 Role of Class 6 Semaphorins in the Vascular System

It has recently been described that Sema6A participates in angiogenesis. One study shows that Sema6A is expressed in ECs during phases of angiogenesis or vascular remodeling, and regulates EC survival and growth by modulating VEGFR2 expression and signaling [149]. Moreover, Sema6A null mice are vital and appear to show reduced developmental angiogenesis associated with increased EC death

[149]. In contrast to a proangiogenic, prosurvival role for Sema6A, another study shows that Sema6A acts as an angiogenesis inhibitor and its expression is controlled by miRNA27a/b [150]. *In vitro*, the extracellular domain of Sema6A is sufficient to inhibit VEGF-mediated Src, Erk, and FAK activation in ECs and inhibit angiogenesis [151].

Apart from Sema6A, not much is known about other members of the Sema6 proteins. However, Sema6B silencing in HUVECs inhibited FGF-induced proliferation, and Sema6D expression levels were increased in gastric carcinoma samples [129, 152]. These results point to a more proangiogenic function of sema6 direction for Sema6 members, which needs to be further evaluated, especially with regard to Sema6A.

11.5.4 Semaphorins and Their Receptors in Pathological Angiogenesis

Different semaphorins have been shown to play a role in cancer, a disease in which they have been mostly studied. Semaphorins are known to affect tumor progression by either acting on tumor cells directly, on the tumor microenvironment, and/or by modulating tumor angiogenesis [153]. Expression of semaphorins and their receptors has been found not only on tumor cells but also on different cellular types of the tumor microenvironment, including ECs. Moreover, the expression of certain semaphorins and their receptors seems to be regulated in response to hypoxia [154]. While certain Semaphorins display proangiogenic and protumorigenic properties (Sema4D), others seem to function as antitumor molecules (Sema3A, Sema3B, Sema3E, Sema3F, and Sema3G) (Table 11.1). Here we focus on describing the main effects of the different semaphorins on tumor ECs and tumor angiogenesis.

Semaphorins with Antiangiogenic Properties in Tumors Similar to developmental angiogenesis, class 3 semaphoring are also the best-studied semaphoring family members in tumor angiogenesis, displaying primarily inhibitory effects (Table 11.1). Using three different tumor models, it was shown that Sema3A is expressed in ECs in premalignant lesions and that it becomes downregulated in the tumor endothelium, correlating with enhanced vascularization and tumor progression [155]. Similarly, downregulation of Sema3A expression levels (as well as Sema3B and Sema3F, see below) correlates with increased invasiveness in breast cancer samples [156], thus suggesting a role for Sema3A as an antiangiogenic molecule in cancer. Indeed, pharmacological inhibition of Sema3A in these tumors led to enhanced angiogenesis and tumor progression. Moreover, ectopic administration of Sema3A decreased neovascularization and tumor development, accompanied by enhanced vascular normalization and pericyte recruitment [155]. In line with this, systemic Sema3A overexpression or tumor-targeted Sema3A delivery was shown to inhibit vessel function and increase tumor hypoxia and necrosis [157]. Anti-Nrp1 treatment in combination with anti-VEGF antibodies consistently shows an additive effect in reducing tumor growth and vascular density with lack of pericyte coverage [137].

Sema3F has also been shown to block tumor angiogenesis. It is markedly downregulated in highly metastatic human cell lines. When metastatic cells are transfected with Sema3F and implanted into mice, the resultant tumors do not metastasize but rather show large areas of cell death and less vascularization [158]. In addition, different tumor cells transfected with Sem3F develop smaller tumors with a significant lower concentration of blood vessels [159, 160]. *In vitro*, tumor cells expressing Sema3F chemorepelled endothelial and lymphatic cells via Nrp2 [158]. Interestingly, expression of Sema3F and Nrp2 are regulated by the tumor suppressor p53 [158].

Sema3G, Sema3B, and Nrp2 expression were found to correlate in glioma patient samples with prolonged survival [161]. Moreover, tumor cells expressing Sema3G or Sema3E, and implanted in the mammary fat pads of mice, develop tumors with a significantly reduced amount of tumor vessels compared with control cell suppression [162, 163]. Exogenous Sema3E consistently inhibits tumor angiogenesis [128].

Semaphorins with Proangiogenic Properties in Tumors Sema3C is the only member of the Sema3 subclass described so far to exert a proangiogenic effect (Table 11.1), even though the amount of literature is limited. Sema3C was found to be highly expressed in neoplastic cells of human gastric cancer samples and to be mainly located at the invasive front [164]. Using an orthotopic model in nude mice gastric cancer cells with silence Sema3C led to reduced primary tumors, reduced metastatic liver tumors, and reduced vessel density. *In vitro*, capillary-like tube formation was reduced by the addition of culture media of Sema3C-deficient tumor cells compared with media of control tumor cells [164].

Class 4 semaphorins are cell membrane anchor proteins mainly characterized to have a proangiogenic function. Among these, Sema4D is the most studied. (Table 11.1). Sema4D is expressed by several carcinoma cells derived, for example, from prostate, colon, breast, and lung tissue, as well as in head and neck squamous cell carcinomas. Its receptor, PlexinB1, is primarily expressed by ECs of these tumors [165]. Sema4D knockout mice show decreased tumor progression when challenged by a tumor-grafting assay [166, 167]. Tumor vasculature increase in oral squamous cell carcinoma xenograft models could be blocked with functional antibodies against Sema4D, thus indicating a proangiogenic role for Sema4D in these tumors [168]. Moreover, it could be shown that Sema4D specifically derived from tumor-associated macrophages (TAMs) appears to be important for neovascularization [166]. Sema4D synergizes the effects of VEGF and possesses a similar expression pattern as VEGF. In addition, its expression is regulated by hypoxia [167, 168]. Mechanistically, it was shown that membrane-bound Sema4D can be proteolytically cleaved by type-1 matrix metalloproteinase MT1-MMP (expressed by several tumor cell lines) to then bind its receptor, PlexinB1, in ECs and initiate the proangiogenic response [169]. Following its binding to PlexinB1, Sema4D promotes the phosphorylation of two receptor tyrosine kinases, Met and Ron, leading to induction of angiogenesis, tumor invasion, and metastasis [170].

Role of Semaphorins in Other Pathological Angiogenesis Conditions: Ischemia Ischemic retinopathies, characterized by increased extraretinal vessels, show (when recapitulated in *in vivo* mouse models) increased levels of PlexinD1 in blood vessels [126]. Consistent with an antiangiogenic role for Sema3E, pathological progression of ischemic retinopathy was reduced by intravitreal administration of Sema3E [126]. Moreover, Sema3E and PlexinD1 are upregulated in ischemic limbs but also in response to hypoxia and diabetes. Consequently, VEGF treatment subtly improved neovascularization in these ischemic models due to intrinsic Sema3E inhibition. However, VEGF treatment with additional blocking of Sema3E/PlexinD1 signaling increased vascular recovery [171]. Sema3A is also secreted by hypoxic neurons in the avascular retina and prevents vascular regeneration of the ischemic retina. Silencing Sema3A expression consistently enhances normal vascular regeneration [172].

Sema4A, as well as its receptors PlexinB2 and PlexinD1, are found on immune cells such as monocytes and macrophages, and macrophage activation leads to an increase in Sema4A, PlexinB2, and PlexinD1 expression [173]. In an *in vivo* cardiac ischemia/reperfusion model, Sema4A is highly expressed in the macrophages recruited to the ischemic area, and sustains macrophage-induced angiogenesis [173]. Acting via PlexinD1, Sema4A leads to enhanced expression of VEGF in macrophages, which in turn acts on ECs to induce an angiogenic response [173].

11.6 Conclusions

Discovery of the molecular link between the neuronal system and the vascular system within the last years has expanded and modified our understanding of angiogenesis. The previous portfolio of angiogenic signaling factors such as VEGF–VEGFR, Dll4–Notch, or angiopoietins–Tie receptors has been substantially enriched with the appearance of a whole set of neuronal molecules that also modulate vascular responses in health and disease. The prior, historically described canonical axon guidance cues (netrins, slits, ephrins, and semaphorins) astonishingly integrate into the concept of angiogenesis. Their overall effects in the vascular system appear to be very similar to their role in a neuronal context (Fig. 11.5). Interestingly, during an angiogenic process, each guidance cue appears to exert a distinct function. The range of action of these canonical axon guidance factors is quite broad, starting from the induction and maintenance of an activated vasculature, including the directed guidance of angiogenic tip-cell ending, and ending in the regulation of vascular-quiescence and crosstalk with other cell types.

The effect of netrin signaling in the angiogenic context is highly dependent on the receptor composition of cells of the vascular compartment. In particular, Netrin/Unc5 interactions show the capacity to stabilize the vasculature (Fig. 11.5). Netrin/Unc5 signaling on vascular sprouts might be seen as a mechanism to strengthen the newly formed vessels and thereafter prevent hypervascularization. Similarly, slits and their Robo receptors stabilize newly grown vessels by affecting stalk cells and regulating mural cell coverage (Fig. 11.5). Could Slit/Robo signaling, in addition to Dll4–Notch, be another set of ligand/receptor pairs that establish and maintain

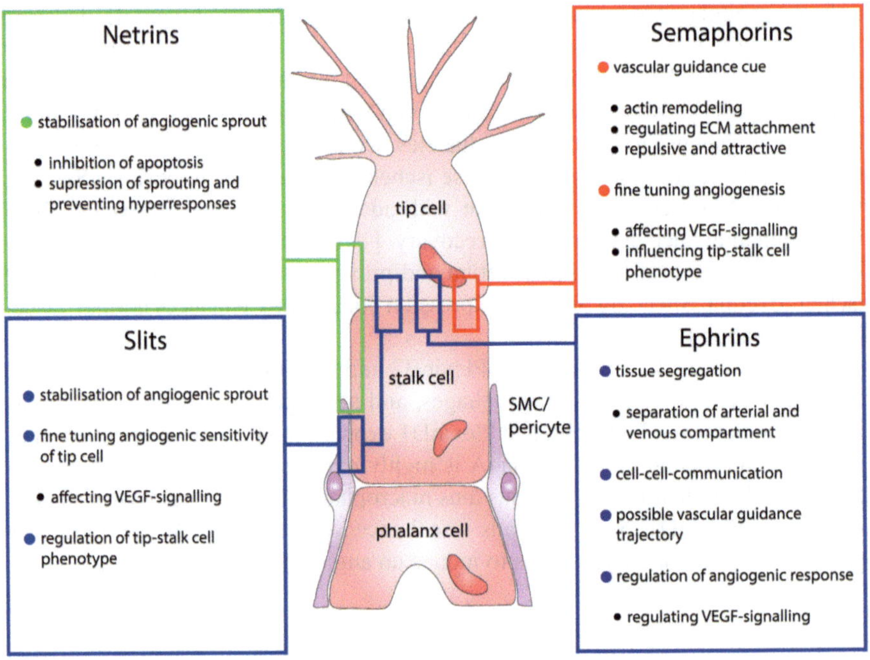

Fig. 11.5 Integrative overview of the role of axon guidance cues in angiogenesis netrins, slits, ephrins, and semaphorins have a defined role in the vascular system. Remarkably, all molecules possess their own functional mechanisms and influence the vascular system at diverse levels, reaching from activated angiogenic sprouts, over the communication between cells within the vascular compartment, and up to the crosstalk of endothelial cells with the surrounding tissue. This scheme summarizes their functions and illustrates the site of their effect. (Adapted with permission from Macmillan Publishers Ltd: Kitajewski [175], copyright 2011)

tip/stalk cells? Moreover, some semaphorins functionally support the induction of the tip-cell phenotype (Sema/Plexin signaling) and mainly affect tip cell sprouting (Fig. 11.5). Similar to descriptions in the axonal growth cone, semaphorin signaling induces remodeling of the actin cytoskeleton that in turn influences the further path of the newly growing vessel. Lastly, membrane-bound ephrins and their Eph receptors function overall as a bidirectional signal transducer between ECs, or between ECs and their surrounding tissue, which serves as a powerful property to pass on signals and affect a larger amount on cells within the vascular/cellular network (Fig. 11.5).

How the molecular mechanisms by which the different canonical axon guidance cues act on blood vessels and how some of them crosstalk and interact with other vascular signals, such as VEGF/VEGFR and Dll4/Notch, is becoming clearer. Moreover, interactions between receptors of netrins and slits have also been shown. Thus, further investigations into the molecular mechanisms that are controlled by these guidance factors and how their signaling is integrated to elicit a specific cellular response will help to expand our understanding of vascular, and perhaps neural, development and tissue homeostasis.

Finally, each of these canonical guidance signals seems to also play a role in pathological angiogenesis, primarily shown so far in a cancer setting. A better understanding of how netrins, slits, semaphorins, and ephrins affect angiogenesis in disease conditions will benefit the future design of therapeutic strategies that might help combat cancer and ischemia.

Acknowledgements This work is supported by the European Research Council (ERC-StG-2012; 311367 NeuroVascular Link).

References

1. Koch M, Murrell JR, Hunter DD, Olson PF, Jin W, Keene DR, et al. A novel member of the netrin family, beta-netrin, shares homology with the beta chain of laminin: identification, expression, and functional characterization. J Cell Biol. 2000;151(2):221–34.
2. Seaman C, Cooper HM. Netrin-3 protein is localized to the axons of motor, sensory, and sympathetic neurons. Mech Dev. 2001;101(1–2):245–8.
3. Nakashiba T, Ikeda T, Nishimura S, Tashiro K, Honjo T, Culotti JG, et al. Netrin-G1: a novel glycosyl phosphatidylinositol-linked mammalian netrin that is functionally divergent from classical netrins. J Neurosci. 2000;20(17):6540–50.
4. Kennedy TE, Serafini T, de la Torre JR, Tessier-Lavigne M. Netrins are diffusible chemotropic factors for commissural axons in the embryonic spinal cord. Cell. 1994;78(3):425–35.
5. Lin JC, Ho WH, Gurney A, Rosenthal A. The netrin-G1 ligand NGL-1 promotes the outgrowth of thalamocortical axons. Nat Neurosci. 2003;6(12):1270–6.
6. Yebra M, Montgomery AM, Diaferia GR, Kaido T, Silletti S, Perez B, et al. Recognition of the neural chemoattractant Netrin-1 by integrins alpha6beta4 and alpha3beta1 regulates epithelial cell adhesion and migration. Dev Cell. 2003;5(5):695–707.
7. Vielmetter J, Kayyem JF, Roman JM, Dreyer WJ. Neogenin, an avian cell surface protein expressed during terminal neuronal differentiation, is closely related to the human tumor suppressor molecule deleted in colorectal cancer. J Cell Biol. 1994;127(6 Pt 2):2009–20.
8. Ly A, Nikolaev A, Suresh G, Zheng Y, Tessier-Lavigne M, Stein E. DSCAM is a netrin receptor that collaborates with DCC in mediating turning responses to netrin-1. Cell. 2008;133(7):1241–54.
9. Kim S, Burette A, Chung HS, Kwon SK, Woo J, Lee HW, et al. NGL family PSD-95-interacting adhesion molecules regulate excitatory synapse formation. Nat Neurosci. 2006;9(10):1294–301.
10. Lai Wing Sun K, Correia JP, Kennedy TE. Netrins: versatile extracellular cues with diverse functions. Development. 2011;138(11):2153–69.
11. Hong K, Hinck L, Nishiyama M, Poo MM, Tessier-Lavigne M, Stein E. A ligand-gated association between cytoplasmic domains of UNC5 and DCC family receptors converts netrin-induced growth cone attraction to repulsion. Cell. 1999;97(7):927–41.
12. Keleman K, Dickson BJ. Short- and long-range repulsion by the Drosophila Unc5 netrin receptor. Neuron. 2001;32(4):605–17.
13. Forcet C, Ye X, Granger L, Corset V, Shin H, Bredesen DE, et al. The dependence receptor DCC (deleted in colorectal cancer) defines an alternative mechanism for caspase activation. Proc Natl Acad Sci U S A. 2001;98(6):3416–21.
14. Llambi F, Causeret F, Bloch-Gallego E, Mehlen P. Netrin-1 acts as a survival factor via its receptors UNC5H and DCC. EMBO J. 2001;20(11):2715–22.
15. Metin C, Deleglise D, Serafini T, Kennedy TE, Tessier-Lavigne M. A role for netrin-1 in the guidance of cortical efferents. Development. 1997;124(24):5063–74.

16. Hakanen J, Duprat S, Salminen M. Netrin1 is required for neural and glial precursor migrations into the olfactory bulb. Dev Biol. 2011;355(1):101–14.
17. Bloch-Gallego E, Ezan F, Tessier-Lavigne M, Sotelo C. Floor plate and netrin-1 are involved in the migration and survival of inferior olivary neurons. J Neurosci. 1999;19(11):4407–20.
18. Colamarino SA, Tessier-Lavigne M. The axonal chemoattractant netrin-1 is also a chemorepellent for trochlear motor axons. Cell. 1995;81(4):621–9.
19. Burgess RW, Jucius TJ, Ackerman SL. Motor axon guidance of the mammalian trochlear and phrenic nerves: dependence on the netrin receptor Unc5c and modifier loci. J Neurosci. 2006;26(21):5756–66.
20. Guthrie S, Pini A. Chemorepulsion of developing motor axons by the floor plate. Neuron. 1995;14(6):1117–30.
21. Guijarro P, Simo S, Pascual M, Abasolo I, Del Rio JA, Soriano E. Netrin1 exerts a chemorepulsive effect on migrating cerebellar interneurons in a Dcc-independent way. Mol Cell Neurosci. 2006;33(4):389–400.
22. Yin Y, Sanes JR, Miner JH. Identification and expression of mouse netrin-4. Mech Dev. 2000;96(1):115–9.
23. Nakashiba T, Nishimura S, Ikeda T, Itohara S. Complementary expression and neurite outgrowth activity of netrin-G subfamily members. Mech Dev. 2002;111(1–2):47–60.
24. Qin S, Yu L, Gao Y, Zhou R, Zhang C. Characterization of the receptors for axon guidance factor netrin-4 and identification of the binding domains. Mol Cell Neurosci. 2007;34(2):243–50.
25. Shimizu A, Nakayama H, Wang P, Konig C, Akino T, Sandlund J, et al. Netrin-1 promotes glioblastoma cell invasiveness and angiogenesis by multiple pathways including activation of RhoA, cathepsin B, and cAMP-response element-binding protein. J Biol Chem. 2013;288(4):2210–22.
26. Lejmi E, Leconte L, Pedron-Mazoyer S, Ropert S, Raoul W, Lavalette S, et al. Netrin-4 inhibits angiogenesis via binding to neogenin and recruitment of Unc5B. Proc Natl Acad Sci U S A. 2008;105(34):12491–6.
27. Lu X, Noble F L, Yuan L, Jiang Q, De Lafarge B, Sugiyama D, et al. The netrin receptor UNC5B mediates guidance events controlling morphogenesis of the vascular system. Nature. 2004;432(7014):179–86.
28. Park KW, Crouse D, Lee M, Karnik SK, Sorensen LK, Murphy KJ, et al. The axonal attractant Netrin-1 is an angiogenic factor. Proc Natl Acad Sci U S A. 2004;101(46):16210–5.
29. Wilson BD, Ii M, Park KW, Suli A, Sorensen LK, Larrieu-Lahargue F, et al. Netrins promote developmental and therapeutic angiogenesis. Science. 2006;313(5787):640–4.
30. Testori J, Schweighofer B, Helfrich I, Sturtzel C, Lipnik K, Gesierich S, et al. The VEGF-regulated transcription factor HLX controls the expression of guidance cues and negatively regulates sprouting of endothelial cells. Blood. 2011;117(9):2735–44.
31. Bouvree K, Larrivee B, Lv X, Yuan L, DeLafarge B, Freitas C, et al. Netrin-1 inhibits sprouting angiogenesis in developing avian embryos. Dev Biol. 2008;318(1):172–83.
32. Navankasattusas S, Whitehead KJ, Suli A, Sorensen LK, Lim AH, Zhao J, et al. The netrin receptor UNC5B promotes angiogenesis in specific vascular beds. Development. 2008;135(4):659–67.
33. Dumartin L, Quemener C, Laklai H, Herbert J, Bicknell R, Bousquet C, et al. Netrin-1 mediates early events in pancreatic adenocarcinoma progression, acting on tumor and endothelial cells. Gastroenterology. 2010;138(4):1595–606, 606 e1–8.
34. Castets M, Mehlen P. Netrin-1 role in angiogenesis: to be or not to be a pro-angiogenic factor? Cell Cycle. 2010;9(8):1466–71.
35. Castets M, Coissieux MM, Delloye-Bourgeois C, Bernard L, Delcros JG, Bernet A, et al. Inhibition of endothelial cell apoptosis by netrin-1 during angiogenesis. Dev Cell. 2009;16(4):614–20.
36. Nguyen A, Cai H. Netrin-1 induces angiogenesis via a DCC-dependent ERK1/2-eNOS feed-forward mechanism. Proc Natl Acad Sci U S A. 2006;103(17):6530–5.

37. Lim AH, Suli A, Yaniv K, Weinstein B, Li DY, Chien CB. Motoneurons are essential for vascular pathfinding. Development. 2011;138(17):3847–57.
38. Larrivee B, Freitas C, Trombe M, Lv X, Delafarge B, Yuan L, et al. Activation of the UNC5B receptor by Netrin-1 inhibits sprouting angiogenesis. Genes Dev. 2007;21(19):2433–47.
39. Xu H, Liu J, Xiong S, Le YZ, Xia X. Suppression of retinal neovascularization by lentivirus-mediated netrin-1 small hairpin. Ophthalmic Res. 2012;47(3):163–9.
40. Li Q, Yao D, Ma J, Zhu J, Xu X, Ren Y, et al. Transplantation of MSCs in combination with netrin-1 improves neoangiogenesis in a rat model of hind limb ischemia. J Surg Res. 2011;166(1):162–9.
41. Liu J, Xia X, Xiong S, Le Y, Xu H. Intravitreous high expression level of netrin-1 in patients with proliferative diabetic retinopathy. Eye Sci. 2011;26(2):35–42.
42. Lambert E, Coissieux MM, Laudet V, Mehlen P. Netrin-4 acts as a pro-angiogenic factor during zebrafish development. J Biol Chem. 2012;287(6):3987–99.
43. Nacht M, Martin TB St, Byrne A, Klinger KW, Teicher BA, Madden SL, et al. Netrin-4 regulates angiogenic responses and tumor cell growth. Exp Cell Res. 2009;315(5):784–94.
44. Larrieu-Lahargue F, Welm AL, Thomas KR, Li DY. Netrin-4 induces lymphangiogenesis in vivo. Blood. 2010;115(26):5418–26.
45. Larrieu-Lahargue F, Welm AL, Thomas KR, Li DY. Netrin-4 activates endothelial integrin {alpha}6{beta}1. Circ Res. 2011;109(7):770–4.
46. Hoang S, Liauw J, Choi M, Choi M, Guzman RG, Steinberg GK. Netrin-4 enhances angiogenesis and neurologic outcome after cerebral ischemia. J Cereb Blood Flow Metab. 2009;29(2):385–97.
47. Eveno C, Broqueres-You D, Feron JG, Rampanou A, Tijeras-Raballand A, Ropert S, et al. Netrin-4 delays colorectal cancer carcinomatosis by inhibiting tumor angiogenesis. Am J Pathol. 2011;178(4):1861–9.
48. Mehlen P, Delloye-Bourgeois C, Chedotal A. Novel roles for Slits and netrins: axon guidance cues as anticancer targets? Nat Rev Cancer. 2011;11(3):188–97.
49. Huminiecki L, Gorn M, Suchting S, Poulsom R, Bicknell R. Magic roundabout is a new member of the roundabout receptor family that is endothelial specific and expressed at sites of active angiogenesis. Genomics. 2002;79(4):547–52.
50. Park KW, Morrison CM, Sorensen LK, Jones CA, Rao Y, Chien CB, et al. Robo4 is a vascular-specific receptor that inhibits endothelial migration. Dev Biol. 2003;261(1):251–67.
51. Guijarro-Munoz I, Cuesta AM, Alvarez-Cienfuegos A, Geng JG, Alvarez-Vallina L, Sanz L. The axonal repellent Slit2 inhibits pericyte migration: potential implications in angiogenesis. Exp Cell Res. 2012;318(4):371–8.
52. Whitford KL, Marillat V, Stein E, Goodman CS, Tessier-Lavigne M, Chedotal A, et al. Regulation of cortical dendrite development by Slit-Robo interactions. Neuron. 2002;33(1):47–61.
53. Shu T, Sundaresan V, McCarthy MM, Richards LJ. Slit2 guides both precrossing and postcrossing callosal axons at the midline in vivo. J Neurosci. 2003;23(22):8176–84.
54. Kidd T, Bland KS, Goodman CS. Slit is the midline repellent for the robo receptor in Drosophila. Cell. 1999;96(6):785–94.
55. Shu T, Richards LJ. Cortical axon guidance by the glial wedge during the development of the corpus callosum. J Neurosci. 2001;21(8):2749–58.
56. Chen Z, Gore BB, Long H, Ma L, Tessier-Lavigne M. Alternative splicing of the Robo3 axon guidance receptor governs the midline switch from attraction to repulsion. Neuron. 2008;58(3):325–32.
57. Stein E, Tessier-Lavigne M. Hierarchical organization of guidance receptors: silencing of netrin attraction by slit through a Robo/DCC receptor complex. Science. 2001;291(5510):1928–38.
58. Jones CA, London NR, Chen H, Park KW, Sauvaget D, Stockton RA, et al. Robo4 stabilizes the vascular network by inhibiting pathologic angiogenesis and endothelial hyperpermeability. Nat Med. 2008;14(4):448–53.
59. Kaur S, Castellone MD, Bedell VM, Konar M, Gutkind JS, Ramchandran R. Robo4 signaling in endothelial cells implies attraction guidance mechanisms. J Biol Chem. 2006;281(16):11347–56.

60. Jones CA, Nishiya N, London NR, Zhu W, Sorensen LK, Chan AC, et al. Slit2-Robo4 signalling promotes vascular stability by blocking Arf6 activity. Nat Cell Biol. 2009;11(11):1325–31.
61. Zhang B, Dietrich UM, Geng JG, Bicknell R, Esko JD, Wang L. Repulsive axon guidance molecule Slit3 is a novel angiogenic factor. Blood. 2009;114(19):4300–9.
62. Marlow R, Binnewies M, Sorensen LK, Monica SD, Strickland P, Forsberg EC, et al. Vascular Robo4 restricts proangiogenic VEGF signaling in breast. Proc Natl Acad Sci U S A. 2010;107(23):10520–5.
63. London NR, Li DY. Robo4-dependent Slit signaling stabilizes the vasculature during pathologic angiogenesis and cytokine storm. Curr Opin Hematol. 2011;18(3):186–90.
64. Koch AW, Mathivet T, Larrivee B, Tong RK, Kowalski J, Pibouin-Fragner L, et al. Robo4 maintains vessel integrity and inhibits angiogenesis by interacting with UNC5B. Dev Cell. 2011;20(1):33–46.
65. Hussain SA, Piper M, Fukuhara N, Strochlic L, Cho G, Howitt JA, et al. A molecular mechanism for the heparan sulfate dependence of slit-robo signaling. J Biol Chem. 2006;281(51):39693–8.
66. Fukuhara N, Howitt JA, Hussain SA, Hohenester E. Structural and functional analysis of slit and heparin binding to immunoglobulin-like domains 1 and 2 of Drosophila Robo. J Biol Chem. 2008;283(23):16226–34.
67. Sheldon H, Andre M, Legg JA, Heal P, Herbert JM, Sainson R, et al. Active involvement of Robo1 and Robo4 in filopodia formation and endothelial cell motility mediated via WASP and other actin nucleation-promoting factors. FASEB J. 2009;23(2):513–22.
68. Wang B, Xiao Y, Ding BB, Zhang N, Yuan X, Gui L, et al. Induction of tumor angiogenesis by Slit-Robo signaling and inhibition of cancer growth by blocking Robo activity. Cancer Cell. 2003;4(1):19–29.
69. Liao WX, Laurent LC, Agent S, Hodges J, Chen DB. Human placental expression of SLIT/ROBO signaling cues: effects of preeclampsia and hypoxia. Biol Reprod. 2012;86(4):111.
70. Wang LJ, Zhao Y, Han B, Ma YG, Zhang J, Yang DM, et al. Targeting Slit-Roundabout signaling inhibits tumor angiogenesis in chemical-induced squamous cell carcinogenesis. Cancer Sci. 2008;99(3):510–7.
71. Yang XM, Han HX, Sui F, Dai YM, Chen M, Geng JG. Slit-Robo signaling mediates lymphangiogenesis and promotes tumor lymphatic metastasis. Biochem Biophys Res Commun. 2010;396(2):571–7.
72. Grone J, Doebler O, Loddenkemper C, Hotz B, Buhr HJ, Bhargava S. Robo1/Robo4: differential expression of angiogenic markers in colorectal cancer. Oncol Rep. 2006;15(6):1437–43.
73. Lemke G. A coherent nomenclature for Eph receptors and their ligands. Mol Cell Neurosci. 1997;9(5–6):331–2.
74. Pasquale EB. Eph-ephrin promiscuity is now crystal clear. Nat Neurosci. 2004;7(5):417–8.
75. Himanen JP, Chumley MJ, Lackmann M, Li C, Barton WA, Jeffrey PD, et al. Repelling class discrimination: ephrin-A5 binds to and activates EphB2 receptor signaling. Nat Neurosci. 2004;7(5):501–9.
76. Triplett JW, Feldheim DA. Eph and ephrin signaling in the formation of topographic maps. Semin Cell Dev Biol. 2012;23(1):7–15.
77. Suetterlin P, Marler KM, Drescher U. Axonal ephrinA/EphA interactions, and the emergence of order in topographic projections. Semin Cell Dev Biol. 2012;23(1):1–6.
78. Jiao JW, Feldheim DA, Chen DF. Ephrins as negative regulators of adult neurogenesis in diverse regions of the central nervous system. Proc Natl Acad Sci U S A. 2008;105(25):8778–83.
79. Hruska M, Dalva MB. Ephrin regulation of synapse formation, function and plasticity. Mol Cell Neurosci. 2012;50(1):35–44.
80. Janes PW, Nievergall E, Lackmann M. Concepts and consequences of Eph receptor clustering. Semin Cell Dev Biol. 2012;23(1):43–50.
81. Zhou N, Zhao WD, Liu DX, Liang Y, Fang WG, Li B, et al. Inactivation of EphA2 promotes tight junction formation and impairs angiogenesis in brain endothelial cells. Microvasc Res. 2011;82(2):113–21.

82. Cheng N, Brantley DM, Liu H, Lin Q, Enriquez M, Gale N, et al. Blockade of EphA receptor tyrosine kinase activation inhibits vascular endothelial cell growth factor-induced angiogenesis. Mol Cancer Res. 2002;1(1):2–11.
83. McBride JL, Ruiz JC. Ephrin-A1 is expressed at sites of vascular development in the mouse. Mech Dev. 1998;77(2):201–4.
84. Chen J, Hicks D, Brantley-Sieders D, Cheng N, McCollum GW, Qi-Werdich X, et al. Inhibition of retinal neovascularization by soluble EphA2 receptor. Exp Eye Res. 2006;82(4):664–73.
85. Okazaki T, Ni A, Baluk P, Ayeni OA, Kearley J, Coyle AJ, et al. Capillary defects and exaggerated inflammatory response in the airways of EphA2-deficient mice. Am J Pathol. 2009;174(6):2388–99.
86. Obi S, Yamamoto K, Shimizu N, Kumagaya S, Masumura T, Sokabe T, et al. Fluid shear stress induces arterial differentiation of endothelial progenitor cells. J Appl Physiol. 2009;106(1):203–11.
87. Hayashi S, Asahara T, Masuda H, Isner JM, Losordo DW. Functional ephrin-B2 expression for promotive interaction between arterial and venous vessels in postnatal neovascularization. Circulation. 2005;111(17):2210–8.
88. Wang Y, Nakayama M, Pitulescu ME, Schmidt TS, Bochenek ML, Sakakibara A, et al. Ephrin-B2 controls VEGF-induced angiogenesis and lymphangiogenesis. Nature. 2010;465(7297):483–6.
89. Zhang XQ, Takakura N, Oike Y, Inada T, Gale NW, Yancopoulos GD, et al. Stromal cells expressing ephrin-B2 promote the growth and sprouting of ephrin-B2(+) endothelial cells. Blood. 2001;98(4):1028–37.
90. Kim YH, Hu H, Guevara-Gallardo S, Lam MT, Fong SY, Wang RA. Artery and vein size is balanced by Notch and ephrin B2/EphB4 during angiogenesis. Development. 2008;135(22):3755–64.
91. Herbert SP, Huisken J, Kim TN, Feldman ME, Houseman BT, Wang RA, et al. Arterial-venous segregation by selective cell sprouting: an alternative mode of blood vessel formation. Science. 2009;326(5950):294–8.
92. Luxey M, Laussu J, Jungas T, Davy A. Generation of transgenic mice overexpressing EfnB2 in endothelial cells. Genesis. 2011;49(10):811–20.
93. Oike Y, Ito Y, Hamada K, Zhang XQ, Miyata K, Arai F, et al. Regulation of vasculogenesis and angiogenesis by EphB/ephrin-B2 signaling between endothelial cells and surrounding mesenchymal cells. Blood. 2002;100(4):1326–33.
94. Adams RH, Diella F, Hennig S, Helmbacher F, Deutsch U, Klein R. The cytoplasmic domain of the ligand ephrinB2 is required for vascular morphogenesis but not cranial neural crest migration. Cell. 2001;104(1):57–69.
95. Sawamiphak S, Seidel S, Essmann CL, Wilkinson GA, Pitulescu ME, Acker T, et al. Ephrin-B2 regulates VEGFR2 function in developmental and tumour angiogenesis. Nature. 2010;465(7297):487–91.
96. Bochenek ML, Dickinson S, Astin JW, Adams RH, Nobes CD. Ephrin-B2 regulates endothelial cell morphology and motility independently of Eph-receptor binding. J Cell Sci. 2010;123(8):1235–46.
97. Noren NK, Lu M, Freeman AL, Koolpe M, Pasquale EB. Interplay between EphB4 on tumor cells and vascular ephrin-B2 regulates tumor growth. Proc Natl Acad Sci U S A. 2004;101(15):5583–8.
98. Gerety SS, Wang HU, Chen ZF, Anderson DJ. Symmetrical mutant phenotypes of the receptor EphB4 and its specific transmembrane ligand ephrin-B2 in cardiovascular development. Mol Cell. 1999;4(3):403–14.
99. Kim JH, Peacock MR, George SC, Hughes CC. BMP9 induces EphrinB2 expression in endothelial cells through an Alk1-BMPRII/ActRII-ID1/ID3-dependent pathway: implications for hereditary hemorrhagic telangiectasia type II. Angiogenesis. 2012;15(3):497–509.
100. Kertesz N, Krasnoperov V, Reddy R, Leshanski L, Kumar SR, Zozulya S, et al. The soluble extracellular domain of EphB4 (sEphB4) antagonizes EphB4-EphrinB2 interaction, modulates angiogenesis, and inhibits tumor growth. Blood. 2006;107(6):2330–8.

101. Nakayama M, Nakayama A, van Lessen M, Yamamoto H, Hoffmann S, Drexler HC, et al. Spatial regulation of VEGF receptor endocytosis in angiogenesis. Nat Cell Biol. 2013;15(3):249–60.
102. Salvucci O, Maric D, Economopoulou M, Sakakibara S, Merlin S, Follenzi A, et al. EphrinB reverse signaling contributes to endothelial and mural cell assembly into vascular structures. Blood. 2009;114(8):1707–16.
103. Foo SS, Turner CJ, Adams S, Compagni A, Aubyn D, Kogata N, et al. Ephrin-B2 controls cell motility and adhesion during blood-vessel-wall assembly. Cell. 2006;124(1):161–73.
104. Erber R, Eichelsbacher U, Powajbo V, Korn T, Djonov V, Lin J, et al. EphB4 controls blood vascular morphogenesis during postnatal angiogenesis. EMBO J. 2006;25(3):628–41.
105. Abengozar MA, de Frutos S, Ferreiro S, Soriano J, Perez-Martinez M, Olmeda D, et al. Blocking ephrinB2 with highly specific antibodies inhibits angiogenesis, lymphangiogenesis, and tumor growth. Blood. 2012;119(19):4565–76.
106. Martiny-Baron G, Korff T, Schaffner F, Esser N, Eggstein S, Marme D, et al. Inhibition of tumor growth and angiogenesis by soluble EphB4. Neoplasia. 2004;6(3):248–57.
107. Ogawa K, Pasqualini R, Lindberg RA, Kain R, Freeman AL, Pasquale EB. The ephrin-A1 ligand and its receptor, EphA2, are expressed during tumor neovascularization. Oncogene. 2000;19(52):6043–52.
108. Fang WB, Brantley-Sieders DM, Parker MA, Reith AD, Chen J. A kinase-dependent role for EphA2 receptor in promoting tumor growth and metastasis. Oncogene. 2005;24(53):7859–68.
109. Brantley-Sieders DM, Fang WB, Hicks DJ, Zhuang G, Shyr Y, Chen J. Impaired tumor microenvironment in EphA2-deficient mice inhibits tumor angiogenesis and metastatic progression. FASEB J. 2005;19(13):1884–6.
110. Brantley DM, Cheng N, Thompson EJ, Lin Q, Brekken RA, Thorpe PE, et al. Soluble Eph A receptors inhibit tumor angiogenesis and progression in vivo. Oncogene. 2002;21(46):7011–26.
111. Dobrzanski P, Hunter K, Jones-Bolin S, Chang H, Robinson C, Pritchard S, et al. Antiangiogenic and antitumor efficacy of EphA2 receptor antagonist. Cancer Res. 2004;64(3):910–9.
112. Pasterkamp RJ, Kolodkin AL. Semaphorin junction: making tracks toward neural connectivity. Curr Opin Neurobiol. 2003;13(1):79–89.
113. Eichmann A, Makinen T, Alitalo K. Neural guidance molecules regulate vascular remodeling and vessel navigation. Genes Dev. 2005;19(9):1013–21.
114. Serini G, Napione L, Arese M, Bussolino F. Besides adhesion: new perspectives of integrin functions in angiogenesis. Cardiovasc Res. 2008;78(2):213–22.
115. Neufeld G, Kessler O. The semaphorins: versatile regulators of tumour progression and tumour angiogenesis. Nat Rev Cancer. 2008;8(8):632–45.
116. Whitaker GB, Limberg BJ, Rosenbaum JS. Vascular endothelial growth factor receptor-2 and neuropilin-1 form a receptor complex that is responsible for the differential signaling potency of VEGF(165) and VEGF(121). J Biol Chem. 2001;276(27):25520–31.
117. van der Zwaag B, Hellemons AJ, Leenders WP, Burbach JP, Brunner HG, Padberg GW, et al. PLEXIN-D1, a novel plexin family member, is expressed in vascular endothelium and the central nervous system during mouse embryogenesis. Dev Dyn. 2002;225(3):336–43.
118. Torres-Vazquez J, Gitler AD, Fraser SD, Berk JD, Van NP, Fishman MC, et al. Semaphorin-plexin signaling guides patterning of the developing vasculature. Dev Cell. 2004;7(1):117–23.
119. Zygmunt T, Gay CM, Blondelle J, Singh MK, Flaherty KM, Means PC, et al. Semaphorin-PlexinD1 signaling limits angiogenic potential via the VEGF decoy receptor sFlt1. Dev Cell. 2011;21(2):301–14.
120. Gay CM, Zygmunt T, Torres-Vazquez J. Diverse functions for the semaphorin receptor PlexinD1 in development and disease. Dev Biol. 2011;349(1):1–19.
121. Gu C, Yoshida Y, Livet J, Reimert DV, Mann F, Merte J, et al. Semaphorin 3E and plexin-D1 control vascular pattern independently of neuropilins. Science. 2005;307(5707):265–8.

122. Zhang Y, Singh MK, Degenhardt KR, Lu MM, Bennett J, Yoshida Y, et al. Tie2Cre-mediated inactivation of plexinD1 results in congenital heart, vascular and skeletal defects. Dev Biol. 2009;325(1):82–93.
123. Vieira JM, Schwarz Q, Ruhrberg C. Selective requirements for NRP1 ligands during neurovascular patterning. Development. 2007;134(10):1833–43.
124. Serini G, Valdembri D, Zanivan S, Morterra G, Burkhardt C, Caccavari F, et al. Class 3 semaphorins control vascular morphogenesis by inhibiting integrin function. Nature. 2003;424(6947):391–7.
125. Meadows SM, Ratliff LA, Singh MK, Epstein JA, Cleaver O. Resolution of defective dorsal aortae patterning in Sema3E-deficient mice occurs via angiogenic remodeling. Dev Dyn. 2013;242(5):580–90.
126. Fukushima Y, Okada M, Kataoka H, Hirashima M, Yoshida Y, Mann F, et al. Sema3E-PlexinD1 signaling selectively suppresses disoriented angiogenesis in ischemic retinopathy in mice. J Clin Invest. 2011;121(5):1974–85.
127. Kim J, Oh WJ, Gaiano N, Yoshida Y, Gu C. Semaphorin 3E-Plexin-D1 signaling regulates VEGF function in developmental angiogenesis via a feedback mechanism. Genes Dev. 2011;25(13):1399–411.
128. Sakurai A, Gavard J, Annas-Linhares Y, Basile JR, Amornphimoltham P, Palmby TR, et al. Semaphorin 3E initiates antiangiogenic signaling through plexin D1 by regulating Arf6 and R-Ras. Mol Cell Biol. 2010;30(12):3086–98.
129. Kigel B, Rabinowicz N, Varshavsky A, Kessler O, Neufeld G. Plexin-A4 promotes tumor progression and tumor angiogenesis by enhancement of VEGF and bFGF signaling. Blood. 2011;118(15):4285–96.
130. Neufeld G, Cohen T, Shraga N, Lange T, Kessler O, Herzog Y. The neuropilins: multifunctional semaphorin and VEGF receptors that modulate axon guidance and angiogenesis. Trends Cardiovasc Med. 2002;12(1):13–9.
131. Lee P, Goishi K, Davidson AJ, Mannix R, Zon L, Klagsbrun M. Neuropilin-1 is required for vascular development and is a mediator of VEGF-dependent angiogenesis in zebrafish. Proc Natl Acad Sci U S .A. 2002;99(16):10470–5.
132. Kawasaki T, Kitsukawa T, Bekku Y, Matsuda Y, Sanbo M, Yagi T, et al. A requirement for neuropilin-1 in embryonic vessel formation. Development. 1999;126(21):4895–902.
133. Gerhardt H, Ruhrberg C, Abramsson A, Fujisawa H, Shima D, Betsholtz C. Neuropilin-1 is required for endothelial tip cell guidance in the developing central nervous system. Dev Dyn. 2004;231(3):503–9.
134. Gu C, Rodriguez ER, Reimert DV, Shu T, Fritzsch B, Richards LJ, et al. Neuropilin-1 conveys semaphorin and VEGF signaling during neural and cardiovascular development. Dev Cell. 2003;5(1):45–57.
135. Fantin A, Vieira JM, Plein A, Denti L, Fruttiger M, Pollard JW, et al. NRP1 acts cell autonomously in endothelium to promote tip cell function during sprouting angiogenesis. Blood. 2013;121(12):2352–62.
136. Appleton BA, Wu P, Maloney J, Yin J, Liang WC, Stawicki S, et al. Structural studies of neuropilin/antibody complexes provide insights into semaphorin and VEGF binding. EMBO J. 2007;26(23):4902–12.
137. Pan Q, Chanthery Y, Liang WC, Stawicki S, Mak J, Rathore N, et al. Blocking neuropilin-1 function has an additive effect with anti-VEGF to inhibit tumor growth. Cancer Cell. 2007;11(1):53–67.
138. Sakurai A, Doci CL, Gutkind JS. Semaphorin signaling in angiogenesis, lymphangiogenesis and cancer. Cell Res. 2012;22(1):23–32.
139. Acevedo LM, Barillas S, Weis SM, Gothert JR, Cheresh DA. Semaphorin 3A suppresses VEGF-mediated angiogenesis yet acts as a vascular permeability factor. Blood. 2008;111(5):2674–80.
140. Guttmann-Raviv N, Shraga-Heled N, Varshavsky A, Guimaraes-Sternberg C, Kessler O, Neufeld G. Semaphorin-3A and semaphorin-3F work together to repel endothelial cells and to inhibit their survival by induction of apoptosis. J Biol Chem. 2007;282(36):26294–305.

141. Fantin A, Schwarz Q, Davidson K, Normando EM, Denti L, Ruhrberg C. The cytoplasmic domain of neuropilin 1 is dispensable for angiogenesis, but promotes the spatial separation of retinal arteries and veins. Development. 2011;138(19):4185–91.
142. Lanahan A, Zhang X, Fantin A, Zhuang Z, Rivera-Molina F, Speichinger K, et al. The neuropilin 1 cytoplasmic domain is required for VEGF-A-dependent arteriogenesis. Dev Cell. 2013;25(2):156–68.
143. Salikhova A, Wang L, Lanahan AA, Liu M, Simons M, Leenders WP, et al. Vascular endothelial growth factor and semaphorin induce neuropilin-1 endocytosis via separate pathways. Circ Res. 2008;103(6):e71–9.
144. Degenhardt K, Singh MK, Aghajanian H, Massera D, Wang Q, Li J, et al. Semaphorin 3d signaling defects are associated with anomalous pulmonary venous connections. Nat Med. 2013;19(6):760–5.
145. Yuan L, Moyon D, Pardanaud L, Breant C, Karkkainen MJ, Alitalo K, et al. Abnormal lymphatic vessel development in neuropilin 2 mutant mice. Development. 2002;129(20):4797–806.
146. Xu Y, Yuan L, Mak J, Pardanaud L, Caunt M, Kasman I, et al. Neuropilin-2 mediates VEGF-C-induced lymphatic sprouting together with VEGFR3. J Cell Biol. 2010;188(1):115–30.
147. Kutschera S, Weber H, Weick A, De Smet F, Genove G, Takemoto M, et al. Differential endothelial transriptomics identifies semaphorin 3G as a vascular class 3 semaphorin. Arterioscler Thromb Vasc Biol. 2011;31(1):151–9.
148. Kessler O, Shraga-Heled N, Lange T, Gutmann-Raviv N, Sabo E, Baruch L, et al. Semaphorin-3F is an inhibitor of tumor angiogenesis. Cancer Res. 2004;64(3):1008–15.
149. Segarra M, Ohnuki H, Maric D, Salvucci O, Hou X, Kumar A, et al. Semaphorin 6A regulates angiogenesis by modulating VEGF signaling. Blood. 2012;120(19):4104–15.
150. Urbich C, Kaluza D, Fromel T, Knau A, Bennewitz K, Boon RA, et al. MicroRNA-27a/b controls endothelial cell repulsion and angiogenesis by targeting semaphorin 6A. Blood. 2012;119(6):1607–16.
151. Dhanabal M, Wu F, Alvarez E, McQueeney KD, Jeffers M, MacDougall J, et al. Recombinant semaphorin 6A-1 ectodomain inhibits in vivo growth factor and tumor cell line-induced angiogenesis. Cancer Biol Ther. 2005;4(6):659–68.
152. Zhao XY, Chen L, Xu Q, Li YH. Expression of semaphorin 6D in gastric carcinoma and its significance. World J Gastroenterol. 2006;12(45):7388–90.
153. Tamagnone L. Emerging role of semaphorins as major regulatory signals and potential therapeutic targets in cancer. Cancer Cell. 2012;22(2):145–52.
154. Neufeld G, Sabag AD, Rabinovicz N, Kessler O. Semaphorins in angiogenesis and tumor progression. Cold Spring Harb Perspect Med. 2012;2(1):a006718.
155. Maione F, Molla F, Meda C, Latini R, Zentilin L, Giacca M, et al. Semaphorin 3A is an endogenous angiogenesis inhibitor that blocks tumor growth and normalizes tumor vasculature in transgenic mouse models. J Clin Invest. 2009;119(11):3356–72.
156. Staton CA, Shaw LA, Valluru M, Hoh L, Koay I, Cross SS, et al. Expression of class 3 semaphorins and their receptors in human breast neoplasia. Histopathology. 2011;59(2):274–82.
157. Casazza A, Fu X, Johansson I, Capparuccia L, Andersson F, Giustacchini A, et al. Systemic and targeted delivery of semaphorin 3A inhibits tumor angiogenesis and progression in mouse tumor models. Arterioscler Thromb Vasc Biol. 2011;31(4):741–9.
158. Futamura M, Kamino H, Miyamoto Y, Kitamura N, Nakamura Y, Ohnishi S, et al. Possible role of semaphorin 3F, a candidate tumor suppressor gene at 3p21.3, in p53-regulated tumor angiogenesis suppression. Cancer Res. 2007;67(4):1451–60.
159. Sabag AD, Bode J, Fink D, Kigel B, Kugler W, Neufeld G. Semaphorin-3D and semaphorin-3E inhibit the development of tumors from glioblastoma cells implanted in the cortex of the brain. Plos One. 2012;7(8):e42912.
160. Bielenberg DG, Wang Y, Fan S, Reighard GL, Scorza R, Abbott AG. A deletion affecting several gene candidates is present in the Evergrowing peach mutant. J Hered. 2004;95(5):436–44.

161. Karayan-Tapon L, Wager M, Guilhot J, Levillain P, Marquant C, Clarhaut J, et al. Semaphorin, neuropilin and VEGF expression in glial tumours: SEMA3G, a prognostic marker? Br J Cancer. 2008;99(7):1153–60.
162. Casazza A, Finisguerra V, Capparuccia L, Camperi A, Swiercz JM, Rizzolio S, et al. Sema3E-Plexin D1 signaling drives human cancer cell invasiveness and metastatic spreading in mice. J Clin Invest. 2010;120(8):2684–98.
163. Kigel B, Varshavsky A, Kessler O, Neufeld G. Successful inhibition of tumor development by specific class-3 semaphorins is associated with expression of appropriate semaphorin receptors by tumor cells. Plos One. 2008;3(9):e3287.
164. Miyato H, Tsuno NH, Kitayama J. Semaphorin 3C is involved in the progression of gastric cancer. Cancer Sci. 2012;103(11):1961–6.
165. Basile JR, Castilho RM, Williams VP, Gutkind JS. Semaphorin 4D provides a link between axon guidance processes and tumor-induced angiogenesis. Proc Natl Acad Sci U S A. 2006;103(24):9017–22.
166. Sierra JR, Corso S, Caione L, Cepero V, Conrotto P, Cignetti A, et al. Tumor angiogenesis and progression are enhanced by Sema4D produced by tumor-associated macrophages. J Exp Med. 2008;205(7):1673–85.
167. Sun Q, Zhou H, Binmadi NO, Basile JR. Hypoxia-inducible factor-1-mediated regulation of semaphorin 4D affects tumor growth and vascularity. J Biol Chem. 2009;284(46):32066–74.
168. Zhou H, Yang YH, Binmadi NO, Proia P, Basile JR. The hypoxia-inducible factor-responsive proteins semaphorin 4D and vascular endothelial growth factor promote tumor growth and angiogenesis in oral squamous cell carcinoma. Exp Cell Res. 2012;318(14):1685–98.
169. Basile JR, Holmbeck K, Bugge TH, Gutkind JS. MT1-MMP controls tumor-induced angiogenesis through the release of semaphorin 4D. J Biol Chem. 2007;282(9):6899–905.
170. Conrotto P, Valdembri D, Corso S, Serini G, Tamagnone L, Comoglio PM, et al. Sema4D induces angiogenesis through Met recruitment by Plexin B1. Blood. 2005;105(11):4321–9.
171. Moriya J, Minamino T, Tateno K, Okada S, Uemura A, Shimizu I, et al. Inhibition of semaphorin as a novel strategy for therapeutic angiogenesis. Circ Res. 2010;106(2):391–8.
172. Joyal JS, Sitaras N, Binet F, Rivera JC, Stahl A, Zaniolo K, et al. Ischemic neurons prevent vascular regeneration of neural tissue by secreting semaphorin 3A. Blood. 2011;117(22):6024–35.
173. Meda C, Molla F, De Pizzol M, Regano D, Maione F, Capano S, et al. Semaphorin 4A exerts a proangiogenic effect by enhancing vascular endothelial growth factor-A expression in macrophages. J Immunol. 2012;188(8):4081–92.
174. Cirulli V, Yebra M. Netrins: beyond the brain. Nat Rev Mol Cell Biol. 2007;8(4):296–306.
175. Kitajewski J. Endothelial laminins underlie the tip cell microenvironment. EMBO Rep. 2011;12(11):1087–8.

Chapter 12
The Vasculature in the Diseased Eye

Hans-Peter Hammes

12.1 Introduction

The retinal vasculature is the only microvascular system that can be assessed non-invasively. While ultrasound and contrast-enhanced angiograms map the body vasculature at a resolution of millimeters, fundus imaging can display the microvasculature in the range of 20 μm of vessel diameter (Fig. 12.1). Although capillaries can only be visualized through contrast enhancement (termed fluorescence angiography) the retinal imaging has gained an enormous interest during recent years because of its mirror function of the body's entire vasculature.

One disease that frequently strikes in the eye is diabetes, and diabetic retinopathy is a frequent encounter in daily medical care. The purpose of this chapter is to a. outline the clinical features of diabetic retinopathy, b. describe model systems which are used to assess the pathogenesis of diabetic retinopathy, and c. to align current knowledge about functional and structural abnormalities caused by factors or pathways discussed elsewhere in this book.

12.2 A Brief Epidemiology of Diabetic Retinopathy

Diabetic retinopathy is a frequent complication of diabetes of both, type 1 and type 2. The overall prevalence of any diabetic retinopathy worldwide is 34.6%, and of sight-threatening diabetic retinopathy (either clinically significant macular edema (CSME) or proliferative diabetic retinopathy) is 10.2% [1]. The highest prevalence rates is in African americans and the lowest is in asians. Retinopathy is more frequent in type 1 than in type 2 diabetes, and increases with

H.-P. Hammes (✉)
5th Medical Department, UMM – University of Heidelberg, Theodor-Kutzer-Ufer 1-3, 68167 Mannheim, Germany
e-mail: hp.hammes@umm.de

M. HH. Schmidt, S. Liebner (eds.), *Endothelial Signaling in Development and Disease*, DOI 10.1007/978-1-4939-2907-8_12

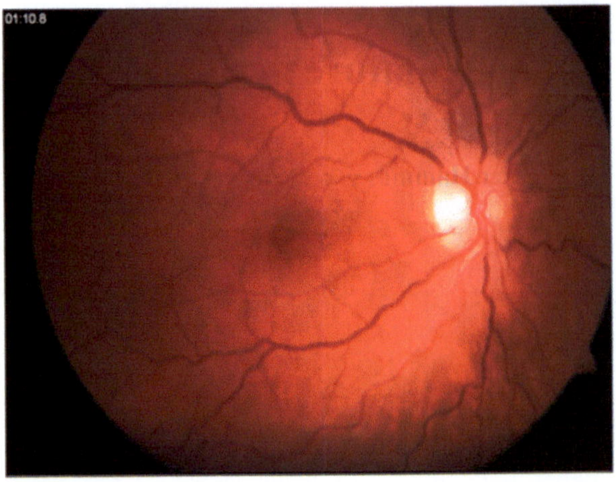

Fig. 12.1 45° Fundus photograph of a 35 year old type 1 diabetic patient after 12 years of disease. No signs of diabetic retinopathy detectable

disease duration, higher glucose/HbA1c levels, blood pressure, and cholesterol. Recent reanalysis of the landmark Wisconsing Epidemiology Study of Diabetic Retinopathy (WESDR) revealed a cumulative incidence of any retinopathy in the cohort surviving 25 years of 97 %, a cumulative rate of retinopathy progression of 83 %, and a rate of progression towards proliferative retinopathy of 43 %. These data confirm, that even after many years of implementation of novel therapies such as intensified insulin therapy, there is still an almost 100 % chance to develop some, and a 50 % chance of developing sight-threatening retinopathy requiring laser therapy. A recent survey of 8700 type 1 diabetic patients from a large data base in Germany and Austria demonstrated a cumulative proportion of any retinopathy of 84.1 %, after 40 years of diabetes, and a cumulative proportion of 50.2 % of advanced retinopathy [2] confirming the above numbers also apply to the local setting. However, advanced proliferative stages appear to decline to some extent in incidence [3]. The follow-up analysis of the Diabetes Control and Complication Trial/Epidemiology (DCCT/EDIC) suggests that the earlier normoglycemia is established, the better the retinal outcome is. Comparing the two treatment groups of DCCT after 30 years of disease, the cumulative incidence of PDR was 21 % in former intensive group, and 50 % in the former conventional group. The latter was comparable with the 47 % cumulative incidence of a population based group of similar diabetes duration. The implementation of strict glycemic control and other measures to normalize the metabolic profile in type 1 diabetes will further decrease the incidence of diabetic retinopathy. This and the translation of DRS and ETDRS findings, which demonstrated that timely laser treatment preserves vision, into treatment of advanced retinopathy stages may also affect visual prognosis in type 1 diabetes [4]. At present, it remains unclear whether the trend of declining retinopathy also applies to type 2 diabetes. On one hand, the numbers of persons with type 2 diabetes increases dramatically (rising the numbers of susceptible patients), on the other hand, implementation of metabolic control, polypharmacy, and timely treatment, in particular for CSME, reduces the numbers of patients with vision threatening retinopathy.

Diabetes is a systemic disease, affecting other organs, such as the kidney. Coincident nephropathy has a major impact on the progression of retinopathy to vision-threatening stages in both, type 1 and type 2 diabetes. Almost all patients with proliferative diabetic retinopathy have some degree of nephropathy. In the WESDR, the odds ratio for having retinopathy in microalbuminuric patients was 3.78 for the younger and 1.8 for the older group [5]. This association is confirmed in a recent study, in which the OR for severe retinopathy in microalbuminuric type 1 patients was 4.1 [95 % CI 3.4–4.9]. In patients with even more progressed renal damage as reflected by the transition to proteinuria, the increase in risk for sight-threatening retinopathy stages multiplies. In Danish type 1 diabetic patients, the 5 year cumulative incidence of PDR was 74 % in patients with gross proteinuria, but only 14 % in patients without proteinuria. The OR for advanced retinopathy in macroalbuminuric patients with type 1 diabetes in a large German study was 8.6 [95 % CI 6.4–11.5].

Patients with CSME may represent a special subgroup with possible shared pathogenesis resulting in a generalized vascular hyperpermeability. In a study of 40 type 2 diabetic patients, Knudsen et al. found a significant correlation of urinary albumin excretion and transcapillary albumin escape rates with diabetic macular edema as measured by optical coherence tomography [6]. Since generalized vascular permeability paralled vascular leakiness in target organs of diabetic injury, the conclusion was that with increasing albuminuria the impact on maculopathy would also increase. As a clinical correlate, a strong association of CSME with proteinuria in type 2 diabetic caucasion patients was found [7]. Together, these data suggest that diabetic retinopathy is part of a systemic disease in which the interaction of eye and kidney plays a particular role. This needs to be considered when diabetic retinopathy is used as a paradigm of a neovascular disease.

12.3 Risk Factors for Diabetic Retinopathy—the Impact of Glucose Revisited

Classical risk factors of diabetic retinopathy are disease duration, quality of glucose control, hypertension, lipids, gender, and hormonal changes. Chronic hyperglycemia is the single most important risk factor causing diabetic tissue damage, but its relative contribution has probably been overestimated since the DCCT and the UKPDS have reported beneficial effects of good glycemic control for the prevention of diabetic retinopathy [8, 9]. Analysis of the DCCT data revealed that improved control (i.e. lowering of HbA1c from 9 to 7 %) reduced sustained retinopathy progression by 73 %. However, in the analysis of the entire group, HbA1c and diabetes duration (i.e. the measure of glycemic exposure) explained only approximately 11 % of the variation in retinopathy risk [10]. From that, it was concluded that the remaining 89 % in the variation in risk is explained by factors related to glycemia, not reflected by HbA1c.

Moreover, using digitized photographs capturing the retinal vasculature in patients whose underlying risk profile differs fundamentally from diabetic patients, it was found that up to 12 % of "healthy" (i.e. non diabetic and non hypertensive patients) develop a retinal phenotype that matches mild stages of diabetic retinopathy. These data indicate that the retina is affected by mechanisms which are unrelated to glycemia, but that hyperglycemia still plays a role in the incipience and propagation of retinal damage.

12.4 Retinopathy as a Risk Indicator for Cardiovascular Disease

It is well documented that kidney disease is associated with excess cardiovascular morbidity and mortality in diabetes. The association between diabetic retinopathy and cardiovascular disease, however, had only been documented for advanced retinopathy stages, suggesting that only the subgroup of patients who developed proliferative diabetic retinopathy had an increase in cardiovascular mortality that was independent of classical CVD risk factors. In fact, the WESDR study was the first large study to demonstrate an excess mortality risk of death in patients with diabetic retinopathy, independent of important CVD risk factors such as age, diabetes duration, glycemia and gender [11]. Interestingly, the retinopathy-CVD association is less consistent in type 1 diabetes, mostly because of the role of age in CVD mortality. According to recent data, however, the risk association between retinopathy and CV mortality extends to any retinopathy which indicates that the development of any retinopathy in a person with diabetes is a biomarker for increased mortality. In this metaanalysis including over 19,000 patients from 20 studies, the OR for the composite outcome in type 2 diabetic patients, when adjusted for CV risk factors, was 1.61 (95 % CI 1.32–1.9) for any retinopathy, and 4.22 (95 % CI 2.81–6.33) for advanced retinopathy. For type 1 diabetes, the corresponding OR for any retinopathy was 4.10 (95 % CI 1.50-11.18), and 7.00 (95 % CI 2.22-20.0) for advanced retinopathy. When limiting the analysis to the studies that reported all cause mortality, the meta-analysis of the association between presence of retinopathy and all cause mortality was approximately 2.5 fold for both, type 1 and type 2 diabetes [12]. This is a strong indicator that the retinal vasculature is a mirror of general vascular risk.

12.5 Retinal Vascular Pathology in Diabetes

The diagnosis and staging of diabetic retinopathy is still based exclusively on signs of vascular dysfunction and structural damage. However, novel insights into the complex cellular interactions in the diabetic retina indicate that retinopathy is not

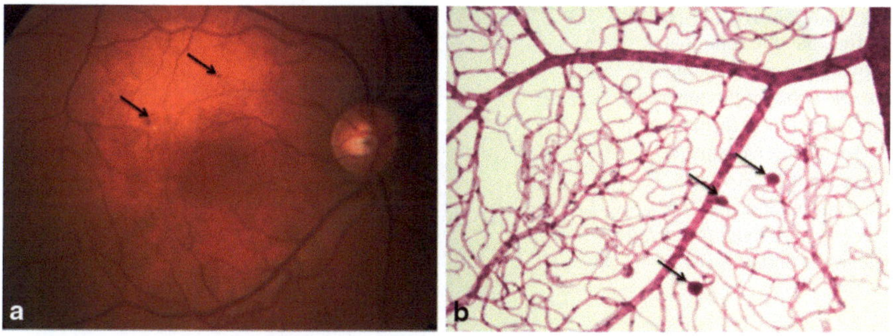

Fig. 12.2 panel **a**: Fundus photograph of a 48 year old type 1 diabetic patients after 23 years of disease. *Arrows*: microaneurysms/dot hemorrages defining mild non-proliferative diabetic retinopathy Panel **b**: Retinal digest preparation showing microaneurysms (*arrows*) in the vicinity of acellular capillaries. Original magnification 200x. Periodic acid Schiff stain

only vascular disease, and that changes of the neural retina may even precede and predict vascular abnormalities (www.laskerfoundation.org/programs/irrf.htm). The implementation of sensitive diagnostic tools into research has opened the view on retinopathy as being the result of damage to multiple different cell types in the retina.

The common characteristics of diabetic vascular damage is increased vascular permeability and progressive vascular occlusion. Clinically, retinopathy is devided into two distinct stages, non-proliferative and proliferative diabetic retinopathy. Only late stages affect vision. Diabetic macular edema is defined as a condition resulting from progressive capillary dropout that causes thickening of the macula through accumulation of extracellular fluid, frequently accompanied by lipoprotein depositions and, if the fovea is involved, vision loss. Any of these advanced stages initiates with vascular changes which are readily detectable by fundus inspection. The earliest and most relevant sign of diabetic retinopathy is the microaneurysm (Fig. 12.2a and b). Clinically, it appears as a red dot of >20 µm in diameter (Fig. 12.2a). Microaneurysms are often hypercellular, indicating angiogenic activity (Fig. 12.2 b). Indeed, microaneurysms are typically located adjacent to areas of capillary dropout, suggesting hypoxic stimulation, and are spherical clusters of cells which stain for endothelial markers, while endothelial cells in capillaries of the vicinity remain mostly negative. This local proliferative response to vascular changes suggests an early abortive attempt of neovascularization amidst vasoregression. Later, microaneurysms can clinically disappear because intraluminar microthrombi occlude them. This can simulate clinical improvement. Both, microaneurysms and dot hemorrhages characterize mild non-proliferative diabetic retinopathy. Around capillaries and microaneurysms, lipoproteins can leak from capillaries as hard exsudates which are yellow, sharply bounded, and variable in size and form. When the macula is not affected, vision remains clinically unaltered. With progressive capillary dropout, avascular areas become confluent, and the retina responds differently to progressive ischemia. On the arteriolar site, microinfarcts of small arterioles disclose as

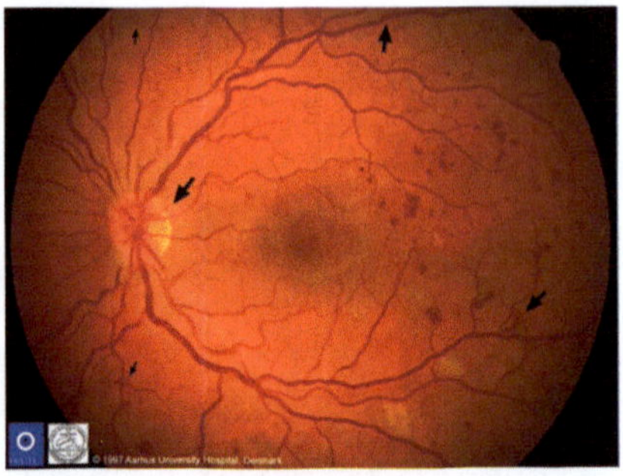

Fig. 12.3 Severe non-proliferative diabetic retinopathy showing venous beading (*arrow upper right* quadrant), intraretinal loop formation (*arrow lower right* panel), intraretinal microvascular abnormalities (IRMA) (*arrows upper* and *lower left* and at the optic disc)

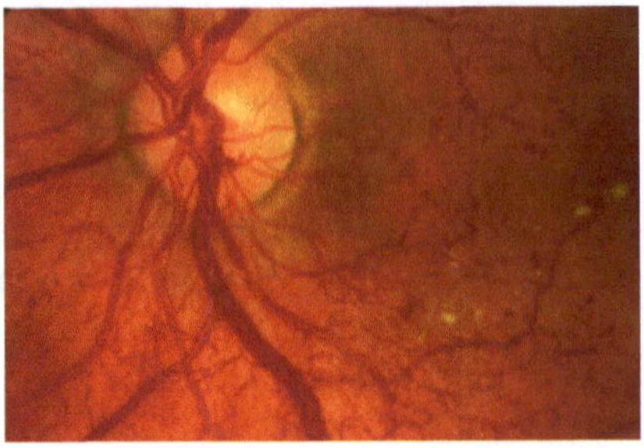

Fig. 12.4 Fundus photograph showing neovascularization from the disc

"cotton-wool-spots", representing nerve fiber swelling in the inner retina. On the venolar site, saccular bulges ("venous beading") develop which are formed by hypercellular venolar segments. Subsequently, new vessel form within the level of the retina, or microaneurysms become numerous and leaky. This stage represents severe, ("preproliferative") non-proliferative diabetic retinopathy at which the risk of blinding proliferative disease is high (Fig. 12.3). Proliferative diabetic retinopathy is defined by growth of new vessels into the vitreous, mainly originating from peripheral venules (NVE, neovascularization elsewhere) or from the optic disc (NVD, neovascularisation from the disc) (Fig. 12.4). Intravitreal hemorrhages cause acute vision loss, while the formation of fibrovascular tissue responding to the intravitreal discharge of inflammatory cells and its contraction is the cause of permanent vision loss through retinal detachment.

Taken together, the clinical distinction between non-proliferative and proliferative diabetic retinopathy is based on clinical decision-making, not on the inherent underlying vascular processes. From a vascular biology point of view, even the initial clinical stages are proliferative in character, suggesting that the retina in diabetes responds with a continuum of proliferative vascular changes.

However, in a subset of patients, the tissue response is not angiogenic, but boosting permeability. When the macular is involved, retinal edema becomes particularly significant, and the resulting entity is diabetic macular edema, and clinically significant, if edema formation involves one disc diameter around the macular. When the edema affects the spatial integrity of the macular area, vision may become impaired. Clinically, macular edema is frequently accompagnied by hard exsudates which represent intraretinal extravasations of lipids at the border between edematous and nonedematous areas. They are considered hallmarks of focal breakdown of the blood retinal barrier, either widespread from large areas of affected capillaries or focal from microaneurysms. The parallel occurrence of leaky and non leaky microaneurysms, and the differences in patient characteristics suggests that local factors determine the clinical phenotype independent of common underlying pathogenetic abnormalities. Notably, the assocation between visual function as measured by acuity, and macular thickness as measured by optical coherence tomography (OCT) is not particularly strong which is of clinical importance when anti-edematous therapies are evaluated.

12.6 Animal Models—do they Reflect Clinical Retinopathy?

Animal models are widely used to study diabetic retinopathy. This is a. the consequence of the paucity of adequate human material, in particular of early stages, or b. the consequence of the availability of models with distinct genetic manipulations, in which the retinal phenotype is studied under diabetic conditions, or as a surrogate. Animal models are a crucial part to understand the pathogenesis, and to identify promising targets for treatment in humans.

Histopathological changes described in animal models of diabetic retinopathy may represent only preclinical human retinopathy, as outlined below, and may vary depending on the model. The only consistent changes that diabetic animals develop are the loss of pericytes and the degeneration of retinal capillaries (also termed vasoregression). Vasoregression and pericyte dropout also occur in the human diabetic retina speaking for the translational validity of animal models. Extensive reviews have highlighted the role, advantages and disadvantages, and other aspects of in vivo models of diabetic retinopathy [13]. In essence, there is no animal model that perfectly reflects human pathology (as there is no animal model of human diabetic nephropathy), and the need for such models has been clearly identified (see www.diamap.eu—5- microvascular complications). It must be emphasized that there is no diabetic animal that develops proliferative retinopathy comparable to humans. The reason for this observation is unclear, and short life span and differences in biochemical and cell biological factors have been cited as underlying causes. Thus, the rationale to study pathogenetic aspects in diabetic animals that develop only preclinical lesions is to identify mechanisms that are involved in the incipient changes in the diabetic retina such as pericyte loss and/or vasoregression for which human material is unavailable.

In chemically induced and spontaneous diabetic rats, pericyte dropout, acellular capillaries and basement membrane thickening consistently develop. Acellular capillaries become significantly more numerous after 6 months of diabetes, with considerable variability in different rat strains. As a consequence, treatment options studied for periods shorter than 6 months may have addressed mechanisms unrelated to the underlying pathogenesis. Pericyte dropout when assessed as pericyte "ghosts" succeeds also at 6 months of diabetes duration, but is identified earlier when quantitative retinal morphometry by image analysis is used [14]. Microaneurysms occasionally develop in the diabetic rat retina, but after extended disease duration (difficult to achieve in diabetic rats). No other lesions develop reproducibly. Diabetic mice of either type of diabetes develop similar lesions within comparable time frames as do rats, but the degree of cellular change is, to some extent, smaller, and gender effects on glycemia and retinopathy development may be stronger.

12.7 Mechanisms of Diabetic Microvascular Damage

The ability of a cell to cope with high ambient glucose is a crucial determinant to explain differential susceptibility of organs towards hyperglycemic damage. Vascular endothelial cells do not show changes in glucose transport rates when ambient glucose is high, which can result in elevated intracellular glucose [15]. The biochemical consequences of this phenomenon have become known as the "unifying hypothesis" about which excellent reviews have been published [16, 17]. For the purpose of this chapter, a brief summary is given here. High intracellular glucose leads to an increase in oxidative stress from different sources, including mitochrondrial overproduction of reactive oxygen species (ROS), enzyme mediated cytoplasmatic ROS production, and receptor mediated ROS generation. As the 50 or so different cell types of the retina are per se variably equipped with glucose transporters responding to the ambient glucose levels, with receptors responding to reactive intermediates and advanced glycation endproducts (AGEs), and enzymes producing ROS in the cytoplasm, it is likely that not one single mechanism is involved. Hyperglycemia activates four major pathogenetic pathways which may affect vascular cells in the diabetic retina: a. increased formation of AGEs and overexpression of its receptor, called RAGE (receptor for AGEs), b. activation of protein kinase C isoforms, c. activation of the polyol pathway, and d. activation of the hexosamine pathway. These mechanisms are the result of a single upstream event, i.e. mitochondrial overproduction of reactive oxygen species. Although some evidence exists that the activation of the polyol pathway is also relevant to microvascular changes in diabetes, glucose in diabetic vascular cells may not be a substrate for aldose reductase, as the Km of aldose reductase for glucose is 100 mM, while the intracellular concentration of glucose endothelial cells exposed to high glucose is in the nanomolar range [18]. In the retina of diabetic rodents, activation of the other pathways have been identified.

In general, AGEs form from glucose, glycolytic intermediates such as dicarbonyls, and from intermediates of free fatty acid oxidation. Extracellular AGEs form from glucose and have a slow formation characteristics, but are found in increased amounts in the extracellular matrix. Intracellular AGEs are rapidly formed from glycolytic intermediates such as methylglyoxal, and contributes to cell damage by three mechanisms: a. modification of protein function, b. alteration of cellular function via receptor recognition, and c. by changes in matrix interactions. AGEs can leak from producing cells to the plasma or accumulate on long-lived matrix molecules [19]. They can interact with the AGE receptor RAGE on various cells which activates the transcription factor NFkB. In the diabetic retina, RAGE has been identified on glial cells [20]. RAGE expression is regulated by high glucose in vitro and in vivo, involving methylglyoxal as a stimulator which increases NFkB and AP-1 binding to the RAGE promoter. AGEs like ROS can induce the overproduction of VEGF by retinal glia. AGEs can cooperate with the hexosamine pathway (see below) to modify transcriptional co-repressors changing transcriptional activation of growth factors [21]. Thus, AGEs can mimick hypoxia in the diabetic retina. On the contrary, AGEs may interfere with transactivation of HIF-1 alpha which mediates hypoxia-stimulated VEGF production. Covalent modification of the transcriptional coactivator p300 by hyperglycemia through methylglyoxal is responsible for the decreased association with HIF-1 alpha leading to reduced VEGF transcription [22]. The net balance of pro- versus antiangiogenic effects of AGEs in the diabetic retina may tip to the antiangiogenic site, since vasoregression dominates in the animal model in which these observations are made.

Several PKC isoforms are involved in cellular dysfunctions of the diabetic retina. Persistent PKC activation results from enhanced de-novo synthesis of diacylglycerol from hyperglycemia-induced triose phosphate which is available because mitochondrial ROS overproduction inhibitor the glycolytic enzyme GAPDH. PKC activation can also result from RAGE activation via AGEs. The role of PKCß activation affecting retinal blood flow and permeability has been established [23]. Specifically, VEGF activates PKC beta in the retina by membrane translocation and subsequent hyperpermeability, which can almost completely be blocked by ruboxistaurin. A novel association has been recently reported between pericyte loss and PKC δ[24]. PKC δ activation leads to increased expression of SHP-1, a protein tyrosine phosphatase affecting PDGF-ß receptor downstream signaling and pericyte apoptosis. PKC activation may also affect VEGF transcription which affects vacular permeability.

The fourth hyperglycemia-driven biochemical abnormality is the activation of the hexosamine pathway. Fructose-6-phosphate, diverted from glycolysis, is the substrate for fructose 6-phosphate amidotransferase (GFAT) and is converted into UDP-N-acetylglucosamine (GlucNac). Specific transferases utilize GlucNac for modification of serine and threonine residues. As mentioned, AGE modification of the transcriptional repressor mSin3A increases the recruitment of such a transferase resulting in the posttranslational modification of SP3, which results in a decreased binding to the glucose-sensitive GC box in the angiopoietin-2 (Ang-2) promoter [21]. The resulting overexpression of Ang-2 precedes pericyte loss in the diabetic retina, and accelerates the formation of capillary dropout. Ang-2 has also been implicated in pericyte migration as an additional mechanism by which incipient pericyte loss in the diabetic retina can occur beyond apoptosis [25]. The hexosamin pathway was also reported to be relevant for TGF-ß1 and PAI-1 expression, however, of which the relevance in the diabetic retina has yet to be determined.

Given the inability to prevent complications by only targeting one of the above described abnormalities, a unifying mechanisms was introduced representing a common denominator of the biochemical abnormalities, and a mechanism by which hyperglycemic memory could be explained. This mechanism conceptually involves the overproduction of superoxide by the mitochondrial electron-transport chain. Intracellular excess glucose induces an abnormally high flux of electron donors into the mitochondrial electron transport chain. The resulting voltage gradient leaks superoxide because electron transfer is blocked at complex III of the electron chain. In endothelial cells exposed to excess glucose in vitro, superoxide production and subsequent ROS formation is increased. ROS overproduction is blunted by collapse of the gradient through UCP-1 or superoxide degradation by dismutase activity. "Rho zero" cells lacking mitochondria fail to produce ROS suggesting that mitochondria are an important source for ROS overproduction in glucose exposed cells. Mitochondrial ROS overproduction causes DNA damage which activates the repair/protection enzyme system called poly-ADP ribose polymerase (PARP). PARP which normaly resides in the nucleus in its inactive form, translocates to the cytoplasm upon ROS activation causing polymers of ADP-ribose to form. These polymers inhibit GAPDH activity which leads to an increase in the glycolytic intermediates which are the substrates for the biochemical pathways and thus lead to an overactivation. In the diabetic retina, ROS overproduction, PARP activation, increased availability of methylglyoxal, activation of the PKC and the hexosamine pathway have been detected [18, 26]. Several of these changes are readily reversible upon glucose normalisation but cannot explain findings from preclinical and clinical studies showing a lasting effect of hyperglycemia despite complete euglycemic restoration. For example, diabetic dogs developed similar degrees of retinopathy when treatment was changed from bad to good glycemic control. Clinically, patients on good glycemic control over 6.5 years in Diabetes Control and Complication Trial persisted to have much lower retinopathy progression during subsequent 14 years of follow-up, although their glycemia had become identical to former control patients whose glycemia had been much worse during the initial 6.5 years [27]. This phenomenon termed "hyperglycemic memory" is found in patients with type 2 diabetes ("hyperglycemic legacy") and only glycemic control, but not blood pressure control produces such memory in the retina [28]. One cause for a memory effect is the covalent attachment of some AGEs to long-lived molecules such as extracellular matrix components. Another cause is the induction of sustained gene expression by glucose-mediated mitochondrial overproduction involving activating methylation of histones associated with critical gene promoters [29]. One example is the activation of the monomethyltransferase Set7 which was shown to be induced by transient high glucose levels. Sixteen hours of high glucose exposure resulted in 6 days of NF-kB transactivation. The Set7 activation led to a transcriptional activation of the p65 subunit of NF-kB (lasting for days) which induced proinflammatory cytokine production both in endothelial cells in vitro and in diabetic tissues in vivo. The permanent activation was inhibitable by superoxided dismutase and, importantly, by overexpression of glyoxalase-1 which is the critical enzyme to degrade methylglyoxal. Similarly, increased H4K3 methylation of the SOD2 gene causes downregulation of the enzyme activity, and failure to normalize the deficit upon euglycemic reentry suggests yet another potential role of epigenetic modifications, caused by ROS and AGEs in hyperglycemic memory. Other posttranslational modifications occurring in the diabetic setting comprise acylation, acetylation, prenylation and ubiquitination, as recently outlined [30].

The validity of the unifying hypothesis has been confirmed in rodents in two ways: parameters such as ROS, AGEs, and products of the hexosamine pathway have been found in the diabetic rat retina, and b. it was shown that both, catalytic antioxidants and metabolic signal blockers are effective in preventing retinal vasoregression. As outlined above, ROS lead to an inhibition of GAPDH activity, and an increased concentration of intermediates upstream the enzyme block. The resultant increased flux through metabolic pathways calls upon stage a strategy that shifts glycolytic intermediates such as fructose-6-phosphate and glyceraldehyde-3-phosphate from the hexosamine and the AGE pathway into alternative, less toxic pathways. Both metabolites are end-products of the non-oxidative part of the pentose-phosphate shuttle of which transketolase is the rate-limiting enzyme. The cofactor of transketolase is thiamine, which can activate the enzyme by 25 %. By contrast, the lipid soluble derivative benfotiamine activates the enzyme by 250 % in endothelial cells. Long-term treatment using benfotiamine in models of diabetic retinopathy largely prevented vasoregression suggesting that metabolic signal blockade is a useful concept for the prevention of diabetic retinopathy [18].

Recently, a link was proposed between metabolism and physiological angiogenesis [31]. The concept implicated that with a reduction of glycolytic flux several important functions involved in the regular formation of retinal blood vessels were impaired, such as endothelial cell migration, tip cell competence and stalk cell proliferation. Whether such changes are relevant for the early diabetic retina is unclear, because a. the findings have been obtained from newborn mice (p5) in which the retinal metabolic demand is different than in the adult retina, b. chronic hyperglycemia affects glycolysis of retinal cells quite variably since some cells can protect themselves from abnormal glucose exposure, while others cannot, c. different cell types modify the microenvironment (including glycolysis) via regulating glucose transporter activity, and d. the endophenotype of physiological angiogenesis is grossly different from diabetic retinopathy.

12.8 Diabetic Pericyte Loss

Pericyte loss is a common and universal finding in diabetic animal models. According to time course studies, pericyte loss in a STZ diabetic rat starts approximately 2 months after onset of hyperglycemia, and thus precedes the development of vasoregression (i.e. consecutive loss of endothelial cells) by several months. The degree of pericyte dropout is equal in the entire retina, suggesting that the level of metabolic stress is linked to the level of pericyte dropout in the diabetic model [32]. The level of pericyte dropout varies between strains and species, ranging between 15 and 50% in diabetic rodents after 6 months duration. The cause of diabetic pericyte dropout remains still undetermined, although more mechanistic insight has been obtained during recent years. The biochemical injury involves primarily the capillary endothelium, but there is some evidence that microvascular pericytes are specifically damaged by hyperglycemia. Activation of pro-apoptotic signalling and apoptosis in pericytes has been observed in vivo, and some of the above outlined biochemical changes are harmful to pericytes when present in abundance. For example, AGEs induce dose- and time dependent apoptosis in pericytes in vitro through activation of the transcription factor FOXO1, mediated in part by p38 and JNK MAP kinases [33]. Repetitive administration of AGEs to non-diabetic animals induces a selective dropout of pericytes after a short period of time [34], and endogenous formation of AGEs which is found in diabetes, accumulate in pericytes as well, suggesting that pericytes hava clearing function [35]. Whether AGEs kill pericytes in situ has not been conclusively shown and the time course of AGE accumulation in pericytes is inconsistent with that of pericyte loss in diabetic animals, suggesting that AGES may not be entirely responsible for pericyte loss. Additionally, the effect of AGE inhibitors on pericyte dropout is modest at best. Alternatively, NF-κB activation and the increase of the extracellular matrix protein βIG-H3 can induce pericyte apoptosis, and more recently, it has been demonstrated that hyperglycemia activates PKC-δ and p38 mitogen-activated protein kinase [MAPK] leading to pericyte apoptosis and vasoregression independently of NF-κB [24]. Thus, a number of mechanisms may exist which operate in parallel to induce pericyte apoptosis in the diabetic retina. From model calculations, however, it is evident that the degree of apoptotic pericytes in retinal digest specimens of diabetic rats is too low to explain the total number of pericytes lost after several months. Alternative mechanisms have therefore to be considered, and it was identified that pericyte loss can also be the result of an active process involving migration of pericytes away from the capillaries, driven by the angiopoietin-Tie system. Gain of function experiments in non-diabetic animals revealed the induction of pericyte dropout in the vicinity of Ang-2 overexpressing sites. When these animals were rendered diabetic, the formation of acellular capillaries was aggravated linking pericyte dropout with vasoregression under specific conditions. Diabetes in mice with a partial Ang-2 deficit yielded the prevention of pericyte dropout and the reduction of acellular capillary formation [25]. In the retina, Ang-2 is expressed in three cell types, (i) in endothelial cells, (ii) in Müller cells, and (iii) in horizontal cells. In diabetes, chronic hyperglycemia increases glucose flux in microvascular endothelial cells and Müller cells, causing modification of the corepressor mSin3A by the reactive intermediate methylglyoxal which results in the recruitment of the enzyme O-GlcNAc transferase to an mSin3A-Sp3 complex. Subsequent modification of Sp3 by O-linked N-acetlyglucosamine decreased its binding to a glucose-responsive GC box in the Ang-2 promoter and the activation of Ang-2 transcription [21]. These data are consistent with the hypothesis, i.e. that pericyte loss in the diabetic retina is actively induced by glial cells overexpressing Ang-2 in response to high glucose.

12.9 Permeability as a Function of Vascular Cells

An important aspect of progression of clinical diabetic retinopathy is the increasingly leaky blood brain barrier. Despite major progress in methodologies, the available in vitro and in vivo model systems only reflects selected aspects of the human disease. A recent review by Klaasen, van Noorden and Schlingemann gives a precise insight into the state-of-the-art, the limitations and the relevance of molecular basis of the blood-retinal barrier for the diabetic retina [36]. Moreover, several chapters in this book provide novel insight into the interaction of cells which determine retinal blood retinal barrier (Chap. 6 by Bravi and Lambugnani and Chap. 7 by Phillipson, Christoffersson, Claesoon-Welsh & Welsh). As noted, the STZ-diabetic retina, although developing signs of early diabetic retinopathy, but not of diabetic macular edema (in particular as these animals do not have a macula). Signs of hyperpermeability in this model are often modest and diffuse, and extrapolation to the pericapillary or perimicroaneurysmatic leakage in DME is probably undue. The best in-vitro system to study the blood retina barrier should be composed of the cells which reflect the complexity of the neurovascular unit (NVU) in the human retina. No in vitro system exists that reflects all aspects of the NVU, so that the diabetic rodent retina is probably the best approximate, but the lack or paucity of data and the gaps in translatability may still be major and explain why some clinical trials in humans with DME failed in the past. Novel models have been generated such as the Akimba mouse which combines hyperglycemia from Akita parents and photoreceptor overexpression of VEGF from the Kimba parents, and their usefulness in closing this experimental gap will be demonstrated [37].

The mechanisms underlying the breakdown of the blood retinal barrier are identical to those discussed for vasoregression, but the cellular events are complex, because not only vascular cells (endothelial cells and pericytes) are involved, but also neuroglial cells. Conceptually, disruption of the blood retinal barrier function in diabetes invokes alterations of junctional proteins, proteins of the transcellular transport maschinery, and changes in the structure and composition of matrix proteins. Hyperglycemia causes downregulation of endothelial tight junction proteins which represents a major mechanism in BRB breakdown. Additionally, junctional molecules are regulated in diabetes by a complex interplay of kinases and phosphatases, which, when removed from tight junctional complexes, result in increased paracellular permeability. Another level of complexity is added by the regulation of VE-cadherin and ß-catenin through hyperglycemia, ROS or shear stress, all of which can be found in the diabetic retina, which play an important role not only in the regulation of adherens junctions, but also in intracellular signalling. Caveolin-1 is a major determinant of a group of molecules involved in the transcellular, but also in the paracellular transport, as it determines the expression of tight junction proteins. These processes are regulated by hyperglycemia, and are possible targets of antipermeability strategies in vivo.

As mentioned, pericytes are selectively and progressively lost in diabetic retinal capillaries. Their physiological role, in particular the contribution to the tightness of the blood-retinal barrier, has been extensively described in a recent review by Armulik et al. [38]. Pericytes are critical in the formation and maintenance of this barrier which is based, among other factors, on the unusual high number of pericytes which cover the retinal capillaries, the high transendothelial electrical resistance that pericytes produce and their ability to regulate endothelial junctions. More recently, pericytes were found to regulate transcytosis although the precise mechanisms and molecules involved are still not known. The pericyte deficient neurovascular unit showed convoluted interendothelial junctions, mislocalized glial water pores, increased endothelial vesicles, and upregulated adhesion molecules promoting leukocyte adherens to the activated endothelium. These abnormalities are largely reminiscent of changes that occur in the diabetic retina.

12.10 The Role of Müller Cells

Another cell of the NVU cooperates in the promotion of altered fluid handling, which is the retinal Müller cell. Müller cells are specialized glial cells which span the entire thickness of the neural retina. They represent a functional and structural interface between neurons and blood vessels, and they guarantee the proper handling of nutrients waste products, water, ions, and neurotransmitters. Thereby, they permit survival and function of multiple neuronal and vascular cells, and they regulate blood flow and barrier properties of small vessels [39]. Virtually all stimuli can activate Müller cells and Müller cell gliosis is a stereotypical response to a plethora of these stimuli. The upregulation of the cytoplasmic glial fibrillary acidic protein (GFAP) is a representative of gliosis, and observed from the early stages of DR on [40]. Although it has been suggested that the Müller cell gliosis in DR is a general change that occurs in response to pathogenic factors such as ischemia-hypoxia, oxidative stress or chronic inflammation [41], the time course of GFAP upregulation in vivo favours reactive oxygen species and hyperglycemia-driven water and ion imbalance as the major trigger. Interestingly, Müller cells which become activated respond by the release of neurotrophic factors and may represent a response to injury to limit damage. However, with persisting stimuli, the same cells progressively interfere with regular tissue repair and thus promote neurodegeneration. Additional contribution to tissue damage can result from the persistence of ion and water dysbalance and impaired neurotransmitter removal. The overexpression of heat shock proteins in Müller cells of diabetic rats may thus represent an adaptive response to maintain homeostasis and to protect neurons, but can also, if persistent, insult vascular cells and cause vasoregression [40, 42]. Similarly, the production of NO from Müller cells can be both, beneficial and harmfull. Müller cells constitutively express NO synthases and upregulate the expression of the inducible form of NO synthase in DR [43]. Low levels of NO are beneficial for neurons, platelet function and retinal perfusion, while high levels are neurotoxic [44, 45]. Thus, the same gliotic reaction may induce beneficial and detrimental effects, depending on the duration and/or amplitude of the trigger. Hyperglycemic NO overproduction also induces cytotoxic prostaglandins from Müller cells, as well the release of VEGF and bFGF, which are neuroprotective, but tend to damage when produced persistently.

Despite the concerted proangiogenic activity that is found in the diabetic rat and mouse retina, the absence of any proliferative retinopathy and the limited intracapillary endothelial proliferative response is surprising and cannot be explained by the short life span of these animals, but should give reason to reconsider both the pro- as well as the less well characterized anti-angiogenic mechanisms that may exist. Müller cells stimulated by hyperglycemia produce proinflammatory cytokines such as interleukin1ß, interleukin 6, and tumor necrosis factor-α. These cytokines have been demonstrated to affect several of the structural and functional parameters, including pericyte dropout, vasoregression, and leakage [41].

12.11 The Role of Microglia

Another cell component of the neurovascular unit is the microglia. These retina-based cells are constituents of the innate immunity system, and respond to danger signals by upregulation of M1 cytokines. A subpopulation of microglial cells is activated early during diabetic retinopathy, and some of the inflammatory signal can be blunted by minocycline, an antibiotic with strong anti-inflammatory properties, and with a peptide that mimicks the tissue-protective properties of erythropoietin [46]. In a rat model of retinal degeneration, we recently indentified activated microglia as one of the mediators of vasoregression. CD-74 was identified as an activation marker of glial cells, and the CD-74 positive microglia was found on degenerating retinal capillaries [47, 48]. Depletion of activated microglia by clodronate coated liposomes alleviated vasoregression. Similar mechanisms have yet to be identified in diabetic models, but preliminary data suggest that both, hyperglycemia and reactive alpha-oxoaldehydes activate retinal microglia (Hammes et al. unpublished).

12.12 Leukostasis, Endothelial Damage and Retinal Inflammation

Leukocytes interact and bind to adhesion molecules which are upregulated in the diabetic retinal vasculature such as ICAM-1 and VCAM, driven by PKC-NF-kB regulatory axis [49]. The major ligand/receptor pair mediating adhesive interactions between leukocytes and retinal endothelial cells is the ß2-integrin-ICAM-1 system. A second pathway that promotes leukocyte-endothelial interaction is the enzyme 6-N-acetyl glucosamine-transferase (core 2 GlcNac-T) in leukocytes which results in glycan modification on the leukocyte surface, promoting leukocyte rolling interactions with endothelial selectins [50]. A third pathway involves the interaction of the receptor for AGEs RAGE and the ß2-integrin Mac-1 [51]. RAGE-mediated leukocyte recruitment in vivo is particularly important in diabetic mice. Consequently, leukostasis may affect the retinal microvasculature mostly by two mechanisms which can be interrelated, one being the occlusion of the microvascular bed, the other one being the direct endothelial cell injury by secretion of cytotoxic factors or by interaction of the Fas-Fas-L ligand receptor system. Apart from technical aspects confounding the results obtained when leukostasis is measured by the ex vivo technique, it has been shown that leukostasis is probably not causal for retinal vasoregression as it was shown that retinal capillary occlusion was not affected although leukostasis was inhibited, and vice versa [26]. Leukostasis may thus not be essential for the development of vasoregression in the early diabetic retina.

12.13 Advanced Retinopathy

With progressing ischemia, the diabetic retina provides evidence first for intraretinal, and then for preretinal neovascularization by sprouting angiogenesis. The predominant but not exclusive growth factor involved is vascular endothelial growth factor [52]. Properties of VEGF comply with pathological angiogenesis and increased permeability. The strongest signal for VEGF regulation is hypoxia, but in the diabetic retina, glucose, glycolytic intermediates, ROS and AGEs can all contribute to upregulation of VEGF. The signalling pathways of VEGF are outlined elsewhere in this book. VEGF may have mechanistic companions which can explain the observation that one third of patients with active proliferative diabetic retinopathy have excess vitreous levels of VEGF while two third have VEGF levels which are comparable to those with quiescent PDR but show active neovascularization. One such candidate that can complement the neovascular potential of VEGF is erythropoietin [53]. Like VEGF, its transcription in the retina is hypoxia-driven, it is found in considerable amounts in the vitreous of patients with proliferative diabetic retinopathy, and its inhibition reduces experimental proliferative retinopathy in newborn mouse oxygen-induced retinopathy (OIR) models. Furthermore, patients who receive erythropoietin in larger doses are at higher risk to develop retinopathy when hypoxia is present in the retina.

Various other growth factors such as TNF-a, Ang-2, erythropoietin, GH/IGF-1, MCP-1, and SDF-1 have been identified in ocular fluids of patients with proliferative diabetic retinopathy and diabetic macular edema, but none of them has a proven clinical role [54, 55]. Novel cytokines continue to be identified from the vitreous of PDR patients, with many of them showing a significant correlation with

VEGF, but with others that do not. These are considered candidates which may be developed for co-administration with anti-VEGF antibodies, or as alternatives when anti-VEGF therapy is ineffective.

12.14 To be Considered

Many of the preclinical compounds which are developed to inhibit proliferative diabetic retinopathy are spin-offs from antiproliferative or tumor therapy. Most of these drugs have targets whose expression as never been studied in a human proliferative diabetic retina, because the appropriate material is not available. Therefore, the mouse OIR model serves a surrogate for proliferative diabetic retinopathy and mimicks essential steps of angiogenesis. The largest difference between this model and the human diabetic disease is defined by the large impact of irreversible glucose-mediated tissue damage of long-standing diabetes in contrast to the developmental angiogenesis of the newborn mouse. As the transcriptom and the make-up of the tissue scaffold in proliferative diabetic retinopathy is likely to be very different from all experimental models developed so far, preclinical data interpretation should focus on the inconsistencies rather than the consistencies between the experimental model and the clinical pattern.

References

1. Yau JW, et al. Global prevalence and major risk factors of diabetic retinopathy. Diabetes Care. 2012;35(3):556–64.
2. Hammes HP, et al. Diabetic retinopathy in type 1 diabetes-a contemporary analysis of 8,784 patients. Diabetologia. 2011;54(8):1977–84.
3. Kyto JP, et al. Decline in the cumulative incidence of severe diabetic retinopathy in patients with type 1 diabetes. Diabetes Care. 2011;34(9):2005–7.
4. Antonetti DA, Klein R, Gardner TW. Diabetic retinopathy. N Engl J Med. 2012;366(13):1227–39.
5. Klein R. Diabetic retinopathy and nephropathy. In: Cortes P, Mogensen CE, editors. The diabetic kidney, Totowa: Humana Press; 2006. p. 473–98.
6. Knudsen ST, et al. Macular edema reflects generalized vascular hyperpermeability in type 2 diabetic patients with retinopathy. Diabetes Care. 2002;25(12):2328–34.
7. Knudsen LL, et al. The North Jutland County Diabetic Retinopathy Study (NCDRS) 2. Non-ophthalmic parameters and clinically significant macular oedema. Br J Ophthalmol. 2007;91(12):1593–5.
8. The Diabetes Control and Complications Trial Research Group. The effect of intensive treatment of diabetes on the development and progression of long-term complications in insulin-dependent diabetes mellitus. N Engl J Med. 1993;329(14):977–86.
9. UK Prospective Diabetes Study (UKPDS) Group. Intensive blood-glucose control with sulphonylureas or insulin compared with conventional treatment and risk of complications in patients with type 2 diabetes (UKPDS 33). The Lancet. 1998;352(9131):837–53.

10. Hirsch IB, Brownlee M. The effect of glucose variability on the risk of microvascular complications in type 1 diabetes. Diabetes Care. 2007;30(1):186-7. author reply 188–9.
11. Klein R, et al. Association of ocular disease and mortality in a diabetic population. Arch Ophthalmol. 1999;117(11):1487–95.
12. Kramer CK, et al. Diabetic retinopathy predicts all-cause mortality and cardiovascular events in both type 1 and 2 diabetes: meta-analysis of observational studies. Diabetes Care. 34(5):1238–44.
13. Robinson R, et al. Update on animal models of diabetic retinopathy: from molecular approaches to mice and higher mammals. Dis Model Mech. 5(4):444–56.
14. Hammes HP, et al. Angiopoietin-2 causes pericyte dropout in the normal retina: evidence for involvement in diabetic retinopathy. Diabetes. 2004;53(4):1104–10.
15. Nishikawa T, et al. Normalizing mitochondrial superoxide production blocks three pathways of hyperglycaemic damage. Nature. 2000;404(6779):787–90.
16. Brownlee M. The pathobiology of diabetic complications: a unifying mechanism. Diabetes. 2005;54(6):1615–25.
17. Giacco F, Brownlee M. Oxidative stress and diabetic complications. Circ Res. 107(9):1058–70.
18. Hammes HP, et al. Benfotiamine blocks three major pathways of hyperglycemic damage and prevents experimental diabetic retinopathy. Nat Med. 2003;9(3):294–9.
19. Brownlee M. Biochemistry and molecular cell biology of diabetic complications. Nature. 2001;414(6865):813–20.
20. Wang Y, et al. Receptor for advanced glycation end product expression in experimental diabetic retinopathy. Ann N Y Acad Sci. 2008;1126:42–5.
21. Yao D, et al. High glucose increases angiopoietin-2 transcription in microvascular endothelial cells through methylglyoxal modification of mSin3A. J Biol Chem. 2007;282(42):31038–45.
22. Thangarajah H, et al. HIF-1alpha dysfunction in diabetes. Cell Cycle. 2010;9(1):75–9.
23. Geraldes P, King GL. Activation of protein kinase C isoforms and its impact on diabetic complications. Circ Res. 2010;106(8):1319–31.
24. Geraldes P, et al. Activation of PKC-delta and SHP-1 by hyperglycemia causes vascular cell apoptosis and diabetic retinopathy. Nat Med. 2009;15(11):1298–306.
25. Pfister F, et al. Pericyte migration: a novel mechanism of pericyte loss in experimental diabetic retinopathy. Diabetes. 2008;57(9):2495–502.
26. Zheng L, Kern TS. Role of nitric oxide, superoxide, peroxynitrite and PARP in diabetic retinopathy. Front Biosci (Landmark. Ed). 2009;14:3974–87.
27. The Diabetes Control and Complications Trial/Epidemiology of Diabetes Interventions and Complications Research Group. Retinopathy and nephropathy in patients with type 1 diabetes four years after a trial of intensive therapy. N Engl J Med. 2000;342(6):381–9.
28. Chalmers J, Cooper ME. UKPDS and the legacy effect. N Engl J Med. 2008;359(15):1618–20.
29. El-Osta A, et al. Transient high glucose causes persistent epigenetic changes and altered gene expression during subsequent normoglycemia. J Exp Med. 2008;205(10):2409–17.
30. Forbes JM, Cooper ME. Mechanisms of diabetic complications. Physiol Rev. 93(1):137–88.
31. De Bock K, et al. Role of PFKFB3-driven glycolysis in vessel sprouting. Cell. 2013;154(3):651–63.
32. Kern TS, Engerman RL. Vascular lesions in diabetes are distributed non-uniformly within the retina. Exp Eye Res. 1995;60(5):545–9.
33. Alikhani M, Roy S, Graves DT. FOXO1 plays an essential role in apoptosis of retinal pericytes. Mol Vis. 2010;16:408–15.
34. Yatoh S, et al. Antioxidants and an inhibitor of advanced glycation ameliorate death of retinal microvascular cells in diabetic retinopathy. Diabetes Metab Res Rev. 2006;22(1):38–45.
35. Behl Y, et al. Diabetes-enhanced tumor necrosis factor-alpha production promotes apoptosis and the loss of retinal microvascular cells in type 1 and type 2 models of diabetic retinopathy. Am J Pathol. 2008;172(5):1411–8.

36. Klaassen I, Van Noorden, CJ, Schlingemann RO. Molecular basis of the inner blood-retinal barrier and its breakdown in diabetic macular edema and other pathological conditions. Prog Retin Eye Res. 2013;34:19–48.
37. Rakoczy EP, et al. Characterization of a mouse model of hyperglycemia and retinal neovascularization. Am J Pathol. 2010;177(5):2659–70.
38. Armulik A, Genove G, Betsholtz C. Pericytes: developmental, physiological, and pathological perspectives, problems, and promises. Dev Cell. 2011;21(2):193–215.
39. Metea MR, Newman EA. Glial cells dilate and constrict blood vessels: a mechanism of neurovascular coupling. J Neurosci. 2006;26(11):2862–70.
40. Hammes HP, Federoff HJ, Brownlee M. Nerve growth factor prevents both neuroretinal programmed cell death and capillary pathology in experimental diabetes. Mol Med. 1995;1(5):527–34.
41. Bringmann A, et al. Muller cells in the healthy and diseased retina. Prog Retin Eye Res. 2006;25(4):397–424.
42. Carlevaro MF, et al. Transferrin promotes endothelial cell migration and invasion: implication in cartilage neovascularization. J Cell Biol. 1997;136(6):1375–84.
43. Mishra A, Newman EA. Inhibition of inducible nitric oxide synthase reverses the loss of functional hyperemia in diabetic retinopathy. Glia. 2010;58(16):1996–2004.
44. Du Y, Sarthy VP, Kern TS. Interaction between NO and COX pathways in retinal cells exposed to elevated glucose and retina of diabetic rats. Am J Physiol Regul Integr Comp Physiol. 2004;287(4):R735–41.
45. Kashii S, et al. Dual actions of nitric oxide in N-methyl-D-aspartate receptor-mediated neurotoxicity in cultured retinal neurons. Brain Res. 1996;711(1–2):93–101.
46. Krady JK, et al. Minocycline reduces proinflammatory cytokine expression, microglial activation, and caspase-3 activation in a rodent model of diabetic retinopathy. Diabetes. 2005;54(5):1559–65.
47. Feng Y, et al. Gene expression profiling of vasoregression in the retina–involvement of microglial cells. PLoS One. 2011;6(2):e16865.
48. Feng Y, et al. Vasoregression linked to neuronal damage in the rat with defect of polycystin-2. PLoS One. 2009;4(10):e7328.
49. Booth G, et al. Mechanisms of amelioration of glucose-induced endothelial dysfunction following inhibition of protein kinase C in vivo. Diabetes. 2002;51(5):1556–64.
50. Chibber R, et al. Activity of the glycosylating enzyme, core 2 GlcNAc (beta1,6) transferase, is higher in polymorphonuclear leukocytes from diabetic patients compared with age-matched control subjects: relevance to capillary occlusion in diabetic retinopathy. Diabetes. 2000;49(10):1724–30.
51. Chavakis T, Bierhaus A, Nawroth PP. RAGE (receptor for advanced glycation end products): a central player in the inflammatory response. Microbes Infect. 2004;6(13):1219–25.
52. Aiello LP, et al. Vascular endothelial growth factor in ocular fluid of patients with diabetic retinopathy and other retinal disorders. N Engl J Med. 1994;331(22):1480–7.
53. Watanabe D, et al. Erythropoietin as a retinal angiogenic factor in proliferative diabetic retinopathy. N Engl J Med. 2005;353(8):782–92.
54. Bromberg-White JL, et al. Identification of VEGF-Independent Cytokines in Proliferative Diabetic Retinopathy Vitreous. Invest Ophthalmol Vis Sci. 2013;54(10):6472–80.
55. Jonas JB, et al. Cytokine concentration in aqueous humor of eyes with diabetic macular edema. Retina. 2012;32(10):2150–7.

Chapter 13
The Angiopoietin—Tie System: Common Signaling Pathways for Angiogenesis, Cancer, and Inflammation

Yvonne Reiss, Alexander Scholz and Karl H. Plate

13.1 Introduction

Angiogenesis involves the complex signaling between multiple angiogenic growth factors, and requires the coordinated interaction between endothelial and adjacent cells. Gene ablation studies in mice identified vascular endothelial growth factor (VEGF) as the principal molecule that drives endothelial cell sprouting and primary network formation [1–4] (see also Chap. 1). In contrast, angiopoietin/Tie signaling is important at subsequent stages of vascular development through the control of vessel remodeling and maturation [5, 6]. Ang1 and Ang2 act as agonistic and antagonistic molecules on the tyrosine kinase with immunoglobulin (Ig) and epidermal growth factor (EGF) homology domains (Tie2). Ang1 has the capability to activate Tie2, while Ang2 has been identified as a context-dependent inhibitory ligand [7–10]. Ang2 overexpression in transgenic mice leads to an embryonic lethal phenotype similar to Ang1 or Tie2 deletion [8, 9, 11, 12]. Those early genetic studies dating back to the 1990s implied that signaling through Tie2 is dependent on the balance of Ang1 and Ang2. In the quiescent vasculature in adults, Ang1 provides a basal signal through constitutive Tie2 phosphorylation to maintain endothelial cell integrity [13] (Fig. 13.1). In contrast, under pathological conditions, Ang2 coun-

Y. Reiss (✉) · A. Scholz · K. H. Plate
Institute of Neurology (Edinger Institute), Goethe University Medical School, Frankfurt, Germany
e-mail: Yvonne.Reiss@kgu.de

Y. Reiss · K. H. Plate
German Cancer Consortium, Partner Site Frankfurt/Mainz, Frankfurt 60528, Germany

German Centre of Cardiovascular Research, Partner Site Rhein/Main, Frankfurt, Germany

A. Scholz
Laboratory of Immunology and Vascular Biology, Department of Pathology, Stanford University School of Medicine, Stanford, CA, USA

M. HH. Schmidt, S. Liebner (eds.), *Endothelial Signaling in Development and Disease*, DOI 10.1007/978-1-4939-2907-8_13

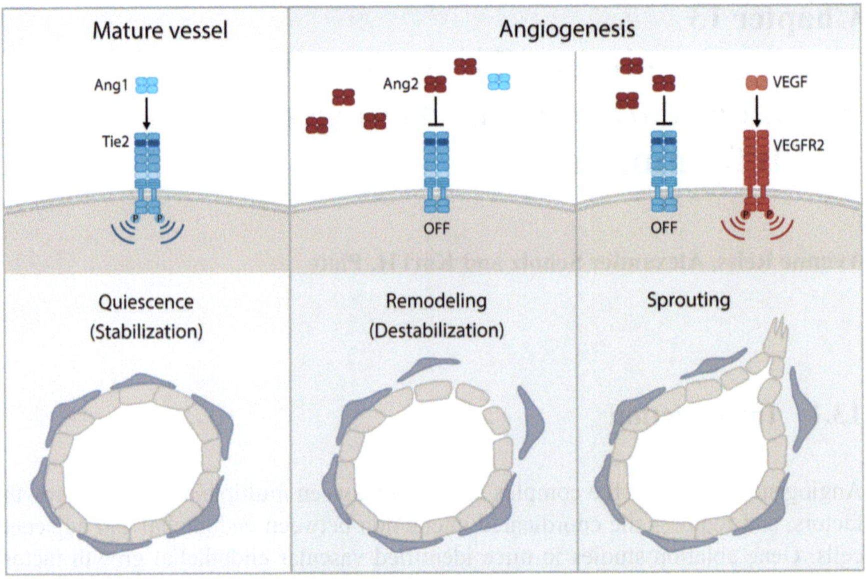

Fig. 13.1 Angiopoietin functions in the vasculature. In the adult vasculature, angiopoietin-1 (Ang1) contributes to the stabilization of pre-existing vessels by providing a basal signal that leads to constitutive Tie2 phosphorylation (quiescence). Under angiogenic conditions, Ang2 interferes with the Ang1-induced Tie2 phosphorylation. As a consequence, the vasculature is destabilized. In conjunction with vascular endothelial growth factor (VEGF), Ang2 induces vascular sprouting

teracts these effects and leads to vessel destabilization, thereby allowing plasticity [14]. Hence, Ang2 is a gatekeeper of vessel growth versus regression and cooperates with additional growth factors, such as VEGF [5, 15] (Fig. 13.1). Similar to their functions in vascular development, Ang1 and Ang2 exert adverse functions during pathological angiogenesis, e.g. in tumors. Constitutive Ang1 signaling promotes the recruitment of pericytes and leads to vessel stabilization [5] (Fig. 13.1). On the contrary, Ang2 is highly upregulated at early stages of tumor progression to drive neovessel formation [14, 16, 17]. In addition to the vascular functions, angiopoietins have also been implicated in the context of inflammation and the recruitment of inflammatory cells [5, 6, 18–20] (Fig. 13.2). Due to these novel functions and to the specific expression in angiogenic, but not normal vessels, angiopoietins have evolved as a potential target for therapeutic intervention in combination with other angiogenic factors [21–24].

13.2 Angiopoietin/Tie Signaling During Development

During embryogenesis, the vascular system is the first organ system that develops through the assembly of endothelial/mesodermal precursor cells (angioblasts) by a process referred to as vasculogenesis [25, 26]. In this context, VEGF signaling

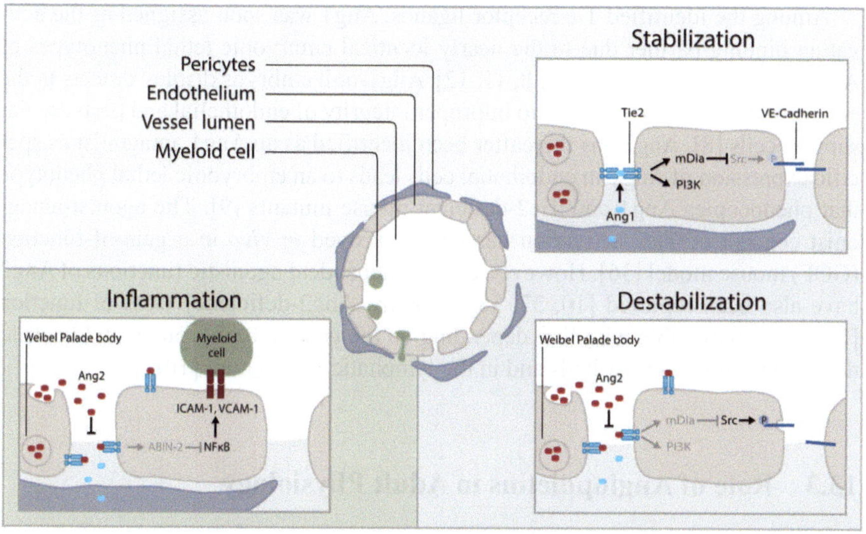

Fig. 13.2 Role of angiopoietins in the regulation of endothelial cell homeostasis. In the quiescent vasculature, paracrine Ang1 signaling leads to the translocation of Tie2 in endothelial cell junctions. Tie2 signal transduction induces survival signals via the PI3K/AKT pathway, and allows vessel stabilization by preventing vascular endothelial (VE)-cadherin phosphorylation via the Src kinase and mammalian diaphanous (*mDia*) [*upper right panel*]. In angiogenic endothelial cells, Ang2 is secreted from Weibel–Palade bodies and acts in an autocrine manner on Tie2, thereby interfering with the Ang1-induced vessel stabilization (*lower right panel*). Furthermore, Ang2 interferes with nuclear factor kappa B (*NF-κB*) signaling via the adaptor protein ABIN-2, and thus promotes the upregulation of adhesion receptors (intercellular adhesion molecule-1 [*ICAM-1*], vascular cell adehsion molecule-1 [*VCAM-1*]) and the recruitment of immune cells

is crucial for the formation of a primary vascular plexus (E8.5) [27, 28] (see also Chap. 1). At later stages (E9.5–12.5), the embryonic vasculature expands by sprouting angiogenesis and maturates through the recruitment of pericytes and smooth muscle cells (SMCs). This process is essentially governed by angiopoietin/Tie2 signaling [5, 6, 27]. Angiopoietins are secreted growth factors that exert downstream signaling through the Tie2 receptor tyrosine kinase. Tie2 is predominantly expressed in endothelial cells. Consequently, deletion of the Tie2 gene leads to embryonic lethality around E10.5–12.5 due to cardiac defects and defective remodeling of the poorly organized primary vascular plexus [11, 12]. This primitive plexus contains fewer endothelial cells that display disrupted interactions with perivascular support cells. Further studies also identified irregular hematopoiesis in the Tie2-null animals [29]. The Tie2 homolog Tie1 receptor tyrosine kinase Tie1 is still considered orphan with no apparent ligand. Ang1 has been described to bind and cluster Tie1 at cell–cell contacts only in the presence of Tie2 [30, 31]. Tie1 deficiency in mice leads to embryonic lethality between E13.5 and birth due to defects in endothelial cell integrity and the formation of pulmonary edema [12, 32]. In addition, deletion of Tie1 affects lymphangiogenesis [33, 34]. However, defects in the lymphatic vasculature were manifested prior to vascular alterations around E12.5. In addition, and in contrast to Tie2, Tie1 deficiency does not effect definitive hematopoiesis [35].

Among the identified Tie receptor ligands, Ang1 was soon assigned as the activating binding partner due to the nearly identical embryonic lethal phenotypes of Ang1 and Tie2 mutant mice [7, 8, 11, 12]. Ang1-null embryos display defects in the endocard and die by E12.5 due to improper integrity of endothelial and perivascular support cells [8]. Ang2 has thereafter been identified as an Ang1 antagonist as specific expression of Ang2 in endothelial cells leads to an embryonic lethal phenotype that phenocopies Ang1 and Tie2-deficient mouse mutants [9]. The agonist-antagonist concept of Tie2 activation has been approved *in vivo* in a gain-of-function (GOF) mouse model [36]. However, context-dependent agonistic functions of Ang2 have also been reported [10, 37, 38]. Notably, Ang2-deficiency (loss-of-function [LOF]) is compatible with life, depending on the genetic background, and leads to defects in some vascular beds and in the lymphatic vasculature [10].

13.3 Role of Angiopoietins in Adult Physiology

Among the four Angiopoietin ligands, Ang1 and Ang2 are characterized best, while Ang3 and Ang4 represent less well-studied orthologs in humans and mice [5, 6, 28]. Ang1 is expressed in numerous cell types, including perivascular cells (SMCs and pericytes), fibroblasts, osteoblasts, or tumor cells, and acts in a paracrine manner on Tie2 [5, 7, 16, 39]. In contrast, Ang-2 is predominantly expressed in endothelial cells and only upregulated at sites of angiogenesis, i.e. during development or pathological angiogenesis [9, 10, 15, 16]. Ang2 exerts its functions on the Tie2 receptor tyrosine kinase in an autocrine manner. Interestingly, Ang2 is not only regulated at the transcriptional level upon hypoxia or by cytokine activation but is also additionally stored in Weibel–Palade bodies (WPB) [5, 40]. Storage in this particular cellular compartment, which is special to endothelial cells, allows the rapid release of its contents (cytokines, growth factors, adhesion receptors). In addition, it elicits immediate responses to stress factors such as histamin, thrombin, etc., and provides a link to the inflammatory cascade [18–20, 40] (Fig. 13.2). Ang1 and Ang2 show high sequence and structure homologies [5, 9, 41]. Angiopoietins consist of an amino-terminal (N) domain, a coiled-coil (C) domain, and a carboxy-terminal fibrinogen-homology (F) domain [41]. While the F domain is responsible for the interaction with the Tie2 receptor, the C-terminal and N-terminal domains are essential for clustering of the ligand. Angiopoietins primarily exist as tetramers and higher order oligomers, and structural analysis revealed that at least one tetramer of Ang1 is necessary to activate endogenous Tie2 [41–43]. Interestingly, Ang2 occurs in the same higher-order multimers as Ang1 but fails to induce receptor activation [41, 42, 44]. Originally, it was identified that Ang1 and Ang2 bind to the same domains of Tie2 with similar kinetics (first Ig-like domain and the EGF-like repeats) [45, 46]. However, more recent evidence suggests the second Ig domain as the primary angiopoietin binding site [42]. Hitherto, it is not entirely resolved how Ang1 and Ang2 achieve the different agonistic versus antagonistic functions, although they form remarkably similar complexes with Tie2 [42, 46, 47]. Recent evidence indicated that specific molecular surface structures are important for the interaction with Tie2 [47]. In particular, three critical residues within the angiopoietin fibrinogen domain are necessary to confer to the differential ligand activities [47].

Tie2 is expressed on vascular and lymphatic endothelial cells and is also present on circulating hematopoietic cells, including megakaryocytes and neutrophils [5, 48]. Additionally, a subpopulation of tumor-infiltrating monocytes has been shown to express Tie2 [49–51]. Tie1 is specifically expressed in endothelial cells but does not directly bind to angiopoietins [5, 6]. Tie1 is abundantly expressed in the embryonic vasculature but its expression subsides with vessel maturation. In the adult, Tie1 is induced in malignant melanoma [6, 52] or in areas of disturbed flow in atherogenic vascular niches [6, 53]. Tie1 and Tie2 form complexes at the endothelial cell surface, which are mainly mediated by electrostatic forces [54]. The association of Tie1 with Tie2 has an inhibitory effect on the receptor, which is independent on the Tie1 kinase domain. Interestingly, stimulation with Ang1, but not Ang2, was able to dissociate Tie1 from Tie2, leading to receptor phosphorylation [54]. Upon stimulation with Ang1, Tie2 not only dissociates from Tie1 but also induces receptor translocation and assembly in distinct signaling complexes [31, 55]. In confluent resting endothelial cells, Ang1 induces the receptor translocation to cell–cell junctions and mediates clustering of Tie2 in trans with neighboring endothelial cells. This shifts intracellular signals to increased Akt signaling and results in endothelial cell quiescence. In contrast, in mobile endothelial cells the interaction of matrix bound Ang1 to Tie2 mediates adhesion and induces a promigratory phenotype mediated by Erk and Dok-R signaling [31, 55]. Interestingly, Ang2 is additionally able to elicit signaling via integrins in Tie2-negative/silenced endothelial cells and thus differentially regulates angiogenesis versus destabilization, depending on the presence of endothelial Tie2 [56, 57].

The deployment of genetic mouse models (GOF/LOF) helped to further understand angiopoietin signaling in the adult. Ang1 signaling is thus dispensable in quiescent vessels in the adult but necessary to modulate vascular response after injury [58]. This finding challenges the concept of Ang1 acting as a survival factor [43]. Nonetheless, Ang1-mediated constitutive Tie2 phosphorylation leads to vessel stabilization in the adult vasculature [13, 36]. This signal is antagonized by Ang2 *in vivo* and thus leads to vessel destabilization [36, 59]. Detailed mechanistic insights on how endothelial cell stabilization versus destabilization are maintained by angiopoietin signaling are rare. Evidence for vascular stabilization and leakage resistance mediated by Ang1 was deducted from studies in GOF mice [60, 61]. Mice overexpressing Ang1 in the skin display more vessels with larger volume, indicative for sprouting [62]. In addition, these mice are leakage-resistant against permeability-inducing agents such as mustard oil or are unreponsive against VEGF-induced permeability [60, 61]. Mechanistically, this effect has been pinpointed to the crosstalk between Ang1 and vascular endothelial (VE)-cadherin signaling via the small protein kinase Src [63, 64]. In detail, Ang1 stimulation of endothelial cells prevents VEGF-induced endothelial cell permeability by sequestering Src through mammalian diaphanous (mDia) [63, 64] (Fig. 13.2). Ang2 has been implicated to antagonize these functions *in vivo* (Fig. 13.2) [for details see also Chaps. 6 and

8]. Intradermal and tracheal permeability was largely attenuated in an Ang2 LOF model [65]. A further study demonstrated direct evidence for the involvement of Ang2 in vascular permeability in the skin of Ang2 GOF animals [66]. Hitherto, the permeability changes induced by Ang2 were demonstrated merely in peripheral vessels. The question arises whether specialized endothelial cells of the brain are similarly affected by Ang2 and whether the blood–brain barrier (BBB) easily opens upon challenge with Ang2. Until now, evidence *in vivo* is rare. However, one study evidenced that Ang2-mediated BBB impairment was antagonized by an Ang2 inhibitory antibody in a brain metastasis model [67].

13.4 Role of Angiopoietins in Tumors

Angiopoietin/Tie signaling has also been implicated to drive tumor progression [22, 68]. Ang2 has been identified to be expressed early during tumor growth similar to the onset of VEGF expression [14, 16, 17]. Ang2 is not expressed in normal vessels and thus appears to be critical for tumor initiation (i.e. by vessel cooption) and tissue remodeling and has been shown to promote new vessel growth in concert with other growth factors [14, 16, 22]. Consequently, high Ang2 levels have been identified in the serum of patients with different neoplasias and Ang2 has thus also been implicated as a biomarker for cancer and other pathologies that involve neovessel growth or vessel permeability [5, 24]. Numerous studies with genetically engineered tumor cells expressing Ang1 or Ang2 provided evidence that both molecules act as agonists/antagonists to induce or inhibit Tie2 downstream signaling [5, 68, 69]. Although outcome on tumor growth upon Ang1 or Ang2 expression was sometimes controversial (see Reiss [68] for review), the net outcome on tumor angiogenesis was similar to findings during developmental angiogenesis, i.e. vessel stabilization mediated by Ang1 versus vessel destabilization mediated by Ang2 (Fig. 13.2) [5, 6]. Ang1 expression leads to an improved vasculature with more pericytes [5, 6, 68]. As a consequence, perfusion of the normally chaotic tumor vasculature is improved upon Ang1 expression, a process denominated as 'vascular normalization' [70–72]. In contrast, Ang2 expression leads to more, but smaller, vessels, which are devoid of mural cells [5, 6, 68]. This finding is indicative of an immature vascular phenotype, which results in improper perfusion. Our own investigations at the ultrastructural level in both mouse glioma and breast cancer models showed similar vascular changes, indicative of vascular stabilization versus destabilization phenotypes upon Ang1 and Ang2 expression [59, 73]. Moreover, studies in transgenic animals supported the finding obtained in angiopoietin-expressing tumor cell lines. Ang2 LOF identifed the rate-limiting role for Ang2 during the early phases of tumor growth, possibly also regulating the angiogenic switch [74]. Ang2 GOF in subcutanous tumors (Lewis lung carcinoma) or in the Rip1Tag2 transgenic mouse model further supported the proangiogenic and destabilizing functions of Ang2 [75, 76].

13.5 Angiopoietin Functions in the Recruitment of Inflammatory Cells

Aside from the vascular functions, angiopoietins have been implicated as gatekeepers of immediate endothelial responses such as permeability, hemostasis (coagulation), and inflammation [5, 20]. Along with other hemostasis-regulating proteins (e.g. von Willebrand factor [vWF], P-selectin, interleukin-8), Ang2 is stored in WPB [40]. Mediators contained in WPB can be deployed rapidly in response to signaling molecules and mechanical stress [5, 40]. Opposite to this, Ang1 prevents endothelial cell activation and inflammation when overexpressed in the skin of transgenic mice [60]. Moreover, Ang1 was capable of counteracting VEGF-mediated skin permeability and inflammation when coexpressed with VEGF in transgenic mice [61]. Those initial genetic studies indicated that Ang1 is able to circumvent inflammatory reactions by stabilization of the vasculature. In fact, Ang1 has been shown to interfere with nuclear factor kappa B (NFκB) signaling through recruitment of the adapter protein A20-binding inhibitor of NF-κB 2 (ABIN-2) [77]. Our own studies demonstrated that the absence of Ang2 delayed the onset of short-term inflammatory reactions in LOF mice, which was restored by administration of recombinant Ang2 [18]. This implicated antagonistic roles of Ang1 and Ang2 in immediate vascular responses, including inflammation (Fig. 13.2) [19]. Mechanistically, the delayed onset of inflammatory responses was attributed to extended rolling and the defective leukocyte adhesion as evidenced by intravital fluorescence video microscopy [18]. Furthermore, by employing a GOF model, i.e. Ang2 expression in endothelial cells, we demonstrated that Ang2 on its own was effective in inducing the recruitment of innate immune cells and thus served as an instigator of inflammation (Fig. 13.2) [78, 79]. In these mice, short-term inflammation was augmented as a consquence of prolonged myeloid cell adhesion [78]. Similarly, Ang2 has been demonstrated to link vascular remodeling and inflammation in the model of *Mycoplasma pulmonis*-induced airway inflammation in a Tie2-dependent manner [80]. Selective targeting of Ang2 was able to restore Tie2 phosphorylation and reduce disease severity [80].

The recruitment of immune cells from the blood is a key cellular response to tissue damage and inflammation [81]. The immigration of immune cells from blood into tissues is a crucial process that not only applies during inflammatory conditions but also in neoplastic diseases as an inflammatory microenvironment is existent in all tumors [82]. The link between inflammation and cancer was recognized by Virchow in the 19th century [83]. Immune cells are recruited to sites of tumor progression where they are able to exert anti- or protumorigenic functions with the latter also accounting for therapeutic resistance [82, 84–86]. The most frequent subsets of immune cells within the tumor microenvironment are tumor-associated macrophages (TAMs) and myeloid-derived suppressor cells [82, 84]. Importantly, growth factors or cytokines that are secreted by tumor-infiltrating immune cells activate key inflammatory transcription factors, e.g. NF-κB [82, 87]. Tumor-infiltrating myeloid cells, also termed accessory cells, have been recognized as major contributors to tumor angiogenesis through the secretion of numerous cytokines or their education by growth factors [86, 88, 89]. High numbers of TAMs have been associated with poor prognosis in numerous cancer entities [90]. They secrete a plethora of cytokines (among which are growth factors (e.g. VEGF), chemokines, and matrix metalloproteinases) that are able to drive angiogenesis in tumors [84, 86]. A subpopulation of these cells expressing the Tie2 receptor tyrosine kinase (Tie2-expressing monocytes [TEM]) has been identified to be crucial for tumor progression [49, 50, 91]. Selective depletion of these TEMs in tumor-bearing mice inhibited tumor angiogenesis and growth, suggesting that they might regulate angiogenic processes in tumors by providing paracrine signals to newly formed blood vessels [49]. Our own studies identified that Ang2 was directly responsible for the recruitment of proangiogenic macrophages in subcutaneous tumors [75] and in settings of inflammation [78]. Furthermore, Ang2 stimulation increased the inherent angiogenic properties of TEMs and led to the upregulation of genes associated with the M2-polarized phenotype of macrophages [75]. In addition, Ang2 transgene expression promoted the expansion of T regulatory cells and thus further contributed to immune suppression [92]. As such, the Ang2–TEM/TAM axis may represent a new dual target for antiangiogenic/anti-immunosuppressive cancer therapy.

13.6 Current Therapeutic Concepts

Antiangiogenic therapy has become a valuable clinical target since the approval of bevacizumab (Avastin®) in 2004 for the treatment of colorectal cancer [93] and thereafter for other cancer entities [72, 94]. However, efficacy of anti-VEGF therapy is rather dismal due to escape mechanisms that involve the presence of proangiogenic innate immune cells or increased invasiveness [84]. Several tumor types, including glioblastoma, are not as responsive or have been shown to develop escape mechansims that lead to further tumor progression despite antiangiogenic therapy [95]. Unresponsiveness to antiangiogenic therapy has, in part, been attributed to the infiltration of tumor-associated macrophages that promote tumor growth by the secretion of proangiogenic cytokines [96–98]. In this regard, novel therapeutic strategies are being explored by pharmaceutical companies. Ang2 emerged as a valuable clinical target as it is solely upregulated under angiogenic conditions, where it is also coexpressed with VEGF [14, 16, 17]. As such, targeting the angiopoietin/Tie signaling pathway opens new avenues for therapeutic inhibition of tumor growth [22–24]. A peptibody targeting Ang1 and Ang2 (Trebananib®/AMG386) [99, 100] is currently being explored in a phase III clinical trial and has met the endpoint of improved progression-free survival [101, 102]. Numerous reagents targeting the angiopoietin/Tie pathway have been developed and were tested in preclinical or clinical models [23, 24, 99, 100, 103–108]. The new drug regimen appears very successful, particularly in combination treatment with drugs that target the VEGF/VEGFR signaling pathway [23, 24]. The combination of VEGF- and angiopoietin-targeting drugs appears to be superior to targeting either pathway alone [23, 24, 38, 109–113].

Table 13.1 Targeting angiogenesis: the angiopoietin versus VEGF axis

	VEGF/VEGFR axis	Ang/Tie2 axis
Major ligands	VEGF A–D, PlGF	Ang1, Ang2, Ang4
Major TK receptors	VEGFR 1–3	Tie2
Co-receptors	Neuropilin	Tie1, integrins
Vascular biology—proof of principle	[122], [123]	[9], [60]
Tumor angiogenesis—proof of principle	[124]	[99]
Major cellular functions	EC proliferation, EC guidance, EC migration, EC survival	Pericyte coverage, EC guidance, vessel maturation, vessel normalization, recruitment of TAMs
First-in-class drug	Bevacizumab (Avastin®)	Trebananib (AMG386)
FDA approval/EU decision	2004/2005	

Ang angiopoietin, *EC* endothelial cell, *PlGF* placental growth factor, *TAMs* tumor-associated macrophages, *TK* tyrosine kinase, *VEGF* vascular endothelial growth factor, *VEGFR* vascular endothelial growth factor receptor

In this regard, it is of interest to note that VEGF blockade has been associated with increased invasiveness and even increased metastasis of the treated tumor. Mechanistically, this phenomenon has been linked to a hypoxia-mediated, c-Met-dependent activation of VEGFR-2 [114]. Although this phenomenon is currently confined to preclinical models as confirmatory data from human cancer are largely missing, the potential risk of increased invasiveness/metastasis remains a major concern. In contrast to VEGF blocklade, inhibition of Ang2 has not been associated with increased metastasis in preclinical cancer models. Blockade of Ang2 resulted in decreased metastasis in a number of preclinical models [105, 106, 113, 115]. The reason why inhibition of one antiangiogenic agent (VEGF) potentially induces invasion/metastasis whereas blocking of another (Ang2) blocks metastasis is not entirely clear but may be related to the extent of therapy-induced hypoxia, as well as to decreased vascular permeability and increased vascular stability as a consequence of Ang2 blockade. Ang1 binding to the Tie2 tyrosine kinase leads to receptor phosphorylation and junctional stabilization in a RhoA kinase-dependent manner [63, 64]. Vice versa, Ang2 is able to block Ang1-induced Tie2 phosphorylation and can reduce cell-matrix interactions, with the consequence of destabilizing vessels [8, 9, 36]. Ang2 inhibitors may therefore counteract Ang2-induced vascular stabilization and thus diminish metastasis.

The dual blockade of VEGF and Ang2 affects numerous cellular compartments, i.e. vascular, perivascular (see Table 13.1), and immune cells (see below), which may explain additive therapeutic results. Myeloid cells have long been known to be responsible for mediating tumor refractoriness [84, 90, 97, 98]. They are associated with poor prognosis and thus their depletion has a beneficial outcome on survival [84, 90, 96]. Continuous expression of Ang2 in endothelial cells (Ang2 GOF mouse model) led to an increased number of tumor-associated myeloid cells that exhibited a proangiogenic/M2 gene signature [75]. In addition, Ang2 GOF increased the frequency of T regulatory cells, which have immunosuppressive capacities and are thus also able to negatively contribute to tumor progression [92]. In addition, high Ang2 levels have recently been associated with therapy resistance [113]. Clearly, these findings are in favor of dual targeting of angiopoietin and VEGF signaling pathways. Interestingly M2-polarized proangiogenic macrophages still reside after Ang2/VEGF combination therapy in a syngeneic glioblastoma model (own unpublished observations). These results demand for additional targeting of the innate immune cell compartment for example by using antibodies against colony-stimulating factor 1 receptor (CSF1R). Proof of principle with drugs targeting tumor-associated macrophages has been established in preclinical models [116, 117]. Moreover, immune cell activation may be an additional option for enhancing efficacy of cancer therapies as an increased number of T lymphocytes/cytotoxic T cells following anti-VEGF or anti-CSF1R therapy have been reported [98, 118]. New avenues targeting the immune system (activation) by the usage of checkpoint inhibitors are currently also being explored [119] and results from preclinical studies will show their efficacy and demonstrate whether they may be useful for future cancer therapies [120, 121].

13.7 Concluding Remarks

Since the discovery of angiopoietins and their cognate Tie receptors, as well as the knowledge regarding their importance in shaping a functional vascular system, the angiopoietin/Tie system has emerged as multifaceted growth factor system with a newly defined link to the immune system. Angiopoietins have wide-ranging effects on the vasculature that include angiogenesis, vascular stabilization and permeability, and the newly defined function in the recruitment of inflammatory cells. The finding that two biological pathways—angiogenesis and inflammation—which are switched on during pathological vessel growth are linked by a common signaling pathway presents a striking opportunity for the development of novel inhibition strategies for cancer treatment by blocking both neovessel formation and inflammatory cell recruitment. Indeed, new promising therapeutic avenues have been explored in recent years, as reviewed in this chapter. Results from ongoing preclinical and clinical studies will prove efficacy and suitablility for future cancer therapy.

Acknowledgments This work was supported by the Collaborative Research Center "Vascular Differentiation and Remodeling" (CRC/Transregio23, project C1), the Cluster of Excellence 147 "Cardiopulmonary System" (ECCPS), Research Area C "Angiogenesis and Alveogenesis" by the German Research Council (DFG) [to YR and KHP]; a fellowship from the German Research Council (DFG) and a start-up grant from the Cluster of Excellence 147 "Cardiopulmonary systems" (ECCPS) [to AS]; a grant from the German Cancer Consortium (DKTK, Partnersite Frankfurt/Mainz) [to KHP]; an Investigator Award (FFF) [to YR] and a research grant of the Goethe University Medical School (to AS).

References

1. Ferrara N, Gerber H-P, LeCouter J. The biology of VEGF and its receptors. Nat Med. 2003;9(6):669–76.
2. Carmeliet P. Angiogenesis in life, disease and medicine. Nature. 2005;438(7070):932–6.
3. Kerbel RS. Tumor angiogenesis. N Engl J Med. 2008;358(19):2039–49.
4. Chung AS, Ferrara N. Developmental and pathological angiogenesis. Annu Rev Cell Dev Biol. 2011;27:563–84.
5. Augustin HG, Koh GY, Thurston G, Alitalo K. Control of vascular morphogenesis and homeostasis through the angiopoietin-Tie system. Nat Rev Mol Cell Biol. 2009;10(3):165–77.
6. Eklund L, Saharinen P. Angiopoietin signaling in the vasculature. Exp Cell Res. 2013;319(9):1271–80.
7. Davis S, Aldrich TH, Jones PF, Acheson A, Compton DL, Jain V, et al. Isolation of angiopoietin-1, a ligand for the TIE2 receptor, by secretion-trap expression cloning. Cell. 1996;87(7):1161–9.
8. Suri C, Jones PF, Patan S, Bartunkova S, Maisonpierre PC, Davis S, et al. Requisite role of angiopoietin-1, a ligand for the TIE2 receptor, during embryonic angiogenesis. Cell. 1996;87(7):1171–80.
9. Maisonpierre PC, Suri C, Jones PF, Bartunkova S, Wiegand SJ, Radziejewski C, et al. Angiopoietin-2, a natural antagonist for Tie2 that disrupts in vivo angiogenesis. Science. 1997;277(5322):55–60.
10. Gale NW, Thurston G, Hackett SF, Renard R, Wang Q, McClain J, et al. Angiopoietin-2 is required for postnatal angiogenesis and lymphatic patterning, and only the latter role is rescued by angiopoietin-1. Dev Cell. 2002;3(3):411–23.
11. Dumont DJ, Gradwohl G, Fong GH, Puri MC, Gertsenstein M, Auerbach A, et al. Dominant-negative and targeted null mutations in the endothelial receptor tyrosine kinase, tek, reveal a critical role in vasculogenesis of the embryo. Genes Dev. 1994;8(16):1897–909.
12. Sato TN, Tozawa Y, Deutsch U, Wolburg-Buchholz K, Fujiwara Y, Gendron-Maguire M, et al. Distinct roles of the receptor tyrosine kinases Tie-1 and Tie-2 in blood vessel formation. Nature. 1995;376(6535):70–4.
13. Wong AL, Haroon ZA, Werner S, Dewhirst MW, Greenberg CS, Peters KG. Tie2 expression and phosphorylation in angiogenic and quiescent adult tissues. Circ Res. 1997;81(4):567–74.
14. Holash J, Maisonpierre PC, Compton D, Boland P, Alexander CR, Zagzag D, et al. Vessel cooption, regression, and growth in tumors mediated by angiopoietins and VEGF. Science. 1999;284(5422):1994–8.
15. Holash J, Wiegand SJ, Yancopoulos GD. New model of tumor angiogenesis: dynamic balance between vessel regression and growth mediated by angiopoietins and VEGF. Oncogene. 1999;18(38):5356–62.
16. Stratmann A, Risau W, Plate KH. Cell type-specific expression of angiopoietin-1 and angiopoietin-2 suggests a role in glioblastoma angiogenesis. Am J Pathol. 1998;153(5):1459–66.
17. Zagzag D. In situ expression of angiopoietins in astrocytomas identifies angiopoietin-2 as an early marker of tumor angiogenesis. Exp Neurol. 1999;159(2):391–400.
18. Fiedler U, Reiss Y, Scharpfenecker M, Grunow V, Koidl S, Thurston G, et al. Angiopoietin-2 sensitizes endothelial cells to TNF-alpha and has a crucial role in the induction of inflammation. Nat Med. 2006;12(2):235–9.
19. Imhof BA, Aurrand-Lions M. Angiogenesis and inflammation face off. Nat Med. 2006;12(2):171–2.
20. Scholz A, Plate KH, Reiss Y. Angiopoietin-2: a multifaceted cytokine that functions in both angiogenesis and inflammation. Ann N Y Acad Sci. Epub 13 Mar 2015. doi: 10.1111/nyas.12726.
21. Huang H, Bhat A, Woodnutt G, Lappe R. Targeting the ANGPT-TIE2 pathway in malignancy. Nat Rev Cancer. 2010;10(8):575–85.

22. Plate KH, Scholz A, Dumont DJ. Tumor angiogenesis and anti-angiogenic therapy in malignant gliomas revisited. Acta Neuropathol. 2012;124(6):763–75.
23. Gerald D, Chintharlapalli S, Augustin HG, Benjamin LE. Angiopoietin-2: an attractive target for improved antiangiogenic tumor therapy. Cancer Res. 2013;73(6):1649–57.
24. Biel NM, Siemann DW. Targeting the Angiopoietin-2/Tie-2 axis in conjunction with VEGF signal interference. Cancer Lett. Epub 12 Oct 2014. doi: 10.1016/j.canlet.2014.09.035.
25. Risau W, Flamme I. Vasculogenesis. Annu Rev Cell Dev Biol. 1995;11:73–91.
26. Risau W. Mechanisms of angiogenesis. Nature. 1997;386(6626):671–4.
27. Adams RH, Alitalo K. Molecular regulation of angiogenesis and lymphangiogenesis. Nat Rev Mol Cell Biol. 2007;8(6):464–78.
28. Jeltsch M, Leppänen V-M, Saharinen P, Alitalo K. Receptor tyrosine kinase-mediated angiogenesis. Cold Spring Harb Perspect Biol. 2013;5(9):pii: a009183.
29. Takakura N, Huang X-L, Naruse T, Hamaguchi I, Dumont DJ, Yancopoulos GD, et al. Critical role of the TIE2 endothelial cell receptor in the development of definitive hematopoiesis. Immunity. 1998;9(5):677–86.
30. Saharinen P, Kerkelä K, Ekman N, Marron M, Brindle N, Lee GM, et al. Multiple angiopoietin recombinant proteins activate the Tie1 receptor tyrosine kinase and promote its interaction with Tie2. J Cell Biol. 2005;169(2):239–43.
31. Saharinen P, Eklund L, Miettinen J, Wirkkala R, Anisimov A, Winderlich M, et al. Angiopoietins assemble distinct Tie2 signalling complexes in endothelial cell-cell and cell-matrix contacts. Nat Cell Biol. 2008;10(5):527–37.
32. Puri MC, Rossant J, Alitalo K, Bernstein A, Partanen J. The receptor tyrosine kinase TIE is required for integrity and survival of vascular endothelial cells. EMBO J. 1995;14(23):5884–91.
33. Qu X, Tompkins K, Batts LE, Puri M, Baldwin HS, Baldwin S. Abnormal embryonic lymphatic vessel development in Tie1 hypomorphic mice. Development. 2010;137(8):1285–95.
34. D'Amico G, Korhonen EA, Waltari M, Saharinen P, Laakkonen P, Alitalo K. Loss of endothelial Tie1 receptor impairs lymphatic vessel development-brief report. Arterioscler Thromb Vasc Biol. 2010;30(2):207–9.
35. Rodewald HR, Sato TN. Tie1, a receptor tyrosine kinase essential for vascular endothelial cell integrity, is not critical for the development of hematopoietic cells. Oncogene. 1996;12(2):397–404.
36. Reiss Y, Droste J, Heil M, Tribulova S, Schmidt MHH, Schaper W, et al. Angiopoietin-2 impairs revascularization after limb ischemia. Circ Res. 2007;101(1):88–96.
37. Daly C, Pasnikowski E, Burova E, Wong V, Aldrich TH, Griffiths J, et al. Angiopoietin-2 functions as an autocrine protective factor in stressed endothelial cells. Proc Natl Acad Sci U S A. 2006 17;103(42):15491–6.
38. Daly C, Eichten A, Castanaro C, Pasnikowski E, Adler A, Lalani AS, et al. Angiopoietin-2 functions as a Tie2 agonist in tumor models, where it limits the effects of VEGF inhibition. Cancer Res. 2013;73(1):108–18.
39. Arai F, Hirao A, Ohmura M, Sato H, Matsuoka S, Takubo K, et al. Tie2/Angiopoietin-1 signaling regulates hematopoietic stem cell quiescence in the bone marrow niche. Cell. 2004;118(2):149–61.
40. Fiedler U, Scharpfenecker M, Koidl S, Hegen A, Grunow V, Schmidt JM, et al. The Tie-2 ligand angiopoietin-2 is stored in and rapidly released upon stimulation from endothelial cell Weibel-Palade bodies. Blood. 2004;103(11):4150–6.
41. Davis S, Papadopoulos N, Aldrich TH, Maisonpierre PC, Huang T, Kovac L, et al. Angiopoietins have distinct modular domains essential for receptor binding, dimerization and superclustering. Nat Struct Biol. 2003;10(1):38–44.
42. Barton WA, Tzvetkova-Robev D, Miranda EP, Kolev MV, Rajashankar KR, Himanen JP, et al. Crystal structures of the Tie2 receptor ectodomain and the angiopoietin-2-Tie2 complex. Nat Struct Mol Biol. 2006;13(6):524–32.
43. Koh GY. Orchestral actions of angiopoietin-1 in vascular regeneration. Trends Mol Med. 2013;19(1):31–9.

44. Kim K-T, Choi H-H, Steinmetz MO, Maco B, Kammerer RA, Ahn SY, et al. Oligomerization and multimerization are critical for angiopoietin-1 to bind and phosphorylate Tie2. J Biol Chem. 2005;280(20):20126–31.
45. Fiedler U, Krissl T, Koidl S, Weiss C, Koblizek T, Deutsch U, et al. Angiopoietin-1 and angiopoietin-2 share the same binding domains in the Tie-2 receptor involving the first Ig-like loop and the epidermal growth factor-like repeats. J Biol Chem. 2003;278(3):1721–7.
46. Barton WA, Tzvetkova D, Nikolov DB. Structure of the angiopoietin-2 receptor binding domain and identification of surfaces involved in Tie2 recognition. Structure. 2005;13(5):825–32.
47. Yu X, Seegar TCM, Dalton AC, Tzvetkova-Robev D, Goldgur Y, Rajashankar KR, et al. Structural basis for angiopoietin-1-mediated signaling initiation. Proc Natl Acad Sci U S A. 2013;110(18):7205–10.
48. Jones N, Iljin K, Dumont DJ, Alitalo K. Tie receptors: new modulators of angiogenic and lymphangiogenic responses. Nat Rev Mol Cell Biol. 2001;2(4):257–67.
49. De Palma M, Venneri MA, Roca C, Naldini L. Targeting exogenous genes to tumor angiogenesis by transplantation of genetically modified hematopoietic stem cells. Nat Med. 2003;9(6):789–95.
50. De Palma M, Venneri MA, Galli R, Sergi Sergi L, Politi LS, Sampaolesi M, et al. Tie2 identifies a hematopoietic lineage of proangiogenic monocytes required for tumor vessel formation and a mesenchymal population of pericyte progenitors. Cancer Cell. 2005;8(3):211–26.
51. Murdoch C, Tazzyman S, Webster S, Lewis CE. Expression of Tie-2 by human monocytes and their responses to angiopoietin-2. J Immunol. 2007;178(11):7405–11.
52. Kaipainen A, Vlaykova T, Hatva E, Böhling T, Jekunen A, Pyrhönen S, et al. Enhanced expression of the tie receptor tyrosine kinase mesenger RNA in the vascular endothelium of metastatic melanomas. Cancer Res. 1994;54(24):6571–7.
53. Porat RM, Grunewald M, Globerman A, Itin A, Barshtein G, Alhonen L, et al. Specific induction of tie1 promoter by disturbed flow in atherosclerosis-prone vascular niches and flow-obstructing pathologies. Circ Res. 2004;94(3):394–401.
54. Seegar TCM, Eller B, Tzvetkova-Robev D, Kolev MV, Henderson SC, Nikolov DB, et al. Tie1-Tie2 interactions mediate functional differences between angiopoietin ligands. Mol Cell. 2010;37(5):643–55.
55. Fukuhara S, Sako K, Minami T, Noda K, Kim HZ, Kodama T, et al. Differential function of Tie2 at cell-cell contacts and cell-substratum contacts regulated by angiopoietin-1. Nat Cell Biol. 2008;10(5):513–26.
56. Felcht M, Luck R, Schering A, Seidel P, Srivastava K, Hu J, et al. Angiopoietin-2 differentially regulates angiogenesis through TIE2 and integrin signaling. J Clin Invest. 2012;122(6):1991–2005.
57. Hakanpaa L, Sipila T, Leppänen V-M, Gautam P, Nurmi H, Jacquemet G, et al. Endothelial destabilization by angiopoietin-2 via integrin β1 activation. Nat Commun. 2015;6:5962.
58. Jeansson M, Gawlik A, Anderson G, Li C, Kerjaschki D, Henkelman M, et al. Angiopoietin-1 is essential in mouse vasculature during development and in response to injury. J Clin Invest. 2011;121(6):2278–89.
59. Reiss Y, Knedla A, Tal AO, Schmidt MHH, Jugold M, Kiessling F, et al. Switching of vascular phenotypes within a murine breast cancer model induced by angiopoietin-2. J Pathol. 2009;217(4):571–80.
60. Thurston G, Suri C, Smith K, McClain J, Sato TN, Yancopoulos GD, et al. Leakage-resistant blood vessels in mice transgenically overexpressing angiopoietin-1. Science. 1999;286(5449):2511–4.
61. Thurston G, Rudge JS, Ioffe E, Zhou H, Ross L, Croll SD, et al. Angiopoietin-1 protects the adult vasculature against plasma leakage. Nat Med. 2000;6(4):460–3.
62. Suri C, McClain J, Thurston G, McDonald DM, Zhou H, Oldmixon EH, et al. Increased vascularization in mice overexpressing angiopoietin-1. Science. 1998;282(5388):468–71.
63. Gavard J, Patel V, Gutkind JS. Angiopoietin-1 prevents VEGF-induced endothelial permeability by sequestering Src through mDia. Dev Cell. 2008;14(1):25–36.

64. Ngok SP, Geyer R, Liu M, Kourtidis A, Agrawal S, Wu C, et al. VEGF and Angiopoietin-1 exert opposing effects on cell junctions by regulating the Rho GEF Syx. J Cell Biol. 2012;199(7):1103–15.
65. Benest AV, Kruse K, Savant S, Thomas M, Laib AM, Loos EK, et al. Angiopoietin-2 is critical for cytokine-induced vascular leakage. PLoS One. 2013;8(8):e70459.
66. Ziegler T, Horstkotte J, Schwab C, Pfetsch V, Weinmann K, Dietzel S, et al. Angiopoietin 2 mediates microvascular and hemodynamic alterations in sepsis. J Clin Invest. Epub 1 Jul 2013. doi: 10.1172/JCI66549
67. Avraham HK, Jiang S, Fu Y, Nakshatri H, Ovadia H, Avraham S. Angiopoietin-2 mediates blood-brain barrier impairment and colonization of triple negative breast cancer cells in brain. J Pathol. 2014;232(3):369–81.
68. Reiss Y. Angiopoietins. Recent Results Cancer Res. 2010;180:3–13.
69. Reiss Y, Machein MR, Plate KH. The role of angiopoietins during angiogenesis in gliomas. Brain Pathol. 2005;15(4):31–17.
70. Winkler F, Kozin SV, Tong RT, Chae S-S, Booth MF, Garkavtsev I, et al. Kinetics of vascular normalization by VEGFR2 blockade governs brain tumor response to radiation: role of oxygenation, angiopoietin-1, and matrix metalloproteinases. Cancer Cell. 2004;6(6):553–63.
71. Jain RK. Normalization of tumor vasculature: an emerging concept in antiangiogenic therapy. Science. 2005;307(5706):58–62.
72. Carmeliet P, Jain RK. Molecular mechanisms and clinical applications of angiogenesis. Nature. 2011;473(7347):298–307.
73. Machein MR, Knedla A, Knoth R, Wagner S, Neuschl E, Plate KH. Angiopoietin-1 promotes tumor angiogenesis in a rat glioma model. Am J Pathol. 2004;165(5):1557–70.
74. Nasarre P, Thomas M, Kruse K, Helfrich I, Wolter V, Deppermann C, et al. Host-derived angiopoietin-2 affects early stages of tumor development and vessel maturation but is dispensable for later stages of tumor growth. Cancer Res. 200969(4):1324–33.
75. Coffelt SB, Tal AO, Scholz A, De Palma M, Patel S, Urbich C, et al. Angiopoietin-2 regulates gene expression in TIE2-expressing monocytes and augments their inherent proangiogenic functions. Cancer Res. 2010;70(13):5270–80.
76. Fagiani E, Lorentz P, Kopfstein L, Christofori G. Angiopoietin-1 and -2 exert antagonistic functions in tumor angiogenesis, yet both induce lymphangiogenesis. Cancer Res. 2011;71(17):5717–27.
77. Hughes DP, Marron MB, Brindle NPJ. The antiinflammatory endothelial tyrosine kinase Tie2 interacts with a novel nuclear factor-kappaB inhibitor ABIN-2. Circ Res. 2003;92(6):630–6.
78. Scholz A, Lang V, Henschler R, Czabanka M, Vajkoczy P, Chavakis E, et al. Angiopoietin-2 promotes myeloid cell infiltration in a βâ-integrin-dependent manner. Blood. 2011;118(18):5050–9.
79. Kim H, Koh GY. Ang2, the instigator of inflammation. Blood. 2011;118(18):4767–8.
80. Tabruyn SP, Colton K, Morisada T, Fuxe J, Wiegand SJ, Thurston G, et al. Angiopoietin-2-driven vascular remodeling in airway inflammation. Am J Pathol. 2010;177(6):3233–44.
81. Butcher EC. Leukocyte-endothelial cell recognition: three (or more) steps to specificity and diversity. Cell. 1991;67(6):1033–6.
82. Grivennikov SI, Greten FR, Karin M. Immunity, inflammation, and cancer. Cell. 2010;140(6):883–99.
83. Balkwill F, Mantovani A. Inflammation and cancer: back to Virchow? Lancet. 2001;357(9255):539–45.
84. Noy R, Pollard JW. Tumor-associated macrophages: from mechanisms to therapy. Immunity. 2014;41(1):49–61.
85. Neutralizing tumor-promoting chronic inflammation: a magic bullet? 2013;339(6117):286–91.
86. Murdoch C, Muthana M, Coffelt SB, Lewis CE. The role of myeloid cells in the promotion of tumour angiogenesis. Nat Rev Cancer. 2008;8(8):618–31.
87. Karin M. Nuclear factor-kappaB in cancer development and progression. Nature. 2006;441(7092):431–6.

88. Avraham-Davidi I, Yona S, Grunewald M, Landsman L, Cochain C, Silvestre JS, et al. On-site education of VEGF-recruited monocytes improves their performance as angiogenic and arteriogenic accessory cells. J Exp Med. 2013;204(5):1057.
89. Grunewald M, Avraham I, Dor Y, Bachar-Lustig E, Itin A, Jung S, et al. VEGF-induced adult neovascularization: recruitment, retention, and role of accessory cells. Cell. 2006;124(1):175–89.
90. Biswas SK, Allavena P, Mantovani A. Tumor-associated macrophages: functional diversity, clinical significance, and open questions. Semin Immunopathol. 2013;35(5):585–600.
91. Venneri MA, De Palma M, Ponzoni M, Pucci F, Scielzo C, Zonari E, et al. Identification of proangiogenic TIE2-expressing monocytes (TEMs) in human peripheral blood and cancer. Blood. 2007;109(12):5276–85.
92. Coffelt SB, Chen Y-Y, Muthana M, Welford AF, Tal AO, Scholz A, et al. Angiopoietin 2 stimulates TIE2-expressing monocytes to suppress T cell activation and to promote regulatory T cell expansion. J Immunol. 2011;186(7):4183–90.
93. Hurwitz H, Fehrenbacher L, Novotny W, Cartwright T, Hainsworth J, Heim W, et al. Bevacizumab plus irinotecan, fluorouracil, and leucovorin for metastatic colorectal cancer. N Engl J Med. 2004;350(23):2335–42.
94. Welti J, Loges S, Dimmeler S, Carmeliet P. Recent molecular discoveries in angiogenesis and antiangiogenic therapies in cancer. J Clin Invest. 2013;123(8):3190–200.
95. Batchelor TT, Reardon DA, de Groot JF, Wick W, Weller M. Antiangiogenic therapy for glioblastoma: current status and future prospects. Clin Cancer Res. 2014;20(22):5612–9.
96. Bergers G, Hanahan D. Modes of resistance to anti-angiogenic therapy. Nat Rev Cancer. 2008;8(8):592–603.
97. Shojaei F, Wu X, Malik AK, Zhong C, Baldwin ME, Schanz S, et al. Tumor refractoriness to anti-VEGF treatment is mediated by CD11b+Gr1+ myeloid cells. Nat Biotechnol. 2007;25(8):911–20.
98. Chung AS, Wu X, Zhuang G, Ngu H, Kasman I, Zhang J, et al. An interleukin-17-mediated paracrine network promotes tumor resistance to anti-angiogenic therapy. Nat Med. 2013;19(9):1114–23.
99. Oliner J, Min H, Leal J, Yu D, Rao S, You E, et al. Suppression of angiogenesis and tumor growth by selective inhibition of angiopoietin-2. Cancer Cell. 2004;6(5):507–16.
100. Coxon A, Bready J, Min H, Kaufman S, Leal J, Yu D, et al. Context-dependent role of angiopoietin-1 inhibition in the suppression of angiogenesis and tumor growth: implications for AMG 386, an angiopoietin-1/2-neutralizing peptibody. Mol Cancer Ther. 2010;9(10):2641–51.
101. Monk BJ, Poveda A, Vergote I, Raspagliesi F, Fujiwara K, Bae D-S, et al. Anti-angiopoietin therapy with trebananib for recurrent ovarian cancer (TRINOVA-1): a randomised, multicentre, double-blind, placebo-controlled phase 3 trial. Lancet Oncol. 2014;15(8):799–808.
102. Liontos M, Lykka M, Dimopoulos M-A, Bamias A. Profile of trebananib (AMG386) and its potential in the treatment of ovarian cancer. Onco Targets Ther. 2014;7:1837–45.
103. Falcón BL, Hashizume H, Koumoutsakos P, Chou J, Bready JV, Coxon A, et al. contrasting actions of selective inhibitors of angiopoietin-1 and angiopoietin-2 on the normalization of tumor blood vessels. Am J Pathol. 2009;175(5):2159–70.
104. Huang H, Lai J-Y, Do J, Liu D, Li L, Del Rosario J, et al. Specifically targeting angiopoietin-2 inhibits angiogenesis, Tie2-expressing monocyte infiltration, and tumor growth. Clin Cancer Res. 2011;17(5):1001–11.
105. Mazzieri R, Pucci F, Moi D, Zonari E, Ranghetti A, Berti A, et al. Targeting the ANG2/TIE2 axis inhibits tumor growth and metastasis by impairing angiogenesis and disabling rebounds of proangiogenic myeloid cells. Cancer Cell. 2011;19(4):512–26.
106. Holopainen T, Saharinen P, D'Amico G, Lampinen A, Eklund L, Sormunen R, et al. Effects of angiopoietin-2-blocking antibody on endothelial cell-cell junctions and lung metastasis. J Natl Cancer Inst. 2012;104(6):461–75.

107. Leow CC, Coffman K, Inigo I, Breen S, Czapiga M, Soukharev S, et al. MEDI3617, a human anti-angiopoietin 2 monoclonal antibody, inhibits angiogenesis and tumor growth in human tumor xenograft models. Int J Oncol. 2012;40(5):1321–30.
108. Thomas M, Kienast Y, Scheuer W, Bähner M, Kaluza K, Gassner C, et al. A novel angiopoietin-2 selective fully human antibody with potent anti-tumoral and anti-angiogenic efficacy and superior side effect profile compared to Pan-Angiopoietin-1/-2 inhibitors. PLoS One. 2013;8(2):e54923.
109. Hashizume H, Falcon BL, Kuroda T, Baluk P, Coxon A, Yu D, et al. Complementary actions of inhibitors of angiopoietin-2 and VEGF on tumor angiogenesis and growth. Cancer Res. 2010;70(6):2213–23.
110. Koh YJ, Kim HZ, Hwang S-I, Lee JE, Oh N, Jung K, et al. Double antiangiogenic protein, DAAP, targeting VEGF-A and angiopoietins in tumor angiogenesis, metastasis, and vascular leakage. Cancer Cell. 2010;18(2):171–84.
111. Brown JL, Cao ZA, Pinzon-Ortiz M, Kendrew J, Reimer C, Wen S, et al. A human monoclonal anti-ANG2 antibody leads to broad antitumor activity in combination with VEGF inhibitors and chemotherapy agents in preclinical models. Mol Cancer Ther. 2010;9(1):145–56.
112. Kienast Y, Klein C, Scheuer W, Raemsch R, Lorenzon E, Bernicke D, et al. Ang-2-VEGF-A CrossMab, a novel bispecific human IgG1 antibody blocking VEGF-A and Ang-2 functions simultaneously, mediates potent antitumor, antiangiogenic, and antimetastatic efficacy. Clin Cancer Res. 2013;19(24):6730–40.
113. Rigamonti N, Kadioglu E, Keklikoglou I, Wyser Rmili C, Leow CC, De Palma M. Role of angiopoietin-2 in adaptive tumor resistance to VEGF signaling blockade. Cell Rep. 2014;8(3):696–706.
114. Lu KV, Chang JP, Parachoniak CA, Pandika MM, Aghi MK, Meyronet D, et al. VEGF inhibits tumor cell invasion and mesenchymal transition through a MET/VEGFR2 complex. Cancer Cell. 2012;22(1):21–35.
115. Srivastava K, Hu J, Korn C, Savant S, Teichert M, Kapel SS, et al. Postsurgical adjuvant tumor therapy by combining anti-angiopoietin-2 and metronomic chemotherapy limits metastatic growth. Cancer Cell. 2014;26(6):880–95.
116. Pyonteck SM, Akkari L, Schuhmacher AJ, Bowman RL, Sevenich L, Quail DF, et al. CSF-1R inhibition alters macrophage polarization and blocks glioma progression. Nat Med. 2013;19(10):1264–72.
117. Ries CH, Cannarile MA, Hoves S, Benz J, Wartha K, Runza V, et al. Targeting tumor-associated macrophages with anti-CSF-1R antibody reveals a strategy for cancer therapy. Cancer Cell. 2014;25(6):846–59.
118. Strachan DC, Ruffell B, Oei Y, Bissell MJ, Coussens LM, Pryer N, et al. CSF1R inhibition delays cervical and mammary tumor growth in murine models by attenuating the turnover of tumor-associated macrophages and enhancing infiltration by CD8(+) T cells. Oncoimmunology. 2013;2(12):e26968.
119. Zhu Y, Knolhoff BL, Meyer MA, Nywening TM, West BL, Luo J, et al. CSF1/CSF1R blockade reprograms tumor-infiltrating macrophages and improves response to T-cell checkpoint immunotherapy in pancreatic cancer models. Cancer Res. 2014;74(18):5057–69.
120. Huang Y, Yuan J, Righi E, Kamoun WS, Ancukiewicz M, Nezivar J, et al. Vascular normalizing doses of antiangiogenic treatment reprogram the immunosuppressive tumor microenvironment and enhance immunotherapy. Proc Natl Acad Sci U S A. 2012;109(43):17561–6.
121. Huang Y, Goel S, Duda DG, Fukumura D, Jain RK. Vascular normalization as an emerging strategy to enhance cancer immunotherapy. Cancer Res. 2013;73(10):2943–8.
122. Carmeliet P, et al. Abnormal blood vessel development and lethality in embryos lacking a single VEGF allele. Nature. 1996;380:435–9.
123. Ferrara N, et al. Heterozygous embryonic lethality induced by targeted inactivation of the VEGF gene. Nature. 1996;380:439–42.
124. Kim KJ, et al. Inhibition of vascular endothelial growth factor-induced angiogenesis suppresses tumour growth in vivo. Nature. 1993;362:841–4.

Chapter 14
Oxygen Signaling in Physiological and Pathological Angiogenesis

Anne-Theres Henze and Massimiliano Mazzone

14.1 Overview of Oxygen-Mediated Pathways

Due to its high reduction potential, oxygen enables organisms to generate large amounts of energy when serving as a terminal electron acceptor during aerobic respiration. Thus, the evolution from unicellular to more complex multicellular organisms with an increased energy demand has been tightly linked to the rise in atmospheric oxygen and implementation of oxygen as a final step in the electron transport chain in order to increase the efficiency in energy production [1]. However, as outlined in more detail below, the indispensability of oxygen is rather a double-edged sword. The role of oxygen in metabolic reactions of aerobic organisms has become detrimental but has also lead to a complete and deadly dependency.

Numerous pathophysiological conditions such as stroke, myocardial infarction, neurological disorder, ischemia, reperfusion of transplanted organs, chronic lung disease, and cancer have been linked to perturbed oxygen supply, doubtlessly attesting that oxygen homeostasis is indispensible for the mammalian system [2–7]. Inevitably connected to aerobic respiration is the production of reactive oxygen species (ROS) [8]. Cells are equipped with a defense system to scavenge these toxic byproducts; however, oxidative stress occurs whenever an imbalance between ROS production and the detoxification capacity of the cell has been reached. Under these circumstances, ROS cause sweeping damage to RNA, DNA and proteins, hence constricting cellular functions [9].

A.-T. Henze (✉)
Department of Oncology, University of Leuven, Laboratory of Molecular Oncology and Angiogenesis, Vesalius Research Center, VIB, Herestraat 49, Box 912, 3000 Leuven, Belgium
e-mail: AnneTheres.Henze@vib-kuleuven.be

M. Mazzone
VIB Vesalius Research Center, KU Leuven, O&N 4, 9e verd, Campus Gasthuisberg, Herestraat 49, bus 912, 3000 Leuven, Belgium
e-mail: Massimiliano.Mazzone@vib-kuleuven.be

M. HH. Schmidt, S. Liebner (eds.), *Endothelial Signaling in Development and Disease,* DOI 10.1007/978-1-4939-2907-8_14

Given the hazardous effects associated with unbalanced oxygen availability, it is easy to conceive that oxygen-dependent mechanisms have evolved to sense oxygen availability, and thus to assure a tight regulation of oxygen supply. On the whole-body level, a vascular infrastructure enables fast distribution of oxygen bound to heme proteins over distances beyond the diffusion limit. Sensory structures such as the carotid body exert control functions to ensure a fast response to disturbed oxygen tensions, commonly resulting in pulmonary vasoconstriction and dilation of systemic vessels [10]. Within these sensory structures, chemoreceptor cells possess the ability to sense oxygen via oxygen-sensitive ion channels, allowing an acute response to modulations in oxygen tension [11]. A more chronic adaption to changes in the oxygen supply is accomplished by oxygen-regulated transcription factors [12]. One of the key transcription factors responsible for adaptive responses to low oxygen pressure is the hypoxia-inducible transcription factor (HIF) [13] (Fig. 14.1a). Inseparably linked to HIFs and their pivotal role in oxygen homeostasis are the prolyl hydroxylase domain proteins (PHDs), which regulate the turnover of HIFs and, strictly speaking, are actual oxygen sensors of the cell since their enzymatic function depends on the availability of oxygen [14, 15]. Factor-inhibiting HIF (FIH), likewise belonging to the iron- and 2-oxoglutarate-dependent enzymes, regulates the transcriptional activity of HIF, and thus represents another component of the oxygen-sensing machinery [16, 17] (Fig. 14.1b). There is a growing body of evidence for other oxygen-sensitive processes. Numerous iron- and 2-oxoglutarate-dependent enzymes have been implicated in serving important functions in oxygen sensing [18]. Of note, JmjC domain-containing proteins, also belonging to the family of iron- and 2-oxoglutarate-dependent enzymes and responsible for epigenetic modifications, have been associated with oxygen sensing and angiogenesis [19]. Mitochondria are also intertwined in oxygen sensing through inhibition of PHD function [20, 21]. However, how the mechanisms impact on PHD activity is controversially discussed. It is believed that under moderate hypoxia, ROS generated from complex III of the electron transport chain inhibits PHD activity and consequentially stabilizes HIF [21]. However, in a different scenario one might imagine that mitochondria diminish PHD activity via oxygen consumption, therefore indirectly decreasing PHD enzymatic function. Nonetheless, cells with a mutant cytochrome b, which produce ROS but do not consume oxygen, provide evidence for mitochondrial-mediated ROS production in HIF stabilization [21].

Thus, when considering the oxygen-related responses on the cellular level, it is interesting to note in this context that the vasculature itself does not only serve as a well-defined oxygen distribution network but is also highly responsive to differences in oxygen tension itself. It has been reported that endothelial cells in the growing vasculature are functionally distinct and exhibit a specific molecular signature [22–25]. These findings are particularly interesting in light of the notion that a growing vessel is exposed to an oxygen gradient with the lowest oxygen level at the forefront of the growing sprout, generally known as the tip cells. In comparison to that, the neighboring stalk cell experiences higher oxygen levels, exemplifying how oxygen signaling contributes to determine the endothelial cell fate and thereby impacts on angiogenic processes. Mechanistically, initiation of the oxygen-sensing

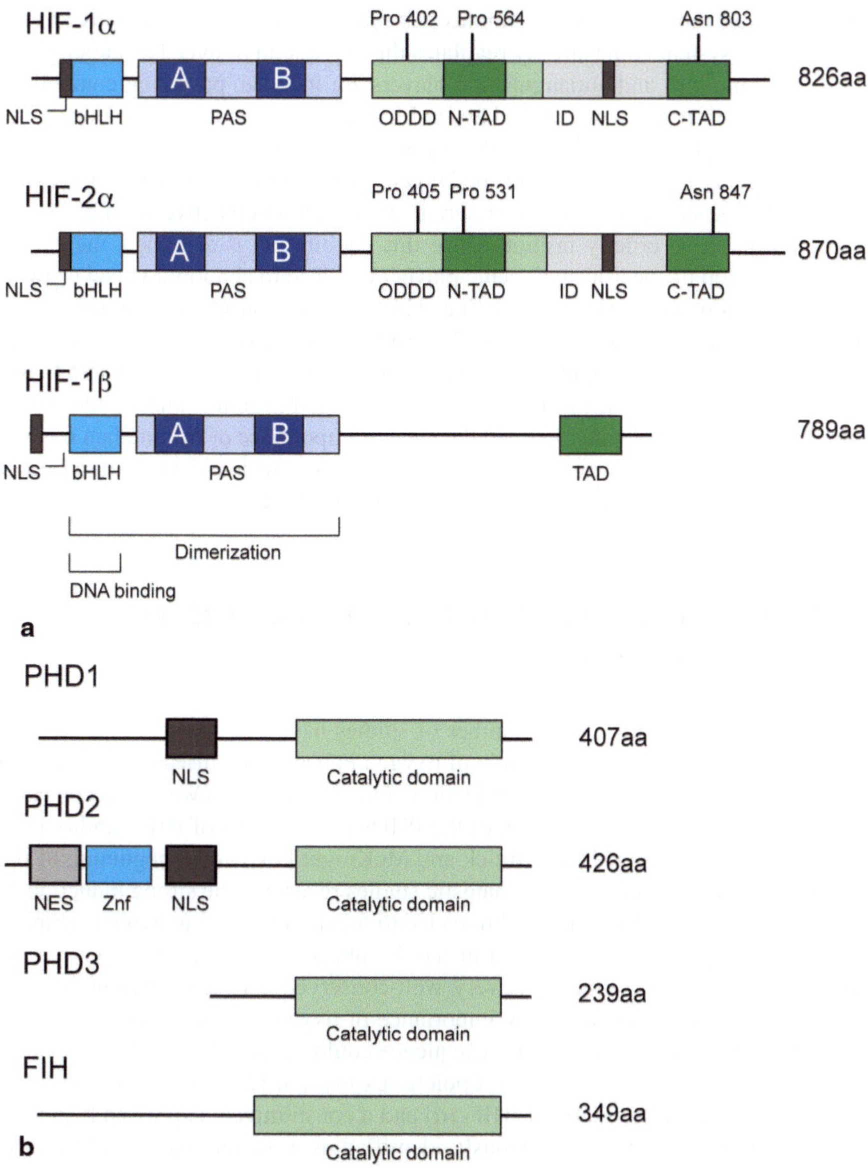

Fig. 14.1 Domain structure for hypoxia-inducible factors (*HIFs*), prolyl hydroxylase domain proteins (*PHDs*), and factor-inhibiting *HIF* (*FIH*). **a** Schematic diagram of the HIF-1α, HIF-2α, and HIF-1β proteins and their functional domains. HIFs contain two Per-ARNT-Sim (*PAS*) domains (*PAS-A* and *PAS-B*), which are required for heterodimerization and DNA binding. HIF-1α and HIF-2α have an N- and C-terminal transactivation domain (*N-TAD* and *C-TAD*), which are linked by an inhibitory domain (*ID*). Two prolyl hydroxylation sites (*prolines 402* and *564* in HIF-1α and *prolines 405* and *531* in HIF-2α) are located within the oxygen-dependent degradation domain (*ODD*) and an asparaginyl hydroxylation site (asparagine 803 in *HIF*-1α and *847* in *HIF*-2α within the C-terminal transactivation domain. **b** Schematic diagram of *PHD* and *FIH* proteins and their functional domains. All hydroxylases contain a conserved catalytic dioxygenase domain at their C-terminal ends. *PHD1* and *PHD2* have a nuclear localization signal (*NLS*) and *PHD2* also contains a nuclear export signal (*NES*). Additionally, *PHD2* has a zink-finger domain (ZnF)

machinery encompasses a very complex and broad response; it is generally known that the HIF system alone already regulates the expression of over 100 target genes [26]. Multiple pro- and antiangiogenic players are induced, providing compelling evidence that hypoxia is one of the key environmental cues that trigger blood vessel growth under physiological and pathophysiological conditions [27]. Physiological angiogenesis is an accurately controlled process where pro- and antiangiogenic factors need to be strictly balanced to assure that the interaction between different cell types occurs in an orderly fashion. Once this equilibrium is disturbed, the excessive production of specific angiogenic players will lead to abnormal vessel growth, with long-lasting consequences, as evinced by the chaotic structure of tumor vessels [28]. Gaining further mechanistic insights from physiological angiogenic processes such as embryonic development, reproduction, or wound healing might help to uncover to what extent these mechanisms are coopted and exploited under pathophysiological conditions. Having outlined the general importance of mammalian oxygen sensing, in the following section we will proceed to provide a detailed description regarding the components of the oxygen-sensing machinery.

14.2 The Role of Hypoxia-Inducible Factors (HIFs) in Oxygen Signaling

In the last two decades a large number of studies have been centered on the HIF system and its implication in the role of oxygen in health and disease. This marked interest was captivated by the identification of the HIF in 1995 (Wang and contributors [29]), and the characterization of the PHDs as regulators of HIF stability (Epstein and co-workers [15], and Bruick and McKnight [14]) at the beginning of the twenty-first century. Initially, mechanistic studies regarding the strong induction of erythropoietin under hypoxic conditions led to the discovery of an hypoxic responsive element (HRE; 5′-RCGTC-3′) in the 3′ enhancer of the erythropoietin gene [30]. This consensus sequence is very well-conserved between different species, pointing towards the evolutionary importance of oxygen sensing. Subsequently, a protein binding to this specific DNA sequence could be identified as the HIF, regulating the hypoxic-dependent erythropoietin expression [29]. HIF is composed of the hypoxia-inducible α-subunit (HIF-1α) and a constitutively expressed β-subunit (HIF-1β) [29]. HIF-1β was previously identified as a binding partner of the aryl hydrocarbon receptor, thus being referred to as aryl hydrocarbon nuclear translocator (ARNT) [31]. Succeeding the cloning of HIF-1α, a closely related protein (HIF-2α), sharing 48% of amino acid identity, was discovered in 1997 [32–35]. To date, it is generally accepted that there are three HIF-α proteins. HIF-1α and HIF-2α are positively associated with HIF target gene expression, whereas the inhibitory Per-ARNT-Sim (PAS) domain (IPAS), an HIF-3α isoform, is reported to serve as a negative regulator of HIF signaling [36, 37]. HIF-1α and β contain a basic helix-loop-helix (bHLH) and PAS domain in the N-terminal half of the protein [29]. These domains are important for its dimerization, and the downstream basic region enables specific DNA binding to the HRE sequence [29, 38]. Subsequently, a C- and N-terminal transactivation domain could be located within the C-terminal half of the HIF-α subunit, specifying the transcriptional activity of HIF [39–41] (Fig. 14.1a). Of note, an important transcriptional regulation is accomplished by FIH-dependent asparaginyl hydroxylation within the C-terminal transactivation domain, sterically inhibiting the recruitment of transcriptional coactivators such as cyclic adenosine monophosphate (cAMP) response element-binding protein CBP/p300 [17, 35, 42] (Fig. 14.2).

There are a number of post-translational regulations of the HIF protein, such as hydroxylation, phosphorylation, acetylation, and sumoylation [14, 15, 42–49]. Hydroxylation within the oxygen-dependent degradation domain (ODD) by PHDs is an important determinant of HIF protein stability. Hydroxyproline HIF is recognized by the von Hippel Lindau protein (pVHL), which is part of an E3 ubiquitin ligase complex subsequently leading to ubiquitination and proteasomal degradation of HIF [50–52] (Fig. 14.2).

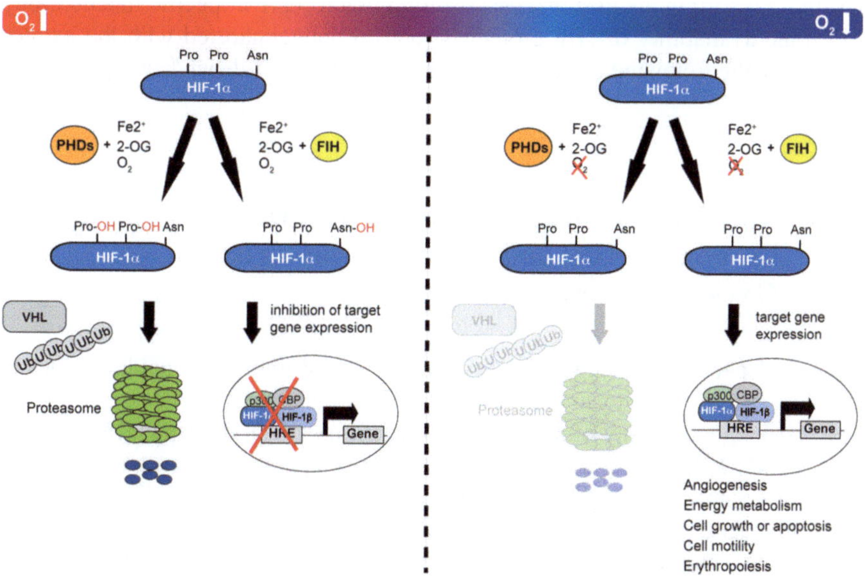

Fig. 14.2 Regulation of hypoxia-inducible factor (*HIF*) degradation and transcriptional activity by prolyl hydroxylase domain proteins (*PHDs*) and factor-inhibiting HIF (*FIH*), respectively. In the presence of divalent iron (*Fe^{2+}*), 2-oxoglutarate, and oxygen, PHDs catalyze the hydroxylation of HIF at two prolyl residues, which leads to their recognition by the E3 ubiquitin ligase von Hippel Lindau protein (*pVHL*), and consequently to ubiquitination and proteasomal degradation of HIF. FIH hydroxylates HIF at an asparagine residue, which hinders the interaction between FIH and its coactivators CBP and p300, and thus leads to transcriptional inhibition. If oxygen levels drop, PHDs and FIH become partly inactive and HIF-α is stabilized, translocates to the nucleus where it heterodimerizes with nuclear HIF-1β heterodimer, and binds with its coactivators to the hypoxic responsive element (*HRE*) of target genes, initiating target gene expression

As aforementioned, HIF-1β messenger RNA (mRNA) and protein are constitutively expressed; however, in contrast, the HIF-1α protein is very unstable and its availability strongly depends on the oxygen tension [53]. In this context, it should be noticed that the transcriptional processes regarding HIF-1α are not strictly regulated by oxygen [54]. Importantly though, the rate of HIF turnover functions as an oxygen-dependent checkpoint. HIF is rapidly degraded under normoxic conditions ($t_{½}$ = 5 min) and accumulates when oxygen levels drop, since degradation via the proteasomal pathway is hampered [55]. This allows HIF to translocate to the nucleus, where it binds as a heterodimer consistent of HIF-1α and HIF-1β to the consensus sequence within the hypoxic response element of target genes [29, 31]. Recruitment of coactivators enables the initiation of the transcriptional complex, resulting in the expression of a number of target genes [56, 57] (Fig. 14.2)

Interestingly HIF-1α and HIF-2α possess overlapping and nonredundant functions [58, 59]. Spatial-temporal differences in the expression exist [60]. Whereas HIF-1α is almost ubiquitously expressed, the abundance of HIF-2α is more restricted to specific cell types, among which are endothelial cells and the carotid body [35, 61, 62]. HIF-2α was also shown to already accumulate under higher oxygen tension, supporting the idea that each isoform might serve a specific function [60]. Since the availability of HIF protein is robustly interconnected with the oxygen tension, in the following section we will focus on the PHD-dependent regulation of HIF turnover.

14.3 The Function of Prolyl Hydroxylase Domain Proteins (PHDs) and Factor-Inhibiting HIF as Oxygen Sensors

Oxygen-dependent HIF regulation takes place at post-transcriptional level. One of the main factors involved in the turnover of HIF proteins are the PHDs (Fig. 14.1b). PHDs belong to a family of nonheme iron- and 2-oxoglutarate-dependent enzymes [14, 15, 51, 52]. The hydroxylation reaction encompasses oxygen and 2-oxoglutarate as cosubstrates, and ferrous iron and ascorbate as cofactors. During the enzymatic reaction, one oxygen atom is utilized to form HIF hydroxyproline, while the other is required for the decarboxylation of 2-oxoglutarate which results in the formation of succinate and CO_2. Fe^{2+} is bound to the active site of the PHD protein, and oxidized during the enzymatic reaction. Ascorbate is needed to reduce the iron during the reaction cycles; however, the consumption does not occur stoichiometrically, therefore leaving room for further investigation [63].

Hydroxylation occurs on Pro402 and Pro564 within the LXXLAP sequence of the ODD in human HIF-1α, and Pro405 and Pro530 in human HIF-2α [64, 65]. Corresponding to their relative proximity to N- and C-termini, these residues are denoted N-ODD or C-ODD, despite their location within the C-terminal halves of the protein. There are three PHD proteins (PHD1–3), and hydroxylation activity on proline 564 is higher for all PHDs; notably, PHD3 is principally inactive at proline 402 [66, 67]. Interestingly, PHDs also differ in their selectivity of HIF-1α versus HIF-2α hydroxylation. While PHD2 exhibits a higher hydroxylation activity on HIF-1α than on HIF-2α, PHD1 and PHD3 preferentially hydroxylate HIF-2α in comparison to HIF-1α [66, 67]. Additionally, FIH was shown to more potently hydroxylate HIF-1, thus contributing to HIF-1- and HIF-2-specific functions [68].

PHD2 is known to be the main HIF regulator under normoxic conditions [69]. Nevertheless there is a growing body of evidence that, dependent on their availability, the relevance of the respective PHD protein might shift. Their functionality might be designated by cell-type specificity, HIF isoform availability, and environmental cues. For instance, PHD2 and, more prominently, PHD3 are induced in an HIF-dependent manner in hypoxia. HIF binding to the HRE could be manifested for both proteins [70, 71]. Interestingly though, HIF-1α and HIF-2α differentially induce PHD2 and PHD3. Whereas HIF-1α induces both PHD2 and PHD3, HIF-2α only enhances PHD3 expression [72, 73].

It is noteworthy that PHDs themselves suppress HIF transcriptional activity, further adding to the complexity of the system [74, 75]. When associating with inhibitor of growth family, member 4, (ING4), a tumor suppressor gene, PHD2 was shown to impede HIF-dependent expression of angiogenic cytokines via transcriptional repression [76]. Thus, in addition to the well-described oxygen-dependent control of HIF and PHDs, other mechanisms are involved in a regulatory function of the HIF response, among them, for instance, receptor-mediated pathways [77]. The knowledge regarding oxygen-independent HIF and PHD regulation will be briefly summarized in the following section.

14.4 Oxygen-Independent HIF and PHD Regulation

There is evidence that HIF target gene expression is enhanced by two well-characterized signaling pathways, the PI3K-AKT-FRAP and Ras-MEK-MAPK cascade [77–79]. First evidence of HIF-induced target gene expression through such a mechanism stems from studies on insulin and insulin-like growth factor (IGF)-1 [80]. Subsequently, an increasing number of growth factors, cytokines, and circulating factors such as platelet-derived growth factor (PDGF), epidermal growth factor (EGF), fibroblast growth factor (FGF)-2, IGF-2, transforming growth factor (TGF)-1β, hepatocyte growth factor (HGF), tumor necrosis factor (TNF)-α, interleukin (IL)-1β, angiotensin-2, and thrombin were found to enhance HIF1α-dependent target gene expression [81–87]. In addition, oncogenes (HER2/neu, Ras, v-Src) and mutations in the tumor suppressor *PTEN* modulate HIF activity [88–91]. Overall, integration of growth factor signaling occurs at the transcriptional as well as post-transcriptional level. For instance, phosphorylation by mitogen-activated protein kinases (MAPK) enhances HIF transcriptional activity, possibly via derepression of the HIF inhibitory domain [45]. This phosphorylation might hamper recognitions and subsequent binding of FIH, consequently enhancing HIF transcriptional activation. On the post-transcriptional level, induction of the PI3K-AKT-FKBP12 rapamycin-associated protein (FRAP)/mammalian target of rapamycin (mTOR) pathway by various cytokines enhances HIF mRNA translation [92]. Given the complexity of growth factor signaling, activation of a kinase cascade, and the subsequent HIF response is profoundly dependent on the cellular context.

Nitric oxide (NO) is known to be a mediator of angiogenesis. Due to its implication in serving a prosurvival and proangiogenic function and its involvement in vasodilation [93], the impact of NO on the PHD/HIF signaling axis will be discussed. NO elicits opposing effects on HIF stability, depending on the oxygen state of the cell. Under normoxia, NO exhibits an inhibitory function on PHD activity, possibly by chelating Fe(II) consequently reducing HIF turnover [94]. Mechanistically, NO inactivates cytochrome c oxidase, impedes ROS production, and causes a redistribution of the available oxygen to the cytosol. However, since oxygen is not a limiting factor under normoxic conditions, this redistribution remains without consequences. However, under hypoxic conditions, these effects gain in importance and lead to activation of PHDs and concomitantly diminished HIF abundance [95]. However, HIF modulation by NO seems to be more complex as it has been shown that the specific experimental conditions critically influence the impact on the HIF signaling axis [96]. Finally, metabolites are also associated with an HIF modulatory function. HIF stability is enhanced by pyruvate, lactate, and oxalacetate, presumably through inhibiting of PHD function [97–99].

The regulation of PHD1 and PHD3 protein stability is again linked to an hypoxic response. The RING finger proteins seven in absentia (Drosophila) homologs 1/2 (SIAH 1/2) were shown to target PHD1 and, most efficiently, PHD3 for polyubiquitination and proteasomal degradation, thereby accomplishing, at least to some extent, PHD1/3 protein destruction. Hypoxia induces SIAH 2 accumulation, most likely in an HIF-independent manner, and thus SIAH 2 controls hypoxic PHD3 abundance in a negative feedback regulatory loop [100]. PHD2 protein destruction is positively regulated by FKBP38, a peptidyl prolyl cis/trans isomerase. Silencing of FKBP38 does not affect PHD2 mRNA but enhances its protein stability [101].

Having obtained a comprehensive understanding of oxygen sensing on a molecular level in the last two sections, we will zoom out and discuss in the next section the functional relevance of these processes during physiological blood vessel remodeling.

14.5 Role of Oxygen Signaling on Physiological and Pathophysiological Angiogenesis

In a very simplified view, the establishment of a vascular system can be subdivided into two processes, vasculogenesis, which refers to the *de novo* formation of blood vessels from endothelial precursor cells during embryonic development, and angiogenesis, which specifies the process of new blood vessel emergence from pre-existing ones [102, 103]. In the embryo, angiogenesis occurs after the primary capillary plexus has been formed. In the adult, angiogenesis is a relatively scarce event, almost exclusively constrained to the ovarian cycle or remodeling processes after injury such as wound healing. A number of very well coordinated events guarantee the establishment of new blood vessels from an existing vascular network. This involves destabilization of vessels by loosening the attachment of pericytes, the subsequent digestion of the basement membrane as well as the extracellular matrix surrounding the blood vessels and its remodeling, and new matrix synthesis. It further includes the ensuing activation of endothelial cell proliferation and migration stimulated by the newly formed matrix until endothelial cells arrest and form tube-like structures. They are successively covered by pericytes to ensure the establishment of a tight network, allowing an efficient blood flow. During this process, endothelial cells are exposed to gradients of different pro- and antiangiogenic factors as well as gradients of oxygen levels [28].

Exemplified by vascular endothelial growth factor (VEGF), one of the most characterized angiogenic factors, it becomes apparent how oxygen sensing is intertwined with the angiogenic process. VEGF abundance is regulated by oxygen tension, as low local oxygen abundance leads to an HIF-driven VEGF expression [104, 105].

This soluble factor exerts a number of different functions during angiogenesis. To name a few, it participates in the destabilization of vessels as it impairs the physical barrier formed by tight junctions between endothelial cells [106, 107]. Furthermore it contributes to endothelial proliferation and migration via binding to VEGF receptor-2 (VEGFR2) expressed at the forefront of the endothelial sprouts, the tip cells. As aforementioned, the cells that trail behind, the stalk cells, differ in their gene expression profile. The tip versus stalk cell phenotype is among others specified by VEGF-mediated activation of VEGFR2 that leads to upregulation of the Notch ligand Dll4 in tip cells and subsequently enhanced Notch signaling and downregulation of VEGFR2 in adjacent stalk cell [108] (Fig. 14.3). However, in addition to regulating the production of angiogenic factors, and thus subsequently triggering the activation of a specific signaling cascade, oxygen levels might also more directly influence the endothelial cell fate.

In this context, it is of interest that Dll4 is regulated in an HIF-dependent manner in response to hypoxia [109], and that Notch intracellular domain (NICD) activity is negatively controlled via FIH [110]. Moreover, FIH exhibits even higher affinity for Notch ICD than for HIF [111]. Hydroxylation of Notch ICD at specific asparagine residues leads to negative regulation of its transcriptional activity. In addition, Notch ICD was shown to positively regulate HIF target gene expression, most likely by competing with HIF for FIH hydroxylation [112]. Thus, the Notch–FIH–HIF signaling interplay might be an interesting determinant for the endothelial cell fate.

It has already been shown that reduced activity of the HIF prolyl hydroxylase PHD2 in endothelial cells evokes a quiescent, 'phalanx' phenotype that involves a closer alignment of cells and thus fosters a more mature and better-perfused vascular network [113]. Growing vessels are exposed to oxygen gradients, with higher oxygen concentrations at the side of the stalk cells and subsequently decreasing

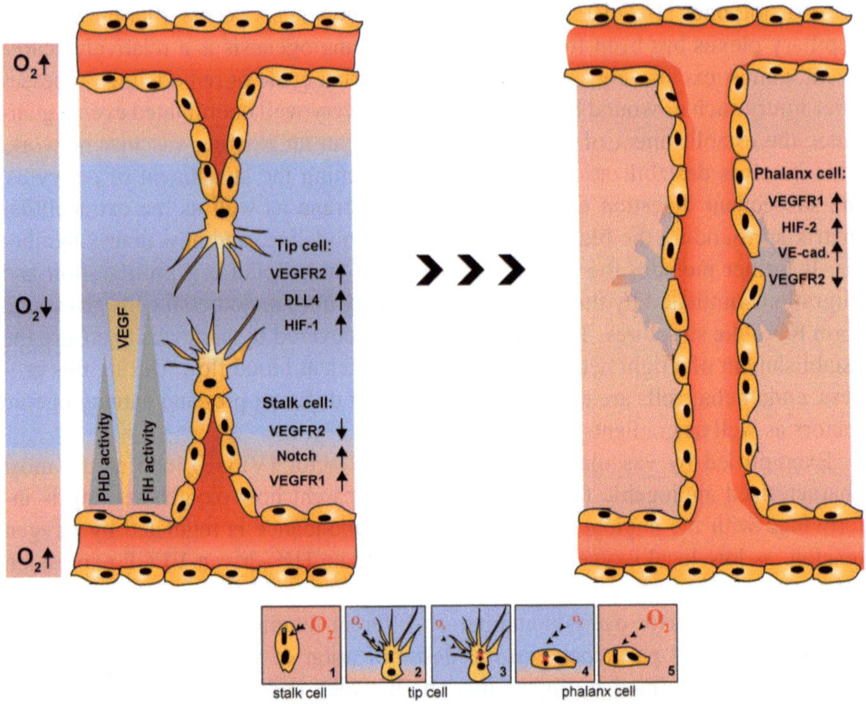

Fig. 14.3 Prolyl hydroxylase domain protein-2 (*PHD2*) is an important determinant of the switch from a motile tip cell to a quiescent phalanx cell phenotype. Outgrowth of vessel sprouts is triggered by proangiogenic growth factors such as vascular endothelial growth factor (VEGF). Cells exposed to the highest concentration of proangiogenic factors are selected to become a tip cell with filopodia invading their surroundings. Proliferating stalk cells that trail behind elongate the sprout. While the growing vessel strikes out alongside a gradient of increasing proangiogenic factors, it encounters decreasing concentrations of oxygen, which concomitantly inhibit PHD and factor-inhibiting hypoxia-inducible factor (*FIH*) activity (*1* and *2*). Two vessel sprouts connect via tip cell fusion in highly hypoxic/anoxic regions. Inactivation of PHD2 function by oxygen scarcity (*3*) favors the switch from a motile tip cell phenotype to a quiescent phalanx phenotype (*4*). Thus, perfusion is established and PHDs are reactivated (*5*); however, the absence of proangiogenic players (following tissue reoxygenation) guarantees a quiescent phenotype will be maintained

oxygen levels towards the tip cells. However, the gradient of angiogenic factors is opposite to the oxygen distribution as outgrowth of vessel sprouts is triggered by proangiogenic growth factors such as VEGF produced by the hypoxic tissue. Tip cell selection occurs where cells are exposed to the highest concentration of proangiogenic factors and proliferating stalk cells elongate the sprout alongside increasing VEGF and decreasing oxygen concentration. At the point where two vessel branches connect, PHD activity is, at least in part, inhibited due to oxygen deprivation. This favors the switch from a motile tip cell phenotype to a quiescent phalanx phenotype (Fig. 14.3).

FIH was shown to exhibit lower affinity for oxygen than PHD2 [66, 114], thus being operational under lower oxygen concentration and inactive at nearly anoxic conditions. Therefore, and with respect to the HIF–FIH–Notch interplay, it will be of particular interest for future studies to dissect the physiological relevance of FIH in endothelial cell types that experience severely hypoxic/anoxic conditions such as the migratory tip cells, and to further define whether the endothelial cell fate (tip vs. stalk cell) is interconnected with FIH functionality.

Evidence from genetic studies has formed the general opinion that HIF-1α and HIF-2α hold nonredundant functions in angiogenesis. Specific deletion of HIF-1 in endothelial cells results in a malfunctioning vascular network under hypoxic conditions such as in tumors [115]. HIF-1 deficiency attenuated hypoxia-induced target gene expression, and this disturbed hypoxic response is not compensated by HIF-2α underlining their distinct roles. Functionally, the genetic loss of HIF-1α hampers endothelial proliferation and migration, and interferes in an autocrine VEGF/VEGFR2 loop since the expression of both is under the control of HIF-1α. While the HIF-1α isoform is ubiquitously expressed, HIF-2α expression is restricted to a subset of cell types, among which are endothelial cells. In particular, HIF-2α is highly expressed in endothelial cells during development, pointing towards a functional relevance in this cell type [32]. Endothelial-specific deletion of HIF-2α disrupts vessel functionality, resulting in smaller and more hypoxic tumors, or impaired and aberrant revascularization of ischemic tissues. Mechanistically, loss of HIF-2α in endothelial cells impairs cell adhesion and thus impacts on vessel functionality [116].

In line with a role of this isoform in vessel integrity, in a different study HIF-2α deletion in endothelial cells caused aberrant tumor vessel growth via reduced expression of ephrin A1 [117]. When considering HIF-2α as an important player in vessel normalization, remodeling, and maturation, it should also be pointed out that endothelial NO sunthase (eNOS), as well as the junctional protein vascular endothelial (VE)-cadherin, are under the specific control of HIF-2α and not HIF-1α [118, 119]. Interestingly, the metastatic spread of tumor cells seems to be differentially governed by HIF-1α and HIF-2α. While endothelial-specific deletion of HIF-1α diminishes lung metastasis, HIF-2α loss enhances the metastatic spread. This effect can be attributed to HIF isoform-specific regulation of NO homeostasis, with HIF-1α loss resulting in reduced NO release and HIF-2α deletion enhancing it. The contrasting HIF-1α/HIF-2α-dependent regulation of NO release results in opposite effects on endothelial tumor cell transmigration. An HIF-1α-deficient endothelium enhances, and an endothelium with disrupted HIF-2α expression inhibits, the transmigration of tumor cells [120].

As outlined above, genetic deletion of PHD2 results in tumor vessel normalization, namely improved endothelial lining, barrier, and stability, carried out through an HIF-2α-driven response. These structural changes warrant a better vessel perfusion and thus tumor oxygenation. As a consequence, tumor cell intravasation and metastasis are reduced [113]. To date, there are no reports for a causal link between endothelial-specific PHD1 and/or PHD3 deletion and vessel malfunction. Overall, inhibition of PHDs in endothelial cells promises a beneficial outcome for cancer treatment. In which way the oxygen-sensing machinery might be exploited for future therapeutic approaches will be the focus of the following section.

Given the fact that oxygen is the main substrate of aerobic organism, it follows that PHDs, and thus oxygen sensing, are closely intertwined with cellular metabolic responses [121–123]. In this respect, a lot of effort has recently been undertaken to understand in which way a metabolic switch can be a cause rather than a consequence of a phenotypic change. A rather new observation suggests the involvement of endothelial cell metabolism in vessel sprouting [124]. In this study, the authors showed that the hypoxia-driven release of the proangiogenic factors VEGF and FGF2 trigger 6-phosphofructo-2-kinase/fructose-2,6-bisphosphatase 3 (PFKFB3) -mediated glycolysis and favor the tip cell phenotype. Strikingly, PFKFB3 signaling even predominates Notch signaling, as evinced by overexpression of both PFKFB3 and NICD in a mosaic-sprouting assay, resulting in a preferred tip cell positioning of these cells. Interestingly, PFKFB3 silencing or genetic deletion does not only impair the tip cell behavior but also affects the stalk cell function by evoking a hypoproliferative stalk cell phenotype [124]. Thus, given the oxygen control of cellular metabolism, these findings uncover an additional control mechanism of vascular sprouting.

14.6 Oxygen-Sensing Pathways as Future Therapeutic Targets

Due to the interrelation between disturbed oxygen homeostasis and pathophysiological conditions such as stroke or cancer, it stands to reason that oxygen signaling pathways are coopted to facilitate disease progression. Interfering in these signaling pathways holds the promise of new and beneficial treatment options. First, evidence for potentially improved curing prospects when targeting oxygen-sensing enzymes stems from drug-mediated inhibition of PHDs, which stimulates angiogenesis and generates a more mature vascular network [125, 126]. However, genetic studies allow a more profound understanding of how interfering in the function of one specific oxygen-sensing enzyme might alter disease progression.

Of utmost importance for cancer therapeutics are the findings that chemotherapeutic treatments in combination with genetic deletion of PHD2 leads to a beneficial outcome [127]—on the one hand, because tumor growth and metastasis are reduced and, on the other hand, as chemotherapeutic side effects on healthy organs are diminished (Fig. 14.4). Endothelial-specific PHD2 deletion alone was previously shown to normalize the tumor vasculature and reduce metastatic tumor cell spread, but without influencing the primary tumor growth [113]. Of note, the combinatory approach of disrupted PHD2 expression and the use of suboptimal doses of chemotherapeutics results in a self-reinforcing antitumor and antimetastatic effect, holding the promise for improved chemotherapeutic regimens. Normalization of the tortuous tumor vasculature through inhibition of PHD2 function allows a better distribution of chemotherapeutics within the cancerous tissue, thus leading to more effective elimination of malignant cells. Importantly, the presence or absence of PHD2 in cancer cells does not alter the response underlining the translational potential of

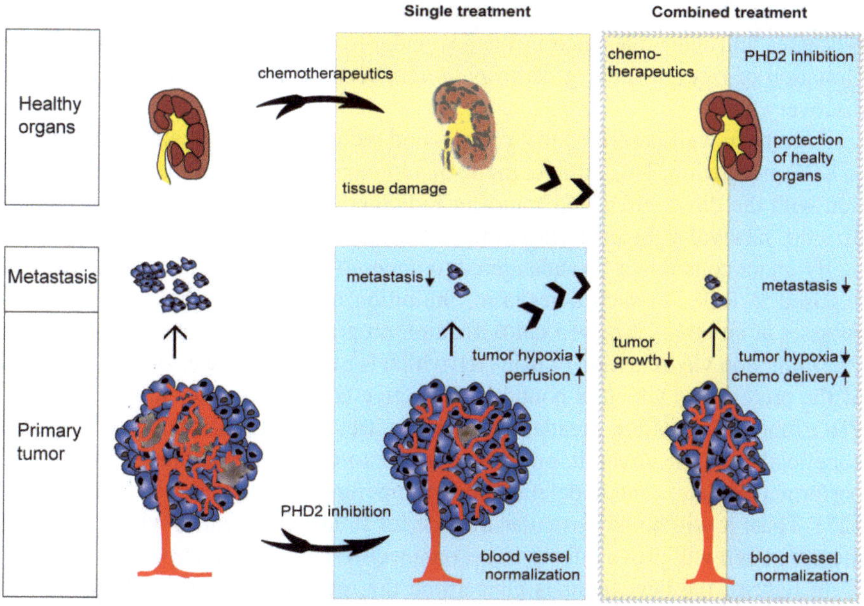

Fig. 14.4 Prolyl hydroxylase domain protein-2 (*PHD2*) inhibition in combination with chemotherapy is beneficial for cancer treatment. Endothelial deletion of PHD2 normalizes chaotic tumor vessels, decreases tumor hypoxia (*brown areas*), and inhibits metastatic tumor cell spread, but does not affect primary tumor growth (box with *blue* background). Suboptimal doses of chemotherapy do not inhibit primary tumor growth but result in cytotoxic side effects, which compromise organ functions (box with *yellow* background). A combinatory approach of suboptimal doses of chemotherapeutics and inhibition of PHD2 further diminishes metastatic tumor cell dissemination but also decreases primary tumor growth as better perfused vessels allow chemotherapeutics to reach their target cells and execute their functions. Importantly, PHD2 deletion protects healthy organs from adverse side effects (box with *half blue, half yellow* background) because of detoxifying responses in normal cells

these findings as drug-mediated inhibition would likewise target different cell types at the same time [127]. Nevertheless, medication-mediated intervention on the FIH/PHD/HIF axis warrants caution. One prerequisite is the development of specific inhibitors for the oxygen-sensing enzymes. FIH and the respective PHD isoforms have been shown to differentially regulate HIF-1α and HIF-2α (see Sect. 3). Given the nonoverlapping and even opposing functions of HIF-1α and HIF-2α in endothelial cells under pathophysiological conditions (e.g. their isoform-specific regulation of metastatic tumor cell spread [120], inhibitors with unique selectivity need to be applied.

This might not only bypass unwanted side effects engendered by induction of unexpected HIF isoform-mediated functions but might also circumvent effects evoked by interference in HIF-independent PHD- or FIH-specific signaling pathways as there is increasing evidence for HIF-independent interaction partners of oxygen-sensing enzymes that might critically alter the response [128].

However given the exceeding importance of oxygen signaling in angiogenic processes and their relevance in therapy induced responses, pharmacological exploitation of oxygen-sensing pathways might serve as a goldmine for future drug discovery.

It has been well-described that chemo- and radiotherapeutic treatment are linked to ROS production. This indirect mode of action might kill cancer cells in conjunction with the direct effect of chemo- or radiotherapy; adversely, ROS also jeopardize the survival of healthy tissue [129–131]. By experiencing DNA damage, formerly intact cells might be endangered to undergo oncogenic transformation when exposed to ROS. In such a scenario, inhibition of PHD2 has been shown to be an asset in an HIF-1/2-driven detoxification program. ROS are assumed to inhibit PHD function via oxidation of iron (II), which is needed for the enzymatic function of the protein and which is bound in the active core of the protein. By inhibiting PHD function, cells are preadapted to these stress conditions, thus resulting in a beneficial outcome. Overall, one has to take into consideration that, in the studies performed to date, genetic deletion occurred before metastatic dissemination [113, 127]. Thus, it will be of particular interest for future therapeutics whether a treatment regimen will prove effectiveness on a well-established (metastatic) tumor.

In light of the new findings concerning the regulatory function of endothelial metabolism on vascular sprouting [124], this control mechanism might hold the promise of new antiangiogenic treatment options. Partial blockage of PFKFB3 with the small molecule 3-(3-pyridinyl)-1- (4-pyridinyl)-2-propen-1-one (3PO) inhibited pathological angiogenesis in ocular and inflammatory models [132].

Summarizing, exploring the function of oxygen-sensing enzymes in the vasculature might hold promise for future drug discovery, thus giving hope for new breakthroughs in the treatment of diseases related to oxygen disturbances.

References

1. Hedges SB, Blair JE, Venturi ML, Shoe JL. A molecular timescale of eukaryote evolution and the rise of complex multicellular life. BMC Evol Biol. 2004;4:2.
2. Bazan NG, Palacios-Pelaez R, Lukiw WJ. Hypoxia signaling to genes: significance in Alzheimer's disease. Mol Neurobiol. 2002;26(2–3):283–98.
3. Bergeron M, Yu AY, Solway KE, Semenza GL, Sharp FR. Induction of hypoxia-inducible factor-1 (HIF-1) and its target genes following focal ischaemia in rat brain. Eur J Neurosci. 1999;11(12):4159–70.
4. Dachs GU, Patterson AV, Firth JD, Ratcliffe PJ, Townsend KM, Stratford IJ, et al. Targeting gene expression to hypoxic tumor cells. Nat Med. 1997;3(5):515–20.
5. Maxwell PH, Dachs GU, Gleadle JM, Nicholls LG, Harris AL, Stratford IJ, et al. Hypoxia-inducible factor-1 modulates gene expression in solid tumors and influences both angiogenesis and tumor growth. Proc Natl Acad Sci U S A. 1997;94(15):8104–9.
6. Carmeliet P, Dor Y, Herbert JM, Fukumura D, Brusselmans K, Dewerchin M, et al. Role of HIF-1alpha in hypoxia-mediated apoptosis, cell proliferation and tumour angiogenesis. Nature. 1998;394(6692):485–90.
7. Semenza GL. Oxygen sensing, homeostasis, and disease. N Engl J Med. 2011;365(6):537–47.

8. Guzy RD, Schumacker PT. Oxygen sensing by mitochondria at complex III: the paradox of increased reactive oxygen species during hypoxia. Exp Physiol. 2006;91(5):807–19.
9. Cooke MS, Evans MD, Dizdaroglu M, Lunec J. Oxidative DNA damage: mechanisms, mutation, and disease. FASEB J. 2003;17(10):1195–214.
10. Lopez-Barneo J, Pardal R, Ortega-Saenz P. Cellular mechanism of oxygen sensing. Annu Rev Physiol. 2001;63:259–87.
11. Peers C. Oxygen-sensitive ion channels. Trends Pharmacol Sci. 1997;18(11):405–8.
12. Cummins EP, Taylor CT. Hypoxia-responsive transcription factors. Pflugers Arch. 2005;450(6):363–71.
13. Adams JM, Difazio LT, Rolandelli RH, Lujan JJ, Hasko G, Csoka B, et al. HIF-1: a key mediator in hypoxia. Acta Physiol Hung. 2009;96(1):19–28.
14. Bruick RK, McKnight SL. A conserved family of prolyl-4-hydroxylases that modify HIF. Science. 2001;294(5545):1337–40.
15. Epstein AC, Gleadle JM, McNeill LA, Hewitson KS, O'Rourke J, Mole DR, et al. C. elegans EGL-9 and mammalian homologs define a family of dioxygenases that regulate HIF by prolyl hydroxylation. Cell. 2001;107(1):43–54.
16. Mahon PC, Hirota K, Semenza GL. FIH-1: a novel protein that interacts with HIF-1alpha and VHL to mediate repression of HIF-1 transcriptional activity. Genes Dev. 2001;15(20):2675–86.
17. Lando D, Peet DJ, Gorman JJ, Whelan DA, Whitelaw ML, Bruick RK. FIH-1 is an asparaginyl hydroxylase enzyme that regulates the transcriptional activity of hypoxia-inducible factor. Genes Dev. 2002;16(12):1466–71.
18. Loenarz C, Schofield CJ. Physiological and biochemical aspects of hydroxylations and demethylations catalyzed by human 2-oxoglutarate oxygenases. Trends Biochem Sci. 2011;36(1):7–18.
19. Boeckel JN, Guarani V, Koyanagi M, Roexe T, Lengeling A, Schermuly RT, et al. Jumonji domain-containing protein 6 (Jmjd6) is required for angiogenic sprouting and regulates splicing of VEGF-receptor 1. Proc Natl Acad Sci U S A. 2011;108(8):3276–81.
20. Kaelin WG Jr. ROS: really involved in oxygen sensing. Cell Metab. 2005;1(6):357–8.
21. Klimova T, Chandel NS. Mitochondrial complex III regulates hypoxic activation of HIF. Cell Death Differ. 2008;15(4):660–6.
22. Gerhardt H, Golding M, Fruttiger M, Ruhrberg C, Lundkvist A, Abramsson A, et al. VEGF guides angiogenic sprouting utilizing endothelial tip cell filopodia. J Cell Biol. 2003;161(6):1163–77.
23. Claxton S, Fruttiger M. Periodic Delta-like 4 expression in developing retinal arteries. Gene Expr Patterns. 2004;5(1):123–7.
24. Lu X, Noble F L, Yuan L, Jiang Q, De Lafarge B, Sugiyama D, et al. The netrin receptor UNC5B mediates guidance events controlling morphogenesis of the vascular system. Nature. 2004;432(7014):179–86.
25. Tammela T, Zarkada G, Wallgard E, Murtomaki A, Suchting S, Wirzenius M, et al. Blocking VEGFR-3 suppresses angiogenic sprouting and vascular network formation. Nature. 2008;454(7204):656–60.
26. Wenger RH, Stiehl DP, Camenisch G. Integration of oxygen signaling at the consensus HRE. Sci STKE. 2005;2005(306):re12.
27. Pugh CW, Ratcliffe PJ. Regulation of angiogenesis by hypoxia: role of the HIF system. Nat Med. 2003;9(6):677–84.
28. Papetti M, Herman IM. Mechanisms of normal and tumor-derived angiogenesis. Am J Physiol Cell Physiol. 2002;282(5):C947–70.
29. Wang GL, Jiang BH, Rue EA, Semenza GL. Hypoxia-inducible factor 1 is a basic-helix-loop-helix-PAS heterodimer regulated by cellular O2 tension. Proc Natl Acad Sci U S A. 1995;92(12):5510–4.
30. Semenza GL, Wang GL. A nuclear factor induced by hypoxia via de novo protein synthesis binds to the human erythropoietin gene enhancer at a site required for transcriptional activation. Mol Cell Biol. 1992;12(12):5447–54.

31. Wood SM, Gleadle JM, Pugh CW, Hankinson O, Ratcliffe PJ. The role of the aryl hydrocarbon receptor nuclear translocator (ARNT) in hypoxic induction of gene expression. Studies in ARNT-deficient cells. J Biol Chem. 1996;271(25):15117–23.
32. Ema M, Taya S, Yokotani N, Sogawa K, Matsuda Y, Fujii-Kuriyama Y. A novel bHLH-PAS factor with close sequence similarity to hypoxia-inducible factor 1alpha regulates the VEGF expression and is potentially involved in lung and vascular development. Proc Natl Acad Sci U S A. 1997;94(9):4273–8.
33. Flamme I, Frohlich T, von Reutern M, Kappel A, Damert A, Risau W. HRF, a putative basic helix-loop-helix-PAS-domain transcription factor is closely related to hypoxia-inducible factor-1 alpha and developmentally expressed in blood vessels. Mech Dev. 1997;63(1):51–60.
34. Hogenesch JB, Chan WK, Jackiw VH, Brown RC, Gu YZ, Pray-Grant M, et al. Characterization of a subset of the basic-helix-loop-helix-PAS superfamily that interacts with components of the dioxin signaling pathway. J Biol Chem. 1997;272(13):8581–93.
35. Tian H, McKnight SL, Russell DW. Endothelial PAS domain protein 1 (EPAS1), a transcription factor selectively expressed in endothelial cells. Genes Dev. 1997;11(1):72–82.
36. Makino Y, Cao R, Svensson K, Bertilsson G, Asman M, Tanaka H, et al. Inhibitory PAS domain protein is a negative regulator of hypoxia-inducible gene expression. Nature. 2001;414(6863):550–4.
37. Makino Y, Kanopka A, Wilson WJ, Tanaka H, Poellinger L. Inhibitory PAS domain protein (IPAS) is a hypoxia-inducible splicing variant of the hypoxia-inducible factor-3alpha locus. J Biol Chem. 2002;277(36):32405–8.
38. Kinoshita K, Kikuchi Y, Sasakura Y, Suzuki M, Fujii-Kuriyama Y, Sogawa K. Altered DNA binding specificity of Arnt by selection of partner bHLH-PAS proteins. Nucleic Acids Res. 2004;32(10):3169–79.
39. Jiang BH, Zheng JZ, Leung SW, Roe R, Semenza GL. Transactivation and inhibitory domains of hypoxia-inducible factor 1alpha. Modulation of transcriptional activity by oxygen tension. J Biol Chem. 1997;272(31):19253–60.
40. Pugh CW, O'Rourke JF, Nagao M, Gleadle JM, Ratcliffe PJ. Activation of hypoxia-inducible factor-1; definition of regulatory domains within the alpha subunit. J Biol Chem. 1997;272(17):11205–14.
41. O'Rourke JF, Tian YM, Ratcliffe PJ, Pugh CW. Oxygen-regulated and transactivating domains in endothelial PAS protein 1: comparison with hypoxia-inducible factor-1alpha. J Biol Chem. 1999;274(4):2060–71.
42. Lando D, Peet DJ, Whelan DA, Gorman JJ, Whitelaw ML. Asparagine hydroxylation of the HIF transactivation domain a hypoxic switch. Science. 2002;295(5556):858–61.
43. Brahimi-Horn C, Mazure N, Pouyssegur J. Signalling via the hypoxia-inducible factor-1alpha requires multiple posttranslational modifications. Cell Signal. 2005;17(1):1–9.
44. Jeong JW, Bae MK, Ahn MY, Kim SH, Sohn TK, Bae MH, et al. Regulation and destabilization of HIF-1alpha by ARD1-mediated acetylation. Cell. 2002;111(5):709–20.
45. Richard DE, Berra E, Gothie E, Roux D, Pouyssegur J. p42/p44 mitogen-activated protein kinases phosphorylate hypoxia-inducible factor 1alpha (HIF-1alpha) and enhance the transcriptional activity of HIF-1. J Biol Chem. 1999;274(46):32631–7.
46. Sodhi A, Montaner S, Patel V, Zohar M, Bais C, Mesri EA, et al. The Kaposi's sarcoma-associated herpes virus G protein-coupled receptor up-regulates vascular endothelial growth factor expression and secretion through mitogen-activated protein kinase and p38 pathways acting on hypoxia-inducible factor 1alpha. Cancer Res. 2000;60(17):4873–80.
47. Minet E, Michel G, Mottet D, Raes M, Michiels C. Transduction pathways involved in Hypoxia-Inducible Factor-1 phosphorylation and activation. Free Radic Biol Med. 2001;31(7):847–55.
48. Salceda S, Beck I, Srinivas V, Caro J. Complex role of protein phosphorylation in gene activation by hypoxia. Kidney Int. 1997;51(2):556–9.
49. Wang GL, Jiang BH, Semenza GL. Effect of protein kinase and phosphatase inhibitors on expression of hypoxia-inducible factor 1. Biochem Biophys Res Commun. 1995;216(2):669–75.

50. Maxwell PH, Wiesener MS, Chang GW, Clifford SC, Vaux EC, Cockman ME, et al. The tumour suppressor protein VHL targets hypoxia-inducible factors for oxygen-dependent proteolysis. Nature. 1999;399(6733):271–5.
51. Ivan M, Kondo K, Yang H, Kim W, Valiando J, Ohh M, et al. HIFalpha targeted for VHL-mediated destruction by proline hydroxylation: implications for O2 sensing. Science. 2001;292(5516):464–8.
52. Jaakkola P, Mole DR, Tian YM, Wilson MI, Gielbert J, Gaskell SJ, et al. Targeting of HIF-alpha to the von Hippel-Lindau ubiquitylation complex by O2-regulated prolyl hydroxylation. Science. 2001;292(5516):468–72.
53. Salceda S, Caro J. Hypoxia-inducible factor 1alpha (HIF-1alpha) protein is rapidly degraded by the ubiquitin-proteasome system under normoxic conditions. Its stabilization by hypoxia depends on redox-induced changes. J Biol Chem. 1997;272(36):22642–7.
54. Gradin K, McGuire J, Wenger RH, Kvietikova I, fhitelaw ML, Toftgard R, et al. Functional interference between hypoxia and dioxin signal transduction pathways: competition for recruitment of the Arnt transcription factor. Mol Cell Biol. 1996;16(10):5221–31.
55. Jewell UR, Kvietikova I, Scheid A, Bauer C, Wenger RH, Gassmann M Induction of HIF-1alpha in response to hypoxia is instantaneous. FASEB J. 2001;15(7):1312–4.
56. Kallio PJ, Okamoto K, O'Brien S, Carrero P, Makino Y, Tanaka H, et al. Signal transduction in hypoxic cells: inducible nuclear translocation and recruitment of the CBP/p300 coactivator by the hypoxia-inducible factor-1alpha. EMBO J. 1998;17(22):6573–86.
57. Arany Z, Huang LE, Eckner R, Bhattacharya S, Jiang C, Goldberg MA, et al. An essential role for p300/CBP in the cellular response to hypoxia. Proc Natl Acad Sci U S A. 1996;93(23):12969–73.
58. Hu CJ, Wang LY, Chodosh LA, Keith B, Simon MC. Differential roles of hypoxia-inducible factor 1alpha (HIF-1alpha) and HIF-2alpha in hypoxic gene regulation. Mol Cell Biol. 2003;23(24):9361–74.
59. Sowter HM, Raval RR, Moore JW, Ratcliffe PJ, Harris AL. Predominant role of hypoxia-inducible transcription factor (Hif)-1alpha versus Hif-2alpha in regulation of the transcriptional response to hypoxia. Cancer Res. 2003;63(19):6130–4.
60. Holmquist-Mengelbier L, Fredlund E, Lofstedt T, Noguera R, Navarro S, Nilsson H, et al. Recruitment of HIF-1alpha and HIF-2alpha to common target genes is differentially regulated in neuroblastoma: HIF-2alpha promotes an aggressive phenotype. Cancer Cell. 2006;10(5):413–23.
61. Tian H, Hammer RE, Matsumoto AM, Russell DW, McKnight SL. The hypoxia-responsive transcription factor EPAS1 is essential for catecholamine homeostasis and protection against heart failure during embryonic development. Genes Dev. 1998;12(21):3320–4.
62. Wiesener MS, Jurgensen JS, Rosenberger C, Scholze CK, Horstrup JH, Warnecke C, et al. Widespread hypoxia-inducible expression of HIF-2alpha in distinct cell populations of different organs. FASEB J. 2003;17(2):271–3.
63. Hewitson KS, Granatino N, Welford RW, McDonough MA, Schofield CJ. Oxidation by 2-oxoglutarate oxygenases: non-haem iron systems in catalysis and signalling. Philos Trans A Math Phys Eng Sci. 2005;363(1829):807–28 (discussion 1035–40).
64. Masson N, Willam C, Maxwell PH, Pugh CW, Ratcliffe PJ. Independent function of two destruction domains in hypoxia-inducible factor-alpha chains activated by prolyl hydroxylation. EMBO J. 2001;20(18):5197–206.
65. Huang LE, Gu J, Schau M, Bunn HF. Regulation of hypoxia-inducible factor 1alpha is mediated by an O2-dependent degradation domain via the ubiquitin-proteasome pathway. Proc Natl Acad Sci U S A. 1998;95(14):7987–92.
66. Hirsila M, Koivunen P, Gunzler V, Kivirikko KI, Myllyharju J. Characterization of the human prolyl 4-hydroxylases that modify the hypoxia-inducible factor. J Biol Chem. 2003;278(33):30772–80.
67. Appelhoff RJ, Tian YM, Raval RR, Turley H, Harris AL, Pugh CW, et al. Differential function of the prolyl hydroxylases PHD1, PHD2, and PHD3 in the regulation of hypoxia-inducible factor. J Biol Chem. 2004;279(37):38458–65.

68. Bracken CP, Fedele AO, Linke S, Balrak W, Lisy K, Whitelaw ML, et al. Cell-specific regulation of hypoxia-inducible factor (HIF)-1alpha and HIF-2alpha stabilization and transactivation in a graded oxygen environment. J Biol Chem. 2006;281(32):22575–85.
69. Berra E, Benizri E, Ginouves A, Volmat V, Roux D, Pouyssegur J. HIF prolyl-hydroxylase 2 is the key oxygen sensor setting low steady-state levels of HIF-1alpha in normoxia. EMBO J. 2003;22(16):4082–90.
70. Metzen E, Stiehl DP, Doege K, Marxsen JH, Hellwig-Burgel T, Jelkmann W. Regulation of the prolyl hydroxylase domain protein 2 (phd2/egln-1) gene: identification of a functional hypoxia-responsive element. Biochem J. 2005;387(Pt 3):711–7.
71. Pescador N, Cuevas Y, Naranjo S, Alcaide M, Villar D, Landazuri MO, et al. Identification of a functional hypoxia-responsive element that regulates the expression of the egl nine homologue 3 (egln3/phd3) gene. Biochem J. 2005;390(Pt 1):189–97.
72. Aprelikova O, Chandramouli GV, Wood M, Vasselli JR, Riss J, Maranchie JK, et al. Regulation of HIF prolyl hydroxylases by hypoxia-inducible factors. J Cell Biochem. 2004;92(3):491–501.
73. Henze AT, Riedel J, Diem T, Wenner J, Flamme I, Pouyseggur J, et al. Prolyl hydroxylases 2 and 3 act in gliomas as protective negative feedback regulators of hypoxia-inducible factors. Cancer Res. 2010;70(1):357–66.
74. To KK, Huang LE. Suppression of hypoxia-inducible factor 1alpha (HIF-1alpha) transcriptional activity by the HIF prolyl hydroxylase EGLN1. J Biol Chem. 2005;280(45):38102–7.
75. Hopfer U, Hopfer H, Jablonski K, Stahl RA, Wolf G. The novel WD-repeat protein Morg1 acts as a molecular scaffold for hypoxia-inducible factor prolyl hydroxylase 3 (PHD3). J Biol Chem. 2006;281(13):8645–55.
76. Ozer A, Wu LC, Bruick RK. The candidate tumor suppressor ING4 represses activation of the hypoxia inducible factor (HIF). Proc Natl Acad Sci U S A. 2005;102(21):7481–6.
77. Treins C, Giorgetti-Peraldi S, Murdaca J, Semenza GL, Van Obberghen E. Insulin stimulates hypoxia-inducible factor 1 through a phosphatidylinositol 3-kinase/target of rapamycin-dependent signaling pathway. J Biol Chem. 2002;277(31):27975–81.
78. Fukuda R, Hirota K, Fan F, Jung YD, Ellis LM, Semenza GL. Insulin-like growth factor 1 induces hypoxia-inducible factor 1-mediated vascular endothelial growth factor expression, which is dependent on MAP kinase and phosphatidylinositol 3-kinase signaling in colon cancer cells. J Biol Chem. 2002;277(41):38205–11.
79. Minet E, Arnould T, Michel G, Roland I, Mottet D, Raes M, et al. ERK activation upon hypoxia: involvement in HIF-1 activation. FEBS Lett. 2000;468(1):53–8.
80. Zelzer E, Levy Y, Kahana C, Shilo BZ, Rubinstein M, Cohen B. Insulin induces transcription of target genes through the hypoxia-inducible factor HIF-1alpha/ARNT. EMBO J. 1998;17(17):5085–94.
81. Zhong H, Chiles K, Feldser D, Laughner E, Hanrahan C, Georgescu MM, et al. Modulation of hypoxia-inducible factor 1alpha expression by the epidermal growth factor/phosphatidylinositol 3-kinase/PTEN/AKT/FRAP pathway in human prostate cancer cells: implications for tumor angiogenesis and therapeutics. Cancer Res. 2000;60(6):1541–5.
82. Feldser D, Agani F, Iyer NV, Pak B, Ferreira G, Semenza GL. Reciprocal positive regulation of hypoxia-inducible factor 1alpha and insulin-like growth factor 2. Cancer Res. 1999;59(16):3915–8.
83. Hellwig-Burgel T, Rutkowski K, Metzen E, Fandrey J, Jelkmann W. Interleukin-1beta and tumor necrosis factor-alpha stimulate DNA binding of hypoxia-inducible factor-1. Blood. 1999;94(5):1561–7.
84. Richard DE, Berra E, Pouyssegur J. Nonhypoxic pathway mediates the induction of hypoxia-inducible factor 1alpha in vascular smooth muscle cells. J Biol Chem. 2000;275(35):26765–71.
85. Gorlach A, Diebold I, Schini-Kerth VB, Berchner-Pfannschmidt U, Roth U, Brandes RP, et al. Thrombin activates the hypoxia-inducible factor-1 signaling pathway in vascular smooth muscle cells: Role of the p22(phox)-containing NADPH oxidase. Circ Res. 2001;89(1):47–54.

86. Tacchini L, Dansi P, Matteucci E, Desiderio MA. Hepatocyte growth factor signalling stimulates hypoxia inducible factor-1 (HIF-1) activity in HepG2 hepatoma cells. Carcinogenesis. 2001;22(9):1363–71.
87. Thornton RD, Lane P, Borghaei RC, Pease EA, Caro J, Mochan E. Interleukin 1 induces hypoxia-inducible factor 1 in human gingival and synovial fibroblasts. Biochem J. 2000;350 (Pt 1):307–12.
88. Laughner E, Taghavi P, Chiles K, Mahon PC, Semenza GL. HER2 (neu) signaling increases the rate of hypoxia-inducible factor 1alpha (HIF-1alpha) synthesis: novel mechanism for HIF-1-mediated vascular endothelial growth factor expression. Mol Cell Biol. 2001;21(12):3995–4004.
89. Chen C, Pore N, Behrooz A, Ismail-Beigi F, Maity A. Regulation of glut1 mRNA by hypoxia-inducible factor-1. Interaction between H-ras and hypoxia. J Biol Chem. 2001;276(12):9519–25.
90. Jiang BH, Agani F, Passaniti A, Semenza GL. V-SRC induces expression of hypoxia-inducible factor 1 (HIF-1) and transcription of genes encoding vascular endothelial growth factor and enolase 1: involvement of HIF-1 in tumor progression. Cancer Res. 1997;57(23):5328–35.
91. Zundel W, Schindler C, Haas-Kogan D, Koong A, Kaper F, Chen E, et al. Loss of PTEN facilitates HIF-1-mediated gene expression. Genes Dev. 2000;14(4):391–6.
92. Blancher C, Moore JW, Robertson N, Harris AL. Effects of ras and von Hippel-Lindau (VHL) gene mutations on hypoxia-inducible factor (HIF)-1alpha, HIF-2alpha, and vascular endothelial growth factor expression and their regulation by the phosphatidylinositol 3′-kinase/Akt signaling pathway. Cancer Res. 2001;61(19):7349–55.
93. Dimmeler S, Zeiher AM. Nitric oxide-an endothelial cell survival factor. Cell Death Differ. 1999;6(10):964–8.
94. Metzen E, Zhou J, Jelkmann W, Fandrey J, Brune B. Nitric oxide impairs normoxic degradation of HIF-1alpha by inhibition of prolyl hydroxylases. Mol Biol Cell. 2003;14(8):3470–81.
95. Hagen T, Taylor CT, Lam F, Moncada S. Redistribution of intracellular oxygen in hypoxia by nitric oxide: effect on HIF1alpha. Science. 2003;302(5652):1975–8.
96. Berchner-Pfannschmidt U, Tug S, Kirsch M, Fandrey J. Oxygen-sensing under the influence of nitric oxide. Cell Signal. 2010;22(3):349–56.
97. Lu H, Forbes RA, Verma A. Hypoxia-inducible factor 1 activation by aerobic glycolysis implicates the Warburg effect in carcinogenesis. J Biol Chem. 2002;277(26):23111–5.
98. Dalgard CL, Lu H, Mohyeldin A, Verma A. Endogenous 2-oxoacids differentially regulate expression of oxygen sensors. Biochem J. 2004;380(Pt 2):419–24.
99. Lu H, Dalgard CL, Mohyeldin A, McFate T, Tait AS, Verma A. Reversible inactivation of HIF-1 prolyl hydroxylases allows cell metabolism to control basal HIF-1. J Biol Chem. 2005;280(51):41928–39.
100. Nakayama K, Frew IJ, Hagensen M, Skals M, Habelhah H, Bhoumik A, et al. Siah2 regulates stability of prolyl-hydroxylases, controls HIF1alpha abundance, and modulates physiological responses to hypoxia. Cell. 2004;117(7):941–52.
101. Barth S, Nesper J, Hasgall PA, Wirthner R, Nytko KJ, Edlich F, et al. The peptidyl prolyl cis/trans isomerase FKBP38 determines hypoxia-inducible transcription factor prolyl-4-hydroxylase PHD2 protein stability. Mol Cell Biol. 2007;27(10):3758–68.
102. Risau W, Flamme I. Vasculogenesis. Annu Rev Cell Dev Biol. 1995;11:73–91.
103. Flamme I, Frolich T, Risau W. Molecular mechanisms of vasculogenesis and embryonic angiogenesis. J Cell Physiol. 1997;173(2):206–10.
104. Shweiki D, Itin A, Soffer D, Keshet E. Vascular endothelial growth factor induced by hypoxia may mediate hypoxia-initiated angiogenesis. Nature. 1992;359(6398):843–5.
105. Kimura H, Weisz A, Ogura T, Hitomi Y, Kurashima Y, Hashimoto K, et al. Identification of hypoxia-inducible factor 1 ancillary sequence and its function in vascular endothelial growth factor gene induction by hypoxia and nitric oxide. J Biol Chem. 2001;276(3):2292–8.
106. Esser S, Lampugnani MG, Corada M, Dejana E, Risau W. Vascular endothelial growth factor induces VE-cadherin tyrosine phosphorylation in endothelial cells. J Cell Sci. 1998;111(Pt 13):1853–65.

107. Kevil CG, Payne DK, Mire E, Alexander JS. Vascular permeability factor/vascular endothelial cell growth factor-mediated permeability occurs through disorganization of endothelial junctional proteins. J Biol Chem. 1998;273(24):15099–103.
108. Phng LK, Gerhardt H. Angiogenesis: a team effort coordinated by notch. Dev Cell. 2009;16(2):196–208.
109. Diez H, Fischer A, Winkler A, Hu CJ, Hatzopoulos AK, Breier G, et al. Hypoxia-mediated activation of Dll4-Notch-Hey2 signaling in endothelial progenitor cells and adoption of arterial cell fate. Exp Cell Res. 2007;313(1):1–9.
110. Zheng X, Linke S, Dias JM, Gradin K, Wallis TP, Hamilton BR, et al. Interaction with factor inhibiting HIF-1 defines an additional mode of cross-coupling between the Notch and hypoxia signaling pathways. Proc Natl Acad Sci U S A. 2008;105(9):3368–73.
111. Coleman ML, McDonough MA, Hewitson KS, Coles C, Mecinovic J, Edelmann M, et al. Asparaginyl hydroxylation of the Notch ankyrin repeat domain by factor inhibiting hypoxia-inducible factor. J Biol Chem. 2007;282(33):24027–38.
112. Gustafsson MV, Zheng X, Pereira T, Gradin K, Jin S, Lundkvist J, et al. Hypoxia requires notch signaling to maintain the undifferentiated cell state. Dev Cell. 2005;9(5):617–28.
113. Mazzone M, Dettori D, Leite de Oliveira R, Loges S, Schmidt T, Jonckx B, et al. Heterozygous deficiency of PHD2 restores tumor oxygenation and inhibits metastasis via endothelial normalization. Cell. 2009;136(5):839–51.
114. Koivunen P, Hirsila M, Gunzler V, Kivirikko KI, Myllyharju J. Catalytic properties of the asparaginyl hydroxylase (FIH) in the oxygen sensing pathway are distinct from those of its prolyl 4-hydroxylases. J Biol Chem. 2004;279(11):9899–904.
115. Tang N, Wang L, Esko J, Giordano FJ, Huang Y, Gerber HP, et al. Loss of HIF-1alpha in endothelial cells disrupts a hypoxia-driven VEGF autocrine loop necessary for tumorigenesis. Cancer Cell. 2004;6(5):485–95.
116. Skuli N, Liu L, Runge A, Wang T, Yuan L, Patel S, et al. Endothelial deletion of hypoxia-inducible factor-2alpha (HIF-2alpha) alters vascular function and tumor angiogenesis. Blood. 2009;114(2):469–77.
117. Yamashita T, Ohneda K, Nagano M, Miyoshi C, Kaneko N, Miwa Y, et al. Hypoxia-inducible transcription factor-2alpha in endothelial cells regulates tumor neovascularization through activation of ephrin A1. J Biol Chem. 2008;283(27):18926–36.
118. Bras A L, Lionneton F, Mattot V, Lelievre E, Caetano B, Spruyt N, et al. HIF-2alpha specifically activates the VE-cadherin promoter independently of hypoxia and in synergy with Ets-1 through two essential ETS-binding sites. Oncogene. 2007;26(53):7480–9.
119. Coulet F, Nadaud S, Agrapart M, Soubrier F. Identification of hypoxia-response element in the human endothelial nitric-oxide synthase gene promoter. J Biol Chem. 2003;278(47):46230–40.
120. Branco-Price C, Zhang N, Schnelle M, Evans C, Katschinski DM, Liao D, et al. Endothelial cell HIF-1alpha and HIF-2alpha differentially regulate metastatic success. Cancer Cell. 2012;21(1):52–65.
121. Aragones J, Schneider M, Van Geyte K, Fraisl P, Dresselaers T, Mazzone M, et al. Deficiency or inhibition of oxygen sensor Phd1 induces hypoxia tolerance by reprogramming basal metabolism. Nat Genet. 2008;40(2):170–80.
122. Aragones J, Fraisl P, Baes M, Carmeliet P. Oxygen sensors at the crossroad of metabolism. Cell Metab. 2009;9(1):11–22.
123. Fraisl P, Mazzone M, Schmidt T, Carmeliet P. Regulation of angiogenesis by oxygen and metabolism. Dev Cell. 2009;16(2):167–79.
124. De Bock K, Georgiadou M, Schoors S, Kuchnio A, Wong BW, Cantelmo AR, et al. Role of PFKFB3-driven glycolysis in vessel sprouting. Cell. 2013;154(3):651–63.
125. Nangaku M, Izuhara Y, Takizawa S, Yamashita T, Fujii-Kuriyama Y, Ohneda O, et al. A novel class of prolyl hydroxylase inhibitors induces angiogenesis and exerts organ protection against ischemia. Arterioscler Thromb Vasc Biol. 2007;27(12):2548–54.

126. Warnecke C, Griethe W, Weidemann A, Jurgensen JS, Willam C, Bachmann S, et al. Activation of the hypoxia-inducible factor-pathway and stimulation of angiogenesis by application of prolyl hydroxylase inhibitors. FASEB J. 2003;17(9):1186–8.
127. Leite de Oliveira R, Deschoemaeker S, Henze AT, Debackere K, Finisguerra V, Takeda Y, et al. Gene-targeting of Phd2 improves tumor response to chemotherapy and prevents side-toxicity. Cancer Cell. 2012;22(2):263–77.
128. Wenger RH, Camenisch G, Stiehl DP, Katschinski DM. HIF prolyl-4-hydroxylase interacting proteins: consequences for drug targeting. Curr Pharm Des. 2009;15(33):3886–94.
129. Anscher MS, Chen L, Rabbani Z, Kang S, Larrier N, Huang H, et al. Recent progress in defining mechanisms and potential targets for prevention of normal tissue injury after radiation therapy. Int J Radiat Oncol Biol Phys. 2005;62(1):255–9.
130. Berthiaume JM, Wallace KB. Adriamycin-induced oxidative mitochondrial cardiotoxicity. Cell Biol Toxicol. 2007;23(1):15–25.
131. Pabla N, Dong Z. Cisplatin nephrotoxicity: mechanisms and renoprotective strategies. Kidney Int. 2008;73(9):994–1007.
132. Schoors S, De Bock K, Cantelmo AR, Georgiadou M, Ghesquiere B, Cauwenberghs S, et al. Partial and transient reduction of glycolysis by PFKFB3 blockade reduces pathological angiogenesis. Cell Metab. 2014;19(1):37–48.

Chapter 15
The Pulmonary Vasculature in Chronic Obstructive Lung Disease

Michael Seimetz and Norbert Weissmann

15.1 Chronic Obstructive Pulmonary Disease (COPD) and its Association with Pulmonary Hypertension (PH)

Chronic obstructive lung disease, or chronic obstructive pulmonary disease (COPD), a collective term for chronic bronchitis and pulmonary emphysema, is one of the major causes of death worldwide. The World Health Organization (WHO) predicts that by 2030, COPD will rank as the third greatest cause of death worldwide. It is characterized by progressive, poorly reversible airflow limitation associated with an abnormal chronic inflammatory response in the lung. COPD is also viewed as a systemic disease, involving skeletal muscle wasting, diaphragmatic dysfunction, and systemic inflammation [1]. In industrialized countries, tobacco smoke (80–90 %) and air pollution are the major triggers for the development of COPD, whereas in developing countries, exposure to biomass smoke, especially during cooking, is most relevant [2].

On the one hand, COPD is viewed as a small airway disease, with destruction of the elastic architecture of the lung leading to enlargement of distal airspaces [3] associated with chronic inflammation of the airways [4, 5]. Influx of inflammatory cells, imbalance of proteases/antiproteases, increased oxidative stress with the con-

M. Seimetz (✉) · N. Weissmann
Excellence Cluster Cardio-Pulmonary System (ECCPS),
Aulweg 130, ECCPS, 35392 Giessen, Germany
e-mail: Michael.Seimetz@innere.med.uni-giessen.de

Universities of Giessen and Marburg Lung Center (UGMLC), Giessen, Germany

German Center for Lung Research (DZL), Giessen, Germany

Justus-Liebig-University, Giessen, Germany

N. Weissmann
e-mail: Norbert.Weissmann@innere.med.uni-giessen.de

M. HH. Schmidt, S. Liebner (eds.), *Endothelial Signaling in Development and Disease*, DOI 10.1007/978-1-4939-2907-8_15

comitant rise in the number of apoptotic cells, and decreased proliferation are likely important upstream events [4, 6, 7].

On the other hand, there is increasing evidence that COPD is also a vascular disease in which cigarette smoke may directly affect the pulmonary vasculature, leading to vascular remodeling, pulmonary hypertension (PH), and finally to cor pulmonale [8–10]. The mean pulmonary artery pressure (mPAP) in PH related to COPD usually ranges between 25 and 35 mmHg, with nearly normal cardiac output [11]. However, mPAPs higher than 40 mmHg are quite common in patients with severe COPD, especially after at least one previous incident of acute respiratory failure [11, 12]. Furthermore, the occurrence of PH in COPD is of prognostic relevance because the presence of PH clearly increases mortality [13, 14].

The exact prevalence of PH in patients with mild or moderate COPD is not known because of the lack of extensive epidemiologic studies [2]. According to the current literature, 30–70 % of patients with mild or moderate COPD also suffer from PH [15]. There are several reasons for the high variation in the rate of observed PH. The necessary right heart catheterization for detection of PH was/is not routinely used in patients with COPD. In addition, the respective studies used different definitions for the onset of PH and varied in study design (measuring at rest/exercise, severity of COPD/PH, methods to measure PH). Some studies have investigated PH in mild, moderate, or severe COPD, with frequencies of PH of approximately 16–44, 43–56, and 59–84 %, respectively [2, 16, 17].

PH is a common complication in advanced COPD [18] and can occur as a consequence of the hypoxia associated with COPD. However, several publications showed that vascular alterations can occur before alveolar destruction is detectable [19–22], indicating that cor pulmonale and late-stage PH are not necessarily secondary to hypoxia in patients with COPD. For instance, in 2003 Santos and colleagues showed that pulmonary vascular remodeling can be seen in smokers who are not yet suffering from COPD [23].

Thus, although PH can occur prior to lung emphysema, its relevance for the pathogenesis of COPD is not yet fully understood. Against this background, in this chapter we summarize the current findings of vascular alterations and pathobiology in COPD.

15.2 PH as a Cause for Right Ventricular Failure in COPD

The progression of PH in COPD is usually slow and pulmonary artery mean pressure (mPAP) can remain stable over periods of 3–12 years [24–26]. In a study following 93 patients during a mean period of 90 months, the average change in mPAP was only +0.5 mmHg/year, independent of the presence of initial PH (defined by mPAP >20 mmHg) [26]. Another study investigating the development of PH in COPD over time (initial mPAP <20 mmHg) showed that only 33/121 developed PH after 6.8 ± 2.9 years [27]. Nevertheless, approximately 30 % of patients with severe COPD exhibited a remarkable worsening of mPAP during follow-up [26]; these patients were characterized by a progressive worsening of PaO_2 and $PaCO_2$ (partial oxygen/carbon dioxide pressure) during the time course, and there was a significant correlation between the changes in PaO_2 and mPAP [25, 26]. Typically, the development of right heart failure (RHF) in patients with chronic respiratory disease is accompanied with the preceding occurrence of PH. The severity of PH and the development of RHF correlate. PH increases the workload of the right ventricle, leading to right ventricular enlargement (hypertrophy plus dilatation), which can result in right ventricular dysfunction. The RHF is usually accompanied by peripheral edema and can be observed in some COPD patients [28, 29]. Peripheral edema is frequently observed in advanced COPD patients and is considered to reflect RHF, but the possible occurrence of RHF in these patients may simply indicate the presence of secondary hyperaldosteronism induced by functional renal insufficiency [30].

The role of pressure overload in the development of RHF in these patients has also been debated. The comparison of COPD patients with and without clinical (edema) and hemodynamic signs of RHF led to the conclusion that RHF was probably due to causes other than PH [31]. In stable conditions, right ventricular contractility, measured by the end-systolic pressure–volume relation, is near normal in COPD patients with PH, but has been found to be decreased during severe exacerbations with marked peripheral edema [31]. Hence, many patients with advanced COPD will never develop RHF. At least some patients experience sequences of RHF during exacerbations associated with a worsening of PH [32]. The level of mPAP is a good prognostic indicator in COPD [32, 33] and in other categories of chronic respiratory disease, such as idiopathic pulmonary fibrosis [33] (the higher the degree of PH the worse the prognosis). It has been shown that life expectancy is worse in patients with PH compared with patients without PH [32], and is particularly poor in patients with a severe degree of PH [33, 34]. The 5-year survival rate of COPD patients with PH (mPAP > 20 mmHg) is approximately 50 % [26, 33]. Long-term oxygen therapy (LTOT) significantly improves the survival of markedly hypoxemic COPD patients also suffering from PH. Accordingly, it can be expected that the prognosis of PH will improve with such a therapy in these patients. This hypothesis can be explained by the fact that mPAP is a good marker of both the duration and severity of alveolar hypoxia in these patients, assuming hypoxia is the causing factor for the PH [14].

15.3 Vascular Alterations

15.3.1 The Process of Remodeling

Vascular changes can occur in smokers without COPD, can precede the development of emphysema in animal models, and can be seen in COPD patients. The respective vessels, especially in end-stage COPD, have thickened walls or are even occluded as a result of remodeling. Vascular remodeling is characterized by thickening of the arterial wall by which the vascular lumen and the internal diameter are reduced, leading to increased resistance and higher intravascular pressure. Of interest, although apparent in vessels of different sizes, muscular arteries and arterioles (small vessels with a diameter $<500\ \mu m$) are predominantly affected [35, 36]. Pulmonary vascular remodeling has been observed at different degrees of disease severity. Indeed, this phenomenon could not only be seen in patients with mild and severe COPD but also in heavy smokers with normal lung function [37]. The major event is hyperplasia of the intima [35, 36], but the other vessel wall layers, the media and the adventitia, are also involved [38]. Intimal hyperplasia results from the proliferation of α-smooth muscle actin- and vimentin-positive cells (corresponding to smooth muscle cells [SMC] associated with elastic and collagen fiber deposition) [23]. Interestingly, some SMCs in the intima express vimentin but not desmin filaments [23]. The presence of these filaments facilitates the discrimination between differentiated and nondifferentiated SMCs [39, 40]. "Accordingly, vimentin-positive, desmin-negative SMCs represent a subpopulation of less differentiated SMCs that may possess synthetic capacity and take part in an ongoing process of vascular remodeling" [9]. Although not fully understood, the occurrence of these not fully differentiated cells could be explained by (i) differentiation from resident precursor cells; (ii) attraction and differentiation of circulating bone-marrow-derived progenitor cells; (iii) dedifferentiation of mature SMCs from the media which migrate to the intima [41]; and (iv) transdifferentiation of endothelial cells (ECs) to SMCs by endothelial-to-mesenchymal transition [42]. The underlying mechanisms are still not fully resolved and have to be deciphered in future. Interestingly, "bone-marrow-derived progenitor vascular cells might exert a dual, opposite effect, contributing to vascular repair through differentiation into ECs or to vessel remodeling through differentiation into SMC". [42].

The respective pathologic picture can be explained by remodeling and angiogenesis induced by inflammation that is related to chronic infection, hypoxia, repeated stretching of hyperinflated lungs, and toxicity of cigarette smoke [43, 44].

Chronic hypoxia at high altitudes can cause PH, but the associated PH is completely reversible a few weeks after return to sea level [11]. This finding may be explained by the observation that PH caused by high altitude primarily induces medial hypertrophy. The involvement of all vessel layers in PH associated with COPD likely explains why PH in COPD is often not reversible by supplemental oxygen, neither acutely [45] nor chronically [46].

15.3.2 Mechanistic Insights into the Remodeling Process

Hypoxemia/hypoxia, chronic inflammation, oxidative and nitrosative stress, endothelial dysfunction, apoptosis, and altered proliferation have all been suggested as factors in the pathogenesis of airway and vascular remodeling [6, 9, 47–50]. Classically, hypoxia has been considered to be the major pathogenic factor for the development of PH in COPD. However, recent studies have shown that pulmonary vascular remodeling and endothelial dysfunction can occur in animal models of lung emphysema [19, 20], in patients with mild COPD not suffering from hypoxemia, and in smokers with normal lung function [23, 35, 51]. In addition, LTOT is not able to reverse PH in many COPD patients [52]. Nevertheless, there are many studies showing that hypoxia plays a role in COPD, at least in severe forms of the disease. Thus, mechanistic aspects related to hypoxia will be discussed. Furthermore, below we give an overview of the most relevant molecules and associated phenomena in the context of COPD and vascular remodeling/PH.

15.3.3 Hypoxia as a Causing Factor for Vascular Remodeling

In chronic respiratory diseases, PH is a result of increased vascular resistance (PVR), whereas cardiac output is often normal. The factors leading to the increase in PVR are manifold [28, 53] but hypoxia secondary to COPD is thought to be a predominant factor [28, 54, 55]. In terms of hypoxia, two mechanisms should be considered: acute hypoxia causing pulmonary vasoconstriction, and chronic hypoxia leading to structural vascular changes (remodeling) over time. Acute hypoxia induces a rise of PVR and mPAP which is part of hypoxic pulmonary vasoconstriction (HPV). Chronic alveolar hypoxia causes precapillary PH in healthy people living in high altitude, similar to that observed in COPD patients. Additionally, the morphological changes of the pulmonary vascular bed (remodeling) are comparable with those of COPD patients with PH (muscularization of pulmonary arterioles, intima thickening in muscular pulmonary arteries and arterioles) [56], although under hypoxic conditions the media is mainly involved in the remodeling.

With regard to hypoxia, it makes sense to distinguish between generalized hypoxia (e.g. seen in high altitude) due to the low partial pressure of oxygen, and localized, regional alveolar hypoxia [49]. Generalized hypoxia has both systemic and organ-specific effects [57, 58]. It results in peripheral vasodilation, general pulmonary vasoconstriction, and activation of a sympathico-adrenergic stress response to increase cardiac output [59]. Moreover, activation of bone marrow and erythropoietin-stimulated red-cell production occurs. Increased cardiac output and polycythemia are adaptive responses to improve blood and oxygen delivery to the hypoxic tissues. In contrast, local alveolar hypoxia occurs by regional lack of alveolar ventilation, e.g. caused by airway obstruction or even in pulmonary edema. HPV here adapts the local blood perfusion to the alveolar ventilation situation, preventing hypoxemia.

Hypoxia occurring within tumors is critical for tumorigenesis caused by the elevated demand for blood and nutrients of the rapidly growing cancer cells that are supplied by structurally abnormal tumor vessels. A similar impairment of the microenvironment oxygenation can occur in the lung because of parenchyma destruction and compression or loss of small vessels. In tumors and in the damaged and remodeled lung, the systemic arterial blood partial pressure of oxygen is usually, normal whereas the tissue microenvironment can be hypoxic.

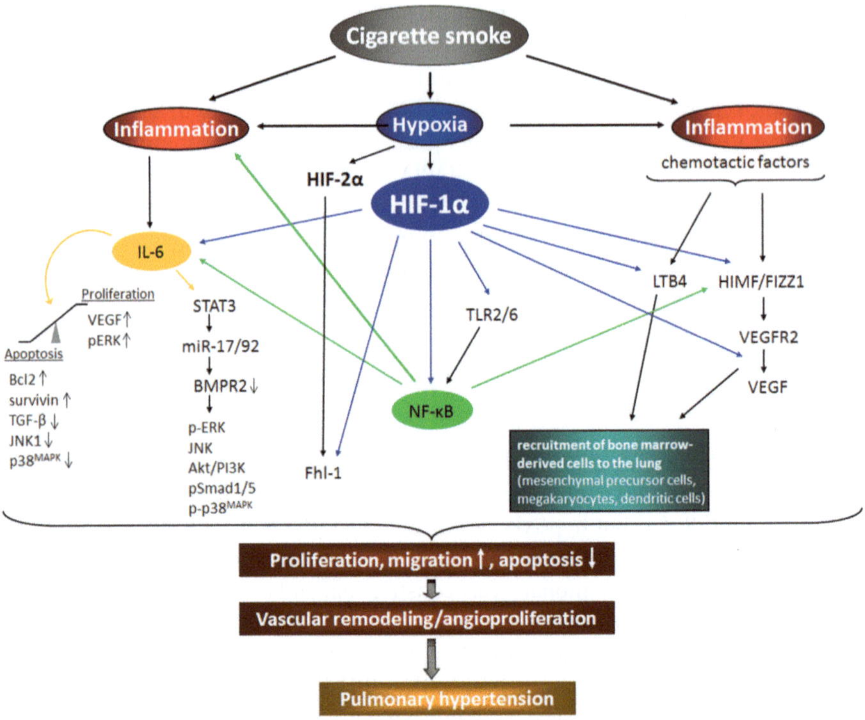

Fig. 15.1 Selection of possible hypoxia-dependent mechanisms contributing to pulmonary vascular remodeling in chronic obstructive pulmonary disease-pulmonary hypertension *(COPD-PH)*. Cigarette smoke-induced airway obstruction and emphysema can result in hypoxia. Hypoxia, amongst others, activates hypoxia-inducible factor *(HIF)-1α*, which can trigger pathways associated with inflammation, the recruitment of bone-marrow-derived cells, and alterations in proliferation/apoptosis balance of vascular endothelial and smooth muscle cells *(SMCs)*. Increased proliferation of SMCs causes narrowing of the vessels, resulting in pulmonary hypertension. All important acronyms of the molecules are explained in the text

Discovery of the oxygen-dependent transcription factor 'hypoxia-inducible factor-1α' (HIF-1α) [60] and subsequent discovery of the closely related HIF-2α was a breakthrough for the understanding of the molecular response of cells to hypoxia (Fig. 15.1, 15.2). HIF controls a variety of hypoxia-dependent genes that are involved in protection from hypoxia or pulmonary vascular remodeling, such as erythropoietin, glucose transporters, vascular endothelial growth factor (VEGF), endothelin-1 (ET-1) and nitric oxide (NO) synthases. HIF-1α and HIF-2α can also link hypoxia and inflammation [61–65]. It has been shown that within 24 h of hypoxia, inflammatory cells accumulate in the lungs, as measured by increased activity of myeloperoxidase. Subsequently, investigations at the molecular level showed that hypoxia-induced mitogenic factor (HIMF; also known as FIZZ1) is released by lung macrophages during stimulation with hypoxia. The proinflammatory and pro-angiogenic effects of HIMF are mediated by VEGF and its receptor VEGFR2 [66]. HIMF is also able to recruit bone-marrow-derived mesenchymal cells to the lung

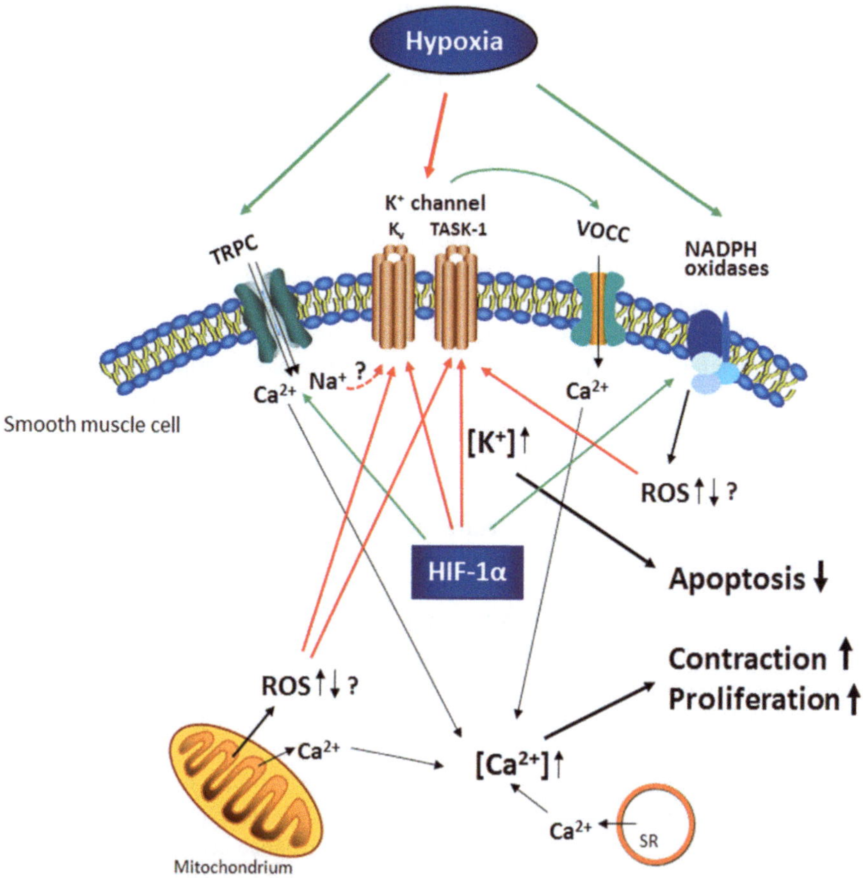

Fig. 15.2 Hypoxia-induced ion channel-mediated increase in proliferation, contraction, and decrease of apoptosis of smooth muscle cells *(SMCs)* contributing to pulmonary vascular remodeling. K_v and TASK-1 channels are downregulated and are less active after chronic hypoxia, leading to accumulation of K^+ within the cell (mediating apoptosis resistance) and membrane potential depolarization of the SMCs. This depolarization causes opening of voltage-operated Ca^{2+} channels *(VOCC)*, especially L-type channels, which mediate Ca^{2+} entry. Hypoxia-dependent reactive oxygen species *(ROS)* regulation derived from nicotinamide adenine dinucleotide phosphate *(NADPH)* oxidases and/or mitochondria is suggested to inhibit the K^+ channels, although it is unclear whether an increase or decrease of *ROS* occurs in hypoxia. Transient receptor potential channel *(TRPC)*-mediated Ca^{2+} or Na^+ influx (speculatively by gating K^+ channels) was also shown to be essential for the intracellular Ca^{2+} increase in at least acute hypoxia. Ca^{2+} release from mitochondria and sarcoplasmic reticulum *(SR)* was shown to additionally increase Ca^{2+} within the cell. Thus, mediated contraction and proliferation of the SMCs can contribute to vascular remodeling. *Colored arrows* depict either activation *(green)* or inhibition *(red)*

in mice [67]. Interestingly, hypoxia can also induce toll-like receptors (TLR) 2 and 6 via HIF-1α [68], which in turn can activate nuclear factor-κB (NF-κB), another proinflammatory transcription factor [69, 70]. The excess of chemotactic factors such as VEGF, HIMF, leukotriene B4 (LTB4), and others induced by hypoxia in the lung can mobilize bone-marrow-derived cells (including mast cells, mesenchymal precursor cells, megakaryocytes, dendritic cells) and their chemotaxis to the lung. Consequently, hypoxic/hypoxemic inflammation is a systemic response [49].

There are many cytokines mediating inflammation but, in the context of PH, special attention was given to HIF-1α-induced interleukin (IL)-6 because its messenger RNA (mRNA), as well as protein levels, were upregulated by chronic hypoxia in mice. Although IL-6-deficient mice showed less inflammation under hypoxic conditions, chronic hypoxic pulmonary hypertension (CHPH) was, however, only marginally reduced [71]. In contrast, IL-6 overexpression, specifically in the lungs, showed enhanced muscularization of small pulmonary vessels and PH under normoxic conditions, indicating that IL-6 may contribute to vascular remodeling in lungs. Of interest, overexpression of IL-6 under hypoxia resulted in severe PH and vascular remodeling up to obliteration [72]. Taken together, IL-6 is upregulated after chronic hypoxia in mouse lungs, but does not seem to be essential for CHPH development. Moreover, the combination of excessive IL-6 and chronic hypoxia seems to change the mode of vascular remodeling towards angioproliferation [49]. Based on recent results, there are currently two theories put forth to explain this observation. The first is that IL-6 is able to exert its effects through modulation of 'bone morphogenetic protein receptor type 2' (BMPR2) through a signal transducer and activator of transcription 3 (STAT3)-microRNA cluster 17/92 pathway [73]. BMPR2 is a member of the transforming growth factor (TGF)-β superfamily of growth factor receptors. It is expressed ubiquitously and is a member of many different signaling pathways, including pERK, JNK, Akt/PI3K [74, 75], pSmad1/5 [76], and p-p38MAPK [77]. Mutations in the BMPR2 gene are shown to be linked to PH development [78], and, moreover, BMPR2 expression is reduced in the pulmonary vasculature in primary PH patients [79]. The other prevailing theory is based on the observation that IL-6 can influence the balance between apoptosis and proliferation in pulmonary arterial SMCs (PASMC) and ECs (PAECs), leading to vascular remodeling. Overexpression of IL-6 induces the angioproliferative growth factor VEGF and intracellular extracellular signal-regulated kinase (ERK), resulting in increased proliferation. Simultaneously, IL-6 downregulates the growth inhibitor TGFβ and proapoptotic mitogen-activated protein kinases (MAPKs) [JNK1, p38MAPK], and upregulates the inhibitors of apoptosis Bcl-2 and survivin, leading to decreased apoptosis [72]. In conclusion, inflammatory cytokines, especially IL-6, can trigger vascular remodeling by influencing signaling pathways which lead to more proliferation and less apoptosis of PASMCs and PAECs (Fig. 15.1).

Furthermore, HIF-1α can induce PASMC depolarization by reduction of K^+ channel expression and activity that is associated with intracellular K^+ accumulation, and increase intracellular Ca^{2+} concentration and pH by upregulating transient receptor potential channels (TRPC) and Na^+/H^+ exchanger [80–83], (Fig. 15.2). It has been shown that the currents of voltage-gated K^+ channels (Kv) are decreased under chronic hypoxic conditions [84–86], most likely mediated by reactive oxygen species (ROS) derived from mitochondria [87–91] and/or nicotinamide adenine dinucleotide phosphate (NADPH) oxidases, such as NOX4 [92], and can be associated with an influx of calcium via TRPC [93]. With regard to ROS in mediating hypoxic responses, there is current controversy as to whether they are up- or downregulated during hypoxia [94, 95].

Moreover, the voltage-independent two-pore-domain K^+ channel, TWIK-related acid-sensitive K^+ channel (TASK)-1 is also inhibited by hypoxia, leading to membrane depolarization and calcium entry through L-type channels [96, 97]. Interestingly, it was recently shown that the inhibition of TASK-1 is also mediated by an ET-1-dependent mechanism [98]. In addition to such channels, L-type Ca^{2+} channels are shown to be a major cellular Ca^{2+} entry pathway [99, 100], (Fig. 15.2).

All these ion alterations are associated with more contractile, apoptosis-resistant, proliferative, and migratory PASMCs under hypoxic conditions [80–83, 100]. The role of HIF-2α in this context is still not clear.

Another recently discovered key player for the development of hypoxia-induced but also nonhypoxia-induced PH, is Fhl1, a protein known to be involved in muscle development (Fig. 15.1). It could be demonstrated that HIF-1α, as well as HIF-2α, induced Fhl-1 expression not only in different animal models of PH but also in human patients with idiopathic pulmonary arterial (PA) hypertension [101]. The increase in Fhl-1 causes elevated proliferation and migration of PASMCs, contributing to vascular remodeling [101].

15.3.4 Hypoxia-Independent Mechanisms Leading to the Development of COPD and PH

Hypoxia has long been thought to be the primary driving force for the development of PH in COPD supported by studies showing a close relationship between mPAP and/or pulmonary resistance and alveolar hypoxia [12, 102, 103]. However, evidence that the causal factors can be hypoxia-independent and more complex is severalfold. (i) It has been shown that oxygen therapy is not able to fully reverse PH in COPD [46, 52]. (ii) Histological investigation in lungs from COPD patients with PH revealed involvement of all vessel layers characterized by prominent intimal thickening, medial hypertrophy, and muscularization of small arterioles [36, 104], whereas hypoxia-induced vascular remodeling is mainly restricted to the vessel media. This finding is supported by the fact that these pulmonary vascular alterations also occurred in nonhypoxic patients with mild airflow obstruction and smokers without any parenchymal disorder, suggesting that vascular changes may be driven by mechanisms independent of hypoxia/hypoxemia [23]. (iii) Consistent with the findings described in humans, a recent study demonstrated that in mice chronically exposed to tobacco smoke, pulmonary vascular remodeling and PH preceded the development of emphysema, and these changes were independent of hypoxia; there was neither hypoxia in the smoking chamber nor hypoxemia in the mice [20]. In addition, this study showed that the expression of genes involved in important pathways associated with PH and COPD, such as apoptosis, proliferation, oxidative stress, extracellular matrix, and inflammation, followed a completely different pattern compared with that observed in chronic hypoxia-induced vascular remodeling, indicating that the smoke- and hypoxia-driven mechanisms were different [20]. The combination of cigarette smoke and hypoxia can have synergistic effects in terms of affecting the vasculature. Guinea pigs exposed to cigarette smoke and subsequently to hypoxia showed a stronger elevation of pulmonary artery pressure and a more pronounced remodeling in small vessels compared with exposure to only one of these stimuli [105].

Additional animal studies reported a direct effect of cigarette smoke on the parenchyma and vasculature. Cigarette smoke can increase the expression of vasoactive mediators in pulmonary arteries [106], affecting the gene expression in pulmonary arteries [107]. The exposure of guinea pigs to chronic cigarette smoke induced emphysema, which was associated with a reduced lung capillary density [108]. It was reported that cigarette smoke extract (CSE) can induce ET-1 in pulmonary artery ECs [109] and reduce prostacyclin synthase expression [110]. Moreover, CSE is able to induce the production of superoxide in ECs, which in turn reacts with NO to peroxynitrite [111]. This radical can inactivate VEGFR2 signaling [112], which is important for EC maintenance and growth. Of interest, CSE-induced EC apoptosis via p53 [113] can be prevented by the phosphodiesterase (PDE)-5 inhibitor sildenafil [114], which leads to increased levels of cyclic guanosine monophosphate (cGMP), an important second messenger. This finding suggests that the cGMP pathway is involved in this context and decreased after smoke exposure.

15.4 Impairment of the Endothelium: Endothelial Dysfunction

New advances regarding the pathogenesis of PH in COPD suggest that an endothelium-derived vasoconstrictor-dilator imbalance caused by endothelial dysfunction associated with decreased expression or uncoupling of endothelial NO synthase (eNOS) could be a driving force for PH development (Fig. 15.3). In this context, expression of VEGF and serotonin transporters seem to be increased [37, 38, 115, 116]. ECs are important for the regulation of vascular homeostasis [117]. They reduce the vascular tone [118] and regulate pulmonary vessel adaptation to increases in blood flow [119] and hypoxia [120, 121]. Endothelial dysfunction has been reported in patients with end-stage COPD who had undergone lung transplantation [115], as well as in patients with mild-to-moderate COPD [37]. The endothelial function is influenced by the expression of vasoreactive mediators controlling vasoconstriction (also proproliferative for SMCs) or vasodilation (also antiproliferative for SMCs). On the one hand, it has been shown that the vasoconstrictive protein ET-1 was increased in patients with primary and secondary PH, including COPD patients [122] and, on the other hand, vasodilative mediators such as eNOS [123, 124] and prostacyclin synthase (PGI2-S) [110] were shown to be downregulated in pulmonary arteries of patients with severe COPD. CSE and/or acrolein (a potent αβ-unsaturated aldehyde contained in cigarette smoke) can decrease the expression of PGI2-S in human pulmonary artery ECs (HPAECs) [110], supporting the hypothesis that its downregulation in COPD patients arises directly from the cigarette smoke ingredients. After exposure to cigarette smoke for 8 months, mice developed

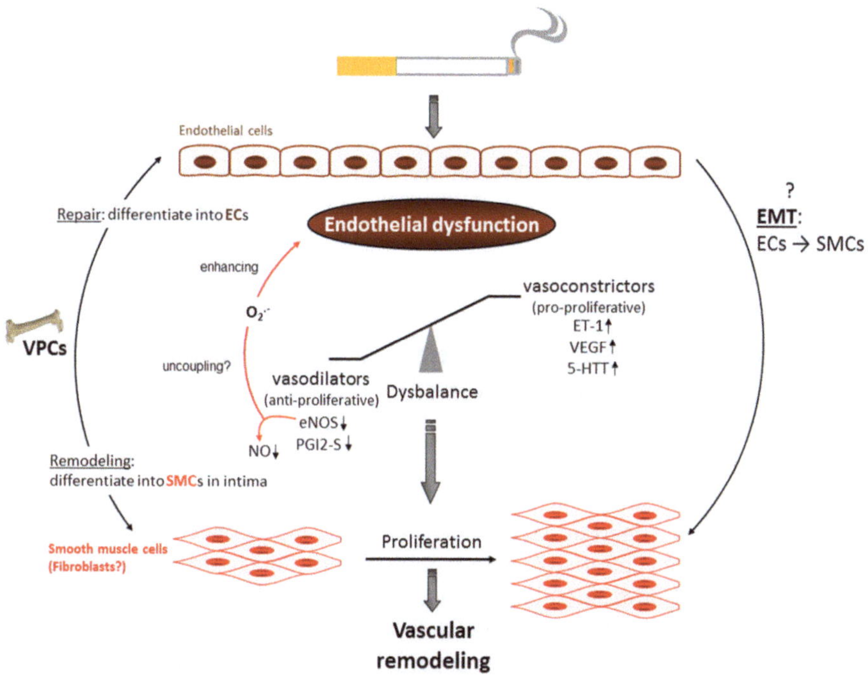

Fig. 15.3 Endothelial dysfunction as a causing factor for the development of vascular remodeling. Cigarette smoke and inflammatory mediators can cause endothelial dysfunction which is triggered by **a** a disbalance of vasodilative and vasoconstrictive molecules towards an excess of vasoconstrictors, and **b** damage/dysregulation of *EC* signaling. Additionally, vascular progenitor cells *(VPCs)* are attracted to the damaged endothelium. Such *VPCs* can either contribute to repair by differentiation into *ECs* or to remodeling by differentiation into *SMCs*. Furthermore, an endothelial-to-mesenchymal transition *(EMT)* may occur, resulting in an SMC phenotype. Vasoconstriction and altered endothelial cell signaling are stimuli for smooth muscle cells *(SMCs)* to proliferate, resulting in vascular remodeling, increased pulmonary artery pressure and, finally, pulmonary hypertension

emphysema and PH, as well as concomitant downregulation of eNOS in lungs and vessels, supporting the observation in humans [20]. Of interest, eNOS-deficient mice developed emphysema and PH upon exposure to cigarette smoke, whereas inducible NO synthase (iNOS)-deficient mice did not [20].

15.5 Inflammatory Cells

Systemic inflammation is a known phenomenon in COPD [125, 126]. Thus, inflammation could also contribute to the pathogenesis of PH in patients with COPD. In fact, the degree of pulmonary vascular remodeling correlates with the amount of inflammatory cells seen in small airways [14, 51]. Compared with nonsmokers, COPD patients show increased numbers of inflammatory cells invading the adventitia of pulmonary muscular arteries [127]. These cells predominantly consist of activated T lymphocytes, especially CD8+T cells [115, 127], which are increased in the arterial adventitia of smokers with normal lung function. The ratio of CD4+/CD8+ is reduced compared with nonsmokers, and is comparable to the situation in patients with mild-to-moderate COPD [127]. The correlation of IL-6 (an inflammatory mediator) expression and elevations in mPAP supports a possible role of inflammation in the pathogenesis of PH (in COPD) [128].

15.6 Oxidative and Nitrosative Stress and the Influence in Vascular (Patho-)Physiology

Evidences for oxidative stress in COPD patients are numerous [7, 129]. Oxidative stress is the result of an imbalance between oxidants and antioxidants in favor of oxidants. Cigarette smoke itself contains high concentrations of ROS [130]. Elevated concentrations of H_2O_2 and 8-isoprostane (oxidative stress markers) can be found in exhaled breath condensate of smokers and ex-smokers, as well as during exacerbations [131–133]. ROS can negatively influence the function of antiproteases, such as α1-antitrypsin and SLPI. This negative influence leads to a protease/antiprotease imbalance accelerating the degradation of elastin in the lung parenchyma, resulting in emphysema [134]. In addition, ROS play an important role in the vasculature. Traditionally, macrophages have been considered as the major ROS source, even in vessel walls [135]. However, several studies showed that all vascular cells (endothelial, smooth muscle, adventitial cells) produce ROS in different amounts, depending on the stimuli to modulate cellular function [136].

ROS, for example produced by macrophages, are normally involved in the elimination of pathogens and as such are critical to the organism. ROS and reactive nitrogen species (RNS) are able to cause protein modification and DNA damage [137]. Indeed, NO, generated by NO synthases, reacts with superoxide (O_2•–) to form the potent oxidant peroxynitrite (ONOO) [138]. This peroxynitrite can react with tyrosine residues from proteins to form nitrotyrosine, shown to be increased in COPD patients [47, 48]. In the vasculature, ROS play an important physiological role, participating as second messengers in endothelial function, in smooth muscle and EC growth and survival, and in the process of remodeling of the vessel wall [139]. Under pathophysiological conditions, these responses are uncontrolled and imbalanced, respectively [140–142]. The main vascular ROS is the superoxide anion, which can inactivate NO, as mentioned above, by generating peroxynitrite. Since NO is an important vasodilator, the reaction with ROS influences the vascular tone [143, 144]. Superoxide can be converted to H_2O_2 by superoxide dismutase (SOD), which is a more stable ROS. The depletion of H_2O_2 is carried out by catalase and glutathione peroxidase producing H_2O. Hydrogen peroxide and other peroxides regulate growth-related signaling in vascular SMCs and inflammatory responses in vascular lesions [141, 145]. ROS can activate several vascular signaling cascades, such as ERKs and MAPKs, which are important in cell growth, proliferation, apoptosis, and differentiation. Furthermore, receptor and nonreceptor tyrosine kinases (shown to be involved in cardiovascular remodeling and damage), and protein tyrosine phosphatases and transcription factors such as NF-κB and AP-1, which induce vascular inflammation [146, 147], are affected by ROS. Several studies reported the role of ROS in growth processes contributing to vascular remodeling and injury. Angiotensin II can lead to NADPH oxidase-generated superoxide production, mediated by the angiotensin type I (ATI) receptor, which is converted to H_2O_2 by SOD acting as a second messenger that results in hypertrophy or hyperplasia of vascular SMCs [148, 149]. This angiotensin II-induced reaction can be inhibited by the flavoprotein inhibitor diphenyleneiodonium (DPI) [148], catalase [149], and knockdown of $p22^{phox}$ [150], supporting the involvement of NADPH oxidases in the vasculature.

ROS are also able to promote vascular remodeling by increasing deposition of extracellular matrix proteins. Collagen and elastic fibers can be degraded by proteinases, especially matrix metalloproteinases (MMPs). These MMPs are secreted by macrophages and vascular SMCs in an inactive form [151]. ROS, for example peroxynitrite, activate MMP2 and 9 in human SMCs, leading to degradation of the basement membrane and elastin [152]. Use of a hypertension model of aldosterone-induced systemic oxidative stress revealed that ET-1-associated processes lead to vascular remodeling [153]. Redox-sensitive inflammatory processes can also contribute to vascular remodeling. Expression of the inflammation-related intracellular adhesion molecule-1 (ICAM-1) is elevated in the aorta from aldosterone-treated rats [153]. Furthermore, Liu et al. [154] reported that angiotensin II-induced oxidative stress caused tissue hypertrophy which was associated with increased ICAM-1 expression. Of interest, NADPH oxidases were involved in this context, producing ROS [154]. It is of note that not only the vascular but also the phagocytic NADPH oxidase is involved in superoxide production in cardiovascular diseases because monocytes and lymphocytes can infiltrate cardiovascular tissues [135].

Taken together, elevated occurrence of superoxide, H_2O_2, and reduced NO bioavailability by reaction with superoxide forming peroxynitrite can contribute fundamentally to vascular remodeling and emphysema development.

15.7 Recent Advances in Molecular Mechanisms of COPD Associated with Vascular Remodeling/PH

15.7.1 Inducible Nitric Oxide Synthase as a Key Player for the Development of Cigarette Smoke-Induced PH and Emphysema

A recent study identified the major NO source in the context of smoke-induced emphysema and PH in mice [20]. Cigarette smoke led to an upregulation of the iNOS mRNA and protein expression, predominantly in small pulmonary vessels, and was associated with increased NO generation. Interestingly, eNOS was downregulated when the disease was established, in mice as well as in COPD patients. It is most probable that the vasodilative NO effect did not appear because of the simultaneous abundance of ROS from both external (cigarette smoke) and internal sources. This study suggested that the subsequent formation of peroxynitrite, most probably increased in this context, had proapoptotic effects (as well as antiproliferative effects) on alveolar epithelial cells type II (AECII) and ECs, promoting emphysema development, vessel loss, and vascular remodeling. Moreover, the level of nitrotyrosine was increased in those mice as well as in smokers without COPD and patients with severe COPD. It is likely that the downregulation of eNOS was associated with uncoupling of this enzyme, leading to the switch of NO to superoxide production, which increased the oxidative stress. Mice deficient in iNOS but not eNOS were protected from vascular remodeling, PH, and emphysema. Treatment with an iNOS inhibitor (L-NIL) could not only prevent disease development but also promote lung regeneration in mice previously exposed to cigarette smoke for 8 months.

Moreover, it has been shown that the development of PH was mediated by iNOS-carrying bone-marrow-derived cells, whereas emphysema development was dependent on iNOS in nonbone-marrow-derived cells [20] (Fig. 15.4). This phenomenon shows that PH and emphysema can develop independently of each other, and could also explain the discrepancy that not all COPD patients also suffer from PH, if such results are transferable to the human situation.

Quite recently, it could be demonstrated in mice and guinea pigs that stimulation of the soluble guanylate cyclase (sGC) could not only prevent the cigarette smoke-induced development of vascular remodeling but also emphysema [155]. The sGC is an enzyme that uses iNOS-generated NO to produce cGMP from guanosine triphosphate (GTP). cGMP acts as a second messenger, amongst others mediating vasodilation, which can decrease the vascular pressure. The combination of the iNOS-related findings [20] with this study demonstrates the important role of the NO–sGC–cGMP axis for the physiology and pathophysiology of the pulmonary vasculature. It has already been suggested that dysregulation of this axis contributes to pulmonary diseases and PH [156–158]. In line with previous findings [20], prevention from vascular remodeling was associated with prevention from emphysema, although causality was not investigated in these studies.

15.7.2 Reactive Oxygen Species as a Trigger for Vascular Remodeling

Although cigarette smoke itself produces ROS, causing oxidative stress, the additional internal ROS sources produced by the organism are still not fully resolved. Depending on the circumstances, several sources such as NADPH oxidases, mitochondria, xanthine oxidase, cyclooxygenases, lipooxygenases, and uncoupled eNOS come into consideration. As mentioned before, uncoupled eNOS can contribute to oxidative stress in terms of COPD [159–163]. Based on experimental and clinical

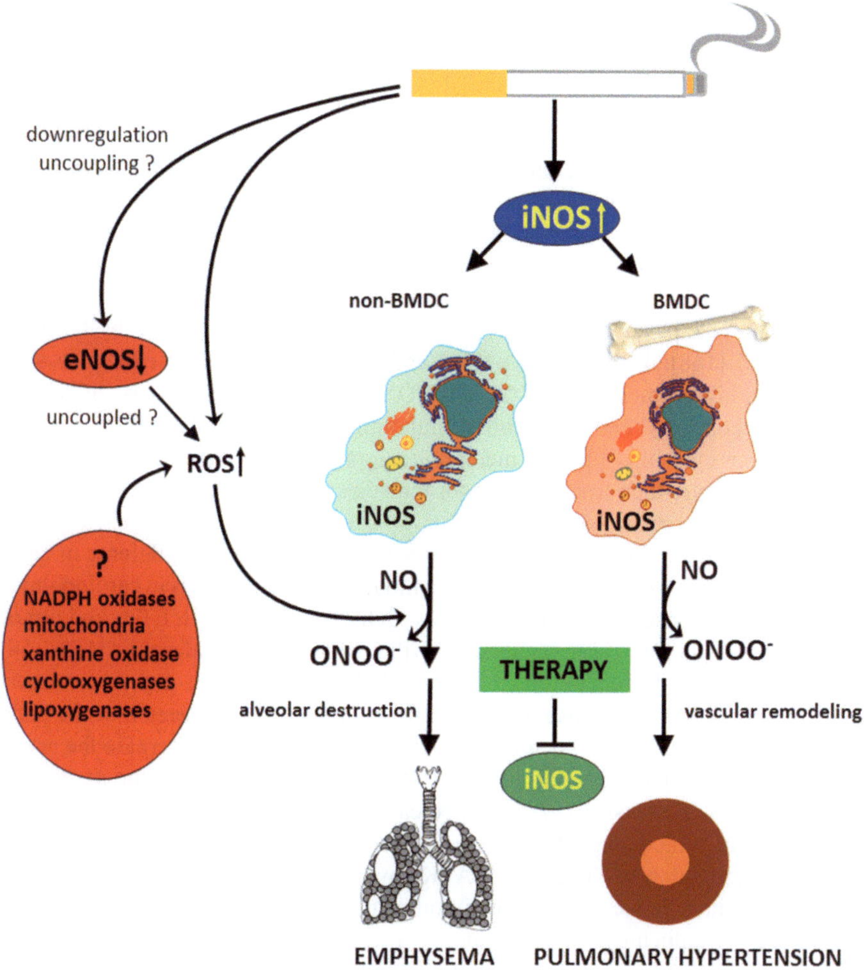

Fig. 15.4 Identification of inducible nitric oxide synthase *(iNOS)* as an essential factor for the development of cigarette smoke-induced emphysema and pulmonary hypertension *(PH)* in mice. Cigarette smoke-mediated upregulation of *iNOS* leads to excessive *NO* production. The formation of peroxynitrite, resulting from the reaction of *NO* with superoxide, was suggested to mediate emphysema and PH development. Superoxide can derive from cigarette smoke itself and/or from uncoupled endothelial *NOS (eNOS)*, nicotinamide adenine dinucleotide phosphate *(NADPH)* oxidases, xanthine oxidases, cyclo- and lipooxidases, and mitochondria. Of interest, *iNOS* generated by nonbone-marrow-derived cells *(NBMDC)*, possibly vascular cells, leads to lung destruction, resulting in emphysema, whereas elevated *iNOS* expression in bone-marrow-derived cells *(BMDC)* causes vascular remodeling. Treatment with the specific *iNOS* inhibitor L-NIL prevents or even reverses pathological alterations

studies, NADPH oxidases are suggested to be the predominant superoxide-producing enzymes in the context of oxidative stress in cardiovascular diseases [164]. The classical NADPH oxidase is expressed in phagocytes; it produces O_2•− as a defense mechanism against bacterial infections [135]. The intrinsic enzyme consists of membrane-bound catalytic subunits that produce the ROS and cytosolic subunits which regulate the function of the enzyme. In phagocytes gp91^{phox} (=NOX2; 'phox' = phagocytic oxidase) is associated with p22^{phox}, being essential for its function [165], and both are located in the membrane. They form a complex with the cytosolic subunits p40^{phox}, p47^{phox}, and p67^{phox}, as well as the G-protein Rac2 [166]. All these subunits, as well as the other membrane-bound enzymes NOX1, NOX4, and NOX5, and homologs to p47^{phox} (NOXO1) and p67^{phox} (NOXA1), are expressed in vascular cells. NOX5 seems to be active without any regulatory subunit. NOX4 is constitutively active, is associated with p22^{phox}, and its function appears to be independent of regulatory subunits, in contrast to NOX1 and NOX2, which need to be activated [167]. The activity of NOX1 is predominantly regulated by NOXO1 and NOXA1 [168]. Activation by p47^{phox} and p67^{phox} is also possible but is less effective than involvement of NOXO1 and NOXA1 [168, 169]. In the case of NOX2, p47^{phox} is phosphorylated upon cell stimulation, followed by complex formation with the other cytosolic subunits, mediating migration to the membrane-bound subunit. Electrons are then transferred from the substrate NADPH to O_2, resulting in superoxide (O_2•–) production [170].

In addition to inflammatory processes often related to immune defense or cigarette smoking, NADPH oxidases can be stimulated by many factors in cardiopulmonary vascular diseases. Vasoactive agonists such as angiotensin II, ET-1, and tumor necrosis factor-α (TNFα) can regulate NADPH oxidases in vascular cells [153, 154, 171]. Even genetic factors might be involved in NADPH oxidase-dependent superoxide production. For instance, polymorphisms in the promoter region of the gene encoding p22^{phox} have been identified. Such polymorphisms can increase the activity of the promoter in vascular SMCs [172], which were shown to be associated with essential hypertension [173] and decreased NO bioavailability [174].

In terms of remodeling, ROS can influence several intracellular signaling cascades, e.g. activation of ERKs and MAPKs, which affect cell growth and differentiation; protein tyrosine phosphatases and transcription factors, for instance NF-κB and AP-1, inducing proinflammatory genes associated with hypertension and atherosclerosis; and receptor and nonreceptor tyrosine kinases, which have been shown to be involved in cardiovascular remodeling and vascular damage [146].

Because of the fact that monocytes and lymphocytes are able to infiltrate cardiovascular tissues and pulmonary vessels, it is important to note that in cardiovascular diseases, such as COPD, not only vascular NADPH oxidase can contribute to ROS formation, and hence to vascular remodeling, but phagocytic oxidase can also contribute [175].

15.7.3 Neprilysin Downregulation Provokes Pulmonary Vascular Remodeling

A recent study demonstrated that neprilysin (neutral endopeptidase [NEP]) might be an important factor for the susceptibility of humans to pulmonary vascular remodeling in response to smoke inhalation and hypoxia [176] (Fig. 15.5). NEP is a trans-

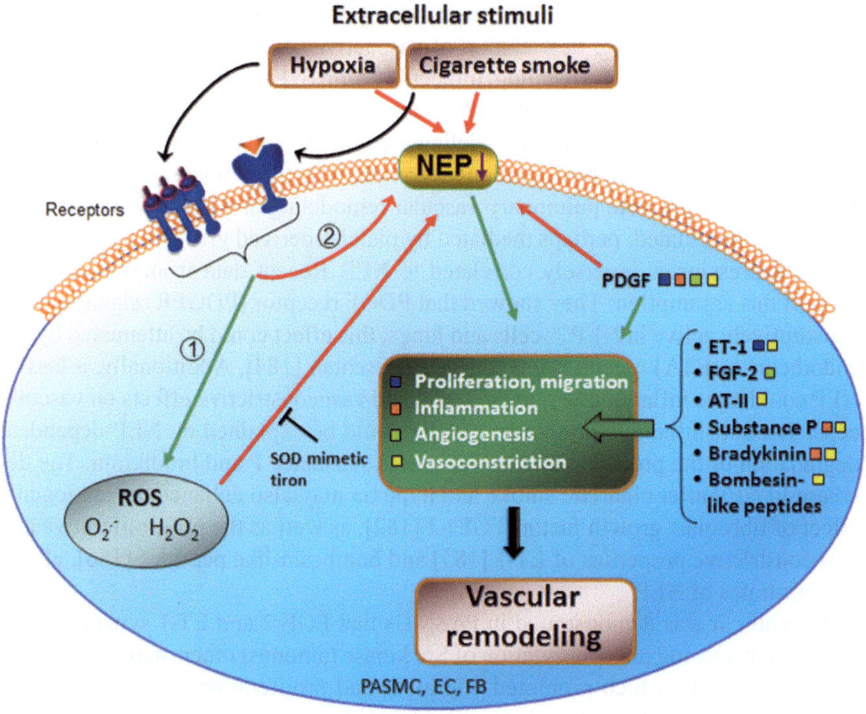

Fig. 15.5 Scheme of proposed neprolysin *(NEP)*-dependent mechanisms leading to vascular remodeling. Extracellular stimuli, such as hypoxia and cigarette smoke, activate pathways in vascular cells, causing downregulation of *NEP* expression and/or activity which was seen in patients with COPD associated with pulmonary hypertension *(PH)*. Stimuli can have direct negative effects on *NEP*, and also indirectly by *1)* increasing reactive oxygen species *(ROS)* and platelet-derived growth factor *(PDGF)*; and *2)* activation of other pathways, in part receptor-mediated. *NEP* downregulation leads to elevated proliferation, migration, inflammation, angiogenesis, and vasoconstriction mediated, amongst others, by depicted molecules, causing vascular remodeling. The *ROS* effect on *NEP* can be inhibited by the superoxide dismutase *(SOD)* mimetic tiron. *Colored boxes* indicate involvement in the respective pathway; *red arrows* represent inhibition/downregulation; *green arrows* represent activation/upregulation. *ET-1* endothelin-1, *FGF-2* fibroblast growth factor, *AT-II* angiotension-II, *PASMC* pulmonary artery smooth muscle cells, *EC* endothelial cells, *FB* fibroblasts

membrane zinc peptidase that is widely expressed, including in PASMCs, PAECs, and fibroblasts [177]. Its activity and expression is decreased by cigarette smoke [178], hypoxia [179, 180], and ROS [181]; it is particularly important in vascular SMCs in terms of remodeling [176]. A genetic depletion of NEP in mice resulted in increased severity of PH associated with stronger proliferation of PASMCs compared with wild-type mice. This finding suggests a protective role of NEP against PH, partly by suppressing proliferation and migration of PASMCs [180].

Interestingly, Wick et al. [176] showed that NEP expression was decreased in lungs from COPD patients with PH and non-COPD PH patients, especially in the distal vasculature where prominent remodeling occurs. NEP is involved in many peptidase-dependent (e.g. degradation of vasoactive neuropeptides) and -independent (e.g. interaction of signaling molecules with NEP's intracellular cytosolic subunit) signaling pathways [182, 183], but its endopeptidase activity-dependent pathways influencing vascular remodeling are not yet well understood. Wick et al. postulated that increased proliferation/migration of dedifferentiated SMCs or myofibroblasts may promote pulmonary vascular remodeling and PH if NEP is less active or downregulated, perhaps mediated by platelet-derived growth factor (PDGF) whose expression is inversely correlated to NEP. Recent data from Karoor et al. support this assumption. They showed that PDGF receptor (PDGFR) signaling was constitutively active in $NEP^{-/-}$ cells and lungs; this effect could be attenuated by the endothelin A (ETA) receptor antagonist ambrisentan [184]. Additionally, a loss of NEP could have inflammatory, angiogenic, and vasoconstrictive effects on vascular cells. The anti-inflammatory action of NEP could be explained by NEP-dependent degradation of the proinflammatory substrates substance P and bradikinin. The decrease in NEP after cigarette smoke and hypoxia may also enhance the angiogenic effect of fibroblast growth factor (FGF)-2 [182], as well as the proproliferative and vasoconstrictive properties of ET-1 [185] and bombesin-like peptides [186], all being substrates of NEP.

Recently, it was demonstrated in PASMCs that FGF-2 and ET-1 synergize with PDGF in increasing phosphorylation of Src kinase (amongst others activating PDGFR) and PDGFR, which promoted migration and proliferation of the cells [184]. The phosphatase PTEN (phosphatase and tensin homolog) also plays an important role in vascular biology. The loss of PTEN results in PH [187]. It is inactivated by phosphorylation (mediated by Src and PDGFR) and downregulated in NEP-deficient PASMCs. This downregulation could be rescued by NEP overexpression in NEP null cells or by downregulation of Src or PDGFR by small interfering RNA (siRNA). These observations suggest that NEP-dependent mechanisms protect against the inactivation of PTEN [184]. Moreover, NEP can be inactivated by ROS, as shown by decreased activity in the presence of H_2O_2 and improved NEP activity when an antioxidant, the SOD mimetic tiron, was present [176].

Of interest, early studies suggest that inhibition of NEP could be beneficial for the treatment of PH [188, 189]. This proposal is based on the fact that NEP can inactivate atrial/brain natriuretic peptides (ANP/BNP), which promote vasodilation by increasing cGMP mediated via natriuretic peptide receptor-A (NPR-A) [190]. cGMP-dependent protein kinase (PKG), cGMP binding PDEs, and cyclic nucleotide-gated ion channels bind cGMP, in which PKG seems to be the main mediator of cGMP signals [191–193]. Binding of ANP/BNP-induced cGMP activates PKG, leading to the catalytic transfer of phosphate from ATP to target proteins. These phosphorylated proteins translate the extracellular stimuli into specific biological outputs [194], such as vasodilation.

It has been shown that NEP antagonists, alone and in combination with angiotensin-converting enzyme (ACE) and endothelin-converting enzyme (ECE) inhibitors, were able to improve cardiac function, decrease systemic blood pressure, and limit cardiac hypertrophy [195–198]. Nevertheless, side effects were observed if single

NEP inhibitors or dual inhibitors (NEP/ACE or NEP/ECE) were used. Nowadays, even triple vasopeptidase inhibitors (NEP/ACE/ECE) are under investigation, with promising preliminary results showing fewer side effects; in particular, the increase in ET-1 can be reversed by simultaneous application of an ECE inhibitor [196]. These combination therapies and the different mechanisms of NEP, ACE, and ECE inhibition have been extensively reviewed by Daull et al. [196].

However, recent observations support the possibility that NEP could be protective against PH [176, 180, 199]. One explanation for the discrepancy between the beneficial cardiac effects and harmful pulmonary effects of NEP inhibition is the well-known phenomenon that the pulmonary and systemic circulations usually respond to hypoxia (a major stimulus for PH) in opposite ways: pulmonary vessels contract to redirect blood flow to better oxygenated areas of the lung, whereas systemic vessels dilate to increase the flow of oxygenated blood to areas of tissue hypoxia or ischemia [180].

In conclusion, in terms of the lung, it is suggested to increase NEP to treat PH, whereas cardiac NEP inhibition could be used for treatment of hypertrophy and improvement of cardiac function. This dichotomy clearly demonstrates that a possible treatment must be dependent not only on the target itself but also on the localization of the target. These data should be kept in mind when considering NEP-related drug therapies.

15.8 Conclusions

The mechanisms of COPD and PH in COPD are still far from being fully understood. Studies primarily from the last decade have shown that vascular remodeling and PH can occur in COPD, not only in severe cases but also in mild-to-moderate forms of the disease, or even in smokers without airflow limitations. Investigations of the molecular mechanisms of COPD and the occurrence of pulmonary vascular remodeling established that COPD associated with PH and pulmonary vascular remodeling is a complicated multifactorial disease involving hypoxia-related and hypoxia-unrelated mechanisms, inflammation, and endothelial dysfunction. Recent advances have changed the long-standing view of the pathobiology of COPD. In the past, vascular alterations (vascular remodeling and PH) have been seen to be secondary events occurring after destruction of the parenchyma, predominantly caused by hypoxia/hypoxemia. However, recent studies have demonstrated that such vascular abnormalities can be early events in COPD, preceding airflow limitations and emphysema, and can be independent of hypoxia. It has been shown, at least in mice, that PH and lung emphysema triggered by tobacco smoke can occur independently, and suggest that vascular molecular alterations can be a trigger for lung emphysema development.

Further elucidation of the contribution of pulmonary vascular changes to COPD development may help to identify new therapeutic concepts for this disease.

References

1. Agusti AG, Noguera A, Sauleda J, Sala E, Pons J, Busquets X. Systemic effects of chronic obstructive pulmonary disease. Eur Respir J. 2003;21(2):347–60.
2. Sertogullarindan B, Gumrukcuoglu HA, Sezgi C, Akil MA. Frequency of pulmonary hypertension in patients with COPD due to biomass smoke and tobacco smoke. Int J Med Sci. 2012;9(6):406–12.
3. Black PN, Ching PS, Beaumont B, Ranasinghe S, Taylor G, Merrilees MJ. Changes in elastic fibres in the small airways and alveoli in COPD. Eur Respir J. 2008;31(5):998–1004.
4. Barnes PJ. Chronic obstructive pulmonary disease. N Engl J Med. 2000;343(4):269–80.
5. Sharafkhaneh A, Hanania NA, Kim V. Pathogenesis of emphysema: from the bench to the bedside. Proc Am Thorac Soc. 2008;5(4):475–7.
6. Churg A, Cosio M, Wright JL. Mechanisms of cigarette smoke-induced COPD: insights from animal models. Am J Physiol Lung Cell Mol Physiol. 2008;294(4):L612–31.
7. Repine JE, Bast A, Lankhorst I. Oxidative stress in chronic obstructive pulmonary disease. Oxidative Stress Study Group. Am J Respir Crit Care Med. 1997;156(2 Pt 1):341–57.
8. Barbera JA, Peinado VI, Santos S. Pulmonary hypertension in chronic obstructive pulmonary disease. Eur Respir J. 2003;21(5):892–905.
9. Peinado VI, Pizarro S, Barbera JA. Pulmonary vascular involvement in COPD. Chest. 2008;134(4):808–14.
10. Wright JL, Levy RD, Churg A. Pulmonary hypertension in chronic obstructive pulmonary disease: current theories of pathogenesis and their implications for treatment. Thorax. 2005;60(7):605–9.
11. Naeije R. Pulmonary hypertension and right heart failure in chronic obstructive pulmonary disease. Proc Am Thorac Soc. 2005;2(1):20–2.
12. Scharf SM, Iqbal M, Keller C, Criner G, Lee S, Fessler HE. Hemodynamic characterization of patients with severe emphysema. Am J Respir Crit Care Med. 2002;166(3):314–22.
13. Cuttica MJ, Kalhan R, Shlobin OA, Ahmad S, Gladwin M, Machado RF, Barnett SD, Nathan SD. Categorization and impact of pulmonary hypertension in patients with advanced COPD. Respir Med. 2010;104(12):1877–82.
14. Oswald-Mammosser M, Weitzenblum E, Quoix E, Moser G, Chaouat A, Charpentier C, Kessler R. Prognostic factors in COPD patients receiving long-term oxygen therapy. Importance of pulmonary artery pressure. Chest. 1995;107(5):1193–8.
15. Minai OA, Chaouat A, Adnot S. Pulmonary hypertension in COPD: epidemiology, significance, and management: pulmonary vascular disease: the global perspective. Chest. 2010;137(6 Suppl):39S-51S.
16. Blanco I, Gimeno E, Munoz PA, Pizarro S, Gistau C, Rodriguez-Roisin R, Roca J, Barbera JA. Hemodynamic and gas exchange effects of sildenafil in patients with chronic obstructive pulmonary disease and pulmonary hypertension. Am J Respir Crit Care Med. 2010;181(3):270–8.
17. Pynnaert C, Lamotte M, Naeije R. Aerobic exercise capacity in COPD patients with and without pulmonary hypertension. Respir Med. 2010;104(1):121–6.
18. Orr R, Smith LJ, Cuttica MJ. Pulmonary hypertension in advanced chronic obstructive pulmonary disease. Curr Opin Pulm Med. 2012;18(2):138–43.
19. Ferrer E, Peinado VI, Diez M, Carrasco JL, Musri MM, Martinez A, Rodriguez-Roisin R, Barbera JA. Effects of cigarette smoke on endothelial function of pulmonary arteries in the guinea pig. Respir Res. 2009;10:76.
20. Seimetz M, Parajuli N, Pichl A, Veit F, Kwapiszewska G, Weisel FC, Milger K, Egemnazarov B, Turowska A, Fuchs B, et al. Inducible NOS inhibition reverses tobacco-smoke-induced emphysema and pulmonary hypertension in mice. Cell. 2011;147(2):293–305.
21. Wright JL, Churg A. Cigarette smoke causes physiologic and morphologic changes of emphysema in the guinea pig. Am Rev Respir Dis. 1990;142(6 Pt 1):1422–8.

22. Wright JL, Churg A. Effect of long-term cigarette smoke exposure on pulmonary vascular structure and function in the guinea pig. Exp Lung Res. 1991;17(6):997–1009.
23. Santos S, Peinado VI, Ramirez J, Melgosa T, Roca J, Rodriguez-Roisin R, Barbera JA. Characterization of pulmonary vascular remodelling in smokers and patients with mild COPD. Eur Respir J. 2002;19(4):632–8.
24. Schrijen F, Uffholtz H, Polu JM, Poincelot F. Pulmonary and systemic hemodynamic evolution in chronic bronchitis. Am Rev Respir Dis. 1978;117(1):25–31.
25. Weitzenblum E, Loiseau A, Hirth C, Mirhom R, Rasaholinjanahary J. Course of pulmonary hemodynamics in patients with chronic obstructive pulmonary disease. Chest. 1979;75(6):656–62.
26. Weitzenblum E, Sautegeau A, Ehrhart M, Mammosser M, Hirth C, Roegel E. Long-term course of pulmonary arterial pressure in chronic obstructive pulmonary disease. Am Rev Respir Dis. 1984;130(6):993–8.
27. Kessler R, Faller M, Weitzenblum E, Chaouat A, Aykut A, Ducolone A, Ehrhart M, Oswald-Mammosser M. "Natural history" of pulmonary hypertension in a series of 131 patients with chronic obstructive lung disease. Am J Respir Crit Care Med. 2001;164(2):219–24.
28. MacNee W. Pathophysiology of cor pulmonale in chronic obstructive pulmonary disease. Part One. Am J Respir Crit Care Med. 1994;150(3):833–52.
29. Richens JM, Howard P. Oedema in cor pulmonale. Clin Sci (Lond). 1982;62(3):255–9.
30. Farber MO, Weinberger MH, Robertson GL, Fineberg NS, Manfredi F. Hormonal abnormalities affecting sodium and water balance in acute respiratory failure due to chronic obstructive lung disease. Chest. 1984;85(1):49–54.
31. MacNee W, Wathen CG, Flenley DC, Muir AD. The effects of controlled oxygen therapy on ventricular function in patients with stable and decompensated cor pulmonale. Am Rev Respir Dis. 1988;137(6):1289–95.
32. Weitzenblum E, Apprill M, Oswald M, Chaouat A, Imbs JL. Pulmonary hemodynamics in patients with chronic obstructive pulmonary disease before and during an episode of peripheral edema. Chest. 1994;105(5):1377–82.
33. Bishop JM, Cross KW. Physiological variables and mortality in patients with various categories of chronic respiratory disease. Bull Eur Physiopathol Respir. 1984;20(6):495–500.
34. Chaouat A, Bugnet AS, Kadaoui N, Schott R, Enache I, Ducolone A, Ehrhart M, Kessler R, Weitzenblum E. Severe pulmonary hypertension and chronic obstructive pulmonary disease. Am J Respir Crit Care Med. 2005;172(2):189–94.
35. Barbera JA, Riverola A, Roca J, Ramirez J, Wagner PD, Ros D, Wiggs BR, Rodriguez-Roisin R. Pulmonary vascular abnormalities and ventilation-perfusion relationships in mild chronic obstructive pulmonary disease. Am J Respir Crit Care Med. 1994;149(2 Pt 1):423–9.
36. Magee F, Wright JL, Wiggs BR, Pare PD, Hogg JC. Pulmonary vascular structure and function in chronic obstructive pulmonary disease. Thorax. 1988;43(3):183–9.
37. Peinado VI, Barbera JA, Ramirez J, Gomez FP, Roca J, Jover L, Gimferrer JM, Rodriguez-Roisin R. Endothelial dysfunction in pulmonary arteries of patients with mild COPD. Am J Physiol. 1998;274(6 Pt 1):L908–13.
38. Eddahibi S, Chaouat A, Morrell N, Fadel E, Fuhrman C, Bugnet AS, Dartevelle P, Housset B, Hamon M, Weitzenblum E, et al. Polymorphism of the serotonin transporter gene and pulmonary hypertension in chronic obstructive pulmonary disease. Circulation. 2003;108(15):1839–44.
39. Owens GK. Regulation of differentiation of vascular smooth muscle cells. Physiol Rev. 1995;75(3):487–517.
40. van der Loop FT, Gabbiani G, Kohnen G, Ramaekers FC, van Eys GJ. Differentiation of smooth muscle cells in human blood vessels as defined by smoothelin, a novel marker for the contractile phenotype. Arterioscler Thromb Vasc Biol. 1997;17(4):665–71.
41. Rabinovitch M. Elastase and the pathobiology of unexplained pulmonary hypertension. Chest. 1998;114(3 Suppl):213S-24S.
42. Barbera JA. Mechanisms of development of chronic obstructive pulmonary disease-associated pulmonary hypertension. Pulm Circ. 2013;3(1):160–4.
43. Naeije R, Barbera JA. Pulmonary hypertension associated with COPD. Crit Care. 2001;5(6):286–9.

44. Wilkinson M, Langhorne CA, Heath D, Barer GR, Howard P. A pathophysiological study of 10 cases of hypoxic cor pulmonale. Q J Med. 1988;66(249):65–85.
45. Lejeune P, Mols P, Naeije R, Hallemans R, Melot C. Acute hemodynamic effects of controlled oxygen therapy in decompensated chronic obstructive pulmonary disease. Crit Care Med. 1984;12(12):1032–5.
46. Timms RM, Khaja FU, Williams GW. Hemodynamic response to oxygen therapy in chronic obstructive pulmonary disease. Ann Intern Med. 1985;102(1):29–36.
47. Ricciardolo FL, Sterk PJ, Gaston B, Folkerts G. Nitric oxide in health and disease of the respiratory system. Physiol Rev. 2004;84(3):731–65.
48. Tsoumakidou M, Tzanakis N, Chrysofakis G, Siafakas NM. Nitrosative stress, heme oxygenase-1 expression and airway inflammation during severe exacerbations of COPD. Chest. 2005;127(6):1911–8.
49. Voelkel NF, Mizuno S, Bogaard HJ. The role of hypoxia in pulmonary vascular diseases: a perspective. Am J Physiol Lung Cell Mol Physiol. 2013;304(7):L457–65.
50. Stockley RA, Mannino D, Barnes PJ. Burden and pathogenesis of chronic obstructive pulmonary disease. Proc Am Thorac Soc. 2009;6(6):524–6.
51. Peinado VI, Barbera JA, Abate P, Ramirez J, Roca J, Santos S, Rodriguez-Roisin R. Inflammatory reaction in pulmonary muscular arteries of patients with mild chronic obstructive pulmonary disease. Am J Respir Crit Care Med. 1999;159(5 Pt 1):1605–11.
52. Weitzenblum E, Sautegeau A, Ehrhart M, Mammosser M, Pelletier A. Long-term oxygen therapy can reverse the progression of pulmonary hypertension in patients with chronic obstructive pulmonary disease. Am Rev Respir Dis. 1985;131(4):493–8.
53. Chaouat A, Naeije R, Weitzenblum E. Pulmonary hypertension in COPD. Eur Respir J. 2008;32(5):1371–85.
54. Fishman AP. Hypoxia on the pulmonary circulation. How and where it acts. Circ Res. 1976;38(4):221–31.
55. Fishman AP. State of the art: chronic cor pulmonale. Am Rev Respir Dis. 1976;114(4):775–94.
56. Weitzenblum E, Chaouat A. Cor pulmonale. Chron Respir Dis. 2009;6(3):177–85.
57. Chao J, Viets Z, Donham P, Wood JG, Gonzalez NC. Dexamethasone blocks the systemic inflammation of alveolar hypoxia at several sites in the inflammatory cascade. Am J Physiol Heart Circ Physiol. 2012;303(2):H168–77.
58. Chao J, Wood JG, Blanco VG, Gonzalez NC. The systemic inflammation of alveolar hypoxia is initiated by alveolar macrophage-borne mediator(s). Am J Respir Cell Mol Biol. 2009;41(5):573–82.
59. Prabhakar NR, Kumar GK, Peng YJ. Sympatho-adrenal activation by chronic intermittent hypoxia. J Appl Physiol. 2012;113(8):1304–10.
60. Wang GL, Semenza GL. Characterization of hypoxia-inducible factor 1 and regulation of DNA binding activity by hypoxia. J Biol Chem. 1993;268(29):21513–8.
61. Archiniegas E, Neves CY, Candelle D, Cardier JE. Thrombin and its protease-activated receptor-1 (PAR1) participate in the endothelial-mesenchymal transdifferentiation process. DNA Cell Biol. 2004;23(12):815–25.
62. Frohlich S, Boylan J, McLoughlin P. Hypoxia-induced inflammation in the lung: a potential therapeutic target in acute lung injury? Am J Respir Cell Mol Biol. 2013;48(3):271–9.
63. Nicolls MR, Voelkel NF. Hypoxia and the lung: beyond hypoxic vasoconstriction. Antioxid Redox Signal. 2007;9(6):741–3.
64. Rius J, Guma M, Schachtrup C, Akassoglou K, Zinkernagel AS, Nizet V, Johnson RS, Haddad GG, Karin M. NF-kappaB links innate immunity to the hypoxic response through transcriptional regulation of HIF-1alpha. Nature. 2008;453(7196):807–11.
65. Taylor CT. Interdependent roles for hypoxia inducible factor and nuclear factor-kappaB in hypoxic inflammation. J Physiol. 2008;586(Pt 17):4055–9.
66. Yamaji-Kegan K, Su Q, Angelini DJ, Champion HC, Johns RA. Hypoxia-induced mitogenic factor has proangiogenic and proinflammatory effects in the lung via VEGF and VEGF receptor-2. Am J Physiol Lung Cell Mol Physiol. 2006;291(6):L1159–68.

67. Angelini DJ, Su Q, Kolosova IA, Fan C, Skinner JT, Yamaji-Kegan K, Collector M, Sharkis SJ, Johns RA. Hypoxia-induced mitogenic factor (HIMF/FIZZ1/RELM alpha) recruits bone marrow-derived cells to the murine pulmonary vasculature. PLoS One. 2010;5(6):e11251.
68. Kuhlicke J, Frick JS, Morote-Garcia JC, Rosenberger P, Eltzschig HK. Hypoxia inducible factor (HIF)-1 coordinates induction of Toll-like receptors TLR2 and TLR6 during hypoxia. PLoS One. 2007;2(12):e1364.
69. van Uden P, Kenneth NS, Rocha S. Regulation of hypoxia-inducible factor-1alpha by NF-kappaB. Biochem J. 2008;412(3):477–84.
70. van Uden P, Kenneth NS, Webster R, Muller HA, Mudie S, Rocha S. Evolutionary conserved regulation of HIF-1beta by NF-kappaB. PLoS Genet. 2011;7(1):e1001285.
71. Savale L, Tu L, Rideau D, Izziki M, Maitre B, Adnot S, Eddahibi S. Impact of interleukin-6 on hypoxia-induced pulmonary hypertension and lung inflammation in mice. Respir Res. 2009;10:6.
72. Steiner MK, Syrkina OL, Kolliputi N, Mark EJ, Hales CA, Waxman AB. Interleukin-6 over-expression induces pulmonary hypertension. Circ Res. 2009;104(2):236–44, (28p following 244).
73. Brock M, Trenkmann M, Gay RE, Michel BA, Gay S, Fischler M, Ulrich S, Speich R, Huber LC. Interleukin-6 modulates the expression of the bone morphogenic protein receptor type II through a novel STAT3-microRNA cluster 17/92 pathway. Circ Res. 2009;104(10):1184–91.
74. Grijelmo C, Rodrigue C, Svrcek M, Bruyneel E, Hendrix A, de Wever O, Gespach C. Pro-invasive activity of BMP-7 through SMAD4/src-independent and ERK/Rac/JNK-dependent signaling pathways in colon cancer cells. Cell Signal. 2007;19(8):1722–32.
75. Tian Q, He XC, Hood L, Li L. Bridging the BMP and Wnt pathways by PI3 kinase/Akt and 14-3-3zeta. Cell Cycle. 2005;4(2):215–6.
76. Rudarakanchana N, Flanagan JA, Chen H, Upton PD, Machado R, Patel D, Trembath RC, Morrell NW. Functional analysis of bone morphogenetic protein type II receptor mutations underlying primary pulmonary hypertension. Hum Mol Genet. 2002;11(13):1517–25.
77. Nohe A, Hassel S, Ehrlich M, Neubauer F, Sebald W, Henis YI, Knaus P. The mode of bone morphogenetic protein (BMP) receptor oligomerization determines different BMP-2 signaling pathways. J Biol Chem. 2002;277(7):5330–8.
78. Rabinovitch M. Molecular pathogenesis of pulmonary arterial hypertension. J Clin Invest. 2012;122(12):4306–13.
79. Atkinson C, Stewart S, Upton PD, Machado R, Thomson JR, Trembath RC, Morrell NW. Primary pulmonary hypertension is associated with reduced pulmonary vascular expression of type II bone morphogenetic protein receptor. Circulation. 2002;105(14):1672–8.
80. Shimoda LA, Fallon M, Pisarcik S, Wang J, Semenza GL. HIF-1 regulates hypoxic induction of NHE1 expression and alkalinization of intracellular pH in pulmonary arterial myocytes. Am J Physiol Lung Cell Mol Physiol. 2006;291(5):L941–9.
81. Shimoda LA, Manalo DJ, Sham JS, Semenza GL, Sylvester JT. Partial HIF-1alpha deficiency impairs pulmonary arterial myocyte electrophysiological responses to hypoxia. Am J Physiol Lung Cell Mol Physiol. 2001;281(1):L202–8.
82. Wang J, Weigand L, Lu W, Sylvester JT, Semenza GL, Shimoda LA. Hypoxia inducible factor 1 mediates hypoxia-induced TRPC expression and elevated intracellular Ca^{2+} in pulmonary arterial smooth muscle cells. Circ Res. 2006;98(12):1528–37.
83. Whitman EM, Pisarcik S, Luke T, Fallon M, Wang J, Sylvester JT, Semenza GL, Shimoda LA. Endothelin-1 mediates hypoxia-induced inhibition of voltage-gated K^+ channel expression in pulmonary arterial myocytes. Am J Physiol Lung Cell Mol Physiol. 2008;294(2):L309–18.
84. Moudgil R, Michelakis ED, Archer SL. The role of k^+ channels in determining pulmonary vascular tone, oxygen sensing, cell proliferation, and apoptosis: implications in hypoxic pulmonary vasoconstriction and pulmonary arterial hypertension. Microcirculation. 2006;13(8):615–32.
85. Remillard CV, Yuan JX. High altitude pulmonary hypertension: role of K^+ and Ca^{2+} channels. High Alt Med Biol. 2005;6(2):133–46.

86. Smirnov SV, Robertson TP, Ward JP, Aaronson PI. Chronic hypoxia is associated with reduced delayed rectifier K+ current in rat pulmonary artery muscle cells. Am J Physiol. 1994;266(1 Pt 2):H365–70.
87. Grimminger F, Weissmann N, Spriestersbach R, Becker E, Rosseau S, Seeger W. Effects of NADPH oxidase inhibitors on hypoxic vasoconstriction in buffer-perfused rabbit lungs. Am J Physiol. 1995;268(5 Pt 1):L747–52.
88. Michelakis ED, Hampl V, Nsair A, Wu X, Harry G, Haromy A, Gurtu R, Archer SL. Diversity in mitochondrial function explains differences in vascular oxygen sensing. Circ Res. 2002;90(12):1307–15.
89. Waypa GB, Schumacker PT. Hypoxic pulmonary vasoconstriction: redox events in oxygen sensing. J Appl Physiol. 2005;98(1):404–14.
90. Weissmann N, Tadic A, Hanze J, Rose F, Winterhalder S, Nollen M, Schermuly RT, Ghofrani HA, Seeger W, Grimminger F. Hypoxic vasoconstriction in intact lungs: a role for NADPH oxidase-derived H(2)O(2)? Am J Physiol Lung Cell Mol Physiol. 2000;279(4):L683–90.
91. Weissmann N, Zeller S, Schafer RU, Turowski C, Ay M, Quanz K, Ghofrani HA, Schermuly RT, Fink L, Seeger W, et al. Impact of mitochondria and NADPH oxidases on acute and sustained hypoxic pulmonary vasoconstriction. Am J Respir Cell Mol Biol. 2006;34(4):505–13.
92. Mittal M, Gu XQ, Pak O, Pamenter ME, Haag D, Fuchs DB, Schermuly RT, Ghofrani HA, Brandes RP, Seeger W, et al. Hypoxia induces Kv channel current inhibition by increased NADPH oxidase-derived reactive oxygen species. Free Radic Biol Med. 2012;52(6):1033–42.
93. Landsberg JW, Yuan JX. Calcium and TRP channels in pulmonary vascular smooth muscle cell proliferation. News Physiol Sci. 2004;19:44–50.
94. Reeve HL, Michelakis E, Nelson DP, Weir EK, Archer SL. Alterations in a redox oxygen sensing mechanism in chronic hypoxia. J Appl Physiol. 2001;90(6):2249–56.
95. Fuchs B, Sommer N, Dietrich A, Schermuly RT, Ghofrani HA, Grimminger F, Seeger W, Gudermann T, Weissmann N. Redox signaling and reactive oxygen species in hypoxic pulmonary vasoconstriction. Respir Physiol Neurobiol. 2010;174(3):282–91.
96. Olschewski A, Li Y, Tang B, Hanze J, Eul B, Bohle RM, Wilhelm J, Morty RE, Brau ME, Weir EK, et al. Impact of TASK-1 in human pulmonary artery smooth muscle cells. Circ Res. 2006;98(8):1072–80.
97. Weir EK, Olschewski A. Role of ion channels in acute and chronic responses of the pulmonary vasculature to hypoxia. Cardiovasc Res. 2006;71(4):630–41.
98. Tang B, Li Y, Nagaraj C, Morty RE, Gabor S, Stacher E, Voswinckel R, Weissmann N, Leithner K, Olschewski H, et al. Endothelin-1 inhibits background two-pore domain channel TASK-1 in primary human pulmonary artery smooth muscle cells. Am J Respir Cell Mol Biol. 2009;41(4):476–83.
99. Herrera GM, Walker BR. Involvement of L-type calcium channels in hypoxic relaxation of vascular smooth muscle. J Vasc Res. 1998;35(4):265–73.
100. Shimoda LA, Polak J. Hypoxia. 4. Hypoxia and ion channel function. Am J Physiol Cell Physiol. 2011;300(5):C951–67.
101. Kwapiszewska G, Wygrecka M, Marsh LM, Schmitt S, Trosser R, Wilhelm J, Helmus K, Eul B, Zakrzewicz A, Ghofrani HA, et al. Fhl-1, a new key protein in pulmonary hypertension. Circulation. 2008;118(11):1183–94.
102. Sims MW, Margolis DJ, Localio AR, Panettieri RA, Kawut SM, Christie JD. Impact of pulmonary artery pressure on exercise function in severe COPD. Chest. 2009;136(2):412–9.
103. Thabut G, Dauriat G, Stern JB, Logeart D, Levy A, Marrash-Chahla R, Mal H. Pulmonary hemodynamics in advanced COPD candidates for lung volume reduction surgery or lung transplantation. Chest. 2005;127(5):1531–6.
104. Wright JL, Petty T, Thurlbeck WM. Analysis of the structure of the muscular pulmonary arteries in patients with pulmonary hypertension and COPD: National Institutes of Health nocturnal oxygen therapy trial. Lung. 1992;170(2):109–24.

105. Ferrer E, Peinado VI, Castaneda J, Prieto-Lloret J, Olea E, Gonzalez-Martin MC, Vega-Agapito MV, Diez M, Dominguez-Fandos D, Obeso A, et al. Effects of cigarette smoke and hypoxia on pulmonary circulation in the guinea pig. Eur Respir J. 2011;38(3):617–27.
106. Wright JL, Tai H, Churg A. Cigarette smoke induces persisting increases of vasoactive mediators in pulmonary arteries. Am J Respir Cell Mol Biol. 2004;31(5):501–9.
107. Wright JL, Tai H, Dai J, Churg A. Cigarette smoke induces rapid changes in gene expression in pulmonary arteries. Lab Invest. 2002;82(10):1391–8.
108. Yamato H, Sun JP, Churg A, Wright JL. Cigarette smoke-induced emphysema in guinea pigs is associated with diffusely decreased capillary density and capillary narrowing. Lab Invest. 1996;75(2):211–9.
109. Lee SD, Lee DS, Chun YG, Shim TS, Lim CM, Koh Y, Kim WS, Kim DS, Kim WD. Cigarette smoke extract induces endothelin-1 via protein kinase C in pulmonary artery endothelial cells. Am J Physiol Lung Cell Mol Physiol. 2001;281(2):L403–11.
110. Nana-Sinkam SP, Lee JD, Sotto-Santiago S, Stearman RS, Keith RL, Choudhury Q, Cool C, Parr J, Moore MD, Bull TM, et al. Prostacyclin prevents pulmonary endothelial cell apoptosis induced by cigarette smoke. Am J Respir Crit Care Med. 2007;175(7):676–85.
111. Peluffo G, Calcerrada P, Piacenza L, Pizzano N, Radi R. Superoxide-mediated inactivation of nitric oxide and peroxynitrite formation by tobacco smoke in vascular endothelium: studies in cultured cells and smokers. Am J Physiol Heart Circ Physiol. 2009;296(6):H1781–92.
112. Edirisinghe I, Arunachalam G, Wong C, Yao H, Rahman A, Phipps RP, Jin ZG, Rahman I. Cigarette-smoke-induced oxidative/nitrosative stress impairs VEGF- and fluid-shear-stress-mediated signaling in endothelial cells. Antioxid Redox Signal. 2010;12(12):1355–69.
113. Damico R, Simms T, Kim BS, Tekeste Z, Amankwan H, Damarla M, Hassoun PM. p53 mediates cigarette smoke-induced apoptosis of pulmonary endothelial cells: inhibitory effects of macrophage migration inhibitor factor. Am J Respir Cell Mol Biol. 2011;44(3):323–32.
114. Milara J, Juan G, Ortiz JL, Guijarro R, Losada M, Serrano A, Morcillo EJ, Cortijo J. Cigarette smoke-induced pulmonary endothelial dysfunction is partially suppressed by sildenafil. Eur J Pharm Sci. 2010;39(5):363–72.
115. Dinh-Xuan AT, Higenbottam TW, Clelland CA, Pepke-Zaba J, Cremona G, Butt AY, Large SR, Wells FC, Wallwork J. Impairment of endothelium-dependent pulmonary-artery relaxation in chronic obstructive lung disease. N Engl J Med. 1991;324(22):1539–47.
116. Santos S, Peinado VI, Ramirez J, Morales-Blanhir J, Bastos R, Roca J, Rodriguez-Roisin R, Barbera JA. Enhanced expression of vascular endothelial growth factor in pulmonary arteries of smokers and patients with moderate chronic obstructive pulmonary disease. Am J Respir Crit Care Med. 2003;167(9):1250–6.
117. Furchgott RF, Zawadzki JV. The obligatory role of endothelial cells in the relaxation of arterial smooth muscle by acetylcholine. Nature. 1980;288(5789):373–6.
118. Stamler JS, Loh E, Roddy MA, Currie KE, Creager MA. Nitric oxide regulates basal systemic and pulmonary vascular resistance in healthy humans. Circulation. 1994;89(5):2035–40.
119. Cremona G, Wood AM, Hall LW, Bower EA, Higenbottam T. Effect of inhibitors of nitric oxide release and action on vascular tone in isolated lungs of pig, sheep, dog and man. J Physiol. 1994;481 (Pt 1):185–95.
120. Barbera JA, Roger N, Roca J, Rovira I, Higenbottam TW, Rodriguez-Roisin R. Worsening of pulmonary gas exchange with nitric oxide inhalation in chronic obstructive pulmonary disease. Lancet. 1996;347(8999):436–40.
121. Greenberg B, Kishiyama S. Endothelium-dependent and -independent responses to severe hypoxia in rat pulmonary artery. Am J Physiol. 1993;265(5 Pt 2):H1712–20.
122. Giaid A, Yanagisawa M, Langleben D, Michel RP, Levy R, Shennib H, Kimura S, Masaki T, Duguid WP, Stewart DJ. Expression of endothelin-1 in the lungs of patients with pulmonary hypertension. N Engl J Med. 1993;328(24):1732–9.

123. Barbera JA, Peinado VI, Santos S, Ramirez J, Roca J, Rodriguez-Roisin R. Reduced expression of endothelial nitric oxide synthase in pulmonary arteries of smokers. Am J Respir Crit Care Med. 2001;164(4):709–13.
124. Giaid A, Saleh D. Reduced expression of endothelial nitric oxide synthase in the lungs of patients with pulmonary hypertension. N Engl J Med. 1995;333(4):214–21.
125. Gan WQ, Man SF, Senthilselvan A, Sin DD. Association between chronic obstructive pulmonary disease and systemic inflammation: a systematic review and a meta-analysis. Thorax. 2004;59(7):574–80.
126. Walter RE, Wilk JB, Larson MG, Vasan RS, Keaney JF Jr, Lipinska I, O'Connor GT, Benjamin EJ. Systemic inflammation and COPD: the Framingham Heart Study. Chest. 2008;133(1):19–25.
127. Saetta M, Baraldo S, Corbino L, Turato G, Braccioni F, Rea F, Cavallesco G, Tropeano G, Mapp CE, Maestrelli P, et al. CD^{8+} ve cells in the lungs of smokers with chronic obstructive pulmonary disease. Am J Respir Crit Care Med. 1999;160(2):711–7.
128. Chaouat A, Savale L, Chouaid C, Tu L, Sztrymf B, Canuet M, Maitre B, Housset B, Brandt C, Le Corvoisier P, et al. Role for interleukin-6 in COPD-related pulmonary hypertension. Chest. 2009;136(3):678–87.
129. MacNee W. Oxidative stress and lung inflammation in airways disease. Eur J Pharmacol. 2001;429(1–3):195–207.
130. Pryor WA, Stone K. Oxidants in cigarette smoke. Radicals, hydrogen peroxide, peroxynitrate, and peroxynitrite. Ann N Y Acad Sci. 1993;686:12–27.
131. Dekhuijzen PN, Aben KK, Dekker I, Aarts LP, Wielders PL, van Herwaarden CL, Bast A. Increased exhalation of hydrogen peroxide in patients with stable and unstable chronic obstructive pulmonary disease. Am J Respir Crit Care Med. 1996;154(3 Pt 1):813–6.
132. Montuschi P, Collins JV, Ciabattoni G, Lazzeri N, Corradi M, Kharitonov SA, Barnes PJ. Exhaled 8-isoprostane as an in vivo biomarker of lung oxidative stress in patients with COPD and healthy smokers. Am J Respir Crit Care Med. 2000;162(3 Pt 1):1175–7.
133. Nowak D, Kasielski M, Antczak A, Pietras T, Bialasiewicz P. Increased content of thiobarbituric acid-reactive substances and hydrogen peroxide in the expired breath condensate of patients with stable chronic obstructive pulmonary disease: no significant effect of cigarette smoking. Respir Med. 1999;93(6):389–96.
134. Taggart C, Cervantes-Laurean D, Kim G, McElvaney NG, Wehr N, Moss J, Levine RL. Oxidation of either methionine 351 or methionine 358 in alpha 1-antitrypsin causes loss of anti-neutrophil elastase activity. J Biol Chem. 2000;275(35):27258–65.
135. Cathcart MK. Regulation of superoxide anion production by NADPH oxidase in monocytes/macrophages: contributions to atherosclerosis. Arterioscler Thromb Vasc Biol. 2004;24(1):23–8.
136. Griendling KK, Sorescu D, Ushio-Fukai M. NAD(P)H oxidase: role in cardiovascular biology and disease. Circ Res. 2000;86(5):494–501.
137. Wink DA, Mitchell JB. Chemical biology of nitric oxide: Insights into regulatory, cytotoxic, and cytoprotective mechanisms of nitric oxide. Free Radic Biol Med. 1998;25(4–5):434–56.
138. Szabo C, Ischiropoulos H, Radi R. Peroxynitrite: biochemistry, pathophysiology and development of therapeutics. Nat Rev Drug Discov. 2007;6(8):662–80.
139. Fortuno A, San Jose G, Moreno MU, Diez J, Zalba G. Oxidative stress and vascular remodelling. Exp Physiol. 2005;90(4):457–62.
140. Griendling KK, Harrison DG. Dual role of reactive oxygen species in vascular growth. Circ Res. 1999;85(6):562–3.
141. Irani K. Oxidant signaling in vascular cell growth, death, and survival: a review of the roles of reactive oxygen species in smooth muscle and endothelial cell mitogenic and apoptotic signaling. Circ Res. 2000;87(3):179–83.
142. Taniyama Y, Griendling KK. Reactive oxygen species in the vasculature: molecular and cellular mechanisms. Hypertension. 2003;42(6):1075–81.

143. Cai H, Harrison DG. Endothelial dysfunction in cardiovascular diseases: the role of oxidant stress. Circ Res. 2000;87(10):840–4.
144. Kojda G, Harrison D. Interactions between NO and reactive oxygen species: pathophysiological importance in atherosclerosis, hypertension, diabetes and heart failure. Cardiovasc Res. 1999;43(3):562–71.
145. Li PF, Dietz R, von Harsdorf R. Differential effect of hydrogen peroxide and superoxide anion on apoptosis and proliferation of vascular smooth muscle cells. Circulation. 1997;96(10):3602–9.
146. Cai H. NAD(P)H oxidase-dependent self-propagation of hydrogen peroxide and vascular disease. Circ Res. 2005;96(8):818–22.
147. Touyz RM, Schiffrin EL. Increased generation of superoxide by angiotensin II in smooth muscle cells from resistance arteries of hypertensive patients: role of phospholipase D-dependent NAD(P)H oxidase-sensitive pathways. J Hypertens. 2001;19(7):1245–54.
148. Griendling KK, Minieri CA, Ollerenshaw JD, Alexander RW. Angiotensin II stimulates NADH and NADPH oxidase activity in cultured vascular smooth muscle cells. Circ Res. 1994;74(6):1141–8.
149. Zafari AM, Ushio-Fukai M, Akers M, Yin Q, Shah A, Harrison DG, Taylor WR, Griendling KK. Role of NADH/NADPH oxidase-derived H2O2 in angiotensin II-induced vascular hypertrophy. Hypertension. 1998;32(3):488–95.
150. Ushio-Fukai M, Zafari AM, Fukui T, Ishizaka N, Griendling KK. $p22^{phox}$ is a critical component of the superoxide-generating NADH/NADPH oxidase system and regulates angiotensin II-induced hypertrophy in vascular smooth muscle cells. J Biol Chem. 1996;271(38):23317–21.
151. Galis ZS, Khatri JJ. Matrix metalloproteinases in vascular remodeling and atherogenesis: the good, the bad, and the ugly. Circ Res. 2002;90(3):251–62.
152. Rajagopalan S, Meng XP, Ramasamy S, Harrison DG, Galis ZS. Reactive oxygen species produced by macrophage-derived foam cells regulate the activity of vascular matrix metalloproteinases in vitro. Implications for atherosclerotic plaque stability. J Clin Invest. 1996;98(11):2572–9.
153. Pu Q, Neves MF, Virdis A, Touyz RM, Schiffrin EL. Endothelin antagonism on aldosterone-induced oxidative stress and vascular remodeling. Hypertension. 2003;42(1):49–55.
154. Liu J, Yang F, Yang XP, Jankowski M, Pagano PJ. NAD(P)H oxidase mediates angiotensin II-induced vascular macrophage infiltration and medial hypertrophy. Arterioscler Thromb Vasc Biol. 2003;23(5):776–82.
155. Weissmann N, Lobo B, Pichl A, Parajuli N, Seimetz M, Puig-Pey R, Ferrer E, Peinado VI, Dominguez-Fandos D, Fysikopoulos A, et al. Stimulation of soluble guanylate cyclase prevents cigarette smoke-induced pulmonary hypertension and emphysema. Am J Respir Crit Care Med. 2014;189(11):1359–73.
156. Dumitrascu R, Weissmann N, Ghofrani HA, Dony E, Beuerlein K, Schmidt H, Stasch JP, Gnoth MJ, Seeger W, Grimminger F, et al. Activation of soluble guanylate cyclase reverses experimental pulmonary hypertension and vascular remodeling. Circulation. 2006;113(2):286–95.
157. Evgenov OV, Pacher P, Schmidt PM, Hasko G, Schmidt HH, Stasch JP. NO-independent stimulators and activators of soluble guanylate cyclase: discovery and therapeutic potential. Nat Rev Drug Discov. 2006;5(9):755–68.
158. Li H, Forstermann U. Nitric oxide in the pathogenesis of vascular disease. J Pathol. 2000;190(3):244–54.
159. Arif E, Ahsan A, Vibhuti A, Rajput C, Deepak D, Athar M, Singh B, Pasha MA. Endothelial nitric oxide synthase gene variants contribute to oxidative stress in COPD. Biochem Biophys Res Commun. 2007;361(1):182–8.
160. Forstermann U. Oxidative stress in vascular disease: causes, defense mechanisms and potential therapies. Nat Clin Pract Cardiovasc Med. 2008;5(6):338–49.
161. Hogg JC. Pathophysiology of airflow limitation in chronic obstructive pulmonary disease. Lancet. 2004;364(9435):709–21.

162. Kharitonov SA. NOS:molecular mechanisms, clinical aspects, therapeutic and monitoring approaches. Curr Drug Targets Inflamm Allergy. 2005;4(2):141–9.
163. Kharitonov SA, Barnes PJ. Nitric oxide, nitrotyrosine, and nitric oxide modulators in asthma and chronic obstructive pulmonary disease. Curr Allergy Asthma Rep. 2003;3(2):121–9.
164. Mueller CF, Laude K, McNally JS, Harrison DG. ATVB in focus: redox mechanisms in blood vessels. Arterioscler Thromb Vasc Biol. 2005;25(2):274–8.
165. Lassegue B, Clempus RE. Vascular NAD(P)H oxidases: specific features, expression, and regulation. Am J Physiol Regul Integr Comp Physiol. 2003;285(2):R277–97.
166. Cross AR, Segal AW. The NADPH oxidase of professional phagocytes–prototype of the NOX electron transport chain systems. Biochim Biophys Acta. 2004;1657(1):1–22.
167. Selemidis S, Sobey CG, Wingler K, Schmidt HH, Drummond GR. NADPH oxidases in the vasculature: molecular features, roles in disease and pharmacological inhibition. Pharmacol Ther. 2008;120(3):254–91.
168. Geiszt M, Lekstrom K, Witta J, Leto TL. Proteins homologous to $p47^{phox}$ and $p67^{phox}$ support superoxide production by NAD(P)H oxidase 1 in colon epithelial cells. J Biol Chem. 2003;278(22):20006–12.
169. Ago T, Kuribayashi F, Hiroaki H, Takeya R, Ito T, Kohda D, Sumimoto H. Phosphorylation of $p47^{phox}$ directs phox homology domain from SH3 domain toward phosphoinositides, leading to phagocyte NADPH oxidase activation. Proc Natl Acad Sci U S A. 2003;100(8):4474–9.
170. Brandes RP, Kreuzer J. Vascular NADPH oxidases: molecular mechanisms of activation. Cardiovasc Res. 2005;65(1):16–27.
171. De Keulenaer GW, Alexander RW, Ushio-Fukai M, Ishizaka N, Griendling KK. Tumour necrosis factor alpha activates a $p22^{phox}$-based NADH oxidase in vascular smooth muscle. Biochem J. 1998;329(Pt 3):653–7.
172. Zalba G, San Jose G, Beaumont FJ, Fortuno MA, Fortuno A, Diez J. Polymorphisms and promoter overactivity of the p22(phox) gene in vascular smooth muscle cells from spontaneously hypertensive rats. Circ Res. 2001;88(2):217–22.
173. Moreno MU, San Jose G, Orbe J, Paramo JA, Beloqui O, Diez J, Zalba G. Preliminary characterisation of the promoter of the human p22(phox) gene: identification of a new polymorphism associated with hypertension. FEBS Lett. 2003;542(1–3):27–31.
174. San Jose G, Moreno MU, Olivan S, Beloqui O, Fortuno A, Diez J, Zalba G. Functional effect of the $p22^{phox}$—930 A/G polymorphism on $p22^{phox}$ expression and NADPH oxidase activity in hypertension. Hypertension. 2004;44(2):163–9.
175. Fortuno A, Olivan S, Beloqui O, San Jose G, Moreno MU, Diez J, Zalba G. Association of increased phagocytic NADPH oxidase-dependent superoxide production with diminished nitric oxide generation in essential hypertension. J Hypertens. 2004;22(11):2169–75.
176. Wick MJ, Buesing EJ, Wehling CA, Loomis ZL, Cool CD, Zamora MR, Miller YE, Colgan SP, Hersh LB, Voelkel NF, et al. Decreased neprilysin and pulmonary vascular remodeling in chronic obstructive pulmonary disease. Am J Respir Crit Care Med. 2011;183(3):330–40.
177. Sunday ME, Hua J, Torday JS, Reyes B, Shipp MA. CD10/neutral endopeptidase 24.11 in developing human fetal lung. Patterns of expression and modulation of peptide-mediated proliferation. J Clin Invest. 1992;90(6):2517–25.
178. Dusser DJ, Djokic TD, Borson DB, Nadel JA. Cigarette smoke induces bronchoconstrictor hyperresponsiveness to substance P and inactivates airway neutral endopeptidase in the guinea pig. Possible role of free radicals. J Clin Invest. 1989;84(3):900–6.
179. Carpenter TC, Stenmark KR. Hypoxia decreases lung neprilysin expression and increases pulmonary vascular leak. Am J Physiol Lung Cell Mol Physiol. 2001;281(4):L941–8.
180. Dempsey EC, Wick MJ, Karoor V, Barr EJ, Tallman DW, Wehling CA, Walchak SJ, Laudi S, Le M, Oka M, et al. Neprilysin null mice develop exaggerated pulmonary vascular remodeling in response to chronic hypoxia. Am J Pathol. 2009;174(3):782–96.

181. Shinall H, Song ES, Hersh LB. Susceptibility of amyloid beta peptide degrading enzymes to oxidative damage: a potential Alzheimer's disease spiral. Biochemistry. 2005;44(46):15345–50.
182. Papandreou CN, Nanus DM. Is methylation the key to CD10 loss? J Pediatr Hematol Oncol. 2010;32(1):2–3.
183. Sumitomo M, Shen R, Nanus DM. Involvement of neutral endopeptidase in neoplastic progression. Biochim Biophys Acta. 2005;1751(1):52–9.
184. Karoor V, Oka M, Walchak SJ, Hersh LB, Miller YE, Dempsey EC. Neprilysin regulates pulmonary artery smooth muscle cell phenotype through a platelet-derived growth factor receptor-dependent mechanism. Hypertension. 2013;61(4):921–30.
185. Lee SH, Channick RN. Endothelin antagonism in pulmonary arterial hypertension. Semin Respir Crit Care Med. 2005;26(4):402–8.
186. Jensen RT, Battey JF, Spindel ER, Benya RV. International Union of Pharmacology. LXVIII. Mammalian bombesin receptors: nomenclature, distribution, pharmacology, signaling, and functions in normal and disease states. Pharmacol Rev. 2008;60(1):1–42.
187. Nemenoff RA, Simpson PA, Furgeson SB, Kaplan-Albuquerque N, Crossno J, Garl PJ, Cooper J, Weiser-Evans MC. Targeted deletion of PTEN in smooth muscle cells results in vascular remodeling and recruitment of progenitor cells through induction of stromal cell-derived factor-1alpha. Circ Res. 2008;102(9):1036–45.
188. Klinger JR, Petit RD, Warburton RR, Wrenn DS, Arnal F, Hill NS. Neutral endopeptidase inhibition attenuates development of hypoxic pulmonary hypertension in rats. J Appl Physiol. 1993;75(4):1615–23.
189. Thompson JS, Morice AH. Neutral endopeptidase inhibitors and the pulmonary circulation. Gen Pharmacol. 1996;27(4):581–5.
190. Casserly B, Klinger JR. Brain natriuretic peptide in pulmonary arterial hypertension: biomarker and potential therapeutic agent. Drug Des Devel Ther. 2009;3:269–87.
191. Feil R, Lohmann SM, de Jonge H, Walter U, Hofmann F. Cyclic GMP-dependent protein kinases and the cardiovascular system: insights from genetically modified mice. Circ Res. 2003;93(10):907–16.
192. Lucas KA, Pitari GM, Kazerounian S, Ruiz-Stewart I, Park J, Schulz S, Chepenik KP, Waldman SA. Guanylyl cyclases and signaling by cyclic GMP. Pharmacol Rev. 2000;52(3):375–414.
193. Silberbach M, Roberts CT Jr. Natriuretic peptide signalling: molecular and cellular pathways to growth regulation. Cell Signal. 2001;13(4):221–31.
194. Hamet P, Tremblay J, Pang SC, Skuherska R, Schiffrin EL, Garcia R, Cantin M, Genest J, Palmour R, Ervin FR, et al. Cyclic GMP as mediator and biological marker of atrial natriuretic factor. J Hypertens Suppl. 1986;4(2):49–56.
195. Corti R, Burnett JC Jr, Rouleau JL, Ruschitzka F, Luscher TF. Vasopeptidase inhibitors: a new therapeutic concept in cardiovascular disease? Circulation. 2001;104(15):1856–62.
196. Daull P, Jeng AY, Battistini B. Towards triple vasopeptidase inhibitors for the treatment of cardiovascular diseases. J Cardiovasc Pharmacol. 2007;50(3):247–56.
197. Maki T, Nasa Y, Yamaguchi F, Yoshida H, Mori M, Takada T, Horikawa E, Okano K, Takeo S. Long-term treatment with neutral endopeptidase inhibitor improves cardiac function and reduces natriuretic peptides in rats with chronic heart failure. Cardiovasc Res. 2001;51(3):608–17.
198. Xu J, Carretero OA, Liu YH, Yang F, Shesely EG, Oja-Tebbe N, Yang XP. Dual inhibition of ACE and NEP provides greater cardioprotection in mice with heart failure. J Card Fail. 2004;10(1):83–9.
199. Sumitomo M, Iwase A, Zheng R, Navarro D, Kaminetzky D, Shen R, Georgescu MM, Nanus DM. Synergy in tumor suppression by direct interaction of neutral endopeptidase with PTEN. Cancer Cell. 2004;5(1):67–78.

Chapter 16
Nitric Oxide Synthesis in Vascular Physiology and Pathophysiology

Huige Li, Ning Xia and Ulrich Förstermann

16.1 Introduction

Nitric oxide (NO), the smallest signaling molecule known, is produced by three isoforms of NO synthase (NOS; EC 1.14.13.39): neuronal NOS (nNOS; NOS I), inducible NOS (iNOS; NOS II), and endothelial NOS (eNOS; NOS III). All NOS isoforms utilize L-arginine and molecular oxygen as substrates, and reduced nicotinamide adenine dinucleotide phosphate (NADPH) as a co-substrate. Flavin adenine dinucleotide (FAD), flavin mononucleotide (FMN), and (6*R*-)5, 6, 7, 8-tetrahydro-L-biopterin (BH_4) are co-factors of all NOS isozymes [22].

nNOS is constitutively expressed in central and peripheral neurons, as well as some other cell types. Its functions include synaptic plasticity in the central nervous system, central regulation of blood pressure, and vasodilatation via peripheral nitrergic nerves. Nitrergic nerves are of particular importance in the relaxation of corpus cavernosum and penile erection.

iNOS is usually not expressed in cells under physiological conditions. Its expression can be induced in many cell types by lipopolysaccharide, cytokines, or other agents. Once expressed, iNOS is constantly active and not regulated by intracellular Ca^{2+}concentrations. When induced in macrophages, iNOS produces large amounts of NO, which represent a major cytotoxic principle of those cells on microbes, intracellular pathogens, and parasitic microorganisms. On the other hand, iNOS has been shown to contribute to a number of diseases, including inflammatory diseases and septic shock.

H. Li (✉) · N. Xia · U. Förstermann
Department of Pharmacology, Johannes Gutenberg University Medical Center, Obere Zahlbacher St. 67, 55131 Mainz, Germany
e-mail: huigeli@uni-mainz.de

N. Xia
e-mail: xianing@uni-mainz.de

U. Förstermann
e-mail: ulrich.forstermann@uni-mainz.de

M. HH. Schmidt, S. Liebner (eds.), *Endothelial Signaling in Development and Disease*, DOI 10.1007/978-1-4939-2907-8_16

eNOS is mostly expressed in endothelial cells. It keeps blood vessels dilated, controls blood pressure, and has numerous other vasoprotective and antiatherosclerotic effects [22].

NO formed by NOS enzymes can act on a number of target enzymes and proteins. The most important physiological signaling pathway stimulated by NO is the activation of soluble guanylyl cyclase and the generation of cyclic guanosine monophosphate (cGMP).

16.2 Nitric Oxide Synthase (NOS) Enzymes in the Vasculature

16.2.1 Endothelial NOS (eNOS)

Under physiological conditions, vascular NO is mainly produced by eNOS. This enzyme is constitutively expressed in the endothelium and is activated by shear stress of the flowing blood or by agonists such as bradykinin, acetylcholine, and vascular endothelial growth factor (VEGF) (Fig. 16.1). Molecular mechanisms underlying eNOS activation include elevation of intracellular Ca^{2+}concentration, post-translational modification of the eNOS enzyme (i.e. phosphorylation and deacetylation), and protein–protein interaction [20].

NO produced by eNOS can diffuse from endothelial cells into the underlying smooth muscle cells (SMCs), and can induce vasodilation by stimulating NO-sensitive guanylyl cyclase. Endothelial NO can also diffuse into the blood and inhibit platelet aggregation and adhesion (Fig. 16.1). In addition to these antihypertensive and antithrombotic actions, eNOS-derived NO also possesses multiple

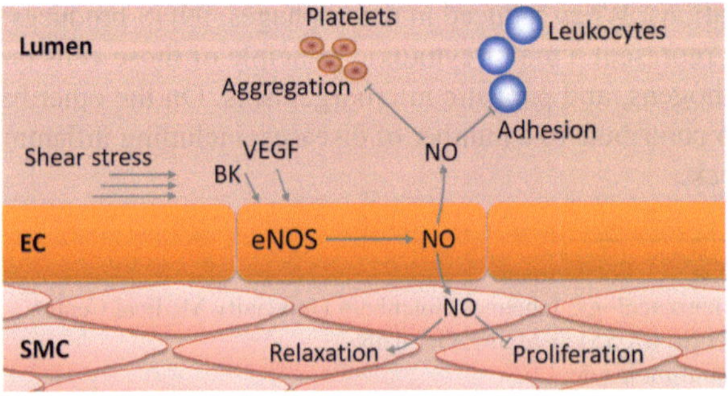

Fig. 16.1 Antihypertensive, antithrombotic, and antiatherosclerotic effects of endothelial nitric oxide synthase (*eNOS*). The eNOS enzyme in endothelial cells (*ECs*) can be activated by shear stress or agonists such as bradykinin (*BK*) and vascular endothelial growth factor (*VEGF*). Endothelial NO diffuses into the blood and inhibits platelet aggregation and adhesion, as well as leukocyte adhesion to the vascular endothelium and leukocyte migration into the vascular wall. NO diffused into smooth muscle cells (*SMC*) induces vasodilation and prevents SMC proliferation

antiatherosclerotic properties, including prevention of leukocyte adhesion to the vascular endothelium and leukocyte migration into the vascular wall, inhibition of low-density lipoprotein (LDL) oxidation, and inhibition of vascular SMC proliferation [49, 51]. Genetic depletion of eNOS exacerbates diet-induced atherosclerosis in the apolipoprotein E-knockout (ApoE-KO) mouse model. The blood pressure of eNOS knockout mice is approximately 30% higher than that of wild-type animals [58]. Recent studies suggest that eNOS is also involved in mitochondrial biogenesis, anti-aging effects, and extension of lifespan in mammals [11, 65].

16.2.2 Inducible NOS (iNOS)

iNOS is normally absent in the vasculature. However, under conditions of inflammation, sepsis, or oxidative stress, iNOS expression can be induced in the endothelium, the media, and/or the adventitia of blood vessels [25, 69].

In contrast to the regulated production of NO by eNOS, iNOS may generate large amounts of NO over long periods of time if substrate and co-factors are not limited. This excessive NO production by iNOS leads to vascular dysfunction, evident as impairment of both vasoconstriction and endothelium-dependent vasorelaxation. Several mechanisms have been proposed by which iNOS impairs contractile responses [105], including continuous activation of the soluble guanylyl cyclase [25], abnormal vascular calcium regulation [10], and oxidative modification of catecholamines [88]. In parallel, the endothelium-dependent, NO-mediated vasodilation response (e.g. to acetylcholine or bradykinin) is impaired by iNOS. This may be the result of reduced NO production from eNOS [25, 40] or enhanced inactivation of eNOS-derived NO by superoxide [100]. Tetrahydrobiopterin (BH_4) is an essential co-factor for NO production by NOS enzymes. iNOS expressed in the endothelium competes with eNOS for BH_4 and reduces NO production from eNOS by limiting BH_4 availability for eNOS [25]. The continuous generation of NO by iNOS induced in the vascular wall can impair the signal transduction cascade that links activation of endothelial receptors to the calcium–calmodulin-dependent activation of eNOS [40]. Moreover, the reduction of endothelium-dependent relaxation may be mediated in part by reduced reactivity of SMCs to NO [17]. The consequence of such dysregulations (impaired vasomotor reactivity to both vasoconstrictor and vasodilator agonists) can be seen, for example, in septic shock. Septic shock is characterized by massive arteriolar vasodilatation, hypotension, and microvascular damage largely mediated by iNOS. Inappropriate vasodilation, abnormal regulation of blood flow to organs, myocardial suppression, and interference with cellular respiration all contribute to hypotension and mortality in septic shock [22, 58].

The induction of iNOS in the vasculature is also associated with enhanced formation of peroxynitrite [59, 100, 106], a key pathogenic mechanism in conditions such as septic shock, stroke, myocardial infarction, chronic heart failure, diabetes, and atherosclerosis [57, 66]. iNOS is present in human atherosclerosis plaque. Genetic deletion of iNOS reduces atherosclerosis in ApoE-KO mice [43]. iNOS also contributes to tissue damage after cerebral ischemia. Inhibition of iNOS by selective pharmacologic inhibitors [34], or gene deletion of iNOS [33], reduces brain damage.

16.2.3 Neuronal NOS (nNOS)

nNOS plays an important role in blood vessels independently of its effects in the central nervous system [61]. nNOS is expressed in perivascular nerve fibers as well as in the vascular wall [84, 96]. Using the selective nNOS inhibitor S-methyl-L-thiocitrulline (SMTC), it has been found that nNOS inhibition leads to a dose-dependent reduction in basal forearm blood flow in human subjects [86]. Importantly, the acetylcholine- [86] and flow-induced [85] vasodilation, effects that are mediated by eNOS, is not affected by SMTC. A similar situation has been observed in human coronary circulation, where SMTC causes a significant reduction in basal coronary blood flow without any effect on increases in flow evoked by intracoronary substance P [85]. In addition, nNOS may play a role in the regulation of mental stress-induced vasodilatation [86] and in the cutaneous blood flow increase in response to whole-body heat stress [39].

In addition to these human studies, animal experiments suggest an atheroprotective role of nNOS in the vasculature. Gene deletion of nNOS in ApoE-KO mice accelerates atherosclerotic plaque formation in the aortic root and descending thoracic aorta [45, 80]. An increase in mortality has also been observed nNOS/ApoE double knockout mice after 24 weeks of Western-type diet [45]. These results are compatible with earlier studies showing accelerated neointimal formation and constrictive vascular remodeling in nNOS-KO mice in a carotid artery ligation model [64, 96].

16.3 Molecular Mechanisms of eNOS Uncoupling

Under physiological conditions, eNOS produces NO, which represents a key element in the vasoprotective function of the endothelium [22, 51, 52]. However, under pathological conditions associated with oxidative stress, eNOS may become dysfunctional [21, 53, 54]. Oxidative stress contributes markedly to endothelial dysfunction, primarily due to rapid oxidative inactivation of NO by excess superoxide. In a second step, the persisting oxidative stress renders eNOS uncoupled (i.e. uncoupling of O_2 reduction from NO synthesis), such that it no longer produces NO but superoxide.

Numerous mechanisms have been proposed to play a role in eNOS uncoupling [22, 51]. Among these, depletion of BH_4, an essential co-factor for the eNOS enzyme, is likely to be a major cause for eNOS uncoupling and endothelial dysfunction. Superoxide can modestly, and peroxynitrite strongly, oxidize BH_4, leading to BH_4 deficiency [48]. ApoE-KO mice show increased oxidative degradation of BH_4 and eNOS uncoupling in cardiovascular tissues [1, 103, 104]. Evidence for BH_4 deficiency and eNOS uncoupling has been obtained in patients with endothelial dysfunction resulting from hypercholesterolemia [93] or diabetes mellitus [28], and in chronic smokers [97].

Another important cause of eNOS uncoupling is a deficiency of L-arginine due to upregulation of arginase expression/activity. In humans and mammals, there are two isoforms of arginases: arginase I (Arg-I) and arginase II (Arg-II), which are encoded by two separate genes [109]. Both Arg-I and Arg-II have been found in the vasculature, with a dependency on species and cell type [70]. Studies performed with vascular endothelial cells suggest that these two isoforms share similar functions, i.e. metabolizing L-arginine to urea and L-ornithine, whereby an upregulation of Arg-I and/or Arg-II limits L-arginine bioavailability for NO production, leading to endothelial dysfunction [109]. Selective endothelial overexpression of Arg-II induces endothelial dysfunction and hypertension and enhances atherosclerosis in mice [99]. Aging is associated with upregulation of Arg-II, enhanced vascular arginase activity, and eNOS uncoupling [41, 111]. The uncoupling of eNOS under conditions of aging can be reversed by arginase inhibitors [41] or by genetic deletion of Arg-II [111]. The expression/activity of vascular arginases is enhanced by a variety of stimuli [70], including angiotensin II [87] or high glucose [76], thrombin [62], and oxidized LDL [79].

Uncoupling of eNOS is a crucial mechanism contributing significantly to endothelial dysfunction and atherogenesis. It not only reduces NO production but also potentiates the pre-existing oxidative stress. The overproduction of reactive oxygen species (ROS) [e.g. superoxide and subsequently peroxynitrite] by uncoupled eNOS in turn enhances oxidation of BH_4 and upregulation of arginase expression/activity [8], creating a vicious circle.

16.4 Uncoupling of eNOS in Diabetes Mellitus

The production of eNOS-derived NO is reduced in diabetes and insulin resistance. The implicated mechanisms include inhibition of eNOS activity by post-translational modification (i.e. enhancement of protein kinase C [PKC]-mediated phosphorylation of eNOS at threonine 495 [94], reduction in phosphorylation of eNOS at serine 1177 [15], and increase in O-linked N-acetylglucosamine modification of eNOS [15]) and eNOS uncoupling due to deficiency of BH_4 or L-arginine [50].

16.4.1 BH_4 Deficiency in Diabetes

In the rat model of streptozotocin-induced type 1 diabetes mellitus, eNOS is uncoupled in blood vessels [29]. The major cause for this eNOS uncoupling is BH_4 oxidation due to PKC-mediated activation and upregulation of NADPH oxidase [29]. Indeed, BH_4 oxidation and BH_4 deficiency is evident in streptozotocin-treated mice [2], as well as in streptozotocin-treated diabetic spontaneously hypertensive rats [19].

Diabetes-induced ROS may cause proteasomal degradation of guanosine 5′-triphosphate cyclohydrolase-1 (GCH1), a rate-limiting enzyme in the synthesis of BH_4 [108], which may subsequently lead to BH_4 deficiency [81, 101, 108]. In addition, S-glutathionylation may represent another important trigger of eNOS uncoupling in the setting of type 1 diabetes [81].

In spontaneously diabetic db/db mice, an animal model of type 2 diabetes that lacks the leptin receptor, the acetylcholine-induced relaxation of isolated small mesenteric arteries is reduced. Although the absolute vascular BH_4 level is unchanged, a relative BH_4 deficiency is evident due to enhanced BH_4 oxidation and a low BH_4:BH_2 ratio [67]. Endothelial dysfunction can be improved by acute *ex vivo* incubation with BH_4 or with sepiapterin, a stable precursor of BH_4 [68]. Fructose-fed rats show insulin resistance with endogenous hyperinsulinemia. Aortic BH_4 contents of fructose-fed rats are decreased, whereas the levels of BH_2 are increased. Impaired endothelial function can be reversed by *ex vivo* preincubation of aortic strips with BH_4 [90] or by oral BH_4 supplementation [91]. These results indicate that BH_4 deficiency represents a major cause for endothelial dysfunction in type 2 diabetes or insulin resistance (Fig. 16.2).

16.4.2 L-Arginine Deficiency in Diabetes

The reduced NO production from eNOS under conditions of diabetes may also be caused by L-arginine deficiency due to induction of arginases. High glucose upregulates Arg-I in (bovine and murine) endothelial cells by stimulating the RhoA-ROCK pathway [76, 110], whereas persistent insulin stimulation (mimicking the hyperinsulinemia condition) in human endothelial cells upregulates the expression and activity of Arg-II via a signaling cascade involving SHP2 phosphorylation and p38 mitogen-activated protein kinase (MAPK) activation [23]. Increased arginase expression/activity decreases L-arginine bioavailability for eNOS on the one hand, and on the other, L-arginine depletion may also lead to eNOS uncoupling (Fig. 16.2).

In aortas from streptozotocin-induced diabetic rats [76] or mice [75, 110], Arg-I expression is enhanced, which is associated with endothelial dysfunction. Aortic Arg-II expression is either barely detectable (rats) or not changed by streptozotocin-induced diabetes (mice). Diabetes-induced coronary vascular dysfunction involves increased arginase activity, with a major involvement of Arg-I [75, 76]. Enhanced arginase activity in the aorta or coronary artery of streptozotocin-induced diabetic animals is believed to induce eNOS uncoupling [75, 76]. Diabetic Arg-I$^{+/-}$Arg-II$^{-/-}$ mice exhibit less arginase activity/expression and less endothelial dysfunction than diabetic wild-type or Arg-I$^{+/+}$Arg-II$^{-/-}$ mice, indicating that Arg-I is likely to be the primary arginase isoform involved in type 1 diabetes-induced vascular dysfunction [75]. In contrast, Arg-II seems to be the primary arginase isoform upregulated in corpus cavernosum tissue of streptozotocin-treated mice, and Arg-II deletion improves corpora cavernosal relaxation in type 1 diabetes [95].

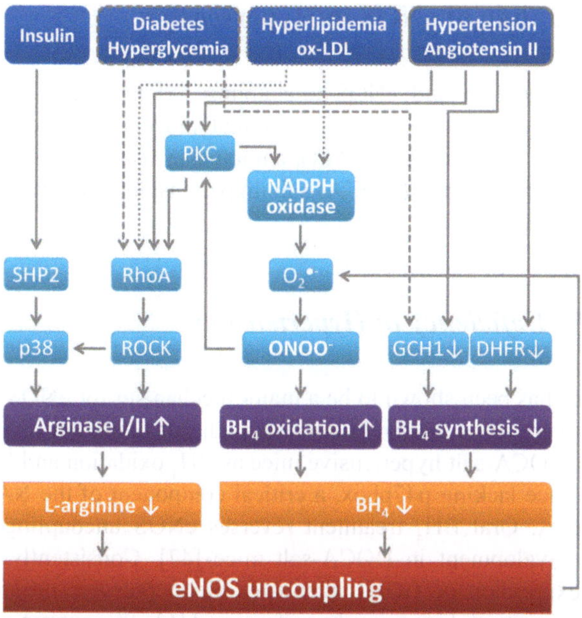

Fig. 16.2 Uncoupling of endothelial NOS (*eNOS*) in cardiovascular disease. Cardiovascular risk factors such as hypertension, hypercholesterolemia, or diabetes mellitus lead to superoxide production by eNOS ('eNOS uncoupling') through two major mechanisms-deficiency of the co-factor tetrahydrobiopterin (*BH_4*) or the substrate L-arginine. Nicotinamide adenine dinucleotide phosphate (*NADPH*) oxidase-derived superoxide (*$O_2\bullet-$*) may react with nitric oxide (*NO*), resulting in peroxynitrite (*ONOO−*). ONOO− oxidizes BH_4 to dihydrobiopterin (*BH_2*). BH_4 deficiency can be exacerbated by downregulation of GTP cyclohydrolase-1 (*GCH1*, the rate-limiting enzyme for BH_4 *de novo* biosynthesis) or dihydrofolate reductase (*DHFR*, the enzyme required for BH_4 regeneration from BH_2). L-arginine deficiency is caused by upregulation of arginase expression and/or activity. Uncoupled eNOS produces reactive oxygen species (e.g. superoxide and subsequently peroxynitrite), which in turn oxidize BH_4 and increase arginase expression and activity, creating a vicious circle.

In type 2 diabetic Goto-Kakizaki rats, coronary artery microvascular dysfunction is associated with increased Arg-II expression. Arginase inhibition restores coronary microvascular function in type 2 diabetic rats by a mechanism related to increased utilization of arginine by NOS and increased NO bioavailability [24].

In patients with type 2 diabetes mellitus, plasma arginase activity is elevated [37]. An upregulation of Arg-I in coronary arterioles of patients with (type 1 or type 2) diabetes mellitus has been shown to contribute to reduced NO production and consequently diminished vasodilation [4]. Arginase inhibition markedly improves endothelium-dependent vasodilatation in the forearm of patients with type 2 diabetes and coronary artery disease, whereas it does not affect endothelial function in healthy controls [89]. This observation indicates a functional role of arginase contributing to endothelial dysfunction in patients with diabetes.

16.5 Uncoupling of eNOS in Hypertension

Endothelial dysfunction and eNOS uncoupling has been demonstrated in different types of hypertension, including animal models of genetic hypertension (spontaneously hypertensive rats, SHR), angiotensin II-induced hypertension, and deoxycorticosterone acetate (DOCA)-salt hypertension (summarized in Li and Forstermann [50]).

16.5.1 BH_4 Deficiency in Hypertension

BH_4 deficiency has been shown to be a major mechanism for eNOS uncoupling in DOCA-salt hypertension. NADPH oxidase-mediated oxidation of BH_4 is evident in the aorta of DOCA-salt hypertensive mice as BH_4 oxidation and BH_4 deficiency are absent in mice lacking p47phox, a critical component of the NADPH oxidase enzyme complex. Oral BH_4 treatment reverses eNOS uncoupling and prevents hypertension development in DOCA-salt mice [47]. Consistently, endothelium-specific overexpression of GCH1 attenuates blood pressure progression in this model of salt-sensitive, low-renin hypertension [16]. In contrast, administration of sepiapterin is not effective in recoupling eNOS in DOCA-salt hypertension. Because of downregulation of endothelial sepiapterin reductase, sepiapterin cannot be converted to BH_4 [112].

In spontaneously hypertensive rats, the expression of vascular NADPH oxidase components is enhanced [50]. BH_4 content is decreased and eNOS in an uncoupled state in SHR aorta [56]. Suppression of NADPH oxidase with gp91ds-tat decreases ROS production in SHR to the level of control Wistar-Kyoto rats (WKY) [113]. Pharmacological reversal of eNOS uncoupling by midostaurin [56] or resveratrol [5] results in a reduction in blood pressure in spontaneously hypertensive rats.

Angiotensin II-induced hypertension is also associated with eNOS uncoupling [63]. Angiotensin II activates vascular PKC, which leads to enhanced expression [63] and activity [6] of vascular NADPH oxidase (Fig. 16.2). Indeed, NADPH oxidases play a crucial role in angiotensin II-induced hypertension. Angiotensin II-induced elevation in blood pressure and production of superoxide are significantly blunted in mice deficient of the NADPH oxidase components Nox1 [60] or p47phox [46], and are potentiated in mice with Nox1 overexpression [13].

BH_4 deficiency represents a major cause for angiotensin II-induced eNOS uncoupling. Angiotensin II reduces the expression of GCH1 and thus BH_4 *de novo* synthesis [82], as well as that of dihydrofolate reductase (DHFR) [7], which catalyzes the regeneration of BH_4 from its oxidized form, BH_2. Oral supplementation with BH_4 restores NO/cGMP signaling in small arteries, and attenuates angiotensin II-induced hypertension in rats [36].

16.5.2 L-Arginine Deficiency in Hypertension

In addition to BH_4 deficiency, the endothelial dysfunction in hypertension may be partially attributable to reduced NO production by eNOS because of L-arginine deficiency. Upregulation of arginase expression/activity in blood vessels has been observed in spontaneously hypertensive rats [12], mineralocorticoid-salt hypertensive rats [74], and Dahl rats with salt-induced hypertension [35]. The mechanisms underlying the upregulation of arginase in hypertension are not completely understood. Angiotensin II has been shown to increase endothelial arginase activity/expression through AT_1 receptors and subsequent activation of RhoA/ROCK/p38 MAPK pathways, leading to endothelial dysfunction [87]. Upregulation of arginase expression/activity is of functional relevance to blood pressure development. Selective endothelial overexpression of Arg-II induces endothelial dysfunction and hypertension in mice [99]. Treatment with arginase inhibitors improves vascular function and lowers blood pressure in spontaneously hypertensive rats [3].

In patients with hypertension, reflex cutaneous vasodilatation is augmented by arginase inhibitors via skin microdialysis catheters [30]. Antihypertensive treatment with the angiotensin-converting enzyme (ACE) inhibitor lisinopril reduced erythrocyte arginase activity in patients with arterial hypertension [42].

16.6 Uncoupling of eNOS in Atherosclerosis

In 1995, it was shown for the first time that LDL enhances superoxide production from eNOS [72]. Thereafter, evidence for eNOS uncoupling has been obtained in hypercholesterolemic mice [48] and patients [93].

16.6.1 BH_4 Deficiency in Atheorsclerosis

Hypercholesterolemic ApoE-KO mice show increased oxidative degradation of BH_4 [1], leading to eNOS uncoupling [1, 103, 104]. Supplementation with BH_4 restores endothelial dysfunction in hypercholesterolemic patients [93]. BH_4 deficiency due to oxidative stress is likely to be a major cause for eNOS uncoupling under these conditions. Diet-induced hypercholesterolemia in rabbits is associated with enhanced NADPH oxidase activity and reduced superoxide dismutase (SOD) activity in the vasculature [98]. Oxidized LDL enhances endothelial superoxide production by stimulating NADPH oxidase [27, 92]. Native LDL and oxidized LDL may also activate endothelial xanthine oxidase [92].

It should be noted that eNOS uncoupling is not an all-or-none phenomenon. Rather, uncoupled eNOS molecules and coupled eNOS proteins may exist in the same cell at the same time [50, 53]. Because of relative BH_4 deficiency, part of the eNOS molecules become uncoupled, while others may still remain coupled. In the hypercholesterolemic ApoE-KO mice, for instance, both superoxide and NO production by eNOS are detectable [71]. Moreover, the damaging effects of superoxide produced by uncoupled eNOS do not overwhelm the protective role of eNOS-derived NO, at least in this mouse model of atherosclerosis [71]. For this reason, genetic deletion of eNOS [9, 44, 71] or pharmacological inhibition [38] of both the coupled and uncoupled eNOS, accelerates atherosclerosis development in ApoE-KO mice.

16.6.2 L-Arginine Deficiency in Atherosclerosis

In addition to BH_4 deficiency, upregulation of arginase expression and/or activity has been shown to contribute to reduced endothelial NO production in experimental models of atherosclerosis, including the ApoE-KO mice (Arg-II) [18, 78] and in hyperlipidemic rabbits (Arg-I and Arg-II) [26]. The aortic arginase activity in ApoE-KO mice is significantly reduced after the removal of the endothelium, suggesting an important contribution from endothelial cells [78]. The functional relevance of arginase upregulation in atherosclerosis has been shown in ApoE-KO mice. Selective endothelial overexpression of Arg-II induces endothelial dysfunction and enhances atherosclerosis in mice [99]. Chronic treatment with an arginase inhibitor for 4 or 8 weeks reduces aortic plaque burden in ApoE-KO mice [78].

The RhoA/ROCK pathway seems to play a central role in the upregulation of arginase expression and activity under conditions of atherosclerosis (Fig. 16.2). Thrombin enhances the activity of Arg-II in human umbilical vein endothelial cells (HUVECs; Arg-I is not detectable in these cells) without changing the protein level of the enzyme [62]. This effect is associated with an upregulation of RhoA and is preventable by statins or ROCK inhibitors [62]. Oxidized LDL not only stimulates arginase enzymatic activity (after 5 min) but also enhances Arg-II protein levels (after 12 h) [79]. The activation of Arg-II by oxidized LDL is mediated by the lectin-like oxidized LDL (LOX-1) receptor and subsequent RhoA/ROCK-dependent microtubule depolymerization, leading to a dissociation of Arg-II from the mitochondria to the cytosol where it limits NO synthesis by eNOS [77]. In bovine aortic endothelial cells, oxidative species (such as peroxynitrite and hydrogen peroxide) increase Arg-I expression and activity through PKC-dependent activation of the RhoA-ROCK pathway [8].

In humans with hypercholesterolemia, the reduced skin blood flow responses are associated with enhanced expression and activity of Arg-I and Arg-II [31]. Statin treatment has no effect of the expression levels of the arginase enzymes but restores the functional vasodilator properties by inhibiting arginase activity [31].

16.7 Pharmacological Prevention of eNOS Uncoupling

Because of eNOS uncoupling under conditions of cardiovascular disease, pharmacological upregulation of eNOS expression alone will not be beneficial because this would lead to enhanced ROS production by the uncoupled eNOS. In fact, many cardiovascular diseases are associated with compensatory upregulation of eNOS expression; however, the compensation is futile, or even harmful, because the eNOS is in an uncoupled state [51, 55].

Therefore, it is essential to prevent eNOS uncoupling, or reverse an existing eNOS uncoupling, for the treatment of cardiovascular disease. Fortunately, eNOS uncoupling is a reversible event. With the growing understanding of molecular mechanisms underlying eNOS uncoupling, numerous pharmacological strategies to prevent eNOS uncoupling have been successfully tested in animal models. These include the ACE inhibitors, AT_1 receptor blockers (ARBs), 3-hydroxy-3-methylglutaryl-coenzyme A (HMG-CoA) reductase inhibitors (statins), the third-generation β-blocker nebivolol, the organic nitrate pentaerythritol tetranitrate (PETN), the plant polyphenolic phytoalexin resveratrol, and some small-molecular-weight eNOS transcription enhancers (for details see our recent review articles [50, 53, 54]). These compounds prevent BH_4 oxidation by inhibiting NADPH oxidase expression/activity. Some drugs additionally improve BH_4 regeneration from BH_2 by upregulating DHFR (e.g. ARBs and PETN), or enhance eNOS enzymatic activity (e.g. ARBs, statins, nebivolol, resveratrol) [50, 53]. Statins [31, 62, 76, 77] and ACE inhibitors [42] also improve eNOS functionality by inhibiting arginase activity. The improvement of NO bioavailability represents part of the pleiotropic effects of these drugs that contribute to their therapeutic benefit.

The long-term use of L-arginine is questionable. A randomized, double-blinded, placebo-controlled study in patients with acute myocardial infarction demonstrated that 6 months of oral L-arginine supplementation (3 g three times daily on top of standard post-infarction therapy) does not have any benefits on vascular stiffness or left ventricular rejection fraction, but increases mortality [83]. In another clinical study, patients with peripheral artery disease receiving L-arginine supplementation (3 g/day for 6 months) show decreased NO production and shortened walking distance [102]. Too much L-arginine may cause harmful effects due to production of other undesired metabolites from L-arginine [14], and chronic L-arginine intake can induce arginase expression/activity, thereby inducing vascular dysfunction [75]. In cultured human endothelial cells, acute L-arginine treatment enhances endothelial NO, whereas chronic L-arginine supplementation (0.5 mmol/L for 7 days) causes Arg-II upregulation, eNOS uncoupling, and endothelial senescence [107]. Therefore, the impact of chronic L-arginine supplementation does not seem beneficial; it is rather detrimental and should not be recommended in the clinical setting [109].

In contrast, promising results have been achieved with arginase inhibitors, although only limited data from clinical studies are currently available. Small-scale 'proof-of-concept' clinical studies have shown that local administration of arginase inhibitors improves vascular function in aged humans [32], as well as in patients with coronary artery disease and type 2 diabetes [89], heart failure [73], and hypertension [30]. Larger clinical studies with systemic arginase inhibition are clearly warranted [70].

16.8 Conclusions

Under conditions of cardiovascular disease, endothelial dysfunction is a result of reduced bioavailability of endothelial NO. The main underlying mechanism is eNOS uncoupling because of a deficiency of the co-factor BH_4 and/or the substrate L-arginine. Some drugs currently in clinical use (e.g. ACE inhibitors, ARBs, statins, nebivolol, and PETN) have the potential to prevent eNOS uncoupling under experimental conditions by elevating BH_4 content. Arginase inhibitors improve eNOS functionality by enhancing L-arginine bioavailability for eNOS, and may have the potential to be used as a novel therapy.

In addition, eNOS expression and activity are also regulated by VEGF. The signal transduction pathways downstream of the VEGF receptor-2 are addressed in Chap. 8 of this book.

References

1. Alp NJ, Mcateer MA, Khoo J, et al. Increased endothelial tetrahydrobiopterin synthesis by targeted transgenic GTP-cyclohydrolase I overexpression reduces endothelial dysfunction and atherosclerosis in ApoE-knockout mice. Arterioscler Thromb Vasc Biol. 2004;24:445–50.
2. Alp NJ, Mussa S, Khoo J, et al. Tetrahydrobiopterin-dependent preservation of nitric oxide-mediated endothelial function in diabetes by targeted transgenic GTP-cyclohydrolase I overexpression. J Clin Invest. 2003;112:725–35.
3. Bagnost T, Berthelot A, Bouhaddi M, et al. Treatment with the arginase inhibitor N(omega)-hydroxy-nor-L-arginine improves vascular function and lowers blood pressure in adult spontaneously hypertensive rat. J Hypertens. 2008;26:1110–8.
4. Beleznai T, Feher A, Spielvogel D, et al. Arginase 1 contributes to diminished coronary arteriolar dilation in patients with diabetes. Am J Physiol Heart Circ Physiol. 2011;300:H777–H83.
5. Bhatt SR, Lokhandwala MF, Banday AA. Resveratrol prevents endothelial nitric oxide synthase uncoupling and attenuates development of hypertension in spontaneously hypertensive rats. Eur J Pharmacol. 2011;667:258–64.
6. Brandes RP, Kreuzer J. Vascular NADPH oxidases: molecular mechanisms of activation. Cardiovasc Res. 2005;65:16–27.
7. Chalupsky K, Cai H. Endothelial dihydrofolate reductase: critical for nitric oxide bioavailability and role in angiotensin II uncoupling of endothelial nitric oxide synthase. Proc Natl Acad Sci U S A. 2005;102:9056–61.
8. Chandra S, Romero MJ, Shatanawi A, et al. Oxidative species increase arginase activity in endothelial cells through the RhoA/Rho kinase pathway. Br J Pharmacol. 2012;165:506–19.
9. Chen J, Kuhlencordt PJ, Astern J, et al. Hypertension does not account for the accelerated atherosclerosis and development of aneurysms in male apolipoprotein e/endothelial nitric oxide synthase double knockout mice. Circulation. 2001;104:2391–4.

10. Chen SJ, Li SY, Shih CC, et al. NO contributes to abnormal vascular calcium regulation and reactivity induced by peritonitis-associated septic shock in rats. Shock. 2010;33:473–8.
11. Csiszar A, Labinskyy N, Pinto JT, et al. Resveratrol induces mitochondrial biogenesis in endothelial cells. Am J Physiol Heart Circ Physiol. 2009;297:H13–H20.
12. Demougeot C, Prigent-Tessier A, Bagnost T, et al. Time course of vascular arginase expression and activity in spontaneously hypertensive rats. Life Sci. 2007;80:1128–34.
13. Dikalova A, Clempus R, Lassegue B, et al. Nox1 overexpression potentiates angiotensin II-induced hypertension and vascular smooth muscle hypertrophy in transgenic mice. Circulation. 2005;112:2668–76.
14. Dioguardi FS. To give or not to give? Lessons from the arginine paradox. J Nutrigenet Nutrigenomics. 2011;4:90–8.
15. Du XL, Edelstein D, Dimmeler S, et al. Hyperglycemia inhibits endothelial nitric oxide synthase activity by posttranslational modification at the Akt site. J Clin Invest. 2001;108:1341–8.
16. Du YH, Guan YY, Alp NJ, et al. Endothelium-specific GTP cyclohydrolase I overexpression attenuates blood pressure progression in salt-sensitive low-renin hypertension. Circulation. 2008;117:1045–54.
17. Eguchi D, D'uscio LV, Wambi C, et al. Inhibitory effect of recombinant iNOS gene expression on vasomotor function of canine basilar artery. Am J Physiol Heart Circ Physiol. 2002;283:H2560–H6.
18. Erdely A, Kepka-Lenhart D, Salmen-Muniz R, et al. Arginase activities and global arginine bioavailability in wild-type and ApoE-deficient mice: responses to high fat and high cholesterol diets. PLoS One. 2010;5:e15253.
19. Faria AM, Papadimitriou A, Silva KC, et al. Uncoupling endothelial nitric oxide synthase is ameliorated by green tea in experimental diabetes by re-establishing tetrahydrobiopterin levels. Diabetes. 2012;61:1838–47.
20. Fleming I. Molecular mechanisms underlying the activation of eNOS. Pflugers Arch. 2010;459:793–806.
21. Forstermann U, Munzel T. Endothelial nitric oxide synthase in vascular disease: from marvel to menace. Circulation. 2006;113:1708–14.
22. Forstermann U, Sessa WC. Nitric oxide synthases: regulation and function. Eur Heart J. 2012;33:829–37, 837a-837d.
23. Giri H, Muthuramu I, Dhar M, et al. Protein tyrosine phosphatase SHP2 mediates chronic insulin-induced endothelial inflammation. Arterioscler Thromb Vasc Biol. 2012;32:1943–50.
24. Gronros J, Jung C, Lundberg JO, et al. Arginase inhibition restores in vivo coronary microvascular function in type 2 diabetic rats. Am J Physiol Heart Circ Physiol. 2011;300:H1174–H81.
25. Gunnett CA, Lund DD, Mcdowell AK, et al. Mechanisms of inducible nitric oxide synthase-mediated vascular dysfunction. Arterioscler Thromb Vasc Biol. 2005;25:1617–22.
26. Hayashi T, Esaki T, Sumi D, et al. Modulating role of estradiol on arginase II expression in hyperlipidemic rabbits as an atheroprotective mechanism. Proc Natl Acad Sci U S A. 2006;103:10485–90.
27. Heinloth A, Heermeier K, Raff U, et al. Stimulation of NADPH oxidase by oxidized low-density lipoprotein induces proliferation of human vascular endothelial cells. J Am Soc Nephrol. 2000;11:1819–25.
28. Heitzer T, Krohn K, Albers S, et al. Tetrahydrobiopterin improves endothelium-dependent vasodilation by increasing nitric oxide activity in patients with type II diabetes mellitus. Diabetologia. 2000;43:1435–8.
29. Hink U, Li H, Mollnau H, et al. Mechanisms underlying endothelial dysfunction in diabetes mellitus. Circ Res. 2001;88:E14–E22.
30. Holowatz LA, Kenney WL. Up-regulation of arginase activity contributes to attenuated reflex cutaneous vasodilatation in hypertensive humans. J Physiol. 2007;581:863–72.

31. Holowatz LA, Santhanam L, Webb A, et al. Oral atorvastatin therapy restores cutaneous microvascular function by decreasing arginase activity in hypercholesterolaemic humans. J Physiol. 2011;589:2093–103.
32. Holowatz LA, Thompson CS, Kenney WL. L-Arginine supplementation or arginase inhibition augments reflex cutaneous vasodilatation in aged human skin. J Physiol. 2006;574:573–81.
33. Iadecola C, Zhang F, Casey R, et al. Delayed reduction of ischemic brain injury and neurological deficits in mice lacking the inducible nitric oxide synthase gene. J Neurosci. 1997;17:9157–64.
34. Iadecola C, Zhang F, Xu X. Inhibition of inducible nitric oxide synthase ameliorates cerebral ischemic damage. Am J Physiol. 1995;268:R286–R92.
35. Johnson FK, Johnson RA, Peyton KJ, et al. Arginase inhibition restores arteriolar endothelial function in Dahl rats with salt-induced hypertension. Am J Physiol Regul Integr Comp Physiol. 2005;288:R1057–R62.
36. Kang KT, Sullivan JC, Spradley FT, et al. Antihypertensive therapy increases tetrahydrobiopterin levels and NO/cGMP signaling in small arteries of angiotensin II-infused hypertensive rats. Am J Physiol Heart Circ Physiol. 2011;300:H718–H24.
37. Kashyap SR, Lara A, Zhang R, et al. Insulin reduces plasma arginase activity in type 2 diabetic patients. Diabetes Care. 2008;31:134–9.
38. Kauser K, Cunha V D, Fitch R, et al. Role of endogenous nitric oxide in progression of atherosclerosis in apolipoprotein E-deficient mice. Am J Physiol Heart Circ Physiol. 2000;278:H1679–H85.
39. Kellogg DL Jr, Zhao JL, Wu Y. Neuronal nitric oxide synthase control mechanisms in the cutaneous vasculature of humans in vivo. J Physiol. 2008;586:847–57.
40. Kessler P, Bauersachs J, Busse R, et al. Inhibition of inducible nitric oxide synthase restores endothelium-dependent relaxations in proinflammatory mediator-induced blood vessels. Arterioscler Thromb Vasc Biol. 1997;17:1746–55.
41. Kim JH, Bugaj LJ, Oh YJ, et al. Arginase inhibition restores NOS coupling and reverses endothelial dysfunction and vascular stiffness in old rats. J Appl Physiol. 2009;(1985) 107:1249–57.
42. Kosenko E, Tikhonova L, Suslikov A, et al. Impacts of lisinopril and lisinopril plus simvastatin on erythrocyte and plasma arginase, nitrite, and nitrate in hypertensive patients. J Clin Pharmacol. 2012;52:102–9.
43. Kuhlencordt PJ, Chen J, Han F, et al. Genetic deficiency of inducible nitric oxide synthase reduces atherosclerosis and lowers plasma lipid peroxides in apolipoprotein E-knockout mice. Circulation. 2001;103:3099–104.
44. Kuhlencordt PJ, Gyurko R, Han F, et al. Accelerated atherosclerosis, aortic aneurysm formation, and ischemic heart disease in apolipoprotein E/endothelial nitric oxide synthase double-knockout mice. Circulation. 2001;104:448–54.
45. Kuhlencordt PJ, Hotten S, Schodel J, et al. Atheroprotective effects of neuronal nitric oxide synthase in apolipoprotein e knockout mice. Arterioscler Thromb Vasc Biol. 2006;26: 1539–44.
46. Landmesser U, Cai H, Dikalov S, et al. Role of p47(phox) in vascular oxidative stress and hypertension caused by angiotensin II. Hypertension. 2002;40:511–5.
47. Landmesser U, Dikalov S, Price SR, et al. Oxidation of tetrahydrobiopterin leads to uncoupling of endothelial cell nitric oxide synthase in hypertension. J Clin Invest. 2003;111:1201–9.
48. Laursen JB, Somers M, Kurz S, et al. Endothelial regulation of vasomotion in apoE-deficient mice: implications for interactions between peroxynitrite and tetrahydrobiopterin. Circulation. 2001;103:1282–8.
49. Li H, Forstermann U. Nitric oxide in the pathogenesis of vascular disease. J Pathol. 2000;190:244–54.
50. Li H, Forstermann U. Pharmacological Prevention of eNOS Uncoupling. Curr Pharm Des. 2014;20:3595–606.
51. Li H, Forstermann U. Prevention of atherosclerosis by interference with the vascular nitric oxide system. Curr Pharm Des. 2009;15:3133–45.

52. Li H, Forstermann U. Structure-activity relationship of staurosporine analogs in regulating expression of endothelial nitric-oxide synthase gene. Mol Pharmacol. 2000;57:427–35.
53. Li H, Forstermann U. Uncoupling of endothelial NO synthase in atherosclerosis and vascular disease. Curr Opin Pharmacol. 2013;13:161–7.
54. Li H, Horke S, Forstermann U. Oxidative stress in vascular disease and its pharmacological prevention. Trends Pharmacol Sci. 2013;34:313–9.
55. Li H, Wallerath T, Munzel T, et al. Regulation of endothelial-type NO synthase expression in pathophysiology and in response to drugs. Nitric Oxide. 2002;7:149–64.
56. Li H, Witte K, August M, et al. Reversal of endothelial nitric oxide synthase uncoupling and up-regulation of endothelial nitric oxide synthase expression lowers blood pressure in hypertensive rats. J Am Coll Cardiol. 2006;47:2536–44.
57. Liaudet L, Rosenblatt-Velin N, Pacher P. Role of peroxynitrite in the cardiovascular dysfunction of septic shock. Curr Vasc Pharmacol. 2013;11:196–207.
58. Liu VW, Huang PL. Cardiovascular roles of nitric oxide: a review of insights from nitric oxide synthase gene disrupted mice. Cardiovasc Res. 2008;77:19–29.
59. Marfella R, Filippo C D, Esposito K, et al. Absence of inducible nitric oxide synthase reduces myocardial damage during ischemia reperfusion in streptozotocin-induced hyperglycemic mice. Diabetes. 2004;53:454–62.
60. Matsuno K, Yamada H, Iwata K, et al. Nox1 is involved in angiotensin II-mediated hypertension: a study in Nox1-deficient mice. Circulation. 2005;112:2677–85.
61. Melikian N, Seddon MD, Casadei B, et al. Neuronal nitric oxide synthase and human vascular regulation. Trends Cardiovasc Med. 2009;19:256–62.
62. Ming XF, Barandier C, Viswambharan H, et al. Thrombin stimulates human endothelial arginase enzymatic activity via RhoA/ROCK pathway: implications for atherosclerotic endothelial dysfunction. Circulation. 2004;110:3708–14.
63. Mollnau H, Wendt M, Szocs K, et al. Effects of angiotensin II infusion on the expression and function of NAD(P)H oxidase and components of nitric oxide/cGMP signaling. Circ Res. 2002;90:E58–E65.
64. Morishita T, Tsutsui M, Shimokawa H, et al. Vasculoprotective roles of neuronal nitric oxide synthase. FASEB J. 2002;16:1994–6.
65. Nisoli E, Tonello C, Cardile A, et al. Calorie restriction promotes mitochondrial biogenesis by inducing the expression of eNOS. Science. 2005;310:314–7.
66. Pacher P, Beckman JS, Liaudet L. Nitric oxide and peroxynitrite in health and disease. Physiol Rev. 2007;87:315–424.
67. Pannirselvam M, Simon V, Verma S, et al. Chronic oral supplementation with sepiapterin prevents endothelial dysfunction and oxidative stress in small mesenteric arteries from diabetic (db/db) mice. Br J Pharmacol. 2003;140:701–6.
68. Pannirselvam M, Verma S, Anderson TJ, et al. Cellular basis of endothelial dysfunction in small mesenteric arteries from spontaneously diabetic (db/db -/-) mice: role of decreased tetrahydrobiopterin bioavailability. Br J Pharmacol. 2002;136:255–63.
69. Pautz A, Art J, Hahn S, et al. Regulation of the expression of inducible nitric oxide synthase. Nitric Oxide. 2010;23:75–93.
70. Pernow J, Jung C. Arginase as a potential target in the treatment of cardiovascular disease: reversal of arginine steal? Cardiovasc Res. 2013;98:334–43.
71. Ponnuswamy P, Schrottle A, Ostermeier E, et al. eNOS protects from atherosclerosis despite relevant superoxide production by the enzyme in apoE mice. PLoS One. 2012;7:e30193.
72. Pritchard KA Jr, Groszek L, Smalley DM, et al. Native low-density lipoprotein increases endothelial cell nitric oxide synthase generation of superoxide anion. Circ Res. 1995;77: 510–8.
73. Quitter F, Figulla HR, Ferrari M, et al. Increased arginase levels in heart failure represent a therapeutic target to rescue microvascular perfusion. Clin Hemorheol Microcirc. 2013;54: 75–85.
74. Rodriguez S, Richert L, Berthelot A. Increased arginase activity in aorta of mineralocorticoid-salt hypertensive rats. Clin Exp Hypertens. 2000;22:75–85.

75. Romero MJ, Iddings JA, Platt DH, et al. Diabetes-induced vascular dysfunction involves arginase I. Am J Physiol Heart Circ Physiol. 2012;302:H159–H66.
76. Romero MJ, Platt DH, Tawfik HE, et al. Diabetes-induced coronary vascular dysfunction involves increased arginase activity. Circ Res. 2008;102:95–102.
77. Ryoo S, Bhunia A, Chang F, et al. OxLDL-dependent activation of arginase II is dependent on the LOX-1 receptor and downstream RhoA signaling. Atherosclerosis. 2011;214:279–87.
78. Ryoo S, Gupta G, Benjo A, et al. Endothelial arginase II: a novel target for the treatment of atherosclerosis. Circ Res. 2008;102:923–32.
79. Ryoo S, Lemmon CA, Soucy KG, et al. Oxidized low-density lipoprotein-dependent endothelial arginase II activation contributes to impaired nitric oxide signaling. Circ Res. 2006;99:951–60.
80. Schodel J, Padmapriya P, Marx A, et al. Expression of neuronal nitric oxide synthase splice variants in atherosclerotic plaques of apoE knockout mice. Atherosclerosis. 2009;206:383–9.
81. Schuhmacher S, Oelze M, Bollmann F, et al. Vascular dysfunction in experimental diabetes is improved by pentaerithrityl tetranitrate but not isosorbide-5-mononitrate therapy. Diabetes. 2011;60:2608–16.
82. Schuhmacher S, Wenzel P, Schulz E, et al. Pentaerythritol tetranitrate improves angiotensin II-induced vascular dysfunction via induction of heme oxygenase-1. Hypertension. 2010;55:897–904.
83. Schulman SP, Becker LC, Kass DA, et al. L-arginine therapy in acute myocardial infarction: the Vascular Interaction With Age in Myocardial Infarction (VINTAGE MI) randomized clinical trial. JAMA. 2006;295:58–64.
84. Schwarz PM, Kleinert H, Forstermann U. Potential functional significance of brain-type and muscle-type nitric oxide synthase I expressed in adventitia and media of rat aorta. Arterioscler Thromb Vasc Biol. 1999;19:2584–90.
85. Seddon M, Melikian N, Dworakowski R, et al. Effects of neuronal nitric oxide synthase on human coronary artery diameter and blood flow in vivo. Circulation. 2009;119:2656–62.
86. Seddon MD, Chowienczyk PJ, Brett SE, et al. Neuronal nitric oxide synthase regulates basal microvascular tone in humans in vivo. Circulation. 2008;117:1991–6.
87. Shatanawi A, Romero MJ, Iddings JA, et al. Angiotensin II-induced vascular endothelial dysfunction through RhoA/Rho kinase/p38 mitogen-activated protein kinase/arginase pathway. Am J Physiol Cell Physiol. 2011;300:C1181–C92.
88. Shelkovnikov S, Gonick HC. Peroxynitrite but not nitric oxide donors destroys epinephrine: HPLC measurement and rat aorta contractility. Life Sci. 2004;75:2765–73.
89. Shemyakin A, Kovamees O, Rafnsson A, et al. Arginase inhibition improves endothelial function in patients with coronary artery disease and type 2 diabetes mellitus. Circulation. 2012;126:2943–50.
90. Shinozaki K, Kashiwagi A, Nishio Y, et al. Abnormal biopterin metabolism is a major cause of impaired endothelium-dependent relaxation through nitric oxide/O2- imbalance in insulin-resistant rat aorta. Diabetes. 1999;48:2437–45.
91. Shinozaki K, Nishio Y, Okamura T, et al. Oral administration of tetrahydrobiopterin prevents endothelial dysfunction and vascular oxidative stress in the aortas of insulin-resistant rats. Circ Res. 2000;87:566–73.
92. Stepp DW, Ou J, Ackerman AW, et al. Native LDL and minimally oxidized LDL differentially regulate superoxide anion in vascular endothelium in situ. Am J Physiol Heart Circ Physiol. 2002;283:H750–H9.
93. Stroes E, Kastelein J, Cosentino F, et al. Tetrahydrobiopterin restores endothelial function in hypercholesterolemia. J Clin Invest. 1997;99:41–6.
94. Taguchi K, Kobayashi T, Matsumoto T, et al. Dysfunction of endothelium-dependent relaxation to insulin via PKC-mediated GRK2/Akt activation in aortas of ob/ob mice. Am J Physiol Heart Circ Physiol. 2011;301:H571–H83.
95. Toque HA, Tostes RC, Yao L, et al. Arginase II deletion increases corpora cavernosa relaxation in diabetic mice. J Sex Med. 2011;8:722–33.
96. Tsutsui M. Neuronal nitric oxide synthase as a novel anti-atherogenic factor. J Atheroscler Thromb. 2004;11:41–8.

97. Ueda S, Matsuoka H, Miyazaki H, et al. Tetrahydrobiopterin restores endothelial function in long-term smokers. J Am Coll Cardiol. 2000;35:71–5.
98. Umeji K, Umemoto S, Itoh S, et al. Comparative effects of pitavastatin and probucol on oxidative stress, Cu/Zn superoxide dismutase, PPAR-gamma, and aortic stiffness in hypercholesterolemia. Am J Physiol Heart Circ Physiol. 2006;291:H2522–H32.
99. Vaisman BL, Andrews KL, Khong SM, et al. Selective endothelial overexpression of arginase II induces endothelial dysfunction and hypertension and enhances atherosclerosis in mice. PLoS One. 2012;7:e39487.
100. Virdis A, Colucci R, Fornai M, et al. Cyclooxygenase-2 inhibition improves vascular endothelial dysfunction in a rat model of endotoxic shock: role of inducible nitric-oxide synthase and oxidative stress. J Pharmacol Exp Ther. 2005;312:945–53.
101. Wenzel P, Daiber A, Oelze M, et al. Mechanisms underlying recoupling of eNOS by HMG-CoA reductase inhibition in a rat model of streptozotocin-induced diabetes mellitus. Atherosclerosis. 2008;198:65–76.
102. Wilson AM, Harada R, Nair N, et al. L-arginine supplementation in peripheral arterial disease: no benefit and possible harm. Circulation. 2007;116:188–95.
103. Wohlfart P, Xu H, Endlich A, et al. Antiatherosclerotic effects of small-molecular-weight compounds enhancing endothelial nitric-oxide synthase (eNOS) expression and preventing eNOS uncoupling. J Pharmacol Exp Ther. 2008;325:370–9.
104. Xia N, Daiber A, Habermeier A, et al. Resveratrol reverses endothelial nitric-oxide synthase uncoupling in apolipoprotein E knockout mice. J Pharmacol Exp Ther. 2010;335:149–54.
105. Xia N, Pautz A, Wollscheid U, et al. Artichoke, cynarin and cyanidin downregulate the expression of inducible nitric oxide synthase in human coronary smooth muscle cells. Molecules. 2014;19:3654–68.
106. Xia Y, Zweier JL. Superoxide and peroxynitrite generation from inducible nitric oxide synthase in macrophages. Proc Natl Acad Sci U S A. 1997;94:6954–8.
107. Xiong Y, Fru MF, Yu Y, et al. Long term exposure to L-arginine accelerates endothelial cell senescence through arginase-II and S6K1 signaling. Aging. 2014;6:369–79.
108. Xu J, Wu Y, Song P, et al. Proteasome-dependent degradation of guanosine 5′-triphosphate cyclohydrolase I causes tetrahydrobiopterin deficiency in diabetes mellitus. Circulation. 2007;116:944–53.
109. Yang Z, Ming XF. Arginase: the emerging therapeutic target for vascular oxidative stress and inflammation. Front Immunol. 2013;4:149.
110. Yao L, Chandra S, Toque HA, et al. Prevention of diabetes-induced arginase activation and vascular dysfunction by Rho kinase (ROCK) knockout. Cardiovasc Res. 2013;97:509–19.
111. Yepuri G, Velagapudi S, Xiong Y, et al. Positive crosstalk between arginase-II and S6K1 in vascular endothelial inflammation and aging. Aging Cell. 2012;11:1005–16.
112. Youn JY, Wang T, Blair J, et al. Endothelium-specific sepiapterin reductase deficiency in DOCA-salt hypertension. Am J Physiol Heart Circ Physiol. 2012;302:H2243–H9.
113. Zhou X, Bohlen HG, Miller SJ, et al. NAD(P)H oxidase-derived peroxide mediates elevated basal and impaired flow-induced NO production in SHR mesenteric arteries in vivo. Am J Physiol Heart Circ Physiol. 2008;295:H1008–H16.

Index

M. HH. Schmidt, S. Liebner (eds.), *Endothelial Signaling in Development and Disease,* DOI 10.1007/978-1-4939-2907-8

MIX
Papier aus verantwortungsvollen Quellen
Paper from responsible sources
FSC® C105338

FSC
www.fsc.org

If you have any concerns about our products,
you can contact us on
ProductSafety@springernature.com

In case Publisher is established outside the EU,
the EU authorized representative is:
Springer Nature Customer Service Center GmbH
Europaplatz 3, 69115 Heidelberg, Germany

Printed by Libri Plureos GmbH
in Hamburg, Germany